LIGHTING for energy-efficient luminous environments

RONALD N. HELMS, PH.D, P.E.
University of Kansas

M. CLAY BELCHER, PH.D, P.E.
University of Kansas

PRENTICE HALL
Englewood Cliffs, New Jersey 07632

Library of Congress Cataloging-in-Publication Data

Helms, Ronald N.
Lighting for energy-efficient luminous environments / Ronald N. Helms, M. Clay Belcher.
p. cm.
Rev. ed. of: Illumination engineering for energy efficient luminous environments. c1980.
Includes bibliographical references and index.
ISBN 0–13–451568–4 : $40.50
1. Electric lighting. 2. Lighting. 3. Electric power—Conservation. I. Belcher, M. Clay. II. Helms, Ronald N. Illumination engineering for energy efficient luminous environments. III. Title.
TK4175.H45 1991
621.32—dc20 90–19471
CIP

Editorial/production supervision
and interior design: Kathryn Gollin Marshak
Manufacturing buyers: Kelly Behr and Susan Brunke

Previously published under the title *Illumination Engineering for Energy Efficient Luminous Environments*

The publisher offers discounts on this book when ordered in bulk quantities. For more information, write:
Special Sales; Prentice-Hall, Inc.
College Technical and Reference Division
Englewood Cliffs, NJ 07632

Printed in the United States of America

10 9 8 7 6 5 4 3 2 1

ISBN 0-13-451568-4

Prentice-Hall International (UK) Limited, *London*
Prentice-Hall of Australia Pty. Limited, *Sydney*
Prentice-Hall Canada Inc., *Toronto*
Prentice-Hall Hispanoamericana, S.A., *Mexico*
Prentice-Hall of India Private Limited, *New Delhi*
Prentice-Hall of Japan, Inc., *Tokyo*
Simon & Schuster Asia Pte. Ltd., *Singapore*
Editora Prentice-Hall do Brasil, Ltda., *Rio de Janeiro*

To
Lawrence Nelson Helms
(1911–1977)

This book is dedicated to my father, Nelson Helms. My dad, who worked for 40 years in a steel mill, wanted a better life for his boys. ''If you want to end up working a swing shift all your life like your dad, don't get a college education. If you want a college education, I'll do everything possible to help.'' My dad passed away in August 1977 of lung cancer at the young age of 66 years.

Ronald N. Helms

To
C. M. ''Sandy'' Belcher
(1924–1990)

Dad exemplified the commandments to ''Love the Lord your God with all your heart and with all your soul and with all your mind'' and ''love your neighbor as yourself.'' It is with deepest love, and appreciation for all he taught me through his words and actions, that I dedicate this book to him.

M. Clay Belcher

contents

preface

The first version of this book was entitled *Illumination Engineering for Energy Efficient Luminous Environments*. The title has been changed to reflect the fact that the book now serves a larger group of students and professionals which includes architects, interior designers, engineers, and optometry students as well as illuminating engineers.

The original book had its beginnings in 1965 when Dr. Ronald N. Helms was invited to join the faculty of the University of Colorado in order ''to develop a program in Illumination-Electrical Systems in Architectural Engineering,'' according to Professor Robert E. Rathburn. It was Bob's dream to make architectural engineering the training ground for all the consulting engineering professions (structural, mechanical, and electrical) as well as construction engineering and management. That book was an outgrowth of lecture notes developed by Dr. Helms and used in the teaching of illumination. The revised book represents a joint venture between Dr. Helms and Dr. M. Clay Belcher. Both authors teach lighting courses in the Architectural Engineering Program at the University of Kansas, located in Lawrence, Kansas.

This new version represents a continuing evolutionary process in response to expanding lighting technology and in response to new insights gained through the teaching and research of Dr. Helms and Dr. Belcher. It has been expanded by the addition of four new chapters, and each of the existing chapters has been updated and amplified. This was done to cover a broader range of design considerations while maintaining a thorough treatment of the fundamentals.

The illuminating engineer may be well grounded in the technical aspects (calculations) while the lighting designer has an understanding of the artistic and aesthetic aspects of lighting. Unfortunately, both groups are usually deficient in one or more of the important fundamental concepts. The key to being a complete lighting professional is to be both an artist and an engineer. The lighting professional must know all aspects of lighting, from the pure art form to the highly complex mathematical technique. Knowing these fundamental concepts, the lighting professional must know how to mix the art and the engineering to make a cohesive, balanced lighting solution.

The task of providing well-balanced lighting solutions has been made more difficult with the realization by most professionals that our nation's natural resources are not limitless. The "energy crunch" of the 1970s was our first alarm. Today the signs of impending energy crisis are even more obvious: utilities operating at peak capacity, environmental concerns, military action in the oil fields. The visibility of lighting makes it a natural target for conservation by those uneducated and undereducated in lighting fundamentals when energy shortages occur. The solution to energy-efficient luminous environments is integral to all chapters in this book. Important fundamental concepts are found in each chapter. All of these concepts must weigh heavily in the balance that must be struck between visibility, visual comfort, productivity/performance/profit, aesthetics, cost, and energy conservation.

The most important chapters of the book are Chapters 1 through 5, which deal with fundamental concepts of importance to all educational and professional fields of lighting study or application. All lighting professionals or students should be thoroughly familiar with these five "basic principles chapters." Most errors made by lighting professionals can be traced back to a lack of understanding of subjects covered in the first five chapters. If the book is to be used as a textbook, a good fundamental course can be built around Chapters 1 through 8. A second, follow-up course in applications could be based on Chapters 8 through 11. An advanced or graduate course which builds off Chapters 1 through 10 could cover the topics in Chapters 11 through 14 as its primary thrust. Any specialized course or seminars utilizing this book should always begin with the "basic principles chapters." In addition, a roadway lighting course or seminar would include Chapters 1 through 4, 6, 8, 11, 12 (Vectors), and 14. A specialized course in daylighting should include Chapters 1 through 8, 12, 13, and 14.

ACKNOWLEDGMENTS

Dr. Ronald N. Helms

In 1982 the Architectural Engineering Program at the University of Kansas was selected as the original beneficiary of the Robert J. Besal Educational Fund. I am most appreciative of the opportunities this has allowed me as the Director of the Besal Lighting Program. In 1984 the Besal Fund made it possible for K.U. to hire a second faculty member in lighting. This was an exciting event in my career. The addition of Clay Belcher to the faculty adds a colleague who "speaks the same language," someone to toss around ideas with, to plan with, to dream with.

I wish to acknowledge Professor Steve Tang and Professor Roland Pierce, two teachers from my undergraduate days at the University of Illinois. Professor Steve Tang, my mentor and advisor, by his example and deeds, planted the seeds that would one day bring me into the teaching profession. Professor Tang was asked by a student why he had given up a highly paid, secure job as chief engineer in a large architectural and engineering firm. Professor Tang's reply was, "the profession has been good to me and I have learned a lot. It was time for me to give something back to the profession. Too many people gain valuable knowledge and experience that they end up taking to the grave. I want to pass some of my experience on to future generations while I am young and capable."

Professor Roland Pierce first stirred my interest in illuminating engineering. He provided additional guidance at the graduate level through individual tutorial sessions and special projects in illumination. Many thanks to both Professor Tang and Professor Pierce for their guidance, encouragement, and untiring dedication to their students—particularly me.

In 1969, I returned to school after five years of teaching to obtain my Ph.D. in Biophysics. My two-year program could not have been possible without the guidance and assistance of Dr. H. Richard Blackwell. I thank Dr. Blackwell for his patience and understanding of the time pressures. I am thankful for the wisdom and guidance of Dr. Glenn Fry and the truly inspirational teaching of Dr. Jack King, both of whom made my educational experience at Ohio State a time to remember.

But above all I want to express my gratitude to my wife, Margo Baron Helms, and my children, Tracey Michelle, Robert Nelson, Laura Elaine, Cathleen Carole, and James Mitchell Helms for their support and love.

Ronald N. Helms, PH.D., P.E., F.I.E.S, F.N.S.A.E.

Dr. M. Clay Belcher

I met Ronald Helms when I was a lighting student in the Architectural Engineering program at the University of Colorado in the mid-1970s. He started out as a teacher, and through the years became mentor, friend, and eventually colleague. I am proud to be associated with an educator who has such unreserved enthusiasm for his subject and especially for his students. When I introduce him to my Introduction to Architectural Engineering class, I tell the students that for anyone serious about a degree in Architectural Engineering, Dr. Helms will give the shirt off his back to make sure they succeed. Thanks for everything, Ron.

Authors always acknowledge the sacrifices made by their families. While preparing the manuscript for this book I learned why. I am deeply appreciative of the loving support given by my wife Beth, and our children Richard Timothy and Anna Caroline.

I am also indebted to all those who have been my students at K.U. You continually teach me about myself, about teaching, and about lighting. Special thanks in this category go to Martin Moeck, a dedicated and gifted student, a valued friend.

Finally, I thank my colleagues in the Illuminating Engineering Society of North America whose dedication to excellence in the art and science of lighting is inspiring. They are too numerous to list, but I am especially appreciative of those who also support me with their prayers.

M. Clay Belcher, P.E., PH.D., M.I.E.S.

1 light, units, and terminology

PHYSICS OF LIGHT

When asked to define poetry, Samuel Johnson replied that it was easier to say what it was not. He continued: "We all *know* what light is; but it is not easy to *tell* what light is." In using this analogy, Dr. Johnson very aptly expressed a problem that has plagued both layperson and physicist through history—the problem of defining light. In fact, light is such a peculiar phenomenon that two theories are required to explain it adequately.

The electromagnetic wave theory of light was the

classical interpretation of the nineteenth century. This perspective gives rise to the following definition:

> Light is visually evaluated radiant energy.

Wave Theory

Scientists' enthusiasm for the wave theory of light was temporarily dampened, however, by the discovery of the photovoltaic effect toward the end of the nineteenth century. A typical apparatus for investigating this effect is shown in a simplified schematic in Figure 1-1. When light strikes the metal plate, some electrons are induced to flow to the other electrode in spite of the "retarding" voltage applied. As that voltage is increased, a level is reached (typically at a few volts) where the electron flow stops.

Why should observation of this phenomenon cause scientists to doubt the wave theory of light? After all, waves can carry energy, presumably enough energy to cause electrons to flow. The problem resides in the time factor. Calculations based on wave theory indicate that it would take about a year for the average electron on the metal plate to receive enough energy for it to migrate to the other electrode. Experiments indicate, however, that electron flow begins *immediately* as light impinges on the plate.

Quantum Theory

In 1905, Albert Einstein showed how an idea that had recently been proposed by Max Planck could be used to explain the photovoltaic effect. From Planck's theory comes the notion that light travels as a stream of little bundles of energy known as quanta. For any frequency of light, v, all these quanta have the same amount of energy, expressed by

$$E = hv$$

where h is known as Planck's constant and has the value of 6.6256×10^{-27} erg/sec.

Though the quantum theory solved the photovoltaic mystery, many physicists

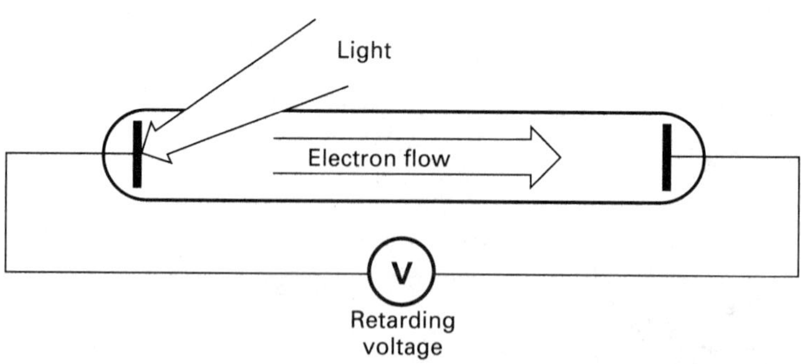

Figure 1-1. Photovoltaic Cell.

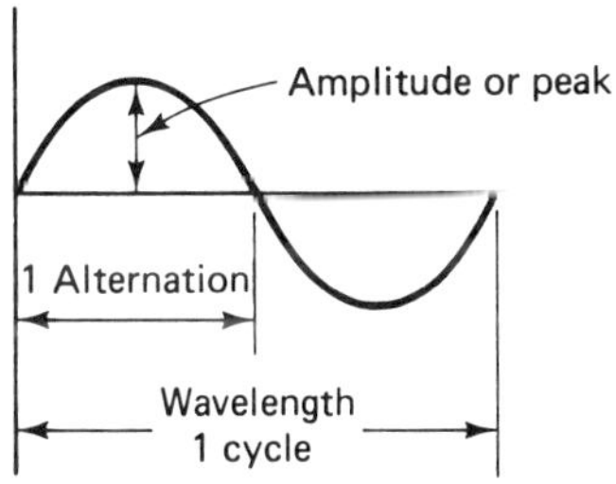

Figure 1-2. Wave Terminology.

were incredulous that light could propagate as a beam of discrete packets of energy instead of like a wave rippling outward from a source.

Duality Theory

In 1929, de Broglie was awarded the Nobel prize for his hypothesis that combined the wave theory and particle (or quantum) theory. Depending on the wavelength (Fig. 1-2), either the wave aspect or the particle aspect may be dominant. For example, radio waves are strictly wavelike, whereas cosmic rays behave like particles. With light, certain of its characteristics are best described by wave theory while others require a quantum approach. Thus, we say light exhibits wave/particle duality.

Electromagnetic Spectrum

Of the wide range of phenomena that exhibit wave characteristics (from cosmic rays to alternating current), light is that portion of the electromagnetic spectrum (Fig. 1-3) which the visual system can detect. This model is very useful for explaining certain aspects of light such as color, interference, and diffraction.

The electromagnetic theory provides the best explanation of the radiant energy characteristics most frequently used by the illuminating engineer. A graphical representation of radiant energy is called a **spectrum.** Different forms of radiant energy are laid out along a continuum from cosmic rays to electricity. The electromagnetic (or radiant energy) spectrum ranges from 3.937×10^{-13} inch (in.) or about 10^{-5} nanometer (nm) for cosmic rays to 3100 miles (4.98×10^{15} nm) for 60-hertz (Hz) electric current.

The dividing line between two forms of radiant energy is not as sharp as Figure 1-3 might indicate but rather there is a gradual transition from one to another. Because of the reciprocal relationship between wavelength and frequency ($\lambda = c/v$), radiant energy can be expressed in terms of its wavelength or frequency along the continuum. Classically, **wavelength** has been used to describe all forms of radiant energy left of the dividing line (Fig. 1-3) between infrared and radar. Radiant energy to the right of this dividing line is described in terms of its **frequency.** This dividing line between infrared and radar is probably due to the physical size of the numbers when expressed in terms of wavelength or frequency. For example, electricity is most commonly designated in terms of its frequency, such as 60 Hz. However, it could also be designated in

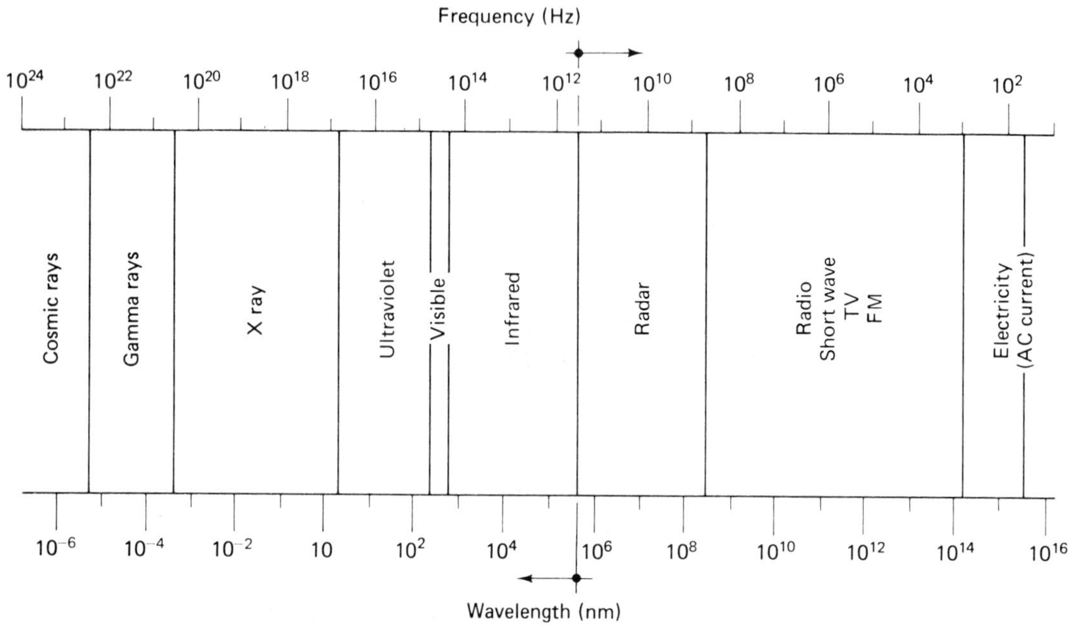

Figure 1-3. Electromagnetic Spectrum.

terms of its wavelength, which is approximately 10^{15} nm. That is a number followed by 15 zeros, which is not a convenient number to work with.

Visible Spectrum

In lighting, we are concerned with a small portion of the electromagnetic spectrum (Fig. 1-3) called the **visible spectrum.** The visible spectrum derives its name from the fact that any energy produced in this narrow band will produce the sensation of vision when it stimulates a normal human eye. The visible spectrum has the following wavelength range (Table 1-2 in the Data Section):

From violet: 0.38×10^{-4} cm $= 0.38 \times 10^{-6}$ m $= 0.38$ μm (micrometers)
= 3800 Å (angstroms) = 15 millionths of an inch
= 380 nm (nanometers)

To red: 0.76×10^{-4} cm $= 0.76 \times 10^{-6}$ m $= 0.76$ μm
= 7600 Å (angstroms) = 30 millionths of an inch
= 760 nm (nanometers)

Each band or part of the visible spectrum will produce a different color sensation when the eye is light adapted. (See Table 1-1.)

Purple exists in the spectrum; however, it occurs because of a red cone–blue

TABLE 1-1 Color Versus Wavelength

Color	Wavelength (nm)
Red	760–630
Orange	630–590
Yellow	590–560
Green	560–490
Blue	490–440
Indigo	440–420
Violet	420–380
Purple	Not a pure spectral color

cone interaction. That is, purple is an artifact of the visual system and not a pure spectral color.

The phenomenon of color sensation that involves the interaction of visible radiant energy and the human visual system is a very important concept. If a "normal" human visual system were stimulated with pure 500 nm (single-wavelength) radiant energy, the light-adapted subject would respond that the light is green. If the wavelength were changed to pure 585 nm radiant energy, the response would be that the light is yellow. Therefore, the wavelength composition of light is important to the sensation of color perception.

LIGHTING UNITS AND TERMINOLOGY

The photometric units and terminology presented in this chapter are the more commonly occurring basic quantities of value to the illuminating engineer or lighting designer. Conversion factors are at the end of this chapter in the data section.

Luminaire

A luminaire is a complete, functional lighting unit, including the lamp(s), housing, electrical components required to power and operate the unit, and control media to direct and enclose the lamp housing. The *old* term for luminaire is a "fixture." A luminaire is a total lighting unit while a fixture is a plumbing unit.

Luminous Energy (Q)

Luminous energy is visually evaluated radiant energy traveling in the form of electromagnetic waves. The unit is the lumen-second (lm-s).

unit: lumen-seconds, lm-sec (= Talbot, T)

Solid Angle (ω)

Solid angle is the ratio of the sphere surface area enclosed to the square of the radius.

$$\omega = \frac{A_{\text{sphere}}}{R^2} = \frac{dA}{R^2}$$

unit: steradian, sr

Luminous Flux (Φ)

Luminous flux is the time rate of flow of luminous energy. For example, the light emitted from a 40 watt fluorescent lamp is at a rate of 3150 lumens.

$$\Phi = \frac{dQ}{dt}$$

unit: $\frac{\text{lm-sec}}{\text{sec}}$ = lumen, lm

The lumen is the flux emitted within a solid angle of 1 steradian by a point source having a luminous intensity of 1 candela.

1 lumen = 1 candela-steradian = 1 cd-sr

Luminous Intensity (*I*)

Luminous intensity is the solid-angular luminous flux density in a given direction.

Luminous intensity is commonly but *incorrectly* referred to as "candlepower." Luminous intensity indicates the ability of a light source to produce illuminance in a given direction.

$$I = \frac{d\Phi}{d\omega}$$

unit: $\frac{\text{lm}}{\text{sr}}$ = candela, cd

Density of Luminous Flux Leaving the Surface

Two concepts are required to describe luminous flux leaving a surface.

Exitance (*M*)

Exitance is defined as the total luminous flux density leaving a surface. It is the "total" luminous flux leaving a surface *without* regard to direction. The *old* term for exitance is "emittance."

$$M = \frac{d\Phi}{dA}$$

$$\textbf{unit: } \frac{\text{lumen}}{\text{ft}^2} = \frac{\text{lm}}{\text{ft}^2}$$

Luminance (L)

Luminance is the luminous flux per unit of projected area (A_θ) per unit solid angle ($d\omega$) leaving a given point in a given direction.

The term used to be called **brightness.** Brightness is used to describe a subjective evaluation of a surface, whereas luminance (or photometric brightness) is used to describe the objective or measured characteristics of a surface.

Luminance is an inherent property of a ray of light.

$$L = \frac{d^2\Phi}{d\omega\, dA_\theta}$$

When the unit of luminous flux is the lumen and the projected area is in square feet (solid angle is unitless), the unit of luminance is the footLambert. Use of this older term is deprecated by the IESNA.

$$\textbf{unit: } \frac{\text{lumen}}{\text{ft}^2} = \text{footLambert, fL}$$

Luminance can also be defined as the luminous intensity (I) of a surface in a given direction (θ) per unit projected area (A_θ) as viewed from that direction.
Since

$$I = \frac{d\Phi}{d\omega}$$

then

$$L = \frac{d^2\Phi}{d\omega\, dA_\theta} = \frac{dI_\theta}{dA_\theta}$$

When the luminous intensity is in candelas and the area is in square inches, the unit of luminance is the candela per square inch. If area is in square feet, the preferred unit is the candela per square foot.

$$\textbf{unit: } 1\,\frac{\text{candela}}{\text{in.}^2} = 1\,\frac{\text{cd}}{\text{in.}^2} = 144\,\frac{\text{cd}}{\text{ft}^2}$$

The conversion from footLamberts to candela per square inch requires the introduction of the constant π. A derivation at the end of this chapter shows how the π factor enters into the following definition:

$$1\ \text{fL} = 1\,\frac{\text{lm}}{\text{ft}^2} \times \frac{1}{\frac{\pi\text{lm}}{\text{cd}}} = \frac{1}{\pi}\,\frac{\text{cd}}{\text{ft}^2} = \frac{1}{144\pi}\,\frac{\text{cd}}{\text{in.}^2}$$

The amount of footLamberts is equal to $1/(144\pi)$ times the amount of candelas per square inch.

$$\text{Conversion formula:} \quad \frac{1\ \text{fL}}{(1/144\pi)(\text{cd/in.}^2)} = \boxed{\frac{144\pi\ \text{in.}^2 \times \text{fL}}{1\ \text{cd}} = 1}$$

For example,

$$3\,\frac{\text{cd}}{\text{in.}^2} \times \frac{144\pi\ \text{in.}^2 \times \text{fL}}{1\ \text{cd}} = 1357.2\ \text{fL}$$

Lambertian Surface (Perfect Diffuser)

A Lambertian surface is a surface that acts like a perfect diffuser. That is, it absorbs all the incidented luminous energy and reemits all the luminous energy in the form of a tangent sphere, according to Lambert's Cosine Law (see Fig. 1-4). Lambert's Cosine Law states that the luminous intensity at any angle (I_θ) is equal to the luminous intensity perpendicular (I_N, normal) times the cosine of angle θ.

$$I_\theta = I_N \cos\theta$$

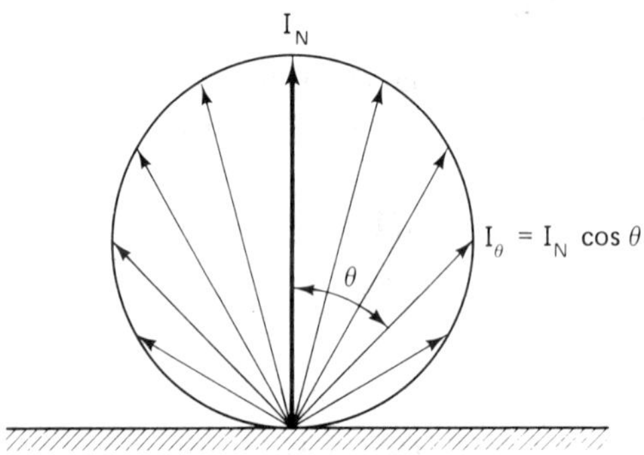

Figure 1-4. Lambert's Cosine Law (Perfect Diffuser).

A Lambertian surface will have a constant luminance irrespective of the direction of view.

$$L = \frac{I_\theta}{A_\theta}$$

Since $A_\theta = A_N \cos\theta$ = projected area (Fig.1-5), and $I_\theta = I_N \cos\theta$ for a Lambertian surface,

$$L = \frac{I_N \cos\theta}{A_N \cos\theta} = \frac{I_N}{A_N} = \text{constant}$$

For a perfect diffuser (Lambertian surface), exitance, M, and luminance, L, are related by the factor of π.

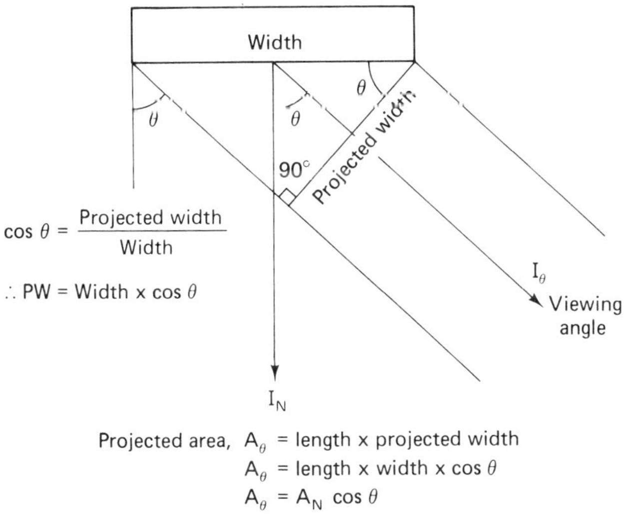

Figure 1-5. Projected Area.

$$M = \frac{d\Phi}{dA} = \pi \times L$$

(Note: This equation is *not* a definition; it is derived. The intensity from a unit area is summed in all directions to determine the total emitted lumens.)

The luminance of a perfectly diffuse (Lambertian) surface is sometimes described by the *old* (obsolete) unit of the footLambert; it is numerically equal to the exitance of the Lambertian surface expressed in lm/ft^2. Thus,

$$L \text{ (fL)} = \pi \times L \text{ (cd/ft}^2\text{)}$$

This equation also is used to convert between fL and cd/ft^2 units even for non-Lambertian surfaces, but in such cases the fL value is no longer equal to the exitance of the surface.

To convert cd/in.^{2} to footLamberts, multiply number of cd/in.^{2} by 144π.

Illuminance (Illumination) (*E*)

The approved terminology is **illuminance,** which has replaced the term **illumination.** The word "illuminance" *sounds* too much *like* the word "luminance" to most individuals. The almost indistinguishable difference in the two words has led some to continue the use of the word "illumination" where illuminance is the correct word. Illuminance is defined as the **incident** luminous flux on a small surface per unit area of the surface (Fig. 1-6).

$$E = \frac{d\Phi}{dA}$$

unit: $\frac{\text{lumen}}{\text{ft}^2} = \frac{\text{lm}}{\text{ft}^2}$ = footcandle, fc

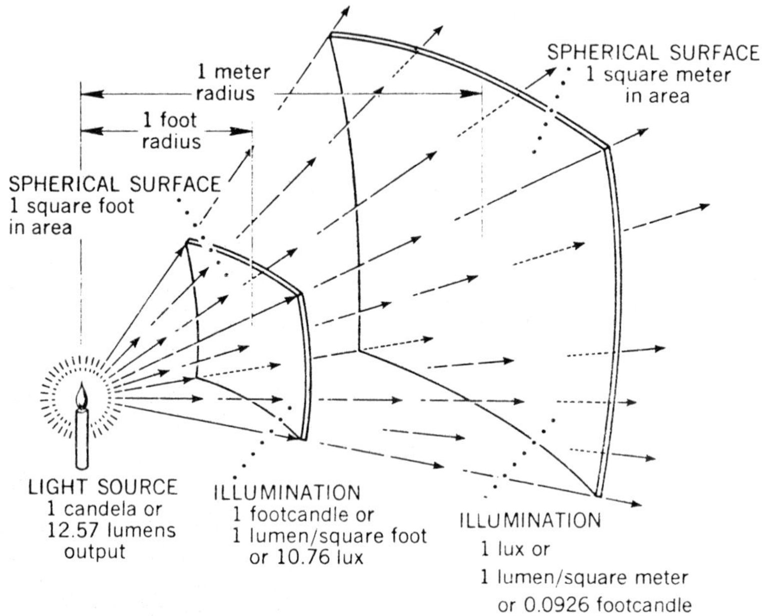

Figure 1-6. Illuminance (1 lm/ft^2 = 1 footcandle).

The **Fundamental Law of Illumination,** known as the **Inverse Square Law** (Fig. 1-7), can be derived.

$$\text{Since } I = \frac{d\Phi}{d\omega} \qquad \text{then } d\Phi = I\, d\omega$$

$$\text{and } d\omega = \frac{dA}{R^2} \qquad \text{then } dA = R^2\, d\omega$$

Therefore,

$$E = \frac{I\, d\omega}{R^2\, d\omega} = \frac{I}{R^2} = \frac{I}{d^2}$$

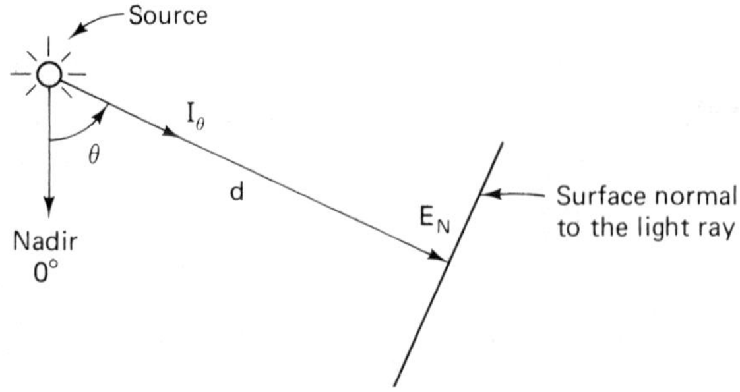

Figure 1-7. Inverse Square Law.

(Since R equals the distance, d, from the light source to the surface area for a unit sphere.)

Reflectance (ρ)

Reflectance is the ability of a surface to reradiate energy. This terminology applies only to a Lambertian surface or perfect diffuser.

$$\rho = \frac{\text{total reflected light (fL)}}{\text{total incident light (fc)}} \times 100$$

$$\text{Since } M = \rho E \qquad \textit{then } \rho = \frac{M}{E} = \frac{\text{emitted}}{\text{incident}} \times 100 = \%$$

units: percent (%)

Most manufacturers (i.e., paint, walls, ceilings) will be able to tell you the reflectance value of their product. This value is usually for some specific light which is incident at a specific angle (i.e., 45 degrees).

Luminance Factor (β)

Luminance factor is a term that applies to the reflectance characteristics of any surface that is non-Lambertian and therefore cannot be described by a single value. Luminance factor is defined as a ratio of the directional light coming off a surface to the directional light coming to a surface. Since

$$L = \beta E$$

$$\text{then } \beta = \frac{\text{emitted (directional)}}{\text{incident (directional)}} \times 100 = \%$$

unit: percent (%)

The amount of light reflecting off a non-Lambertian surface will change depending on

1. The location of the light source
2. The location of the observer
3. The physical characteristics of the surface

The solid (called an **indicatrix**) formed by the various reflected vectors is a function of three variables, h_s, v_o, and v_s. The luminance factor is measured for a fixed v_o angle and an observer located in the $h_o = 0$ plane. The light source is moved about in the hemisphere above the point on the surface (see Fig. 1-8).

If $L = \beta \times E$, then L will change depending on the direction of incident light and the viewing angle, v_o. The β value of the background may change at a *different rate* from the β value of the print. This can cause the *contrast* between the paper and the print to go to 0 (zero), which means that the print can't be seen. Since the location

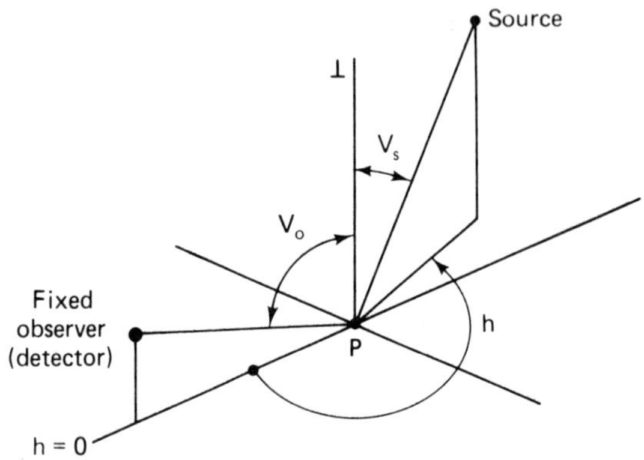

Figure 1-8. Measurement of the Luminance Factor, β. h_s = horizontal angle between the fixed axis h_o and the moving component. v_o = vertical angle from a perpendicular to a point on the surface, *P*, to the line from the observer, *O*, (detector, photometer, photocell), to the point on the surface. v_s = vertical angle from a perpendicular to a point on the surface, *P*, to the line from the source, *s*, to the point on the surface.

of the light source relative to the visual task is so crucial, the lighting should not be placed in spaces until the furniture location and task locations are known.

Reflected Luminous Intensity Indicatrix

Reflected luminous intensity indicatrix (applicable for non-Lambertian surfaces) is a vector function of three variables (h_s, h_o, v_s), as defined. For the RLI indicatrix, h_o = 0 and the v_s angle is fixed for a given indicatrix and the detector moves about in the hemisphere above the point on the surface (see Fig. 1-9).

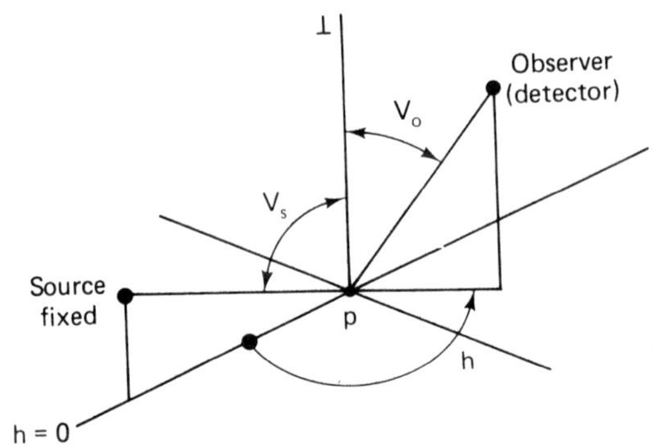

Figure 1-9. Measurement of the Reflected Luminous Intensity Indicatrix.

Transmittance (τ)

Transmittance is a measure of the amount of light transmitted through a material expressed in a percentage. This term applies only to a transparent material.

$$\tau = \frac{\text{total transmitted light (fL)}}{\text{total incident light (fc)}} \times 100$$

$$\text{Since } L = \tau \times E \qquad \text{then } \tau = \frac{L}{E} \times 100 = \%$$

unit: percent (%)

Most glass manufacturers will provide a single value for the transmittance of a type of glass. This value is usually for some specific viewing angle (i.e., 0°) and a specific incident angle (i.e., 0°) for the light ray.

As the angle of incident light increases (Fig. 1-10) the transmittance through the glass will decrease.

If the manufacturer lists the transmittance of $\frac{1}{4}''$ plate glass as 89% which is at 0° incidence angle, the remaining 11% of the incident ray will be absorbed (heat) and reflected.

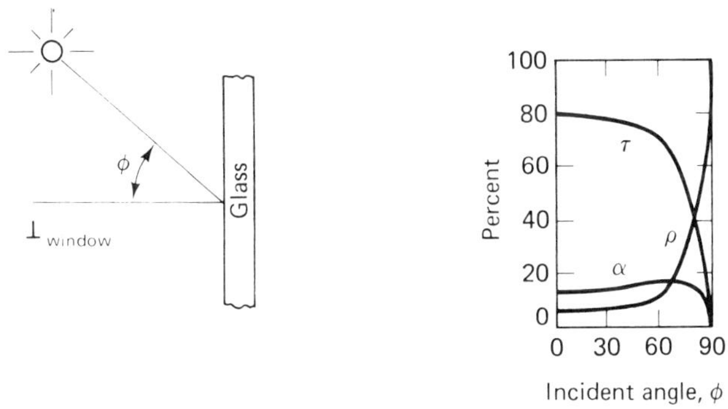

Figure 1-10. Properties of Glass.

Efficacy (*K*)

Luminous efficacy is a ratio of the total luminous flux emitted by a source to the total power input to the source.

$$K = \frac{\text{total luminous flux emitted}}{\text{total input power}} = \frac{\text{light output}}{\text{power input}}$$

$$\textbf{unit:} \frac{\text{lumen}}{\text{watt}} = \frac{\text{lm}}{\text{W}}$$

Efficacy is similar in concept to miles per gallon. The larger the efficacy, the higher the light output with less power consumption.

Example:

A 100 W incandescent lamp produces 1750 lumens while a 40 W fluorescent produces 3150 lumens.

$$\text{incandescent lamp efficacy} = 1750/100 = 17.5 \text{ lm/W}$$

$$\text{fluorescent lamp efficacy} = 3150/40 = 78.8 \text{ lm/W}$$

It will take about two (1.8) of the 100 W incandescent lamps ($1.8 \times 1750 = 3150$ lm) or approximately 200 watts of power (in this example) to produce the same light as one 40 watt fluorescent lamp. Since the heat produced by a light source is a function of the lamp watts (3.41 Btu/hr/W), the incandescent lamps will produce five times the heat for the same light output (lumens).

1-40 W: 40×3.41 Btu/hr/W = 136 Btu/hr

2-100 W: 200×3.41 = 682 Btu/hr

Spectral Luminous Efficiency

Spectral luminous efficiency is the ratio of radiant flux at wavelength λ_m to that at any other wavelength such that both wavelengths produce equally intense visual sensations. λ_m is chosen so that the maximum value of this ratio is 1. The spectral luminous efficiency of all wavelengths defines a function that is the response of the human visual system to radiant energy. Symbols are $V(\lambda)$ for photopic vision and $V'(\lambda)$ for scotopic vision. The function of $V(\lambda)$ is commonly known as the **standard observer curve.**

Unit Sphere

The relationship between candelas, lumens, steradians, and footcandles can be shown by using a sphere 1 ft in radius (unit sphere) with a uniform point source of 1 cd at the center of the sphere (see Fig. 1-11).

For an area of 1 ft^2 on the surface, the solid angle subtended will be 1 sr.

$$\omega = \frac{A}{R^2} = \frac{1\ \text{ft}^2}{1\ \text{ft}^2} = 1\ \text{sr}$$

The 1 candela point source will produce 1 lumen in the unit solid angle.

$$d\Phi = I \times d\omega = 1\ \text{cd} \times 1\ \text{sr} = 1\ \text{lm}$$

The illumination produced on the inside surface of the unit sphere will be 1 lumen on 1 square foot or 1 footcandle.

$$E = \frac{d\Phi}{dA} = \frac{1\ \text{lm}}{1\ \text{ft}^2} = 1\ \text{fc}$$

The total surface area of a sphere is $4\pi R^2$. Therefore, the total surface area of the unit sphere is 4π or 12.57 square feet. If the luminous flux of 1 lumen falls on each square foot, the 1 candela uniform point source will produce a total of 12.57 lumens or 4π lumens.

Mean Spherical Candlepower

Mean spherical candlepower is the average luminous intensity (candlepower) of a light source in all directions in space.

MSCP is equal to the total luminous flux of the light source divided by 4π, which is the total surface area of the unit sphere.

$$MSCP = \frac{\text{total luminous flux}}{\text{total sphere area}} = \frac{\Phi_{\text{total}}}{4\pi}$$

Assumptions. The preceding formula makes two assumptions. First, the light source is *uniform* in its distribution, which means that the luminous intensity vectors

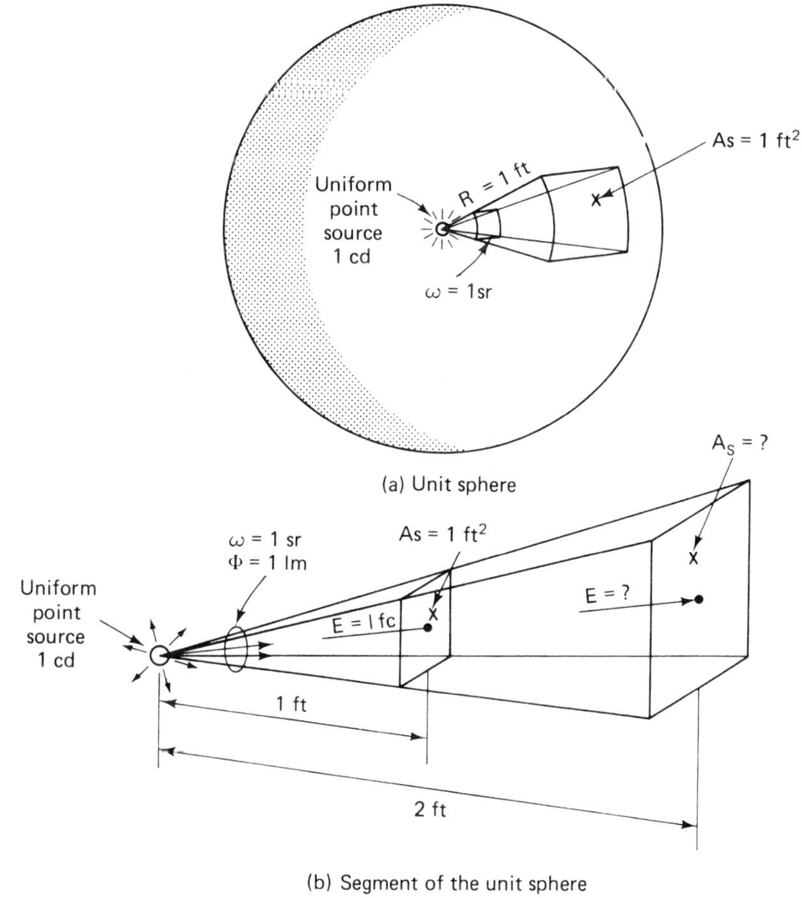

Figure 1-11. Unit Sphere.

radiating from the point source are the same in all directions. This distribution would be for an indicatrix that is a perfect sphere, which cannot exist in any real light source because of the blocking effect of the lamp base. Second, the source of light is a **point source,** which means that the size of the source is very small compared to the distance to the surface.

$$\textbf{unit: } \frac{\text{lumen}}{\text{sr}} = \text{candela, cd}$$

Smashing Point

Smashing point denotes "the time when the lamps have so decreased in light output that it is more economical to break them and replace them with new ones than to continue to burn them" (1).

METRIC SYSTEM

The English system of units will be abandoned in this country in the future. In its place, this country will use the metric system, more properly called the **International System of Units,** abbreviated **SI.** The primary reasons for adopting the SI units are (1) their extensive use in most countries of the world, (2) they are the primary units in the scientific fields, and (3) the need for uniformity between the scientific and engineering fields.

In illuminating engineering only those terms that involve units of length (Table 1-3, Conversion Tables) or areas (Table 1-4, Conversion Tables) will be affected by the conversion. Units of lumens, candela, steradians, and efficacy will remain the same. Therefore, only the units of illuminance and luminance will be affected by the conversion to SI units.

Illuminance (*E*)

The SI unit of illuminance is the lumen per square meter or the **lux;** the amount of lux is the amount of footcandles times 10.76. The constant (10.76) is actually the conversion of square feet to square meters.

$$1 \text{ fc} = 1\frac{\text{lm}}{\text{ft}^2} = 10.76\frac{\text{lm}}{\text{m}^2} = 10.76 \text{ lux}$$

The conversion formula is

$$\frac{1 \text{ fc}}{10.76 \text{ lux}} = \boxed{\frac{\text{fc}}{10.76 \text{ lux}} = 1}$$

For example,

$$30 \text{ fc} \times \frac{10.76 \text{ lux}}{1 \text{ fc}} = 322.8 \text{ lux}$$

Exitance

The SI unit of exitance is lumens per square meter. The amount of lumens per square meter is the amount of lumens per square foot times 10.76. The constant (10.76) is actually the conversion of square feet to square meters.

$$1\frac{\text{lm}}{\text{ft}^2} = 10.76\frac{\text{lm}}{\text{m}^2}$$

The conversion formula is

$$\frac{1 \text{ (lm/ft}^2)}{10.76 \text{ (lm/m}^2)} = \boxed{\frac{\text{m}^2}{10.76 \times \text{ft}^2} = 1}$$

For example,

$$3\,\frac{\text{lm}}{\text{ft}^2} \times \frac{10.76 \times \text{ft}^2}{\text{m}^2} = 32.28\,\frac{\text{lm}}{\text{m}^2}$$

Luminance

The SI unit of luminance (Table 1-5, Conversion Tables) is candelas per square meter. The amount of candelas per square meter is the amount of footLamberts times 3.426. The constant (3.426) contains the conversion of square feet to square meters and includes the factor of π. A candela per square meter is equivalent to a unit called the **nit.**

$$1\text{ fL} = \frac{1\text{ cd}}{\pi\text{ ft}^2} = 3.426\,\frac{\text{cd}}{\text{m}^2} = 3.426\text{ nit}$$

The conversion formula is:

$$\frac{1\text{ fL}}{3.426(\text{cd/m}^2)} = \boxed{\frac{\text{fL} \times \text{m}^2}{3.426\text{ cd}} = 1}$$

For example,

$$452\text{ fL} \times \frac{3.426\text{ cd}}{\text{fL} \times \text{m}^2} = 1548.6\frac{\text{cd}}{\text{m}^2}$$

PHOTOELECTRIC EFFECT

To understand how illumination measuring devices work, one must understand the basic phenomenon known as the **photoelectric effect.** The photoelectric effect is the liberation of electrons from the surface atoms of a metal. If a metal surface is illuminated by a light quantum, electrons will be liberated from the surface. If an electric field exists, the electrons will flow to the anode and create an electric current that can be detected by a galvanometer. A standard photocell will generate a small current when illuminated. The photocell consists of a metal plate coated with a semiconductor. When the photocell is exposed to light, electrons are liberated from the metal surface and are trapped at the interface unless an external circuit is provided.

An illuminance meter (Fig. 1-12) is used to determine the quantity of luminous flux falling on an area (footcandles). Basically, the meter consists of a photocell connected in series with a microammeter calibrated in footcandles. For making interior measurements, the photocell should be cosine and color corrected. The cosine correction is achieved with a specially designed translucent diffusing plate that allows the photocell to correctly evaluate the illuminance regardless of direction. The photocell is also color corrected so that its sensitivity matches the standard observer curve (see Chapter 2) or the photopic luminosity curve ($V\lambda$).

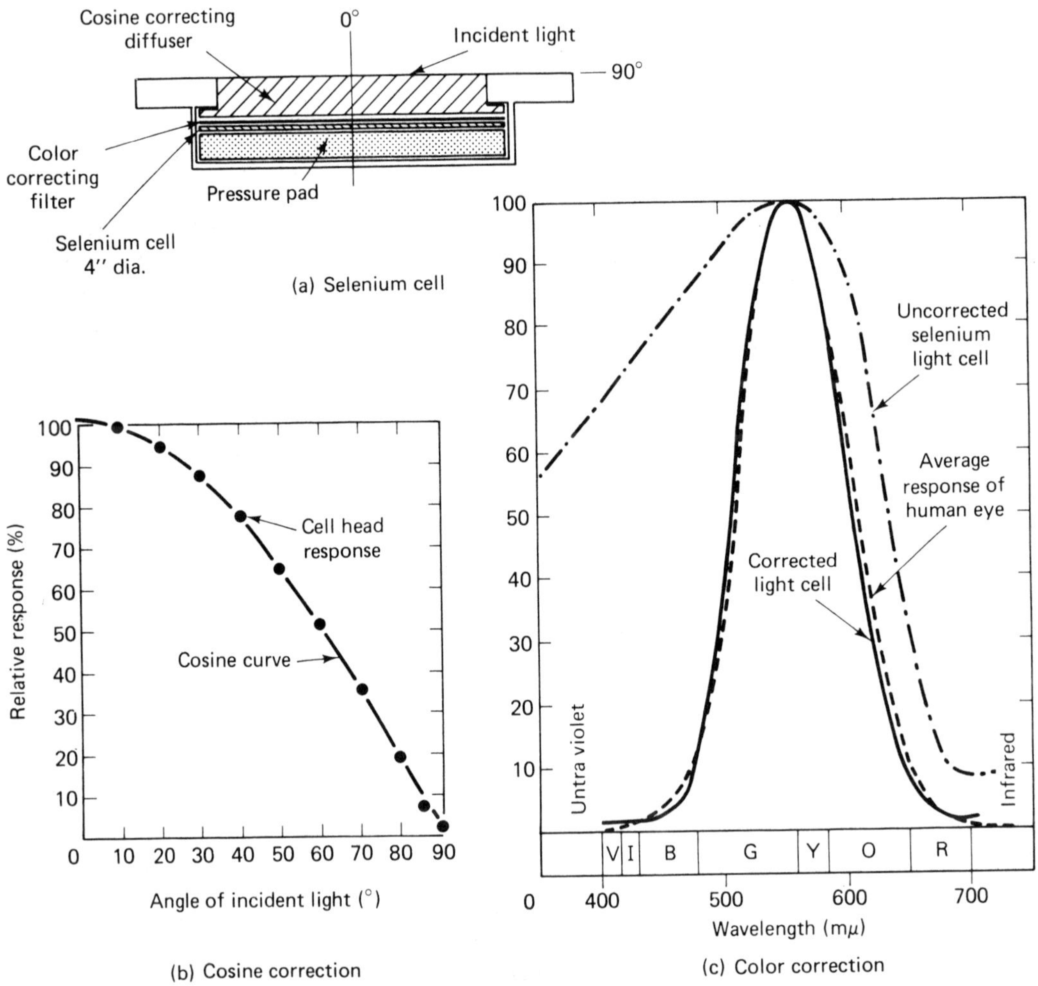

Figure 1-12. Illuminance Meter. (Courtesy of GE Lighting.)

REFERENCES

1. *Electric Lamps*, 1st ed., Part I, p. 12. Scranton, PA: International Correspondence School, 1922.
2. Kaufman, John E., and Jack F. Christensen, eds. *IES Lighting Handbook*, Reference Volume, 1984. New York: Illuminating Engineering Society of North America, 1984.

DATA SECTION

Derivation of Conversion from Luminous Flux (Lumens) to Luminous Intensity (Candelas) (or Where Does the π Come from?)

$$dA = 2\pi r \times R d\theta \qquad \frac{r}{R} = \sin\theta$$

$$r = R\sin\theta, \text{ radius of the zone} \qquad r = R\sin\theta$$

$$dA = 2\pi\, R\sin\theta\, R d\theta = 2\pi R^2 \sin\theta\, d\theta, \text{ area at an infinitesimal zone surface}$$

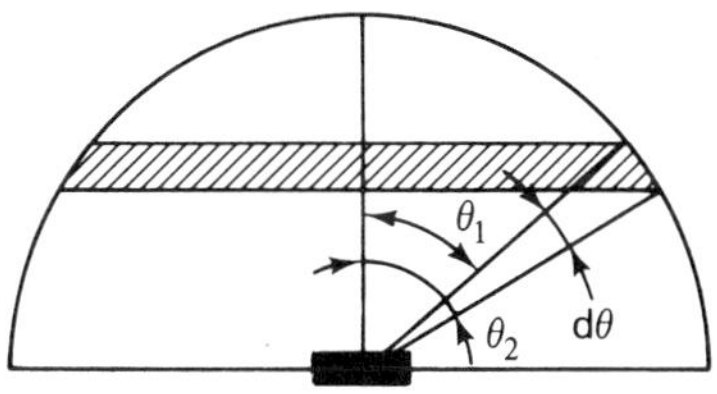

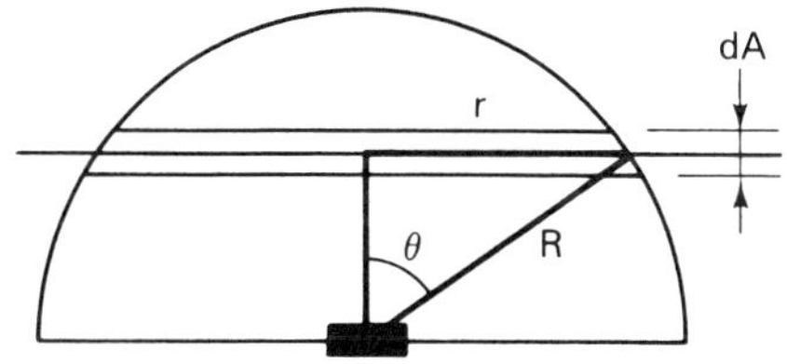

Solid angle subtended: $d\omega = dA/R^2$, by definition. Therefore,

$$d\omega = \frac{2\pi R^2 \sin\theta\, d\theta}{R^2} = 2\pi\sin\theta\, d\theta$$

$$d\phi = I_\theta d\omega$$

General expression for any source: $\phi = 2\pi \int_0^{\pi/2} I_\theta \sin\theta\, d\theta$

$$(I_\theta = I_N \cos\theta, \text{ for Lambertian surface})$$

$$d\phi = I_N \cos\theta\,(2\pi\sin\theta\, d\theta)$$

Therefore,

$$= 2\pi I_N \cos\theta\sin\theta\, d\theta$$

$$\int d\phi = 2\pi I_N \int_0^{\pi/2} \cos\theta\sin\theta\, d\theta$$

$$\phi = 2\pi I_N \left[\frac{1}{2}\sin^2\theta\right]_0^{\pi/2}$$

$$\phi = \pi I_n$$

The amount of lumens is equal to π times the amount of candelas measured normal to the surface. That is, a candela is smaller than a lumen by a factor of π, or 1 cd = π lm.

Definition:

$$1\text{fL} = 1\frac{\text{lm}}{\text{ft}^2} \times \frac{\text{cd}}{\pi\ \text{lm}} = \frac{1\ \text{cd}}{\pi\ \text{ft}^2}$$

Conversion Formulas:

$$\frac{1/\pi\ (\text{cd/ft}^2)}{1\ (\text{lm/ft}^2)} = \frac{1\ \text{cd} \times \text{ft}^2}{\pi\ \text{ft}^2 \times \text{lm}} = \boxed{\frac{\text{cd}}{\pi\ \text{lm}} = 1}$$

TABLE 1-2 Prefixes

Greek prefixes		
Million, mega	1,000,000.	10^6
Thousand, kilo	1,000.	10^3
Hundred, hecto	100.	10^2
Ten, deka	10.	10^1
Latin prefixes		
Tenth, deci	0.1	10^{-1}
Hundredth, centi	0.01	10^{-2}
Thousandth, milli	0.001	10^{-3}
Millionth, micro	0.000001	10^{-6}

TABLE 1-3 Length Conversion Factors

To Obtain \ Multiply Number of ... By	Meters	Centimeters	Millimeters	Micrometers	Nanometers (millimicrons)	Angstroms	Feet	Inches
Meters, m	1	0.01	0.001	10^{-6}	10^{-9}	10^{-10}	3.048×10^{-1}	2.540×10^{-2}
Centimeters, cm	100	1	0.1	10^{-4}	10^{-7}	10^{-8}	3.048×10	2.540
Millimeters, mm	1000	10	1	10^{-3}	10^{-6}	10^{-7}	3.048×10^2	2.540×10
Micrometers, μm	10^6	10^4	10^3	1	0.001	0.0001	3.048×10^5	2.540×10^4
Nanometers, nm (millimicrons)	10^9	10^7	10^6	10^3	1	0.1	3.048×10^8	2.540×10^7
Angstroms, Å	10^{10}	10^8	10^7	10^4	10	1	3.048×10^9	2.540×10^8
Feet, ft	3.281	3.281×10^{-2}	3.281×10^{-3}	3.281×10^{-6}	3.281×10^{-9}	3.281×10^{-10}	1	8.333×10^{-2}
Inches, in.	3.937×10	3.937×10^{-1}	3.937×10^{-2}	3.937×10^{-5}	3.937×10^{-8}	3.937×10^{-9}	12	1

TABLE 1-4 Area Conversion Factors

To Obtain \ Multiply Number of \ By	Square inches	Square feet	Square yards	Square centimeters	Square meters
Square inches, $in.^2$	1	144	1296	0.155	1550
Square feet, ft^2	6.944×10^{-3}	1	9	1.076×10^{-3}	10.76
Square yards, yd^2	7.716×10^{-4}	0.1111	1	1.196×10^{-4}	1.196
Square centimeters, cm^2	6.452	929	8361	1	10^4
Square meters, m^2	6.452×10^{-4}	9.920×10^{-2}	0.8361	10^{-4}	1

TABLE 1-5 Luminance Conversion Factors

To Obtain \ Multiply Number of \ By	Candela/$in.^2$	Candela/cm^2	Candela/m^2	Lambert	MilliLambert	FootLambert
Candela/$in.^2$	1	6	6.45×10^{-4}	2.05×10^{-8}	2.05×10^{-3}	2.21×10^{-3}
Candela/cm^2	0.155	1	10^{-4}	3.183×10^{-7}	3.183×10^{-4}	3.426×10^{-4}
Candela/m^2	1550	10^4	1	3.183×10^{-3}	3.183	3.426
Lamberts	0.487	3.142	3.142×10^{-4}	1	10^{-3}	1.076×10^{-3}
MilliLamberts	487	3142	0.3142	10^3	1	1.076
FootLamberts	452	2919	0.2919	0.929×10^{-3}	0.929	1

2 eye, vision, and visibility

ANATOMY OF THE EYE

Structure of the Eye

The eye or globe is a spherically shaped optical instrument of remarkable design. The globe is made up of three concentric layers (Fig. 2-1) referred to as **tunics.**

Outer Tunic

The **outer tunic** or **fibrous tunic** is divided into two distinctly different parts. The anterior (front) sixth is

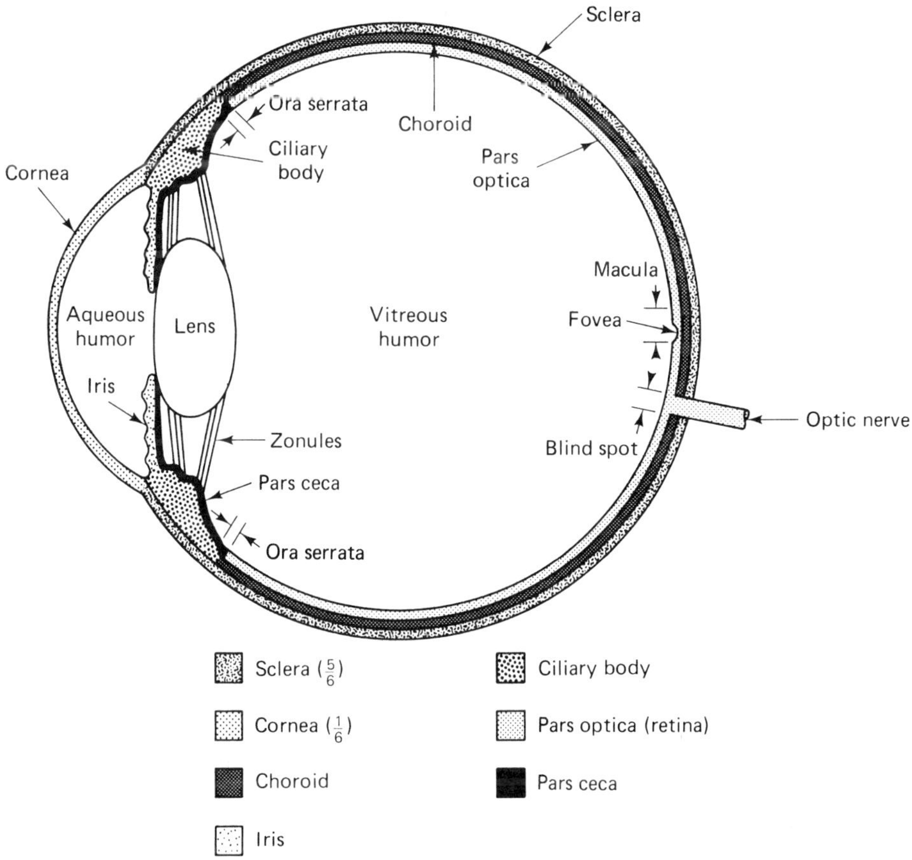

Figure 2-1. Tunics of the Eye.

the **cornea;** the posterior (back) five-sixths is the **sclera.** The cornea is the most important refractive medium in the eye. It is the outer transparent membrane that refracts or bends the incident light so that it falls on the lens and can then be focused on the retina. The sclera, on the other hand, is transluscent and so serves no optical function. It is a tough outer membrane whose purpose is to maintain the shape of the globe.

Middle Tunic

The **middle tunic, vascular tunic,** or **uvea** is subdivided into three parts: choroid, ciliary body, and iris. The most important function of the **choroid** is to provide nourishment for the outer layers of the retina (layers 1 through 4; see Fig. 2-2). The **ciliary body** is important in changing the shape of the lens in **accommodation** or focusing. The ciliary body also produces the aqueous humor that fills the outer chamber of the eye. The aqueous humor is a watery fluid that acts to refract and filter light entering the eye. The **iris** adjusts the amount of light that enters the eye and helps to increase the depth

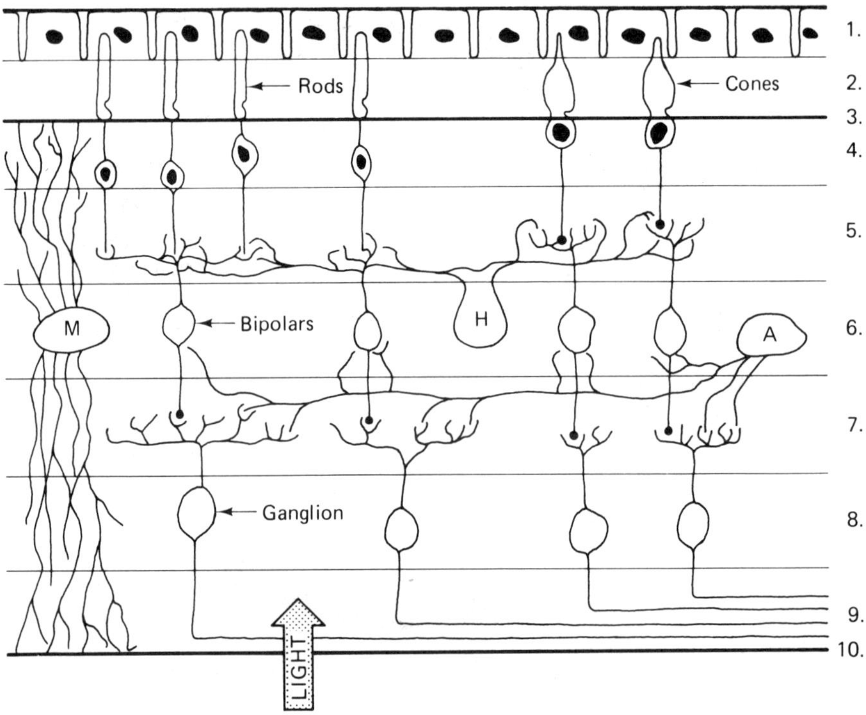

M – Muller cell H – Horizontal cell A – Amacrine cell

1. Pigment epithelium
2. Rod and cone layer
3. Outer limiting membrane
4. Outer nuclear
5. Outer plexiform
6. Inner nuclear
7. Inner plexiform
8. Ganglion cells
9. Optic nerve fibers
10. Inner limiting membrane

Figure 2-2. Layers of the Retina.

of focus. The iris thus performs one component of **adaptation,** or the process of adjusting to ambient light. We see then that the middle tunic is vitally important because it is involved in both of the critical optical functions of the eye, adaptation and accommodation.

The **ciliary zonules** are dense fibrils that run in two sheets from the ciliary body to the capsule of the lens to which they are fused. The ciliary body, the zonules, and the lens work together to perform accommodation. As an object is brought close to the eye, muscles in the ciliary body constrict causing the zonules to slacken and allowing the lens to bulge, thus bringing the nearby object into focus. As the object is moved away from the eye, the body expands, the zonules tighten, and the lens flattens bringing the now distant object into focus. An amazing feature of this process is that it is typically done automatically—we rarely stop to think about focusing.

The eye is said to be functioning in distant vision range when the object is more than 20 feet from the eye. Since the muscles that control accommodation and convergence are in a relaxed state during distant vision, the eye will have less strain and fatigue.

The magnitude of accommodation decreases with age, causing blurred near vision. This decrease in the magnitude of accommodation is believed to be due to a hardening of the lens substance. This condition is known as **presbyopia** and begins at about the age of 40.

Innermost Tunic

The **innermost tunic** or **nervous tunic** is divided into the pars optica (or retina) and the pars ceca. The nervous tunic is an outgrowth of the brain called the **diencephalon.** The pars ceca is the anterior portion of this tunic. It is represented as a thin layer on the innermost aspect of the iris and ciliary body. The pars ceca ends at the **ora serrata.** The ora serrata (see Fig. 2-1) is the transition area between the choroid and ciliary body at the level of the vascular tunic and the transition area between the pars ceca and pars optica at the level of the nervous tunic. There are no visual cells in the ora serrata. The posterior portion of the nervous tunic is the pars optica or retina proper. It is the only light-sensitive portion of the globe.

Two important specialized areas are associated with the retina: the **optic disc** and the **macula.** The optic disc (white spot or blind spot) is the area where the optic nerve leaves the retina (see Fig. 2-1). There are no visual cells at the blind spot. The blind spot appears as a white spot approximately 1.5 millimeters (mm) in diameter when viewed through the front of the eye. The macula (macula lutea or yellow spot) (see Figs. 2-1 and 2-4) is a depression about 1.5 mm in diameter (5.2° external) and is located 3.5 mm temporal (toward the temple) and 0.5 mm inferior to (below) the optic disc. The bottom-most part of the depression is called the fovea centralis or **fovea.** It is an area of about 0.4 mm in diameter (1.4° external). The fovea is the most acute visual area on the retina. The muscles of the eye and the refractive media act to focus light on the fovea. The fovea is the area of clearest vision and best color response.

Retina

The retina, which is the inner lining of the eye, is the receiving or light-sensitive portion. The retina contains delicate light-sensitive nerve fiber endings called **cones** and **rods.**

The retina consists of 10 layers of cells and processes. Figure 2-2 is a schematic cross section of the retina indicating the layers and the names associated with each layer.

Cones

The cones are the photoreceptors that are effective for ''daylight'' vision only. The maximum concentration of cones (Fig. 2-3) is in the fovea. Cone density decreases with increasing distance from the fovea. The fovea is inactive under very dim light (less than 0.003 candela per square foot (cd/ft^2)). The candela per square foot (see Chapter 1) is a unit of luminous energy leaving a surface and arriving at the retina.

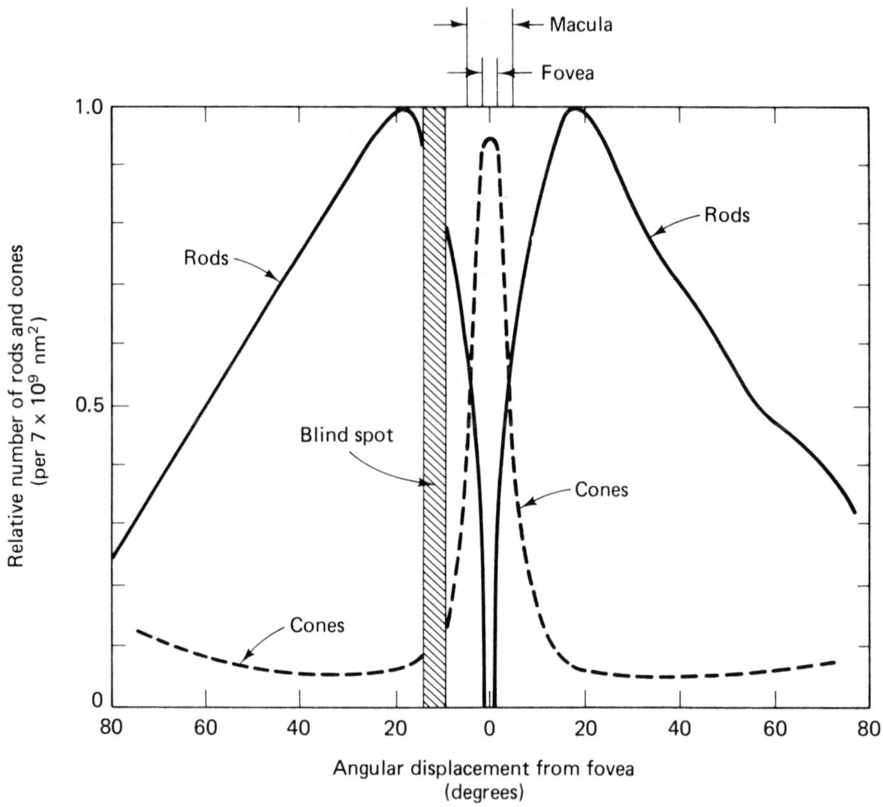

Figure 2-3. Distribution of Rods and Cones.

The rods must take over the visual process during low levels of illuminance. When the eye is receiving light at levels above approximately 0.3 cd/ft^2, the system is said to be operating under **photopic** or pure cone vision.

While moving from a dark environment into a very light environment, the visual system experiences a change in sensitivity. This phenomena is called **light adaptation.** Light adaptation involves primarily the cone system and usually takes less than a minute.

Rods

The rods are more sensitive to light than the cones; they function primarily during "night" vision. There are no rods in the fovea and rod density increases from just outside the fovea outward (Fig. 2-3). Whereas cones are linked one to one with nerve endings, a bundle of rods (many thousand) is served by one nerve ending. These multiple hookups result in very poor visual acuity. However, since each nerve ending serves a relatively large portion of the visual field, the rod system is more likely to detect changes such as motion or flicker. This accounts for the phenomenon of seeing flicker out of the "corner of the eye" which disappears when one turns to face the source, since the periphery of the visual field is where rods are the predominant photoreceptor. Similarly, in dim light situations one is less likely to detect an object by looking straight at it (foveal, cone vision) than by placing the object in the periphery where the more sensitive rods can detect it.

There is no color response with the rod system. Rods produce a black and white response, which is actually a reaction to variations in luminance. Moving from a very light environment into a dark environment results in a change in sensitivity of the visual system due to **dark adaptation.** Dark adaptation is not completely achieved until about 1 hour after light is removed. The initial rapid phase of dark adaptation is due to low-level cone action, while the remaining portion of dark adaptation is due to slower rod action. **Scotopic** or rod vision occurs when the eye has been dark adapted and only the rods are functioning.

Rods and Cones

When both rods and cones are functioning, the visual system is said to be operating in the **mesopic** range. The mesopic range is from 0.003 cd/ft^2 to approximately 0.3 cd/ft^2. The upper limit is dependent on the size, position, and time of exposure of the rod system. This is the range most commonly encountered during nighttime activities such as driving.

Distribution of Cones and Rods

There are 6 to 7 million cones and 75 to 150 million rods distributed across the surface of the retina. Figure 2-3 shows the relative distribution of the cones and rods. The break in the curves (parallel lines) represents the location of the optic disc or blind spot. The largest concentration of both cones and rods is in the region of the macula. Figure 2-4 identifies the area and external angle subtended by areas in the region of the fovea. The external angle represents the projection of an area on the retina out into the external world outside the eye. The external angle is referred to as the **visual angle** (see Visual Acuity section). This angle can be used to determine the physical size of objects at various distances that can be seen by each region (Fig. 2-4) of the retina.

VISUAL PATHWAY

Radiant energy in the visible spectrum passes through the eye and is focused on the retina. The light stimulus causes a photochemical reaction in the rods and cones. This reaction causes nerve impulses to be generated in ganglion cells, which are then transmitted through ganglion cell fibers to the region of the optic disc. The impulse continues along what are known as fibrous processes, the elements which make up the optic nerve (Fig. 2-5). The optic nerve leaves the bony orbit through an opening called the optic foramen and continues to a region where half of the fibers decussate (cross over) in a structure called the **optic chiasma.** The optic fibers leaving the chiasma change in

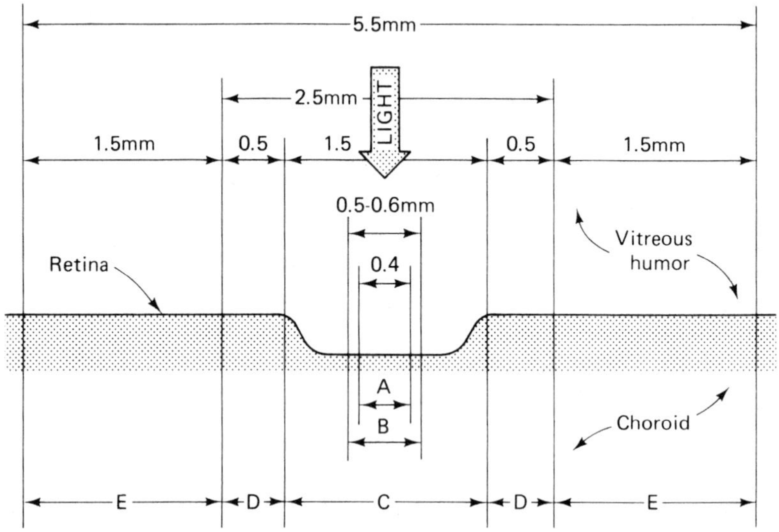

	Region	Diameter		External angle, degrees (object space)
		Microns (μ)	mm	
A.	Fovea	400	0.4	1.4°
B.	Rod free area	500 – 600	0.5 to 0.6	1.7 – 2.0°
C.	Mucula	1500	1.5	5.2°
D.	Parafovea (band approx. 500 μ wide)	2500	2.5	8.6°
E.	Perifovea (band approx. 1500 μ wide)	5500	5.5	19 – 20°

Figure 2-4. Areas Within the Macula Region.

name and make up. Thereafter, the fibers are called the **optic tract.** The impulses travel through the fibers of the optic tract and terminate in the **lateral geniculate body (LGB),** which is located in the thalamus of the brain. The cells of the LGB give rise to new fibers, which form the **optic radiation** fibers that extend out and back through the brain to the cortex (surface of the brain) in a region of the brain called the **occipital lobes** (Fig. 2-6). The area of the cortex that receives the optic radiations surrounds the calcarine fissure and is called the striate area, striate cortex, visual area, or Brodmann's Area 17. The gross area of vision in the brain is called the **occipital cortex.**

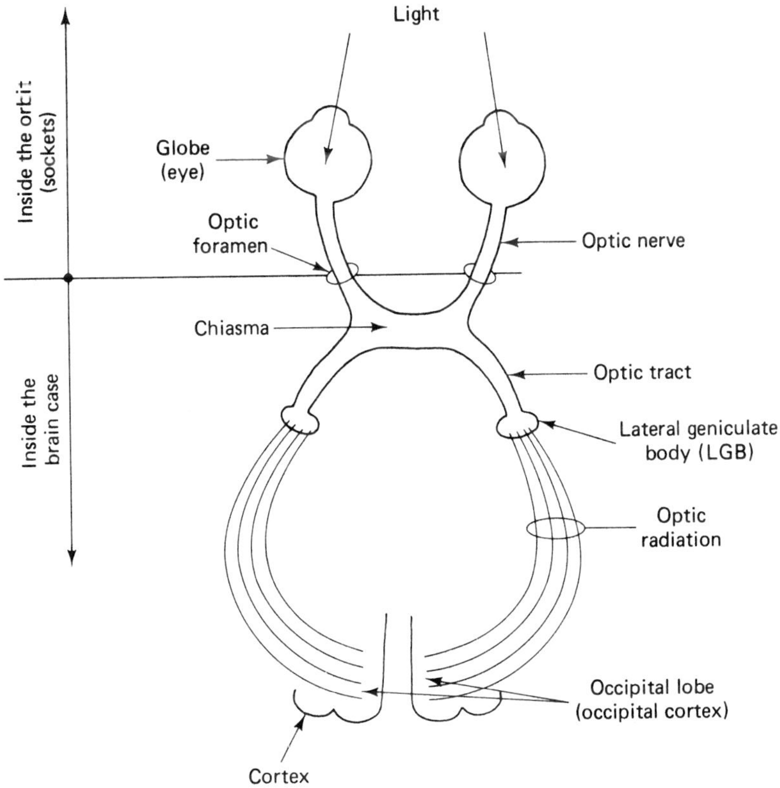

Figure 2-5. Visual Pathway.

SENSITIVITY OF THE VISUAL SYSTEM

The luminous efficiency of radiant energy is dependent on the ability of various receiving and measuring devices. The spectral response characteristics of the human eye vary between individuals so that it is not feasible for any one individual to act as a standard observer. Therefore, experiments had to be conducted on thousands of subjects to determine the average response characteristics of the human visual system.

The spectral sensitivity of the human eye to the entire visible spectrum can be established experimentally. The scotopic (dim-light) experiment was initially conducted by Hecht and Williams in 1922. In 1932 Gibson and Tyndall conducted a bright-light experiment to obtain the photopic curve. These experiments involved measuring the threshold response to stimulus at various wavelengths. The results are indicated in Figure 2-7. Threshold can be thought of as the yes/no point of vision, that is, the point at which some critical detail just disappears, the "I see it/I don't see it" point of vision.

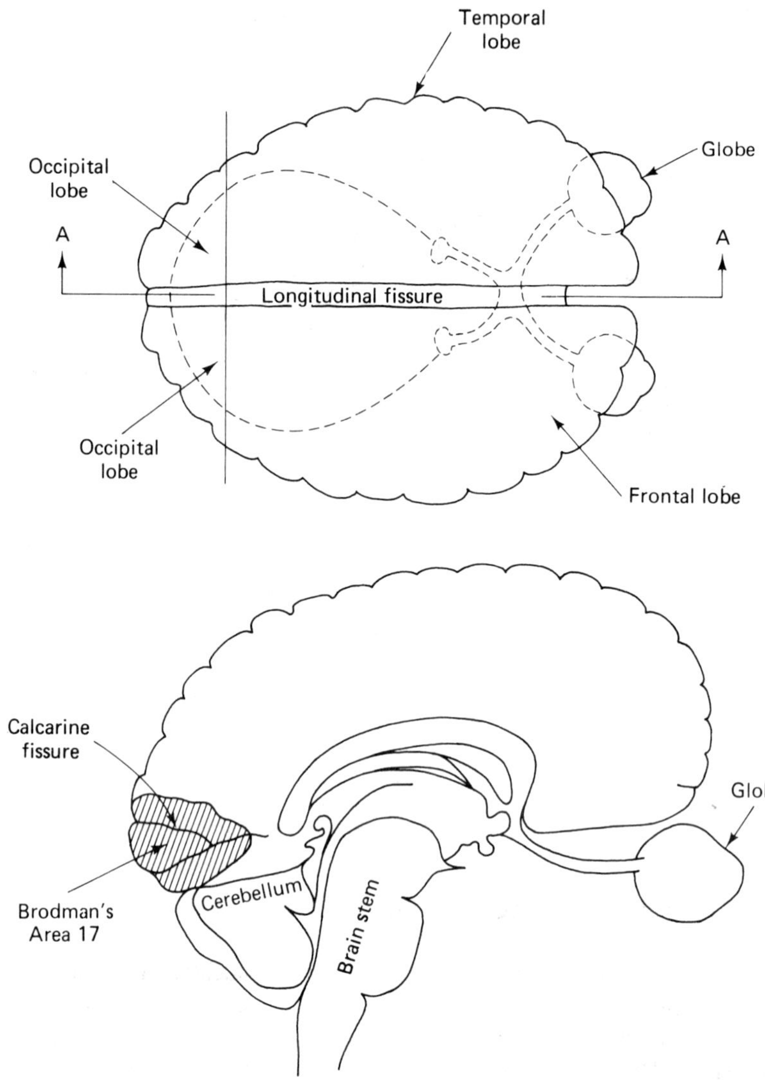

Figure 2-6. Brain and Occipital Cortex.

The interval between the absolute threshold of visibility (scotopic curve) and the initial appearance of hue for a given homogeneous wavelength (photopic curve) is called **photochromatic interval.** This interval is greatest at the shorter wavelengths (blue) and near zero at the longer wavelengths (red). Mesopic vision occurs at intermediate levels of luminance, where rods and cones are believed to work together.

By definition, sensitivity has a reciprocal relationship to threshold. That is,

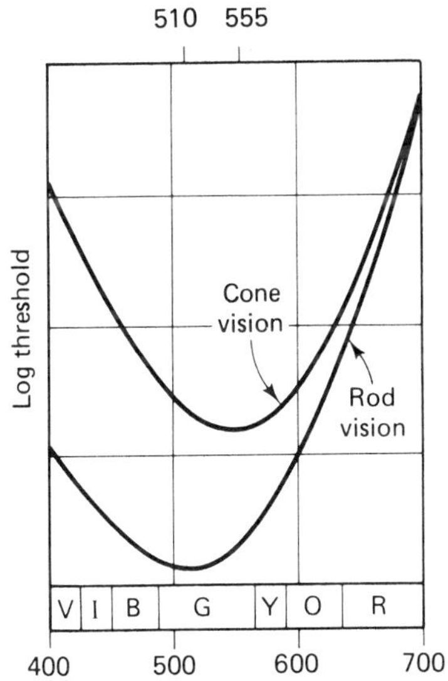

Figure 2-7. Threshold.

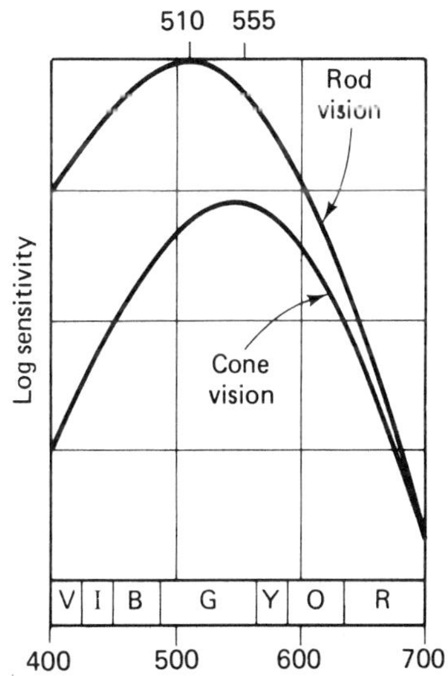

Figure 2-8. Sensitivity.

$$\text{sensitivity} = \frac{1}{T_\lambda}$$

This results in the threshold curve being inverted (Fig. 2-8) and allows one to speak in terms of sensitivity, which is a more meaningful term to most individuals.

Four major facts can be derived from sensitivity curves:

1. Neither the rods nor cones are uniformly sensitive across the visible spectrum.
2. The region of maximal sensitivity is 555 nm for cone vision and 510 nm for rod vision.
3. The rod function lies above the cone function, indicating that throughout most of the spectrum, the rods require less energy for vision (have a greater sensitivity) than the cones.
4. The rods and cones are equally sensitive to radiant energy in the long-wavelength (red) end of the spectrum.

Visual stimulation at scotopic levels does not produce color vision anywhere along the spectrum. Color vision is possible only with light levels of sufficient magnitude to activate the cone system. When only the rods are functioning, all wavelengths are seen as a series of lighter or darker grays. Weak light is visible, but hue is absent.

The previous curves are more commonly presented in the form of photopic and scotopic relative luminosity curves (relative spectral sensitivity curve or standard observer curve). They are shown in Figure 2-9. The notation used is

$$V_\lambda = \text{photopic}$$
$$V'_\lambda = \text{scotopic}$$

Despite the gross difference in absolute sensitivity between rods and cones, the two functions are plotted on the same graph by performing a simple arithmetic adjustment to put them on a relative basis.

$$\text{relative sensitivity} = \frac{1}{T_\lambda} \times T_{\text{max}} \times 100\%$$

The cone sensitivity curve (Fig. 2-8) must be raised a vertical distance of approximately 1.6 log units to make it comparable to the rod sensitivity curve on a relative basis. The relative sensitivity curves are used for photometric problems to represent the response of a standard human observer.

The Purkinje effect is a shift in the maximum sensitivity of the eye from photopic to scotopic vision. The relative spectral sensitivity curve for cones (photopic curve) peaks at 555 nm. During rod or scotopic vision, the relative spectral sensitivity curve shifts 45 nm toward the 400 nm (blue) end of the spectrum so that the peak occurs at 510 nm. This shift results in an increase in sensitivity to shorter wavelengths (400 nm) and a decrease in sensitivity to longer wavelengths (700 nm) for the rod system. Even though objects will be colorless under rod vision, a blue object will appear brighter or more intense during scotopic (rod) vision than a red object of equal reflectance, due to this shift in sensitivity.

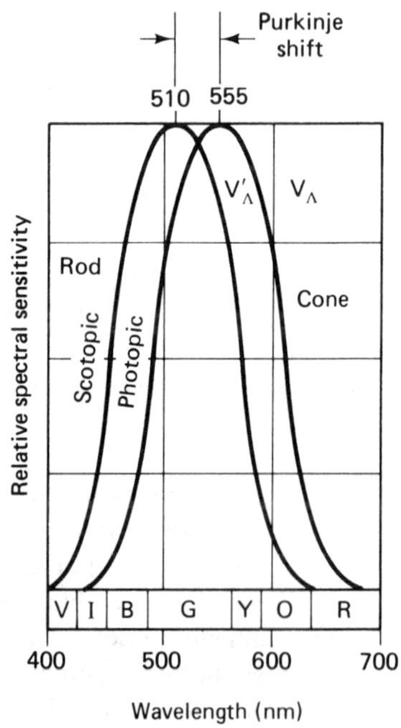

Figure 2-9. Relative Sensitivity.

COLOR VISION

Normal color vision is referred to as **trichromatic** vision, meaning three colors. Color vision is due to a visual system that consists of three different classes of cones. Each cone contains a different pigment, and each pigment has a different spectral response curve. The three spectral response curves for each pigment overlap across the visible spectrum. Each spectral response curve is shifted with respect to the other along the wavelength continuum so that each has a different maximum sensitivity. The peak or maximum sensitivities occur at 570, 535, and 445 nm.

However, as can be seen in Figure 2-10, all cones have some sensitivity at all wavelengths. The color receptors are commonly referred to as the red, green, and blue cones. The designations refer to the wavelengths associated with each cone system. According to this theory, normal color vision is a function of the three kinds of pigments present in the three classes of cones. Thus the normal human visual system consists of four classes of receptors, three associated with the cones and the fourth associated

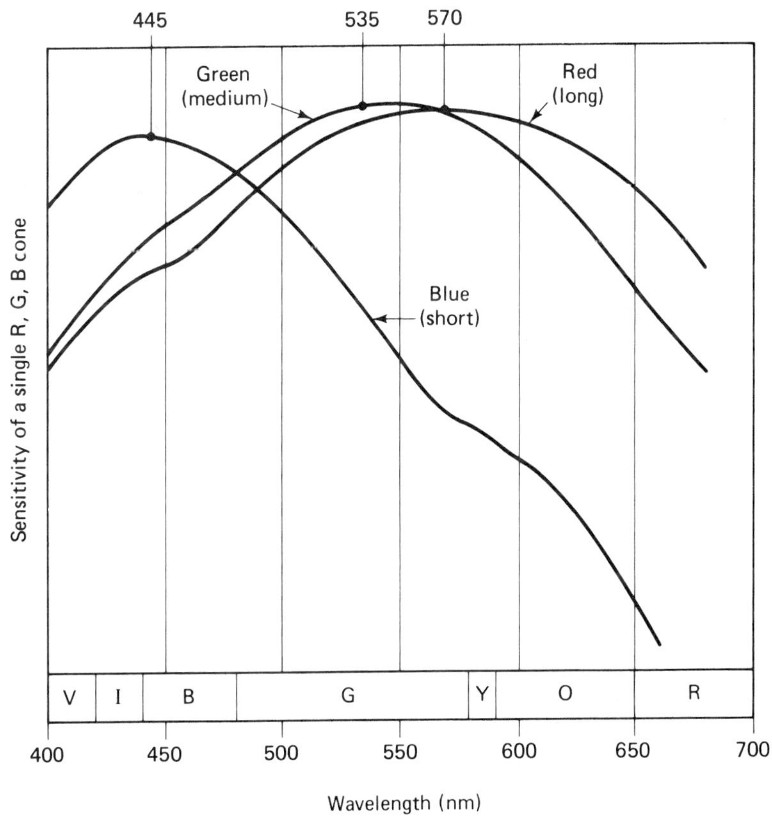

Figure 2-10. Three-Cone Spectral Response. (Original drawing by Dr. Lewis O. Harvey, Jr., based on data from Vos and Walraven.)

with the rod pigment (rhodopsin). Each of the four receptors has its own spectral sensitivity curve, which means that the human visual system is actually tetrachromatic rather than trichromatic.

VISUAL PHENOMENA

Afterimage

Afterimage is the continuation of the visual sensation after the stimulus has ceased. It is caused by a continuation of the photochemical process resulting in nerve impulses, which cause continued object perception for a short period of time. Afterimage effect is proportional to the luminance of the object. Afterimage can be a negative or complementary process. The original color of the object is replaced by its complementary color. Afterimage makes a series of still pictures appear to move, a phenomenon from which motion pictures are derived.

Time Sensitivity

The time sensitivity of the visual system is phenomenal. The human eye can detect a pulse of light that acts for only 1/1000 second. It can detect a second pulse if the time interval between the two pulses is 1/10 s. This latter phenomenon applies only for cone vision, since the cones are much more time sensitive than the rods.

Stimulus Sensitivity

Rods are a thousand times more sensitive to intensity than the cones are at low levels of illuminance. Under scotopic vision, it is said that the rods can detect the flicker of a candle 14 miles away.

BINOCULAR VISION

The binocular vision field is the total vision field seen by both eyes, minus the areas blocked by the nose, eyebrows, and the cheeks (Fig. 2-11). The binocular visual field occurs in the object space. Object space is the real world in front of the observer, as contrasted to the visual space physically inside the brain.

Under normal vision, both eyes focus on an object, and the resulting two images are fused and interpreted by the brain as one. The visual field of both eyes includes more area than the field of one eye alone. The horizontal field subtends an angle of 180°, the average vertical field subtends an angle of 130° (Fig. 2-12).

The central field of distinct vision or foveal vision (Fig. 2-4) subtends an angle of 1.4° in object space. The field surrounding the central visual field (1.4° to 20°) has a level of sensitivity that is about 1% of the sensitivity of foveal vision. Beyond the

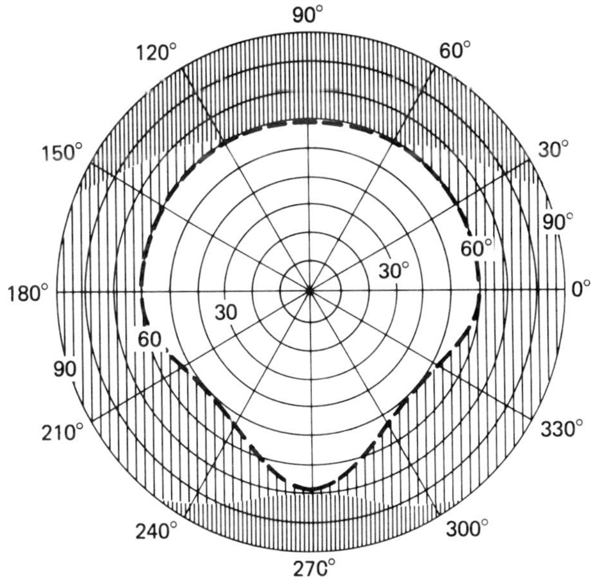

Figure 2-11. Binocular Field of View. (Courtesy of Holophane, Inc., H. L. Logan, 1939.)

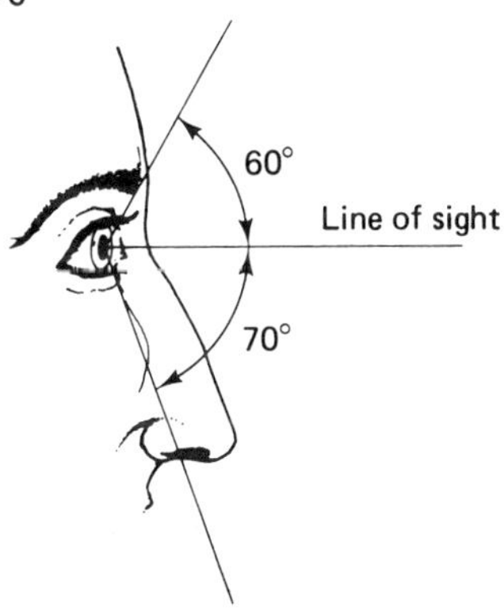

Figure 2-12. Vertical Limits of the Field of View.

20° limit, vision is very indistinct and blurred. However, this outer region is very sensitive to changes in brightness and motion.

VISUAL ACUITY

Visual acuity is defined as a measure of the ability of the eye to distinguish detail. Four factors affect visual acuity: (1) size, (2) luminance, (3) contrast, and (4) time.

Size

Size is the most generally recognized and accepted factor in seeing. Visual acuity is dependent upon the size of an object, which affects the size of the image on the

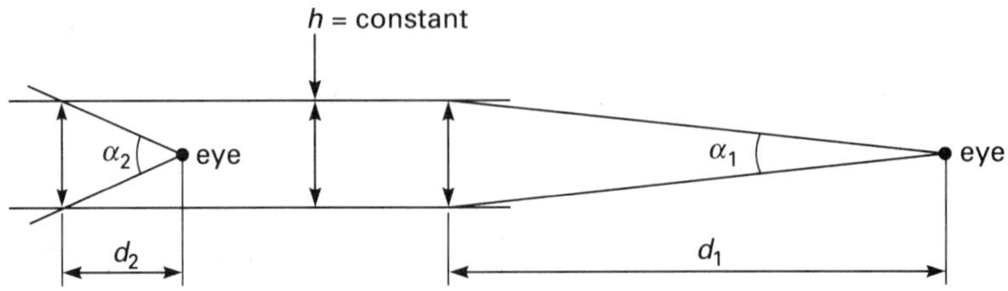

Figure 2-13. Physical Size Versus Visual Size.

retina. The important aspect of size is not the physical size of the object, but the visual angle that the object subtends at the eye. When an object is brought closer to the eye, we are actually increasing the visual angle, thus making the object clearer (see Fig. 2-13).

Visual Angle

The visual axis (*BB*'; see Fig. 2-14) meets the retina at the fovea centralis. To determine the size of the retinal image produced by any external object, we use

$$\tan \alpha = \frac{\overline{AB}}{R}$$

For practical purposes, *R* is measured from the front surface of the cornea. For small visual angles,

$$\tan \alpha = \alpha \text{ (radians)}$$

then

$$\alpha = \frac{\overline{AB}}{R} \text{ (radians)}$$

or

$$\alpha = \frac{180\,\overline{AB}}{\pi R} = \frac{57.30\,\overline{AB}}{R} \text{ (degrees)}$$

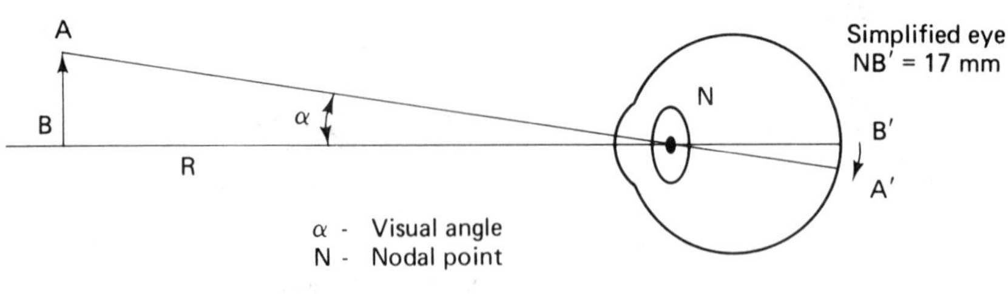

Figure 2-14. Visual Angle.

then

$$\alpha = \frac{180}{\pi} \times 60 \frac{\text{min}}{\text{deg}} \times \frac{\overline{AB}}{R}$$

$$\alpha = \frac{3438\ \overline{AB}}{R} \text{(minutes)}$$

If $AB = h$, height of object, and $R = d$, distance from the object to the front surface of the cornea, then, visual angle

$$\alpha = \frac{h}{d} \times 3438 \text{ (minutes)}$$

Actually, d should be measured to the nodal point, which would add approximately 7 mm to the distance. However, in most cases this distance is ignored.

Visual Acuity

The **visual acuity number** is defined as the reciprocal of the minimum visual angle, in minutes of arc, required to detect some critical detail. The larger the angle required to see an object, the poorer the visual acuity, and hence the smaller the visual acuity number. Increasing the visual angle is achieved by bringing objects closer to the subject.

$$\text{Visual acuity number } \nu = \frac{1}{\alpha} = \frac{0.00029\ d}{h}$$

Snellen found that the threshold size of detail, for most subjects with so-called **normal** vision, was about 1 min of arc for black objects on white background.

The **Landolt ring** or **C** (see Fig. 2-15) has been used as a standard test object for establishing visual acuity. A 1 min Landolt ring will have a critical detail of 1 min with a stroke of 1 min and an overall dimension of 5 min.

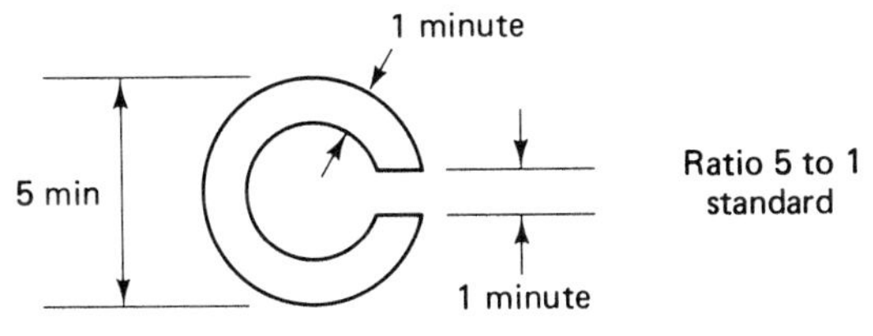

Figure 2-15. Landolt Ring or *C*.

Snellen Fraction

Another measure of acuity, the Snellen fraction is given by

$$\frac{D'}{D} = \frac{\text{actual distance}}{\text{normal distance}}$$

where

D' = distance in feet at which a given line of letters is barely recognizable by any subject
D = distance in feet at which the same line is barely recognizable by a person with normal vision

The standard Snellen test chart is based on a 20 ft viewing distance; therefore, the distance D' from which the test chart is actually viewed is set equal to 20 ft. A Snellen fraction of 20/40 means that the subject can barely read a letter at 20 ft that a person with normal vision can see at 40 ft. Therefore, the person with 20/40 vision has poorer acuity than does the normally sighted person.

Image Height on the Retina

A 1 min object will produce an image height of 0.0049 mm on the retina.

$$h_i = \frac{nh}{d}$$

where

h_i = height of retinal image
n = nodal point in millimeters = 17 mm
h = height of object in millimeters
d = distance from object to nodal point in millimeters

If $\alpha = h/d \times 3438$, then

$$h = \frac{\alpha d}{3438}$$

Therefore,

$$h_i = \frac{n}{d} \times \frac{\alpha d}{3438} = \frac{n\alpha}{3438}$$

For a 1 min test object, $\alpha = 1$ min; then

$$h_i = \frac{17 \times 1}{3438} = 0.0049 \text{ mm} = 4900 \text{ nm}$$

Luminance

Luminance is an objective physical quantity that the layperson often confuses with brightness. It is important, however, for the lighting professional to maintain a distinction between the two. The critical difference lies in the fact that **brightness** is a subjective evaluation or interpretation of the amount of light reaching the visual system. Luminance, on the other hand, is dependent on the amount of light striking a surface and the amount of light being reflected back to the eye. Surfaces with low reflective values require more light than those with high reflective qualities if the luminance is to be equal. For diffuse surfaces,

$$L = \rho \times E \quad (2.1)$$

where

L = luminance
ρ = reflectance
E = illuminance

The visual system sees luminance (the light reflected from surfaces), not illuminance (the light arriving at an object). Therefore, calculational procedures based upon the light reaching a surface will give little indication of how well a task can be seen. Seeing is a function of the light leaving a surface (luminance). The eye receives bundles of light from objects of different luminance in space and must accommodate to these luminance differences by an averaging process.

Comfort and visibility are dependent upon the luminance patterns within the visual field. Comfort is dependent on the variation in luminances in the visual field (luminance ratios). Visibility is affected by the adaptation state, quality of luminance patterns, and clutter within the visual field. Clutter can create an overload of the visual system due to excessive luminance ratios.

Contrast

Contrast is the basic seeing mechanism in vision. **Contrast threshold** is a measure of the ability of an observer to distinguish a minimum difference in luminance between two areas a given percentage of the time. The contrast threshold is expressed as a fraction:

$$C = \left| \frac{L_o - L_b}{L_b} \right| = \left| \frac{\Delta L}{L_b} \right| \quad (2.2)$$

where

L_o = luminance of test object
L_b = luminance of the background
Δ_L = is either an increment or decrement superimposed on the background

Figure 2-16 defines contrast terms.

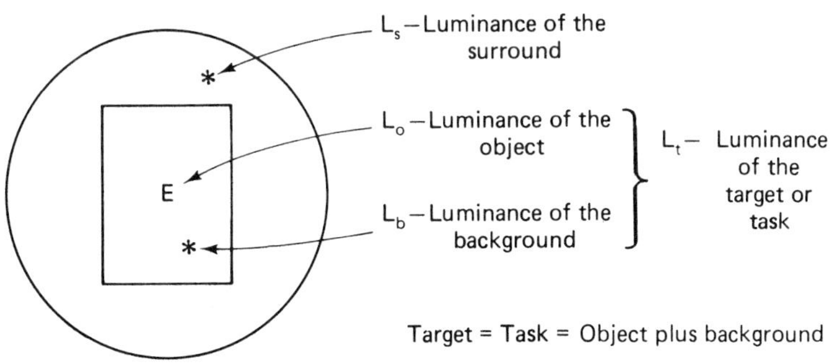

Figure 2-16. Definition of Contrast Terms.

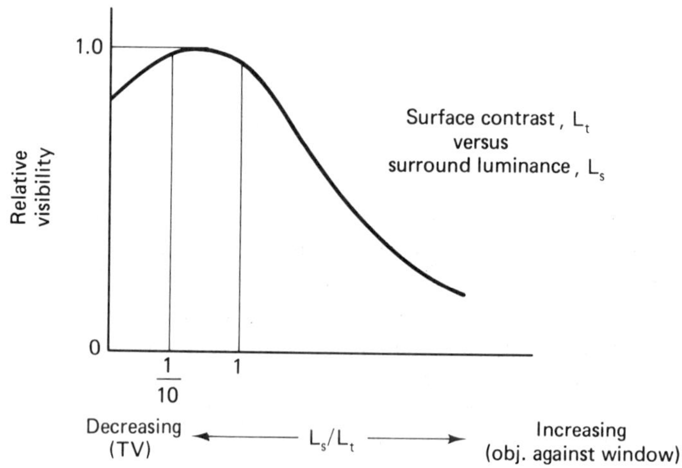

Figure 2-17. Effect of Surround on Surface Contrast.

Surface Contrast

Surface contrast is associated with the contrast between an object (print) and its background (paper). The ability to see surface contrast is affected by physical contrast and the surrounding luminance, L_s.

Outline Contrast

If the surrounding luminance L_s is much greater than the target luminance L_t, the task contrast will be lost and the target will go into silhouette. If surrounding luminance L_s is much less than target luminance L_t, the task contrast may not be affected by excessive luminance but may create discomfort, and hence a reduction in visibility.

To see surface detail, that is, to have good surface discrimination, the luminance ratios in the visual field must be controlled. The maximum visual efficiency occurs when the surrounding luminance L_s, has a ratio to target luminance L_t that ranges from 1/10 to 1.0.

You will notice in Figure 2-17 that visual efficiency falls off rapidly as surrounding luminance exceeds target luminance. Under these conditions the target is in silhouette against the surrounding luminance.

Surface Contrast, Task Contrast, or Contrast

There are two ways of specifying contrast. The first is the classical approach developed by Weston.

$$C = \frac{L_g - L_t}{L_g}$$

where

L_g = the greater luminance
L_l = the lesser luminance

The Weston formula results in contrast limits of C from 0 to 1.

Blackwell found that visibility is not the same if the relationship between object and background is reversed in the Weston formula. For example, if

$$L_o = 5 \text{ fL} \quad \text{and} \quad L_b = 80 \text{ fL}$$

then

$$C = 0.94$$

However, if

$$L_o = 80 \text{ fL} \quad \text{and} \quad L_b = 5 \text{ fL}$$

then the same result is achieved:

$$C = 0.94$$

That is, the contrast values are the same but the visibility is not the same. Therefore, Blackwell developed a formula (eq. 2.2) for contrast which eliminates this problem. That formula, which we saw previously, has become the standard method for calculating contrast.

In equation 2.1 we saw that, for a diffuse surface (e.g., a paper towel), light leaving (luminance, L) is equal to light arriving (illuminance, E) multiplied by reflectance. Thus the Blackwell formula may be written as:

$$C = \left| \frac{E_o\rho_o - E_b\rho_b}{E_b\rho_b} \right|$$

for a diffuse surface. If E is a constant (the same amount of light arrives at the object as arrives at the background), then for a diffuse surface,

$$C = \left| \frac{\rho_o - \rho_b}{\rho_b} \right|$$

The Blackwell formula results in the following limits of contrast. If

$L_b > L_o$, C goes from 0 to +1
(L_o, black print; L_b, white paper)
If $L_o > L_b$, C goes from 0 to ∞
(L_o, white print; L_b, black paper)

The highest visibility occurs when the object is brighter than the background ($L_o > L_b$, white print on black paper). The Blackwell method relates directly to visibility. With this formulation, a task with a contrast of 2 will be twice as visible as a task with a contrast of 1. For example, with white print (80%) on black paper (5%), Blackwell:

$$C = \left| \frac{\rho_o - \rho_b}{\rho_b} \right| = \left| \frac{0.80 - 0.05}{0.05} \right| = \left| \frac{0.75}{0.05} \right| = 15.0$$

Weston:

$$C = \frac{\rho_g = \rho_l}{\rho_g} = \frac{0.80 - 0.05}{0.80} = \frac{0.75}{0.80} = 0.94$$

Contrast Sensitivity

Contrast sensitivity is the level of sensitivity of the visual system to differences in luminance of adjacent surfaces. It is defined as the reciprocal of contrast, and it increases with increasing adaptation luminance. That is, as the total quantity of light reaching the retina increases, the sensitivity to (or the ability to detect) surface contrast will increase (see Fig. 2-18).

If we superimpose the contrast and contrast sensitivity curves, we can see their reciprocal relationship. As the level of contrast sensitivity increases, the visual system needs less contrast for a certain level of visibility (see Fig. 2-19). Points along the contrast curve represent some predetermined level of visibility or equal seeability.

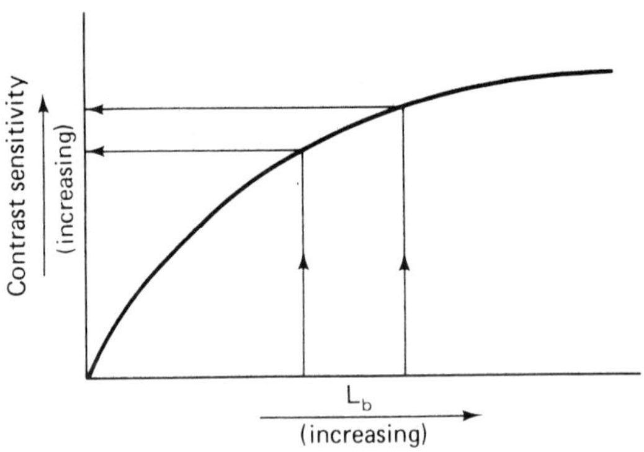

Figure 2-18. Contrast Sensitivity Versus L_b over Normal Indoor Ranges.

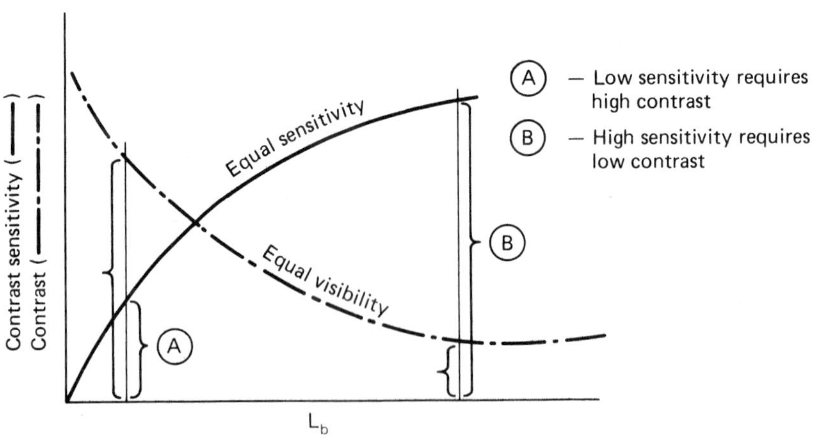

Figure 2-19. Contrast Sensitivity Versus Contrast.

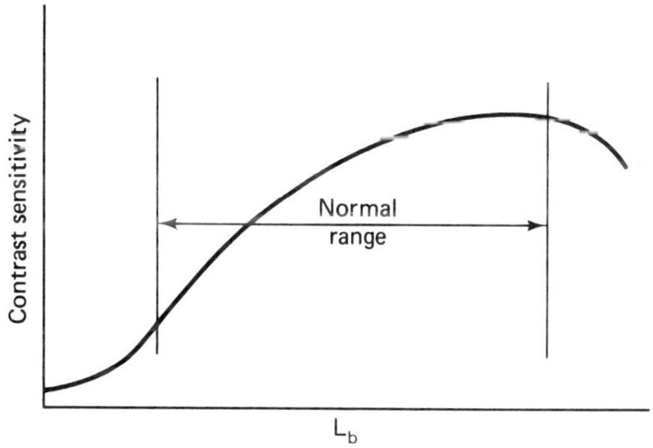

Figure 2-20. Contrast Sensitivity Extended Range.

Over the normal range of background luminances, contrast sensitivity increases with an increase in background luminance (see Fig. 2-20). At very high background luminance, the eye becomes less tolerant of luminance differences between an object (L_o) and its background (L_b), due to an increase in eye sensitivity, and contrast sensitivity may drop off.

Time

Time is the fourth factor that affects visual acuity. Seeing is not instantaneous. A time lag exists in the electrochemical processing of the retinal signal that reaches the brain. As the level of background luminance increases, the time required to interpret details will decrease. Just as the camera requires a longer exposure time in dim light than in bright light, so does the eye. The eye can distinguish and discriminate details at low luminance levels if given enough time.

Figure 2-21 shows that, as luminance increases from 1 to 6 fL, seeing time is

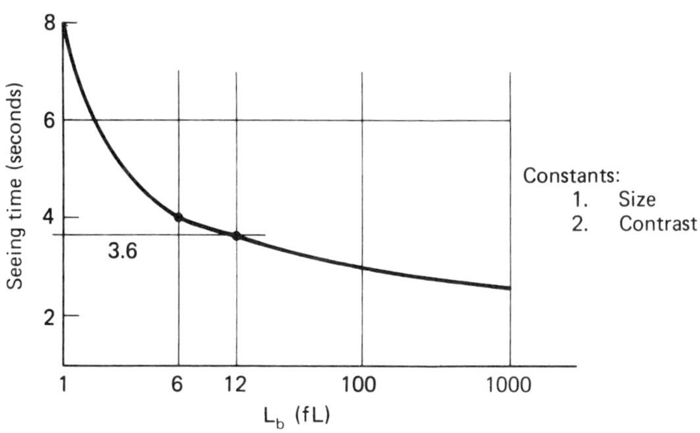

Figure 2-21. Time Versus L_b.

cut in half; an additional increase from 6 to 12 fL will result in an additional decrease in seeing time of 5%.

Conclusions

Visual acuity is affected by four factors: (1) size, (2) luminance, (3) contrast, and (4) time. The lighting design professional usually has little or no control over size or time. If these two variables are fixed, the ability to see surface detail (visual acuity) is dependent on task contrast and background luminance.

For Constant Contrast

As the level of background luminance L_b increases, the sensitivity to contrast increases, which results in an increase in visual acuity (see Fig. 2-22).

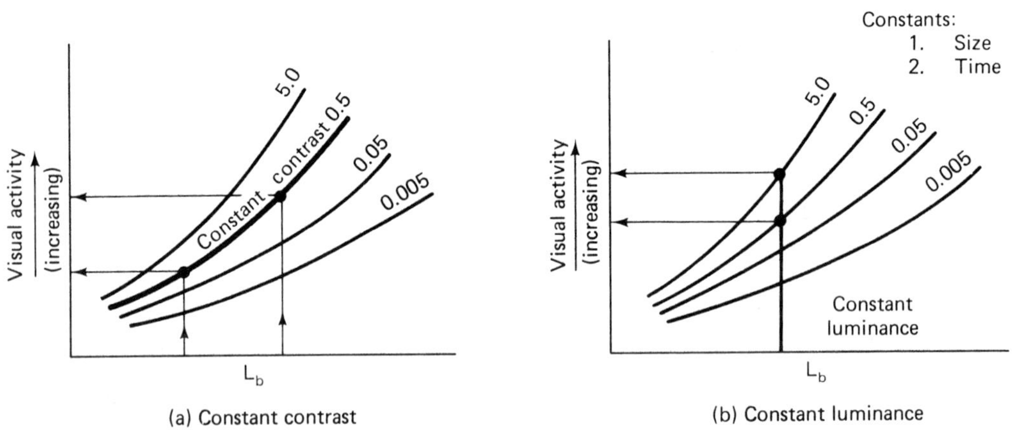

Figure 2-22. Visual Acuity Versus L_b.

For Constant Luminance

At a constant level of background luminance, L_b, the task contrast must be increased to increase the level of visual acuity.

3 color

The subject of color is important to the illumination professional because of the impact of light on color perception. Color is not a property of objects, but a psychological response to the different wavelengths of radiant energy incident on the retina. It is the combination of spectral (color) distribution characteristics of a light source and the reflective properties of an object that produce a specific color appearance for the object. If the spectral characteristics of the light source or object change, the color appearance will change. These interactions must be understood by the lighting designer if the color appearance desired by the architect or interior designer is to be achieved. The lighting professional must also make the interaction known to the architect or interior designer. If color selection is not made with the light source in mind, the brilliant, rich, pure colors chosen by the designer may very well appear dull, drab, and washed out. The time, effort, and cost that went into color selection for room finishes and furnishings will all be wasted.

Color is also important because of its possible effects on visual acuity. Color is considered by some to be the fifth factor that affects visual acuity. Very little research has been done in the United States on the subject of color contrast. It is obvious that color contrast is important and under certain conditions will affect visual acuity. If two objects have exactly the same texture and reflectance properties but differ in color, they can be distinguished from one another because of color contrast. However, if one is composed of "pure" (single-wavelength spectral distribution) red pigment and the other is "pure" green pigment, and they are both illuminated with monochromatic (single wavelength) yellow light, the two objects will appear as a single object since the contrast between the two is zero.

COLOR THEORY

In the seventeenth century, Newton found that "white" light consisted of many different-colored light rays. When he passed sunlight through a prism of glass, the light ray was dispersed into a rainbow of colors he called the **spectrum.** By passing the light spectrum through a second prism, the energy emerged from the second prism as a single ray of "white" light. Newton concluded from his experiments that the seven colors in the spectrum were the fundamental or primary colors.

If a sufficient quantity of light of a single wavelength, such as 650 nm, strikes the retina of the eye, the sensation of distinct color is produced; in this case the stimulus is described as red. Color has been shown to be a sensation or a matter of vision. The sensation of color (color vision) is produced by the action of radiant energy of specific wavelengths acting on the retina of a normal eye. The radiant energy leaves the light source in the form of waves traveling at the speed of light. Variations in the wavelength of the radiant energy will produce different sensations that correspond to different hues or colors.

Newton's experiments showed that the nature of light (its spectral distribution characteristics) influences the perception of color. That is, objects will look different under daylight or incandescent or fluorescent light. Object color is dependent on the nature of the light cast on the object (or incident light), the reflectance of the object, and the response characteristics of the subject's eye. Object color is due to the phenomenon called **selective absorption** (see Fig. 3-1). Selective absorption is the result of object pigment decomposing the incident light, absorbing some of the rays, and reflecting or scattering others. The properties of the material or its pigmentation determines the color appearance by a subtractive or absorptive process. This process allows only certain wavelengths to be reflected, which gives an object its color appearance. A gray appearance

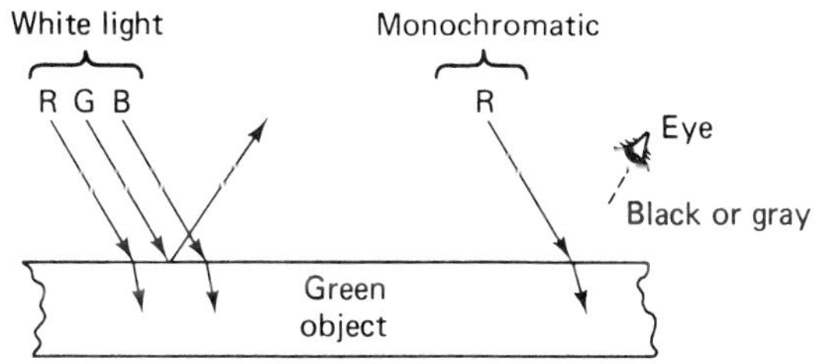

Figure 3-1. Selective Absorption.

is produced by an object absorbing a certain percentage of rays and reflecting the remaining portion without disturbing the relative proportion of energy in the source.

Subtractive Color Mixing Theory

Brewster, experimenting with pigments and dyes, found that three colors existed that could be mixed to form the seven colors Newton had identified in the spectrum. The three colors, which Brewster referred to as the **primaries,** were magenta (red-purple), yellow, and cyan (blue-turquoise). Brewster's theory was in accordance with other research conducted by scientists and colorists.

We refer to Brewster's primaries as the **subtractive primaries,** the primary colors of pigments (Fig. 3-2). By mixing pairs of subtractive primaries, the subtractive **secondaries** (red, blue, and green) are formed: yellow and magenta form red, magenta and cyan form blue, cyan and yellow form green. **Complementary** colors are across from each other in Figure 3-2. For example, blue is referred to as the complement of yellow, or vice versa. A mixture of all three primaries results in black or the absence of color.

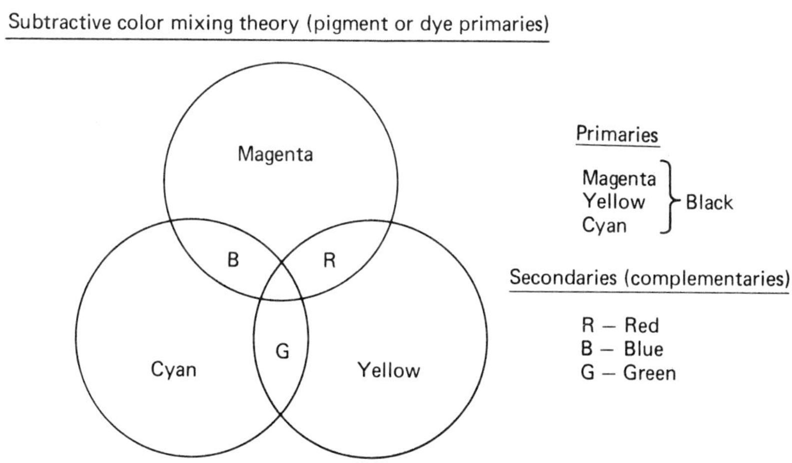

Figure 3-2. Subtractive Primaries.

Additive Color Mixing Theory

In the nineteenth century, Thomas Young formulated the theory that white light is composed of three primary colors (red, green, blue) and that these colors can be mixed in varying proportions to produce the other seven colors of the spectrum. The Young theory was not generally accepted until Helmholtz and Maxwell confirmed it.

Helmholtz, expanding on Young's work, theorized that three groups of nerve fibers or photochemical substances are present in the eye. Each group of nerve fibers is sensitive to one of the three light primaries; that is, one group is sensitive to red, one group to green, and the third group to blue. Helmholtz and Young felt that intermediate hues were derived from action on at least two of the three nerve fiber groups. Therefore, light that simultaneously affects the nerve fibers sensitive to red and green produces the sensation of yellow. The Helmholtz–Young theory was contrary to Brewster's theory, which stated that yellow was a primary color. According to Helmholtz–Young theory, white light would be produced by equal stimulation of all three nerve groups.

The Helmholtz–Young theory still prevails for spectral color mixing. This mixing of colored lights (Fig. 3-3) is called the **additive** color process. When two groups of wavelengths are added, the result is a third color. By mixing the **additive primaries,** the secondary colors are produced. Red and green produce the secondary color yellow; red and blue produce magenta; green and blue produce cyan. Red plus blue plus green will produce white light.

Complementary colors are secondary and across from the primaries on the color circles. Yellow (a secondary produced by mixing red and green) is the complement of blue (a primary). The light secondaries are almost identical with the pigment primaries. White light is achieved by mixing the three primaries of light, and black represents the absence of light, whereas black pigment is a subtraction of the three primaries of light.

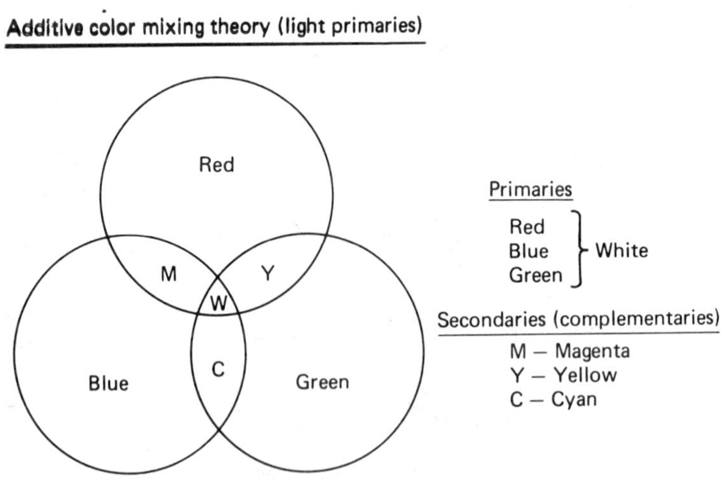

Figure 3-3. Additive Primaries.

Psychological Effects of Color

Colors have long been known to have psychological and emotional effects on people. Some of the psychological and emotional terms that have been used to define these effects and the color associated with them are as follows:

1. Warm, advancing, and stimulating colors: red, orange-red, orange, yellow.
2. Cool, receding, relaxing colors: blue-green, violet.
3. Neutral, tranquil colors: yellow-green, green. Neutral is equidistant between the ends of the spectrum.

Shades, which are achieved by the addition of black, are considered warmer than pure colors. **Tints,** which are achieved by the addition of white, are considered cooler than pure colors. Gray may be neutral, cool, or warm, depending on whether it is tinted or shaded.

Red is associated with health, power, anger, fire, and heat; red-purple with royalty; reddish-brown with harvest and fall. Yellow is associated with the sun (warmth) and is said to be one of the most cheerful colors. Gold is associated with richness and splendor. Green is associated with nature, life, and the out-of-doors. Blue is associated with constancy and fidelity, and is cool. White is associated with light and purity; black is associated with darkness, death, and gloom.

These associations are important facts that should be kept in mind where mood and atmosphere are to be created. The existence of these strong color associations can be proved by conducting a simple word/color association experiment. A subject is asked to respond with the first color name that the subject associates with the sample words shown in Table 3-1.

TABLE 3-1 Color Association Sample Test

Fire (red)*	Death (black)
Richness (gold)	Power (red)
Fall (reddish-brown, orange)	Splendor (gold)
Cool (blue)	Purity (white)
Cheerful (yellow)	Nature (green)
Fidelity (blue)	Warm (red)
Anger (red)	Royalty (red-purple, violet)

* Most common answer is in parentheses.

Color Contrast

Chromatic areas are affected by the adjacent or surrounding area colors. A color patch will appear brighter or less gray if the background color is relatively dark. The same patch will look dimmer or more gray if the background color is relatively light.

COLOR CLASSIFICATION

A number of color classification systems have been proposed over the years. Three of these classification systems will be presented here: (1) Munsell system, (2) Ostwald system, and (3) CIE chromaticity system.

The Munsell and Ostwald systems are used to denote pigment colors for a standard lighting condition. Either of these two systems can be used when ordering paints. A specifier placing an order for a light orange paint could get any one of a thousand tints of orange. However, by using one of the color notation systems, such as 5YR7/8 (Munsell), the specifier would be assured of getting the desired color.

The CIE chromaticity system is used primarily in the fields of color research, manufacturing, processing, and marketing. The CIE chromaticity system has the advantage of combining the effects of object color, source color, and the visual system to determine the actual perceived color under the given conditions. True object color can be predetermined for a given light source color.

Munsell Color System

The Munsell color system was conceived in 1898 by A. H. Munsell, a lecturer in color composition and art at the Normal Art School in Boston. Munsell's basic reason for developing the system was to aid in teaching his color composition class. The system would enable him to describe the colors on his sketches in definite terms. Munsell began by arranging colors on a circle, then on a sphere. From this beginning, the Munsell color system was developed.

The Munsell color system classifies colors on three dimensions: hue, value, and chroma. **Hue** is the name of a color. Technically it is defined as a perceptual attribute which allows us to distinguish from among red, yellow, green, blue, purple, and from among adjacent pairs of these. Thus there are 10 major divisions of hue:

R	Red
YR	Yellow-red
Y	Yellow
GY	Green-yellow
G	Green
BG	Blue-green
B	Blue
PB	Purple-blue
P	Purple
RP	Red-purple

Value is a measure of the lightness or darkness of a color. The higher the value number, the lighter the color and the higher the daylight reflectance. Ten is the highest value possible and represents ideal white. Lower-value numbers indicate darker colors and

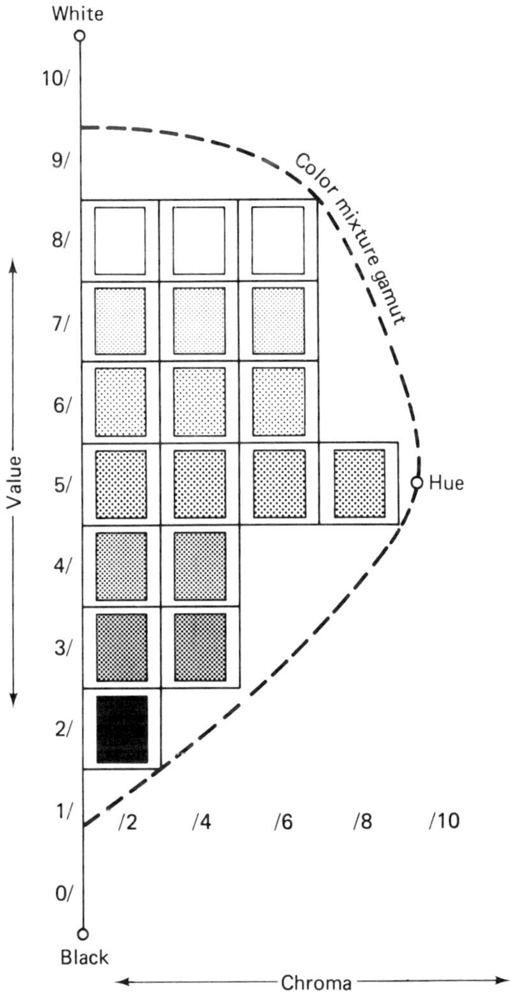

Figure 3-4. Constant Hue Segment. (Courtesy of Munsell Color, Baltimore, Md.)

lower daylight reflectance. Zero is the lowest value possible and represents ideal black. A segment from the system (Fig. 3-4) shows colors progressing from very dark at the bottom of each chart (low value) to very light at the top (high value). They vary by steps that are intended to be visually equal when viewed under the intended illuminant, daylight. Since human visual response is nonlinear, visually equal gradations of lightness are not equal steps of reflectance. Approximate daylight reflectances for the various Munsell values are given in Table 3-2.

Chroma is a measure of the saturation (or purity) of a color. The spectral colors have the maximum chroma, or are 100% pure. By adding gray, a color becomes less saturated until it finally has no color, or is achromatic (gray). Chroma is on the horizontal axis of Figure 3-4. The progression is from achromatic colors (black, gray, or white)

TABLE 3-2 Approximate Daylight Reflectances for Various Munsell Values

Munsell value	Approx. daylight reflectance
10	100%
9	80%
8	60%
7	45%
6	30%
5	20%
4	12%
3	7%
2	3%
1	1%
0	0%

at the left side of each chart to saturated colors on the right side. Thus minimally saturated colors have a chroma designation of two, while the strongest colored specimens producible will be 18 or 20, depending on the hue. Again, the steps are intended to be visually equal.

Constant Hue Chart

The locus of colors producible by mixtures of black, white, and pigment would be curved lines, such as the dashed line of Figure 3-4. The system is built up at each hue by adding progressive columns of colors to the nine neutrals of the same value. Each column of colors departs from the equivalent neutral by two additional steps of chroma. Each row of colors of constant value is extended until the next interval of two chroma steps would extend beyond the colorant-mixture gamut.

Constant Value Chart

The 100-point Munsell hue scale and notation are shown around the outer circle of Figure 3-5. Colors of constant hue are shown by the radial lines intersecting at the point that represents the neutral or gray scale (chroma/0). The hue scale is built up of 10 segments of 10 hues each, such as the segment from 1R to 10R. The fifth hue of each of these segments is noted by the letters alone. Thus 5R is simply R, or red, while 1R is at the extreme red-purple end of the red range and 10R is at the extreme yellow-red end of the red range. The chroma scale up to chroma/10 is shown along any radial line in the figure, and the colors of constant chroma are shown by concentric circles.

Munsell Solid

If the constant hue and constant value charts were shown in three-dimensional form, the resultant figure would be a distorted sphere (Fig. C., color section).

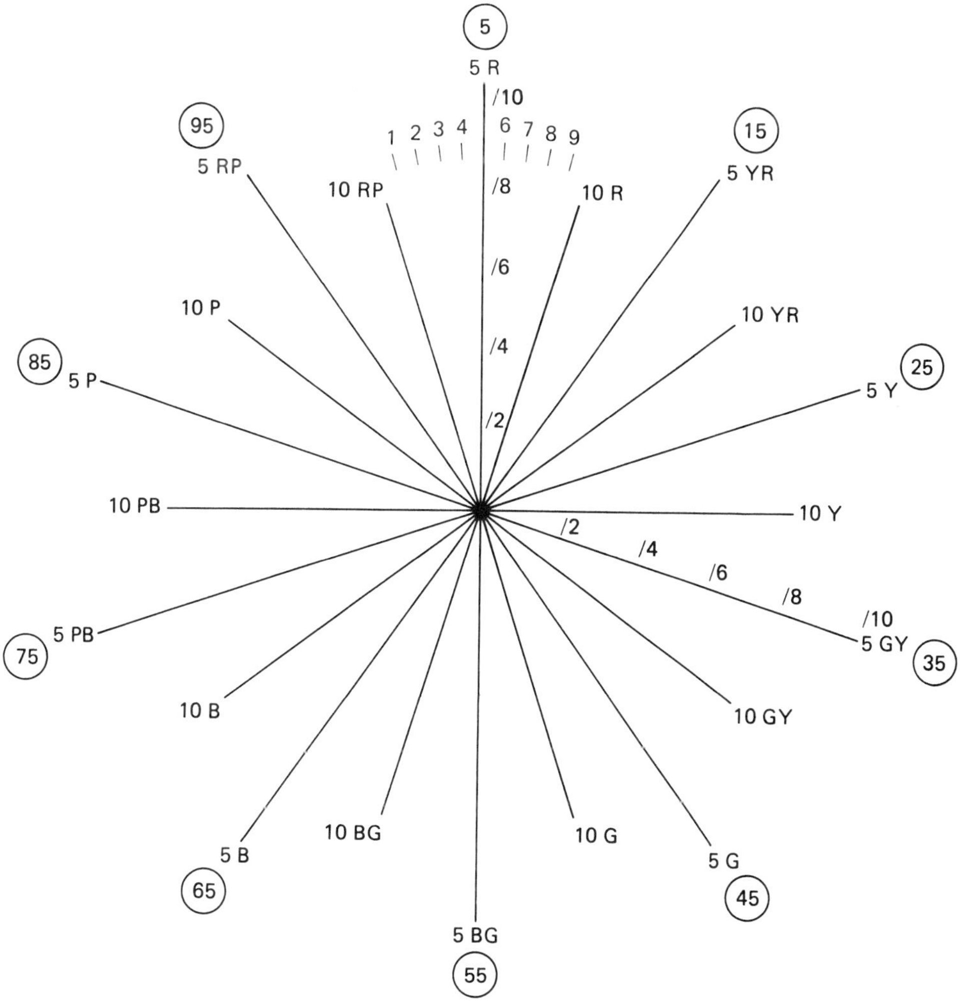

Figure 3-5. Constant Value Segment.

Notation System

Each individual color chip in the Munsell system is denoted by three symbols, for example, 5G 7/4. Here 5G indicates hue-5 green, 7 indicates value, and 4 indicates chroma. The color 5G 7/4 could also be denoted according to the 100-hue divisions, in which case the color would be specified as 45-7/4 (Fig. 3-6). Thus, by using the Munsell color notation system any color can be accurately defined.

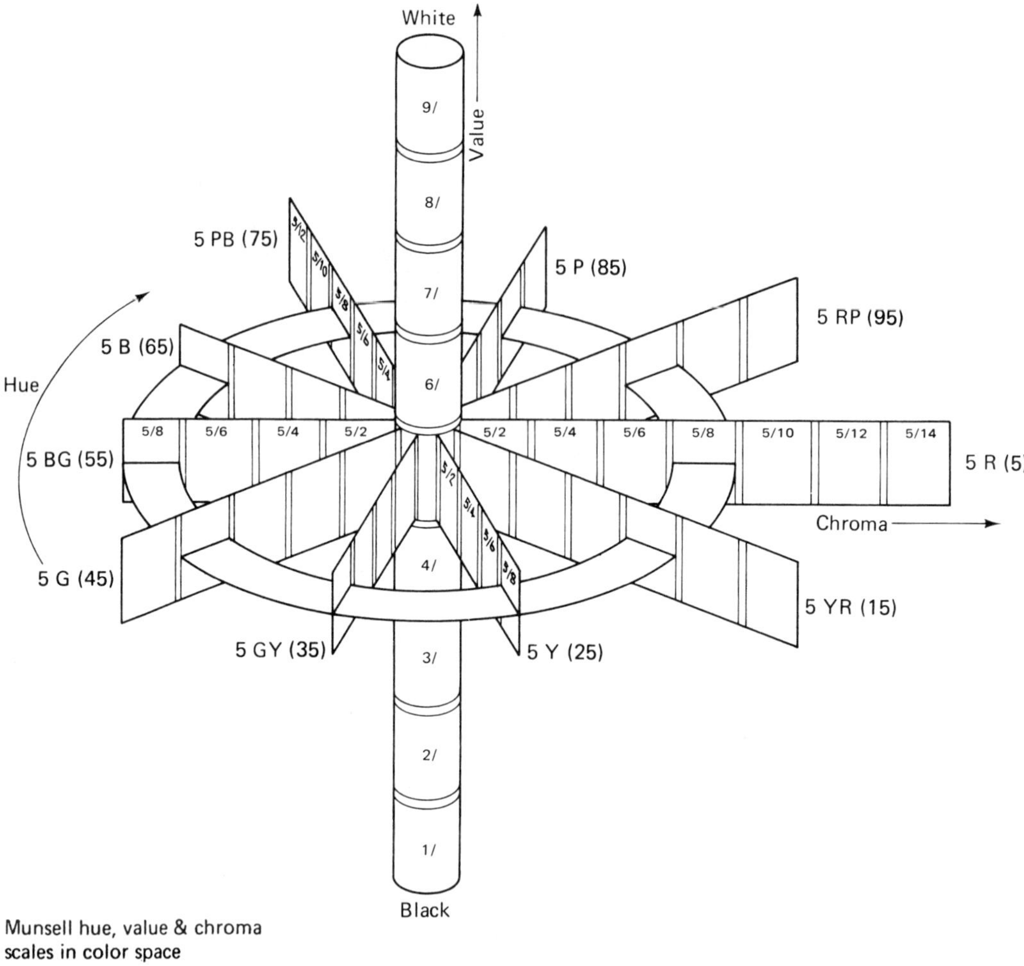

Figure 3-6. Munsell Color Solid. (Courtesy of Munsell Color, Baltimore, Md.)

Color Education

The principal use of the Munsell color system today is for color specification, since the system lends itself so well to this function. However, it is still used for color education as Munsell originally intended. Munsell Color (part of the Macbeth Division of Kollmorgen Corporation) publishes a catalog of available materials and equipment, such as "student sets." The catalog can be obtained by writing to Munsell Color at 2441 North Calvert St., Baltimore, Maryland 21218.

Ostwald Color System

The Ostwald color theory was developed in 1914 by Wilhelm Ostwald. He selected four evenly spaced primaries to develop a color wheel. Ostwald's primaries were lemon-yellow, scarlet, ultramarine, and turquoise green.

Ostwald divided his original color wheel into 100 equally spaced hues. He found this to be impractical, since it was difficult to distinguish one color step from another. He finally decided to divide the color wheel into 4 primary colors with 5 intermediates between each pair of primaries, which formed a color wheel of 24 distinct hues.

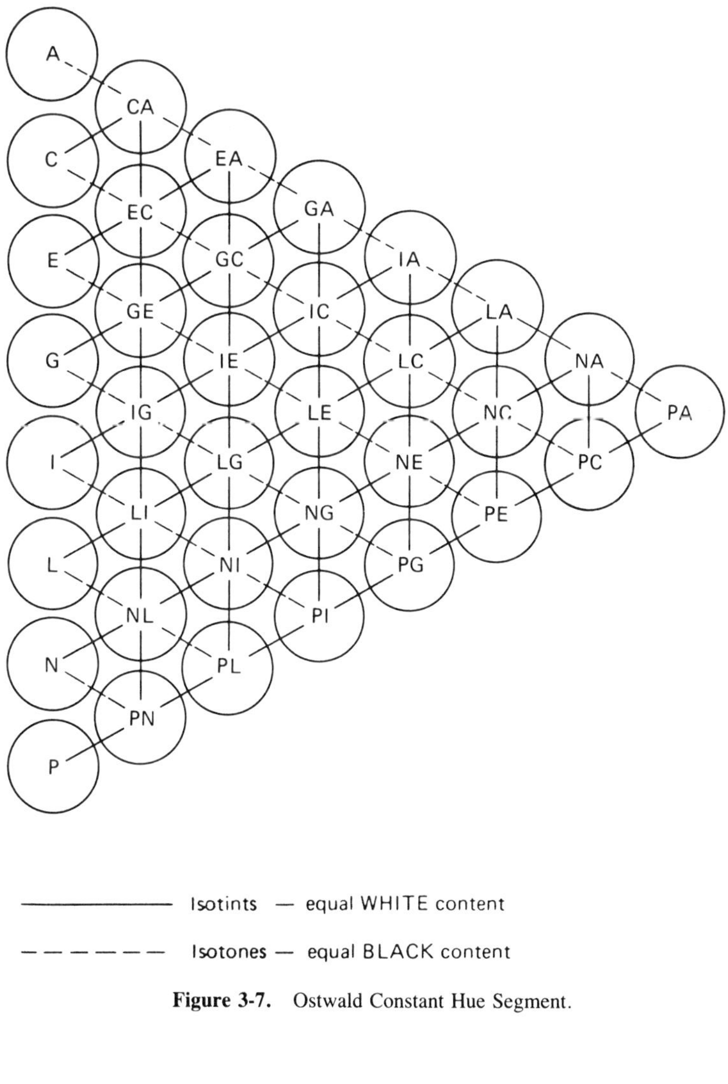

Figure 3-7. Ostwald Constant Hue Segment.

Achromatic Scale

Gustav Fechner, a German scientist, developed experiments that indicated that between black and white our perception is stimulated by a logarithmic progression of gray steps. Based on Fechner's studies, Ostwald organized a gray scale based on a uniformly varying proportion of white and black. His first study consisted of a scale with 25 stops, which he lettered with the alphabet, omitting the letter J. He concluded that a more practical scale needed only 8 gray stops. Each stop is given a letter designation: A, C, E, G, I, L, N, P. These 8 stops represent the **achromatic** scale (or gray scale) in the Ostwald system.

Color Solid

With the 8 grays and 24 hues, the resulting solid contains 680 colors. Each hue segment is an equilateral triangle with each color designated by letter combinations (see Fig. 3-7). The first letter indicates the percentage of **white** content, the second letter the percentage of **black** content. Therefore, colors with like first letters have equal white content and are called **isotints.** Those with like last letters have equal black content and are called **isotones.** Colors around the color solid with the same pair of letters are called **isovalents.** Thus, in Munsell terminology, isovalents have the same value and chroma but differ in hue.

Color Notation

The three-symbol notation system defines the percentages of white, black, and color that each of the color chips contains. Table 3-3 gives the percentages of white and black associated with each letter value.

Example

Ostwald colors 7LE and 20EA.

7 = scarlet
L = first letter = % white = 0.089
E = second letter = % black = 0.645
% color = [1 − (%W + %B)] = 0.266

20 = turquoise green
E = % white = 0.355
A = % black = 0.109
% color = [1 − (%W + %B)] = 0.536

Note that the extreme color PA equals only 0.8558% [1 − (0.0355 + 0.1087)] color saturation. This indicates that, according to the Ostwald theory, full color free from black and white represents an ideal that cannot be realized by any form of pigment or dye.

TABLE 3-3 Ostwald Color System: Percentage of White and Black Associated with Each Letter Value

Ostwald value	White content	Black content
A	0.8913	0.1087
C	0.5623	0.4377
E	0.3548	0.6452
G	0.2239	0.7761
I	0.1413	0.8587
L	0.0891	0.9109
N	0.0562	0.9438
P	0.0355	0.9645

Harmonies

An advantage with the Ostwald system is in its application to color harmony. The Ostwald solid is symmetrical and orderly which makes the Ostwald system a more reliable means of choosing color harmonies. The 680 colors are equally spaced and therefore are geometrically and mathematically related, as well as visually related. The mathematical and geometrical harmony obtainable can be compared to the chords possible in music. That is, equal intervals can be chosen or intervals can be determined by mathematical progression.

CIE Chromaticity System

The International Commission on Illumination (Commission Internationale de l'Eclairage) has chosen as a basis for standardization the response of the three sets of color receptors in the eye. The CIE system uses a tristimulus method of specifying the color of light on a surface in terms of the amounts of the primary colors of light required to match the unknown color in question. The system employs a chromaticity diagram that contains the seven colors of the visible spectrum (Fig. D., color section). The spectral colors are located on the locus (edge) of the chromaticity diagram. A straight line (purple boundary) joins the red and violet ends of the spectral locus to form a closed diagram.

Terms and Definitions

X, Y, and Z are **tristimulus values** that represent the absolute values of the red, green and blue primaries.

x, y, and z are **chromaticity coordinates** or **trichromatic coefficients** that represent the fractional values of the three primaries.

$\bar{x}$, $\bar{y}$, and $\bar{z}$ are **distribution functions** that represent the tristimulus values—the red, green, and blue response of the standard eye.

Standard Light Sources

The CIE system is based on three standard light sources (see Table 3-4).

TABLE 3-4. The Three Standard Light Sources of the CIE System

Source	Description	Chromaticity coordinates
Standard A	Tungsten lamp	$x = 0.4476$ $y = 0.4075$
Standard B	Mean noon sunlight	$x = 0.3485$ $y = 0.3518$
Standard C	Average daylight	$x = 0.3101$ $y = 0.3163$

General

The object color to be specified (unknown color) is subjected to a spectrophotometric measurement to determine its reflectance (or transmittance) properties to each wavelength across the spectrum. The specification of an unknown color is also dependent on the spectral characteristics of the light source under which the object is being viewed.

The distribution functions of the three tristimulus values (x, y, z) have been mathematically transformed so that the tristimulus value y (green) matches the standard luminosity curve (see Chapter 1). The distribution functions are each mathematically combined with the unknown color's spectral reflectance curve and the spectral energy distribution of the light source to define the absolute tristimulus values (X, Y, Z).

The fractional values of red (x) and green (y) of the mixture to match the unknown color can be found by the equations

$$x = \frac{X}{X + Y + Z}, \qquad y = \frac{Y}{X + Y + Z} \qquad \text{(chromaticity coordinates)}$$

Since $x + y + z = 1$, only two fractional values are required to define an unknown color. As a result, the chromaticity diagram (Fig. 3-8), which is used to plot the chromaticity coordinate, can be a two-dimensional diagram. The fractional value of red (x chromaticity coordinate), is plotted along the abscissa, and the fractional value of green (y chromaticity coordinate) is plotted along the ordinate.

The chromaticity coordinates (x, y) of the unknown color (object location) and the coordinates of the light source (source location) are plotted (Fig. 3-8) on the CIE chromaticity diagram. A straight line is drawn *from* the light source location *through* the object location and extended on to the spectral locus of the diagram. The point at which the straight line intersects the spectral locus defines the **dominant wavelength.** The dominant wavelength is the dominant color appearance of the object (unknown

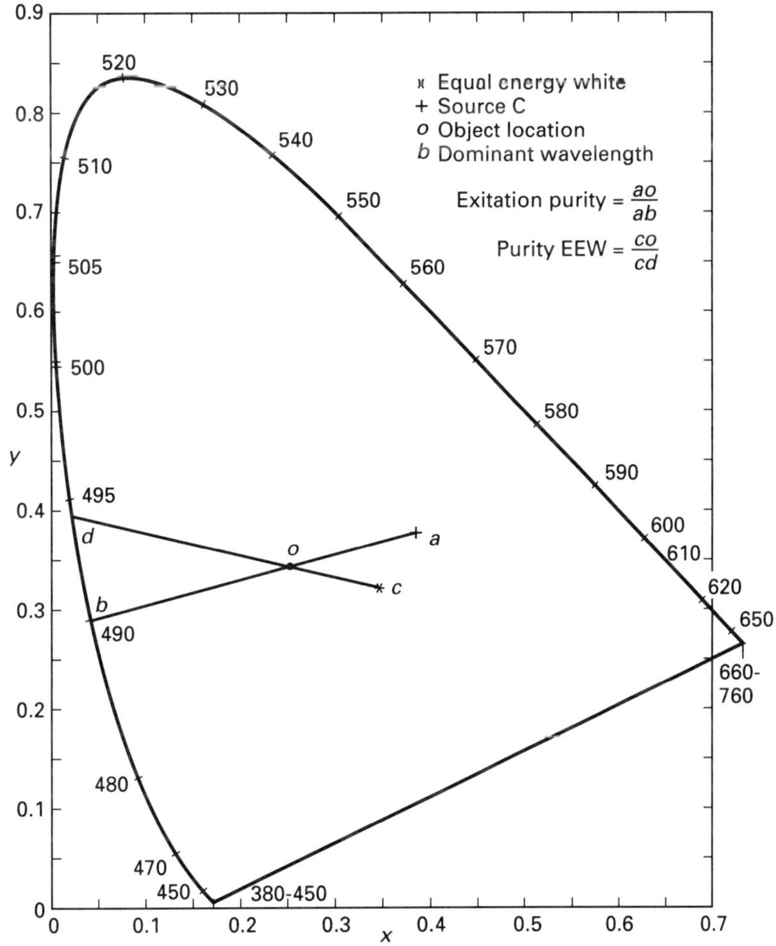

Figure 3-8. Definitions and Terms for CIE Chromaticity Diagram.

color) under that given light source. The excitation purity of the object can be found (Fig. 3-8) by determining the proportional length of the line (*ao*/*ab*) from the light source (*a*) to the object (*o*) and from the light source (*a*) to the locus (*b*) of the diagram.

The maximum color saturation, or 100% purity, occurs on the spectral locus of the diagram; zero saturation or no color occurs at the point defined as **equal energy white** (EEW). Purity relative to EEW is found (Fig. 3-8) by constructing a straight line *from* EEW *through* the object *to* the spectral locus. Purity (EEW) is determined by the proportional length of the line (*co*/*cd*) from EEW (*c*) to the object (*o*) and from EEW (*c*) to the locus (*d*) of the diagram.

Characteristics of CIE (x, y, z) System

1. All chromaticity coefficients are positive.
2. The y value doubles as a measure of luminance (reflectance) and chromaticity coordinate. The x axis represents an axis of constant zero luminance referred to as the Alychne axis, so that y values are luminance values.
3. y (distribution function) is equal to the standard luminosity curve.
4. From 650 to 700 on the spectral locus is a straight line, which represents a z value of zero.

Calculation Methods

There are three methods of obtaining the CIE specifications of an object of unknown color under a given light source: (1) the graphical solution, (2) the weighted ordinate method, and (3) the selected ordinate method (10 or 30 ordinates). The more irregular the curve, the greater is the number of ordinates required for accuracy.

Graphical Solution. A graphical solution is a useful procedure for gaining insight into the workings of the CIE chromaticity system. This is also the most exact procedure and is quite tedious if done longhand. It is the method recommended for the computerized analysis of color using the CIE system.

Example

What is the color appearance (unknown color) of a given object when illuminated by standard source C?

1. Obtain the spectral reflectance distribution data (curve a, Fig. 3-9) for the object of interest.
2. The spectral energy distribution (curve b, Fig. 3-9) of a given light source (this example uses standard source C) is combined with the spectral reflectance distribution of the object (previous). The distribution curves are combined by multiplying the ordinates of the two curves together.
3. Combined distribution curves (curve $a \times b$ = curve c, Fig. 3-9).
4. The ordinates at each wavelength interval along the distribution functions (curve d, Fig. 3-9) are multiplied by the ordinates of the previous combined curve (curve c) at the same wavelength interval.
5. The resulting curves (curve e, Fig. 3-9) represent the absolute values X, Y, and Z for the unknown surface color under the specified light source (standard source C).
6. The chromaticity coordinates x and y are obtained by dividing the area under the X or Y by the sum of the areas under the three curves (e). The x and y values can then be plotted on the chromaticity diagram (see Fig. 3-8, position o) to determine the CIE specification of the object.
7. The chromaticity diagram can be used to determine the following:
   a. The reflectance of the object under the light source, which is equal to the total area under the Y curve (curve e).

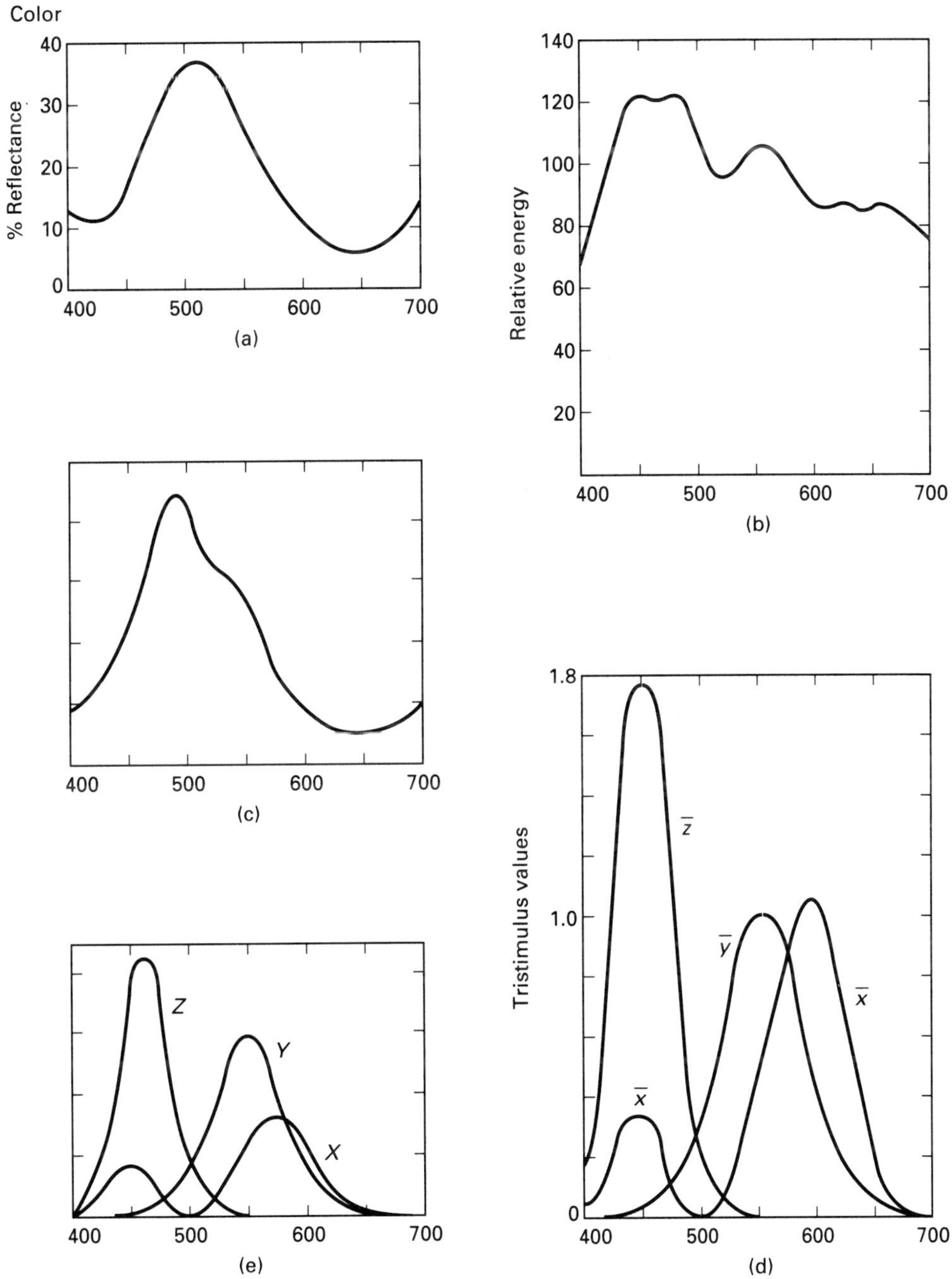

Figure 3-9. Graphical Procedure: Curve (a), Spectral Reflectance Distribution of the Object; Curve (b), Spectral Energy Distribution for Source *C*; Curve (c), Combined Curve; Curve (d), Distribution Functions; and Curve (e), Absolute Values of R, G, B Primaries (Tristimulus Values).

b. The dominant wavelength or predominant color appearance of the object under the light source.
c. The excitation purity relative to the light source.
d. The purity of the object color relative to EEW.
e. The required mixture of two or three light sources to reproduce the same color appearances as the object.

Method of Ten Selected Ordinates. A procedure has been developed to determine rapidly the CIE specification of an object under the three standard illuminants (source A, B, or C). Preselected ordinates (wavelengths) and weighting functions (column factors) have been selected by the color committee of the CIE. An example problem utilizing the method of 10 selected ordinates is given at the end of this chapter.

Source Blending

The matching of a lamp color can be done by mixing the light from two or more other light sources. Any two points joined together with a straight line that passes through the source to be matched (*G*) can be combined in the suitable proportions to give the desired matching color (see Fig. 3-10). Thus the straight line *AB* through *G* shows that wavelengths of 460 nm (blue) and 574 nm (yellow) can be mixed in the proportion of *AG*/*GB* of 574 nm light plus *GB*/*AB* of 460 nm light give the color *G*.

COLOR TEMPERATURE

Color temperature is a method for describing certain color characteristics of light sources. In order to define color temperature we must first understand the term **blackbody radiator.** A blackbody is a theoretical object which is a perfect radiator. As the temperature of a blackbody is raised, it radiates energy in the visible range, first red, changing to orange, yellow, bluish white, and finally white. These colors plot as a curved line on the CIE chromaticity diagram of Fig. 3-11. The curve is known as the blackbody locus.

Color temperature, then, is used to describe the color of a light source by comparing it to the color of a blackbody radiator. For example, the color appearance of an incandescent lamp is similar to a blackbody radiator heated to about 3000 degrees Kelvin. Therefore it is said that the incandescent lamp has a color temperature of 3000K. (The Kelvin temperature scale uses the same size degree as the centigrade scale, but its zero point is at absolute zero, or −273°C).

Note that the color temperature is not a measure of the physical temperature of the light source. It is the temperature of the blackbody radiator when the color appearance is the same as the source being tested. Ironically, low-color temperature corresponds to what is considered a warm appearance (rich in red and orange) while high-color temperature sources are thought of as having a cool appearance (bluish light).

To assign a color temperature to a light source, the source must match the blackbody radiator in terms of chromaticity (color appearance) as well as have a similar **spectral power distribution (SPD)** curve. An incandescent lamp has an SPD curve close to

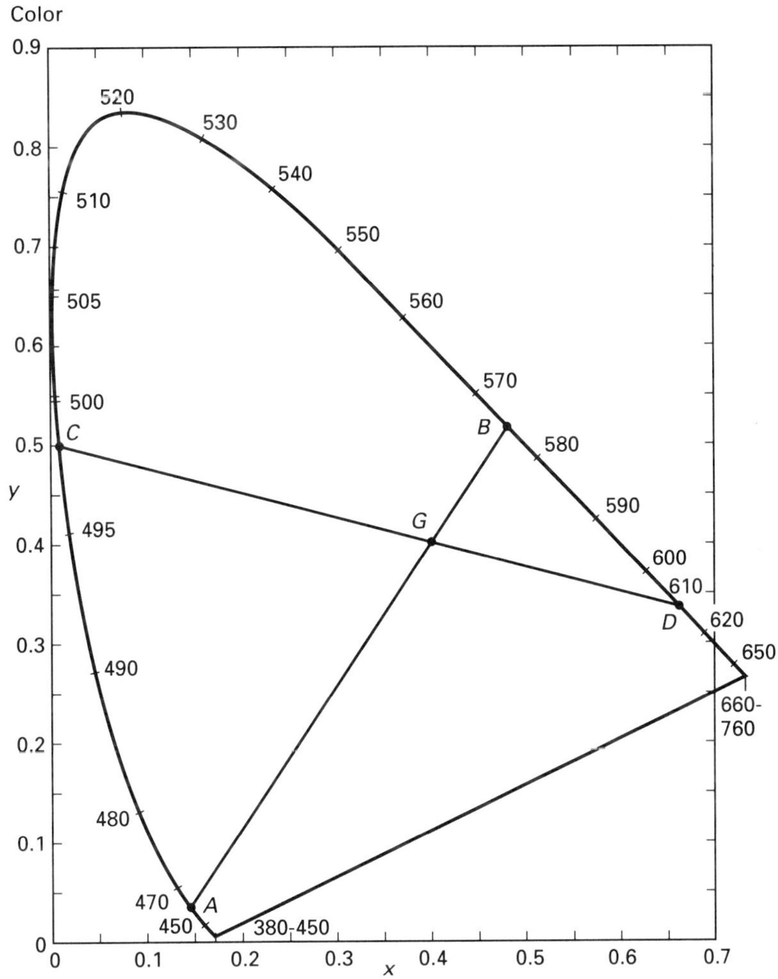

Figure 3-10. Examples of Source Blending.

that of the blackbody radiator; a chromaticity match is also possible. Therefore, color temperature can be properly assigned to incandescent sources.

Most white fluorescent lamps will have a chromaticity match with a blackbody, but they will not have an SPD curve similar to a blackbody radiator. Thus the lamp will not fall on the locus of the blackbody radiator. Therefore, it is improper to assign a color temperature to a fluorescent lamp. Instead, we assign to such lamps a **correlated color temperature.** A correlated color temperature implies a chromaticity match but not an SPD match, since the source falls off the blackbody locus. Isotemperature lines that cross the locus of the blackbody radiator curve are used to determine the correlated color temperature (see Fig. 3-11).

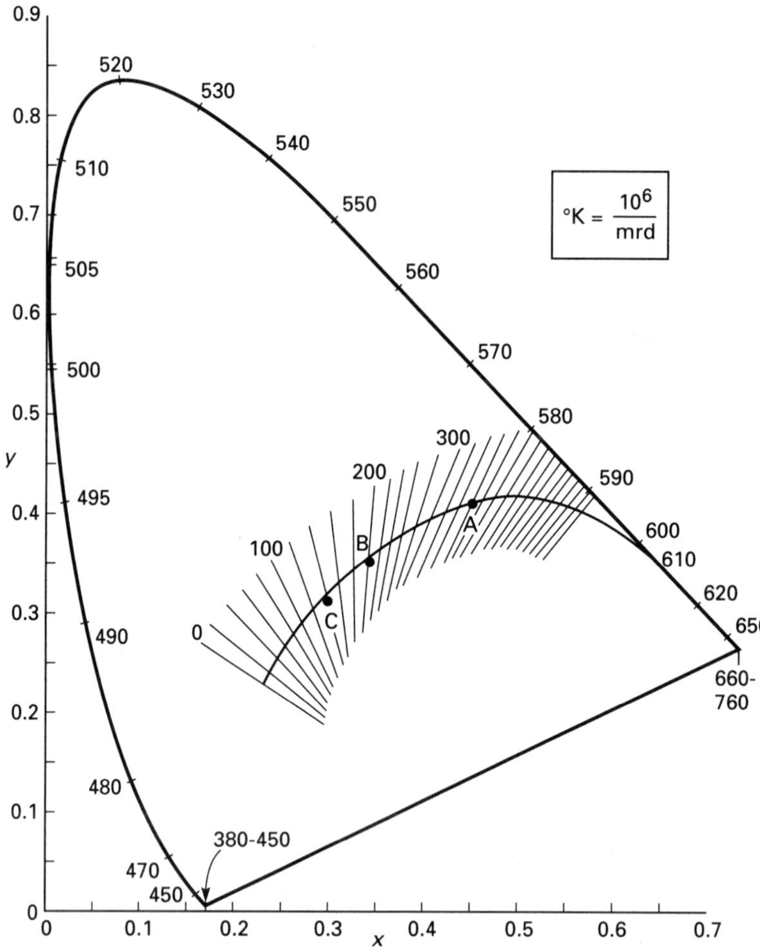

Figure 3-11. Blackbody Locus and Correlated Color Temperature.

Sources that match the blackbody radiator in neither chromaticity nor SPD cannot be properly assigned a color temperature or a correlated color temperature.

COLOR RENDERING INDEX

Another metric used in considering the color properties of a light source is the color rendering index, or CRI. Like color temperature, CRI compares the light source in question to a reference source. Unlike color temperature, which is concerned with the color appearance of light produced, CRI deals with the appearance of colored objects illuminated by a light source and how that appearance compares with the color appearance of objects illuminated by a reference source.

The impetus for development of a color rendering measurement system was the rapid market growth of the fluorescent lamp in the 1940s and 1950s. Manufacturers found that they could make extremely efficient fluorescent lamps, but objects viewed under the light of such lamps had a ghastly green appearance. Fluorescent lamps could also be made which rendered colors more normally, but at the expense of reduced efficiency. In seeking to quantify this effect, the Color Committee of the IES developed the CRI method in 1962. The Commission Internationale de l'Eclairage officially adopted this procedure in 1964 and defined the testing procedure to be used.

In assigning a numerical value to the color rendering properties of a light source, the CRI system compares the shift in object color that occurs when the object is illuminated by the test light source with the shift that occurs under the reference source. The explicit procedure with detailed formulas can be found in the 1974 CIE publication No. 13.2 (1). A simplified outline of this procedure follows.

1. The CIE chromaticity coordinates of eight reference colors are determined for the source for which the CRI is to be determined (the test source) and for the reference source using spectroradiometric measurements. The eight reference colors used are hues of moderate intensity each having the same lightness. The reference source used depends on the (correlated) color temperature of the test source. It is the blackbody radiator for test sources with color temperature below 5000K. A daylight reference is used for test sources with color temperature of 5000K and above.

2. After various transformations and corrections are applied to these raw data, the distance or shift between the test source coordinates and reference source coordinates is calculated for each of the eight colors.

3. A CRI is calculated for each of the eight shifts found in step 2.

4. The overall CRI of the test source is found as the arithmetic mean of the eight individual CRIs calculated in step 3. This overall value is called the general CRI.

Many misconceptions and misapplications of CRI have arisen. For example, since the operating characteristics of incandescent lamps approximate a blackbody, the CRI of these lamps is very high, typically approaching 100. There is a tendency to interpret this as meaning that the color rendering of incandescent is somehow "best." What the actual interpretation should be is that incandescent lamps have very similar color rendering characteristics as their reference source.

Another potential problem with CRI concerns the fact that only sources having the same correlated color temperature can be compared. It is often tempting to think of a 5000K lamp with CRI of 80 as having color rendering that is superior to that of a 3000K lamp with CRI of 60. As the CRI system is defined, however, the only comparative statement we can make about the color rendering of these two lamps is that this 5000K lamp is more similar to its reference source than is the 3000K lamp to its reference source.

Lighting designers should also keep in mind that CRI does not assess subjective impressions such as pleasantness of color appearance. Such evaluations would require a color preference index which, although advocated by some, does not currently exist.

These drawbacks do not mean that the CRI system is without value. The eight individual CRIs tell us if any specific color will be distorted under the test source. The general CRI (usually the only number reported by lamp manufacturers in consumer literature) can serve as a warning to the lighting professional: a low CRI indicates that at least some colors will be distorted.

REFERENCES

1. Commission Internationale de l'Eclairage, "Method of Measuring and Specifying Colour Rendering Properties of Light Sources," CIE publication no. 13.2 (TC −3.2), 1974.
2. Jerome, C. W., "Color Rendering Properties of Light Sources," *Color Research and Application*, Vol. 1, No. 37 (1976).
3. Kaufman, John E., and Jack F. Christensen, eds., *IES Lighting Handbook,* Reference Volume, 1984. New York: Illuminating Engineering Society of North America, 1984.

DATA SECTION

TABLE 3–4 Ten Selected Ordinates for Sources A, B, and C

Source A: Tungsten, 2854K, x = 0.448, y = 0.408			
Ordinate number	X	Y	Z
1	517	508	425
2	561	530	436
3	576	544	444
4	587	555	451
5	597	566	457
6	605	577	463
7	614	588	469
8	623	600	477
9	635	615	488
10	656	640	508
Source B: Noon sunlight, 4800K, x = 0.349, y = 0.352			
1	442	495	423
2	528	520	433
3	559	534	440
4	573	545	446
5	585	556	452
6	595	566	458
7	605	577	464
8	615	589	471
9	628	605	480
10	649	631	499

TABLE 3–4 Cont.

Source C: Average daylight, 6740K, x = 0.310, y = 0.316			
1	436	489	422
2	461	515	432
3	544	530	439
4	564	541	444
5	577	552	450
6	589	562	456
7	600	573	462
8	611	585	469
9	624	601	478
10	644	627	495
Column factors	X	Y	Z
Source A	0.110	0.100	0.036
Source B	0.099	0.100	0.085
Source C	0.098	0.100	0.118

EXAMPLE PROBLEM

Method of 10 Selected Ordinates

Illuminant A

Spectral reflectance for Munsell 5R4/14

	X		*Y*		*Z*	
Ordinate number	λ	λρ	λ	λρ	λ	λρ
1	517	0.040	508	0.040	425	0.050
2	561	0.040	530	0.040	436	0.048
3	576	0.060	544	0.040	444	0.048
4	587	0.062	555	0.040	451	0.045
5	597	0.280	566	0.040	457	0.045
6	605	0.350	577	0.065	463	0.043
7	614	0.550	588	0.165	465	0.043
8	623	0.600	600	0.300	477	0.040
9	635	0.640	615	0.555	488	0.040
10	656	0.675	640	0.650	508	0.040
Sum		3.397		1.935		0.442
x Column factor		0.110		0.100		0.036
	X =	0.374	*Y* =	0.194	*Z* =	0.016

$$x = \frac{X}{X+Y+Z} = \frac{0.374}{0.584} = 0.64$$

Dominant wavelength = 615 nm

$$y = \frac{Y}{X+Y+Z} = \frac{0.194}{0.584} = 0.33$$

Excitation purity $= \frac{ao}{ab} = 84.9\%$

Reflectance $= Y = 0.194 = 19.4\%$

Purity EEW $= \frac{co}{cd} = 93.6\%$

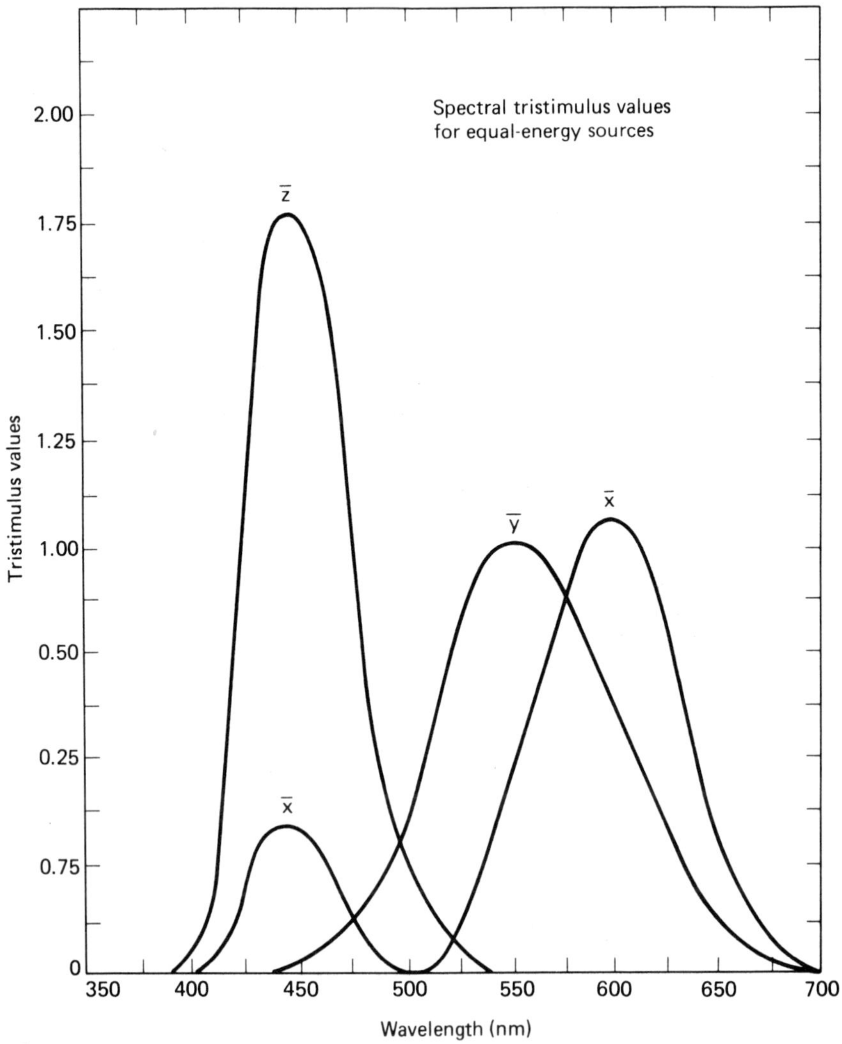

Figure 3-12. Tristimulus Values.

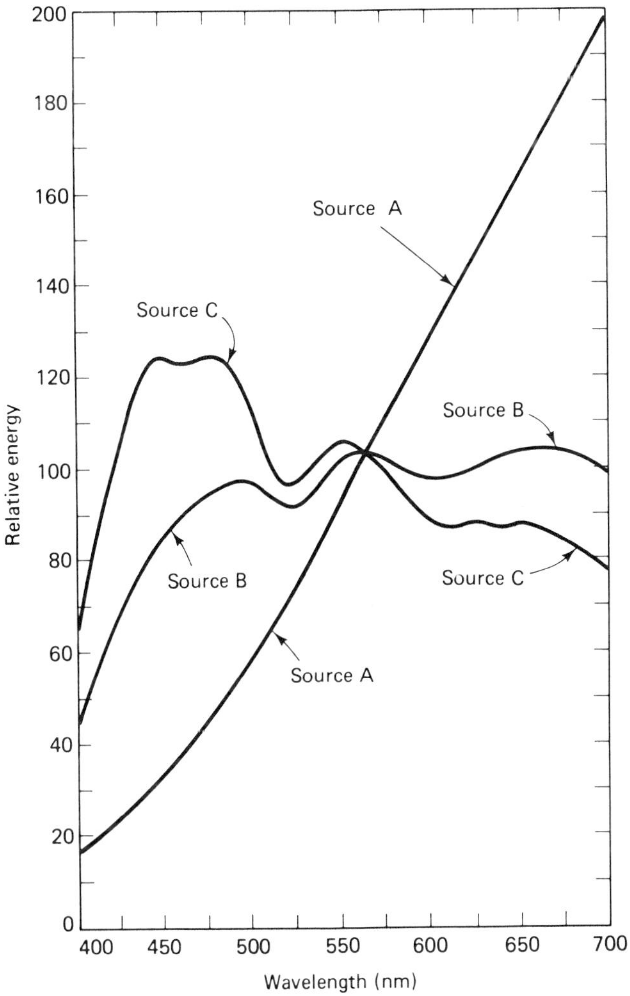

Figure 3-13. Three Standard Light Sources.

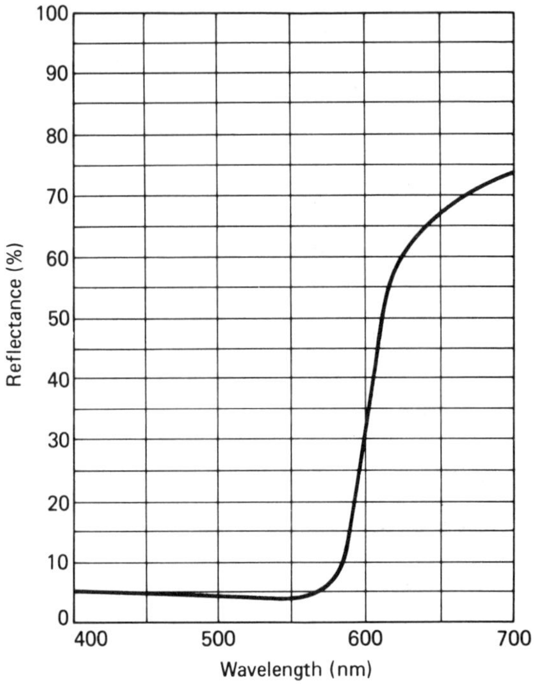

Figure 3-14. Spectral Reflectance for Munsell 5R4.

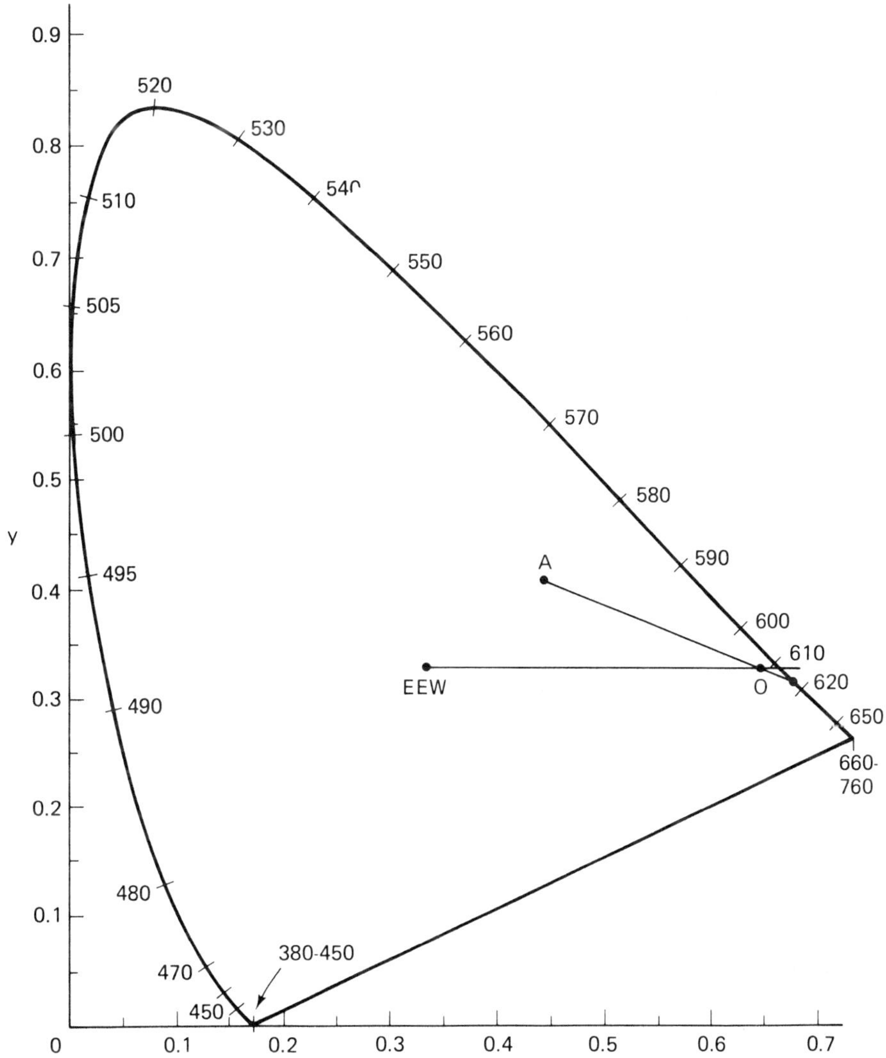

Figure 3-15. Example of CIE Chromaticity Plot.

4 light sources

HISTORY OF THE ELECTRIC LAMP

The success of the electric lamp was dependent on the development of an economical source of electricity. Therefore, the history of the electric lamp is the history of electricity.

About 25 centuries ago a Greek philosopher, Thales, found that a fossil resin called amber became electrically charged when rubbed. The Greek name for amber is *elektron,* which is the basis for the word "electricity." Numerous investigators developed crude devices for generating electricity. In 1745, E. G. von Kleist invented the Leyden jar. The Leyden jar is the forerunner of the condenser. He showed that he could store small quantities of electricity generated by friction machines. Benjamin Franklin connected several Leyden jars in a parallel arrangement (parallel circuits) that produced a strong electrical discharge. His numerous experiments led to the establishment of the principles of parallel and series connections.

Based on Galvani's theory of animal electricity, Alessandro Volta developed the voltaic pile. The voltaic pile consisted of a pile of alternate silver and zinc discs with a piece of cloth soaked in salt water between each metal layer. He showed that electricity could be produced with this arrangement. This was the beginning of the battery. In 1820, a physicist named Hans Oersted discovered that a compass needle could be deflected by current flowing in a wire. Based on Oersted's original findings, André Marie Ampere developed fundamental laws related to current flow in a wire. Ampere showed that current flowing in a wire coil produced a magnetic field, which established a connection between magnetism and electricity.

In 1825, Georg Simon Ohm conducted his important experiments in electricity, which led to the law that states that current flow is directly proportional to the voltage and inversely proportional to the resistance; this is called Ohm's law ($I = E/R$). The invention of the galvanometer by Johann Schweigger made it possible to measure the current flow accurately in a wire. In 1825, William Sturgeon developed the first electromagnet. This was an important step in the process of developing an electric power source that used the principles of magnetism to produce electricity.

In 1831, Michael Faraday showed that a copper disc rotated between the poles of a permanent magnet would produce an electric current that could be detected on a galvanometer. This was the beginning of many years of investigation of the dynamo (electric generator), which is the foundation of the electric lighting industry.

Prior to 1878, all methods of distributing electric current involved constant current dynamos (series circuits). This method was satisfactory for arc lamps since they are constant current devices. However, a series system would be an impractical one for distributing power to electric lamps since it would be impossible to turn off one lamp without turning off all the lamps on the circuit. Thomas Edison knew that if a practical lamp were to be developed, a new distribution system would have to be designed to allow individual control of each lamp. Edison developed a constant voltage dynamo that allowed his incandescent lamps to be operated in parallel. He called this system of distribution a "multiple" system. With the invention of the constant voltage dynamo, Edison began work on a new high-resistance incandescent lamp. In 1880, Edison was granted a patent on a new, high-resistance, carbon-filament vacuum lamp. Edison demonstrated his new incandescent lighting system at Menlo Park, New Jersey. Although Edison did not invent the first incandescent lamp, he did invent the first practical incandescent lamp and power system that was suitable for multiple distribution over large areas.

The first attempt at making an incandescent lamp was made by Warren de la Rue in 1820. The lamp consisted of a glass tube with brass endcap and a coil of platinum wire for a filament. The lamp had a very short burning life, and the operating cost was very high because of the number of batteries required. Between 1820 and

1878 numerous types, sizes, and configurations of incandescent sources were developed and patented.

The first commercial application of the incandescent lamp was on the steamship *Columbia.* Four dynamos were started May 2, 1880 and operated 115 lamps on the *Columbia.* The first application of Edison's system was in the lithography shop of Hinds, Ketchum, and Company in New York City in 1881. Over 150 systems were installed in the next two years utilizing over 30,000 incandescent lamps. The world's largest incandescent electric lamp was built in 1954 to commemorate the seventy-fifth anniversary of Edison's invention. This lamp, on display in Cleveland, stands 42 inches high and uses 75,000 watts of electricity. It weighs 50 pounds and produces as much light as 2875 60-watt lamps.

LIGHT SOURCES

Light Source Terminology

Burning Cycle

The life rating of a light source is given in burning hours based on measurement of a large representative group of lamps under laboratory conditions. The life of a lamp is usually affected by the number of starts. The average rated life is based on lamp industry standards which specify the **burning cycle per start**, a 50% mortality, and approximately a minimum of 70% rated lumens.

Operating Hours per Year

The lamp life alone has little practical meaning for the lighting designer. The important consideration is how long the lamp will be in service for the particular application. The **operating hours per year** for the lighting application can be determined from the following formula.

$$\text{oper. hr/yr} = \frac{\text{wk}}{\text{yr}} \times \frac{\text{days}}{\text{wk}} \times \frac{\text{hr}}{\text{day}} = \frac{\text{hr}}{\text{yr}}$$

For example if the building is open 52 wk/yr for 6 days/wk and 11 hr/day, the operating hours per year for the building will be

$$\text{oper. hr/yr} = 52 \times 6 \times 11 = 3432 \text{ hr/yr}$$

Color Temperature and Correlated Color Temperature

Color temperature and correlated color temperature were defined in Chapter 3. Since incandescent lamps have radiating characteristics ''similar'' to the blackbody radiator (Chapter 3), they can be assigned a ''color temperature.'' However, gaseous discharge lamps do not have radiating characteristics that match the blackbody radiator. Therefore, gaseous discharge lamps may be assigned a ''correlated color temperature'' if they fall within the boundaries established by the isotemperature line shown in Figure 3–11. The isotemperature lines were developed experimentally and should not be extended beyond the limits shown on the diagram. Equal energy white, EEW, on the CIE chromaticity diagram, corresponds to a color temperature of 5451K.

Light Source Selection

TABLE 4-1 Light Source Selection Parameters

1. *Amount of control desired*—a point source will have more accurate control than an area source.
2. *Amount of diffusion desired*—an area source is more diffuse than a point source.
3. *Consistency and reliability of supply voltage*—HID lamps are more sensitive to voltage variation and requires restrike time.
4. *Ambient temperature and humidity*—fluorescent lamps are very sensitive to temperature and humidity.
5. *Air-conditioning load*—the lower the efficacy, the more watts for equal illumination; therefore, more heat is added to the space.
6. *Color rendition*—most sources are available in different SPDs.
7. *Cost*—initial cost versus life-cycle cost.
8. *Efficacy, life, lumen depreciation.*
9. *Control*—on/off versus variable control.
10. *Energy-saving devices.*

Light Source Categories

Light sources (lamps) used today in artificial lighting can be divided into two main categories: incandescent and gaseous discharge. The gaseous discharge type of lamp is either low or high pressure. Low-pressure gaseous discharge sources are the fluorescent and low-pressure sodium lamps. Mercury vapor, metal halide, and high-pressure sodium lamps are considered high-pressure gaseous discharge sources.

These are the most common light sources used in the field of illuminating engineering. Each light source will be described in terms of its three primary components: (1) light-producing element, (2) enclosure (bulb), and (3) electrical connection. The chapter is divided into three sections: (1) incandescent sources, (2) gaseous discharge sources, and (3) ballasts.

INCANDESCENT SOURCES

Standard Incandescent Lamps

Light-Producing Element

Light is produced in the incandescent lamp (Fig. 4-1) by heating a wire or filament to high temperatures, which causes the wire to incandesce (glow). The wire is called a **filament.** The incandescence of the wire is a result of its resistance to the flow of

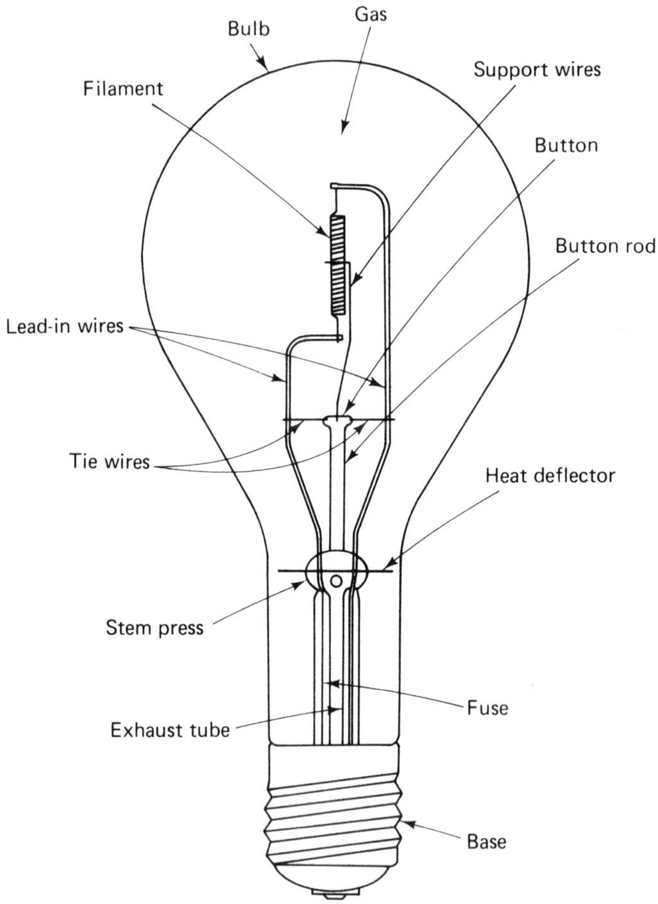

Figure 4-1. Incandescent Lamp. (Courtesy of GTE Products Corp.)

electrical current through the wire. Tungsten is used as a filament material. No other substance is as efficient in converting electrical energy into light on the basis of life and cost. Tungsten has four important characteristics:

1. High melting point
2. Low evaporation
3. High strength and ductility
4. Favorable radiation characteristics

The most common filament letter designations are s, straight; c, coiled; cc, coiled coil; and r, ribbon or flat. Coiled-coil filaments are the most efficient and widely used in the lamps encountered in illuminating engineering. The resistance of cold tungsten is low compared to its operational resistance; therefore, there is a large initial inrush of a current to a cold lamp.

Enclosure

The enclosure or glass envelope is called the **bulb** and serves two primary functions. First, the enclosure keeps air (oxygen) away from the filament. When the filament is exposed to air, evaporation occurs more rapidly. Second, the enclosure maintains a constant environment for the filament to retard evaporation of the tungsten. The enclosure is filled with an inert gas of argon and nitrogen to retard evaporation of the filament. Gas-filled lamps designated type C are 40 W and higher. Lamps of 25 W and less are vacuum lamps, which are designated type B.

The enclosure comes in a variety of shapes, sizes, and wattages.

Lamp Shape Designation

Lamp Shapes. See Figure 4-2.

F = flame	R = reflector
G = globular (round)	PAR = parabolic
PS = pear straight	A = arbitrary (or standard)
T = tubular	S = straight

Lamp Shape Designation. In addition to a letter designation (shape), the bulbs also have a number designation, which represents the maximum diameter of the enclosure in eighths of an inch.

1. Letter(s) represent the *shape*.
2. Number(s) represent the *diameter in* $\frac{1}{8}$*th of an inch*.

For example, a PS48 = pear straight—$\frac{48}{8}$th in. in diameter or 6 in.

Lamp Finishes. Bulb surface treatments may be clear, surface treated, tinted, or built-in silvered control surfaces. The standard lamps on the market are clear, inside frosted (or acid etched), white (silica coating), and white bowl or silver bowl. Colored or tinted bulbs are natural-colored glass, outside spray, ceramic glazing (enameled), or fused-on color filters.

Lamp Order Code (ANSI Code). The ANSI Lamp Ordering Code should be used when it is available to maintain an open specification of a light source.

Incandescent

Shape, "Arbitrary"
|
100A
|
Watts

Shape, "Parabolic"
|
150PAR/FL
| |
Watts Flood pattern

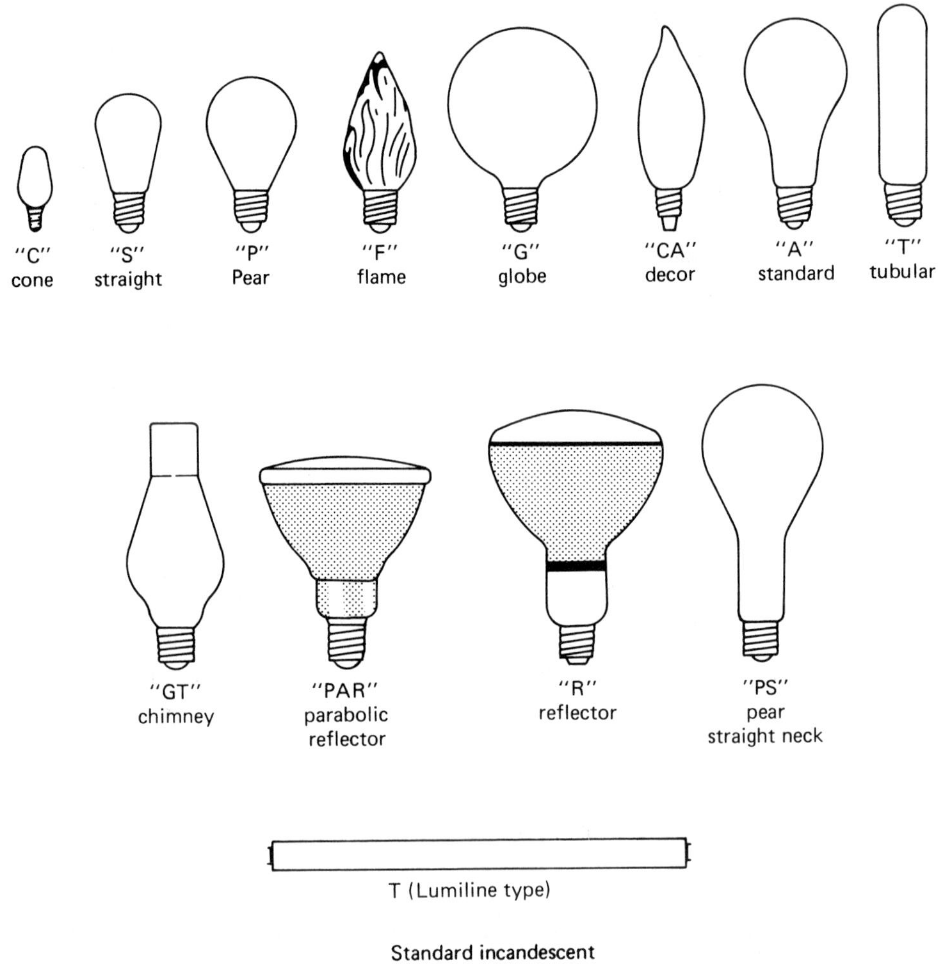

Figure 4-2. Incandescent Lamp Shapes. (Courtesy of GTE Products Corp.)

Electrical Connection

The electrical connection is called the **base.** Some bases are also used to position or align the filament in an optical system.

There are eight types of bases: (1) screw, (2) screw with ring contacts (three-way), (3) skirted screw, (4) bipost, (5) prefocus, (6) disc, (7) bayonet, and (8) prong. The most common base is the screw base: specifically, the medium screw base. Other screw bases include the mogul, medium, intermediate, candelabra, and miniature.

The general-service lamps of 300 W and below normally use the medium screw base; 300 to 500 W lamps use the mogul screw base.

Operating Characteristics

Efficacy. A 100A lamp, which is rated at 1750 lumen, has an efficacy of

$$K_{\text{system}} = \frac{1750}{100} = 17.5 \frac{\text{lm}}{\text{W}}$$

Life. To measure the "life" of a lamp; place the lamp in testing lab with prespecified conditions (temperature, humidity, etc.). The lamp is cycled on/off with a three hour on cycle. When 50% of the lamps have burned out, the burning time is recorded. That is, at 750 hr, 50% of the 100A lamps in the test lab have burned out so the life rating of the lamp is 750 hours.

A 100A lamp has a rated life of 750 hours. Therefore, number of years the 100A lamp will operate in a building that operates for 3432 hr/yr will be

$$\text{no. yr} = \frac{\text{rated life}}{\text{oper. hr/yr}} = \frac{750}{3432} = 0.2 \text{ yr } (2\tfrac{1}{2} \text{ mo})$$

Operating Variables

Voltage Variation. Varying the voltage of an incandescent lamp above or below the rated voltage will affect the characteristics of the lamp. For example, if a 120 V rated lamp is operated at 125 V (104%), the lamp will produce 16% more lumens, 7% more watts, and 38% less life. A 120 V rated lamp operated at 115 V (96%) will provide 13% less lumens, 6% less watts, and 62% more life (see Fig. 4-3).

Lumen Depreciation. The resistance of the filament increases with time owing to evaporation and accompanying decrease in the diameter of the filament. This increase in filament resistance causes a decrease in lumens, amperes, and watts. Further reduction in lumen output is due to the absorption of light by the deposited tungsten on the inside surface of the lamp which is called **bulb wall blackening** (see Fig. 4-4).

SPD curve. The human visual system responds differently to the different wavelengths of radiation. Our mind interprets these different wavelengths as color. Light sources are important in color vision because they provide radiant energy and thus color response. The distribution of wavelengths emitted by a source is known as the **spectral power distribution (SPD).** The SPD of an incandescent lamp is shown in Figure 4-5. Note the tremendous amount of red or long wavelengths present; this is to be expected from a source producing luminous energy by heat.

Energy Distribution. Of the total input energy into the incandescent lamp, only 10% ends up as visible light (see Fig. 4-6).

Energy-Saving Devices

1. *Switching*: On/off control of the lighting
2. *Dimming*: Modern SCR dimmers that reduce energy consumption as the lights are dimmed.

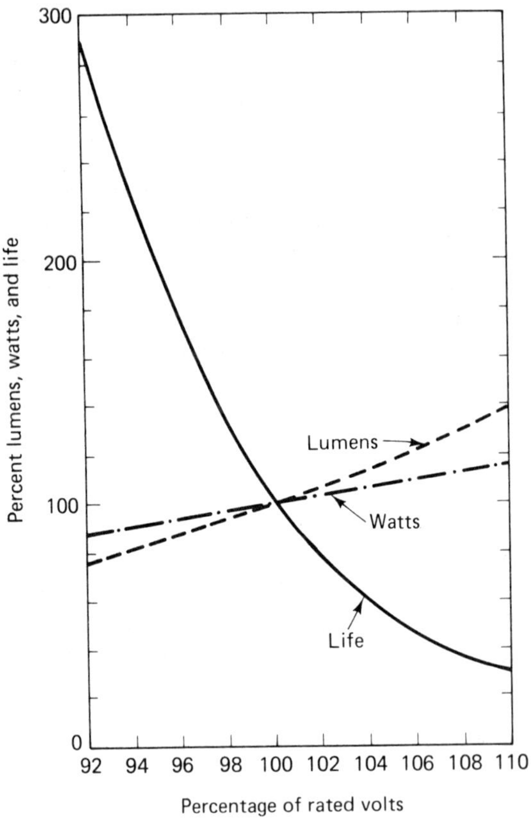

Figure 4-3. Voltage Effect on Lamp Life and Lumen Output. (Courtesy of GTE Products Corp.)

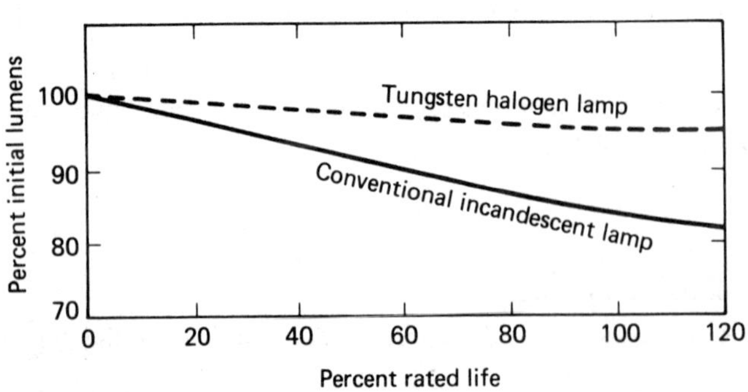

Figure 4-4. Lumen Depreciation of Incandescent Lamps. (Courtesy of GTE Products Corp.)

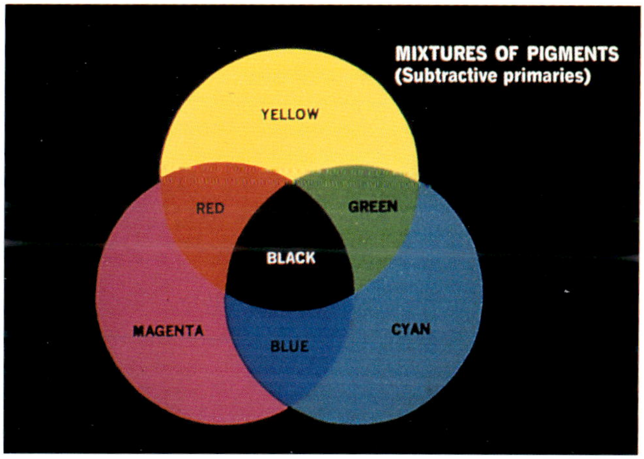

Figure A. Mixtures of Pigments (Subtractive Primaries). (Courtesy of G.E. Lighting.)

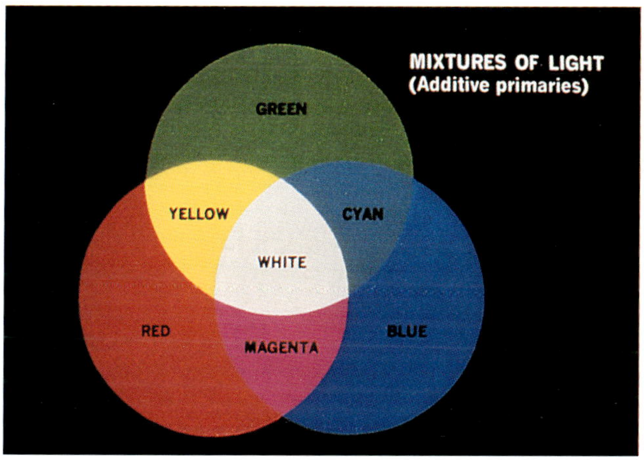

Figure B. Mixtures of Light (Additive Primaries). (Courtesy of G.E. Lighting.)

Figure C. Munsell Color Tree. (Courtesy of Munsell Color, Baltimore, Md.)

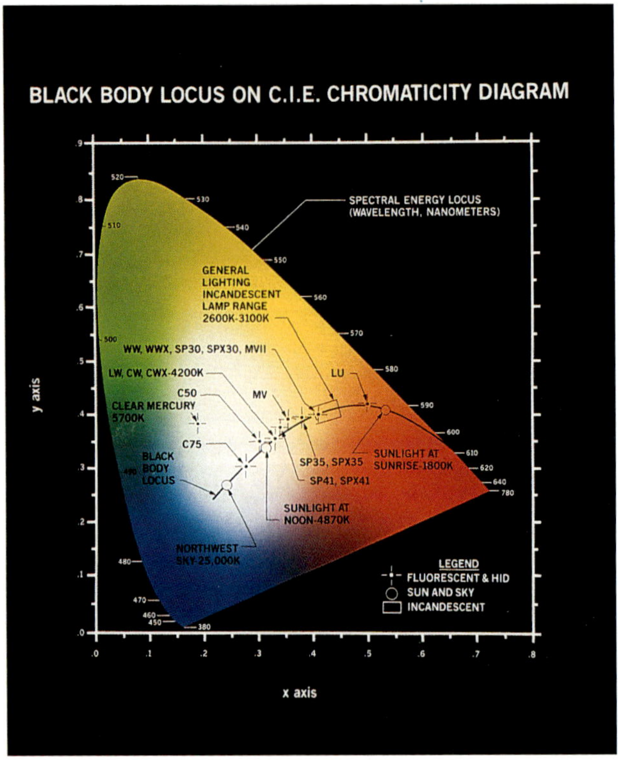

Figure D. Black Body Locus on C.I.E. Chromaticity Diagram. (Courtesy of G.E. Lighting.)

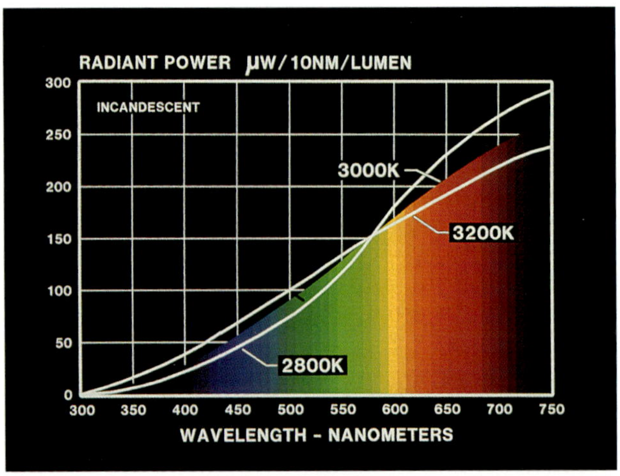

Figure E. Spectral Power Distribution for Incandescent Lamps. (Courtesy of G.E. Lighting.)

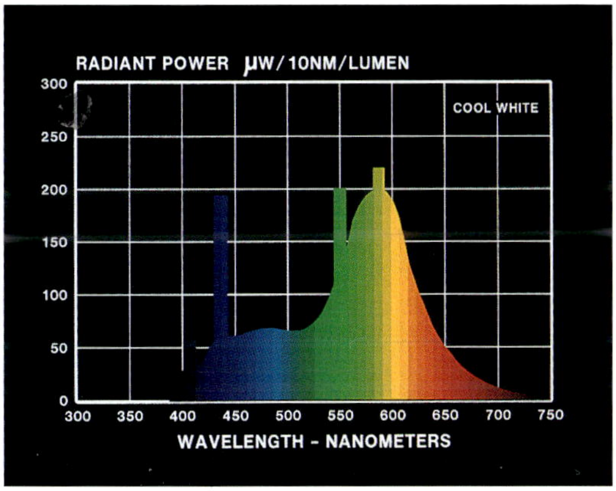

Figure F. Spectral Power Distribution for Cool White Fluorescent Lamps. (Courtesy of G.E. Lighting.)

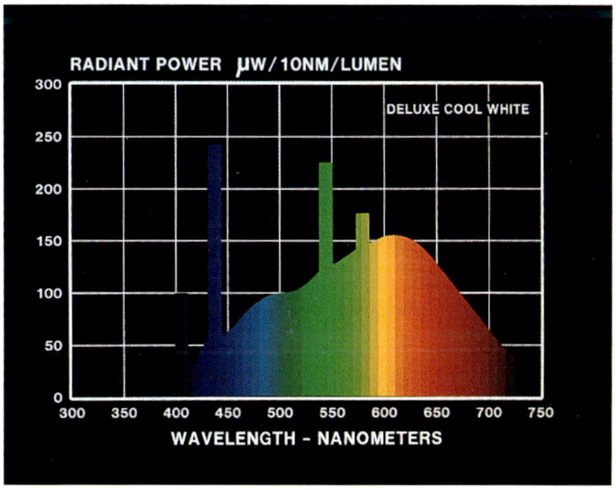

Figure G. Spectral Power Distribution for Deluxe Cool White Fluorescent Lamps. (Courtesy of G.E. Lighting.)

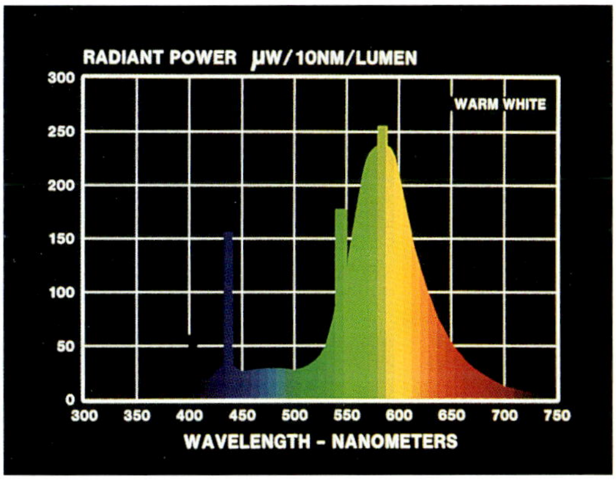

Figure H. Spectral Power Distribution for Warm White Fluorescent Lamps. (Courtesy of G.E. Lighting.)

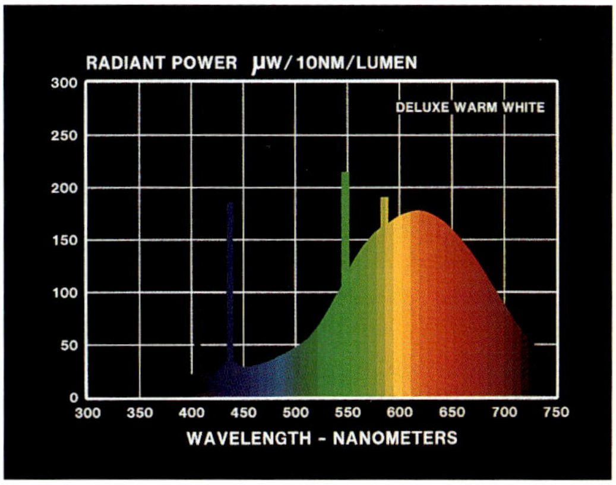

Figure I. Spectral Power Distribution for Deluxe Warm White Fluorescent Lamps. (Courtesy of G.E. Lighting.)

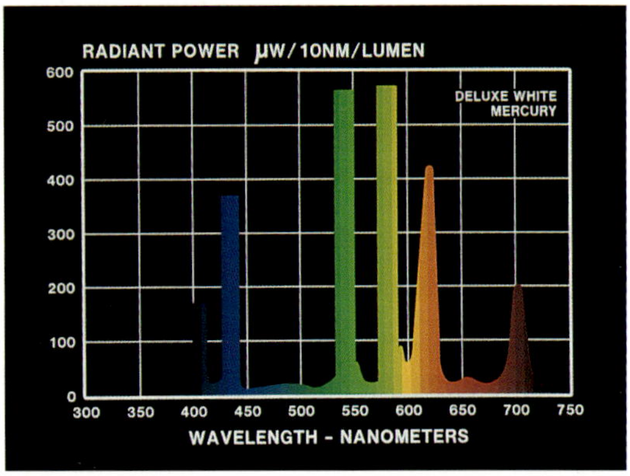

Figure J. Spectral Power Distribution for Deluxe White Mercury Vapor Lamps. (Courtesy of G.E. Lighting.)

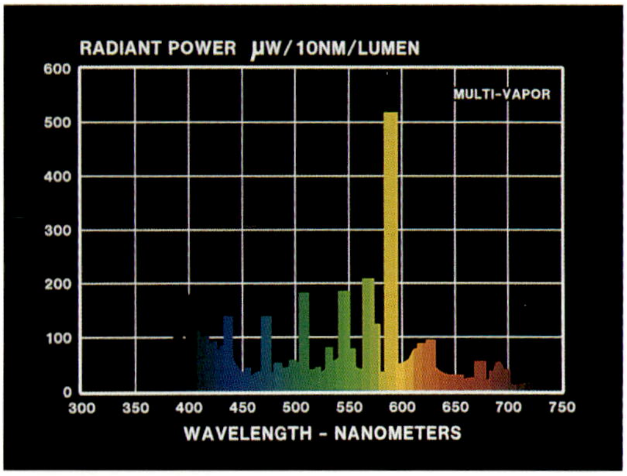

Figure K. Spectral Power Distribution for Metal Halide Lamps. (Multi-Vapor is G.E. Lighting's Tradename.) (Courtesy of G.E. Lighting.)

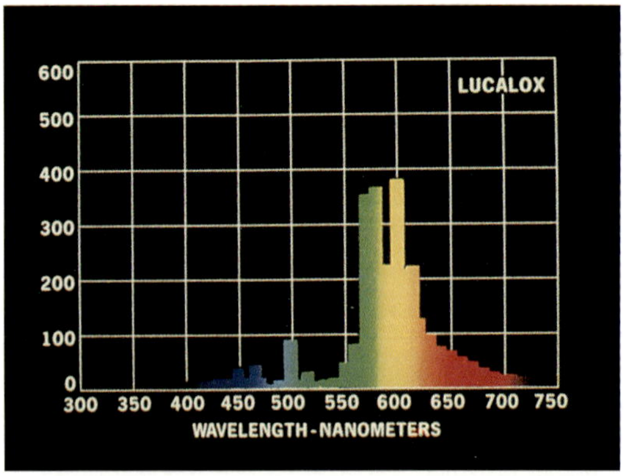

Figure L. Spectral Power Distribution for High-Pressure Sodium Vapor Lamps. (Lucalox is G.E. Lighting's Tradename.) (Courtesy of G.E. Lighting.)

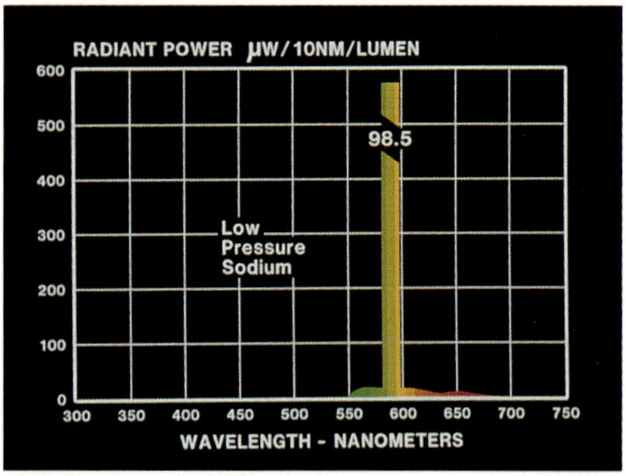

Figure M. Spectral Power Distribution for Low-Pressure Sodium Vapor Lamps. (Courtesy of G.E. Lighting.)

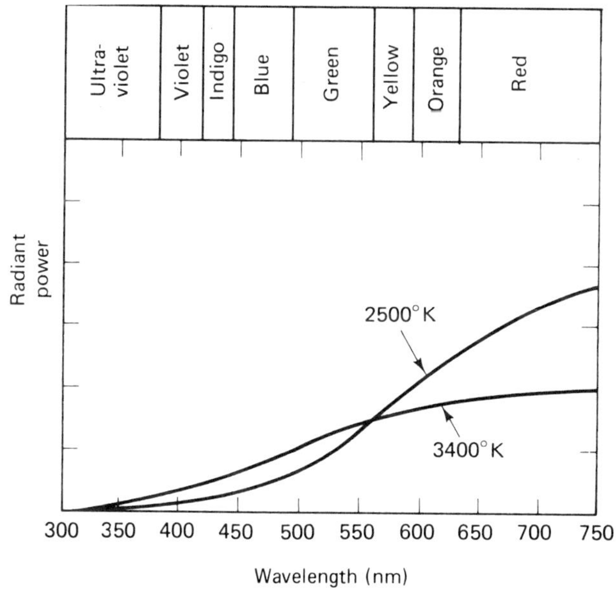

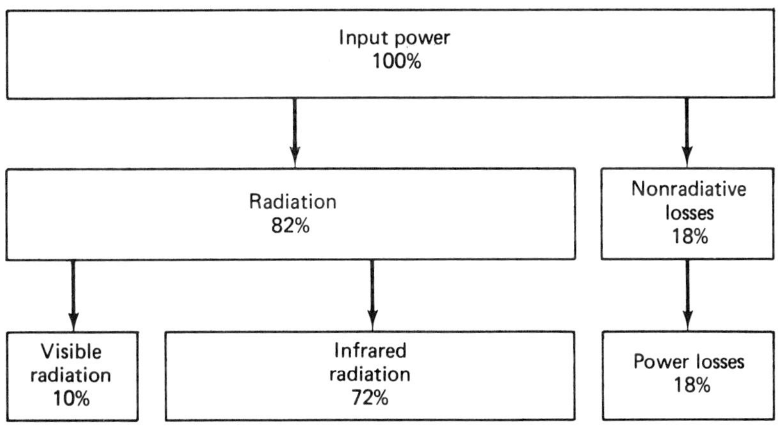

Figure 4-6. Energy Distribution of an Incandescent Lamp, 17.5 lm/W (100 W produces 10% visible radiation).

3. *Energy-saving disc*: A diode that chops half of the sine wave. For a 60 W lamp:

Volts: 120 V to 82 V

Lumens: 870 to 238 lm (73% reduction)

Watts: 60 to 32.6 W (46% reduction)

Efficacy: 14.5 to 7.3 lm/W (50% reduction)

4. *Energy-saving lamps*: Lamps that make use of different fill gases to increase efficacy. These lamps use krypton rather than argon gas used in conventional lamps. The result is a decrease in wattage without a decrease in efficacy. As an added benefit, the life is increased slightly.

Summary

The incandescent lamp has a short life, which increases maintenance cost, and a low efficacy, which increases operating cost. The heat produced by the lighting system is a function of the *watts* consumed by the system. The relationship between heat and watts is expressed in a constant, which is 3.41 Btu/hr/watt. The 100A lamp produces 341 Btu/hr. The incandescent lamp will produce more heat for an equal amount of lumens (light output); hence, incandescent lighting has the potential for a greater impact on the building air conditioning system.

Among the advantages of the "standard" incandescent lamp is the low initial cost of the lamp. The lamp also requires no auxiliary equipment (no ballast), which reduces the initial system cost. Color rendition is acceptable. Often an incandescent system is chosen because it is the easiest and cheapest to dim which may be an important consideration in many designs.

Tungsten-Halogen Lamps

A deficiency of standard incandescent lamps has been lumen depreciation over lamp life. When the filament is heated, it slowly evaporates and redeposits on the inside of the bulb wall (bulb wall blackening). This layer of tungsten deposit then acts like a filter, absorbing some light and lowering the light output. This was overcome by the development of the **halogen regenerative cycle lamp,** which is also called a **tungsten halogen** lamp, **quartz iodine** lamp, **tungsten iodine** lamp, or **quartz** lamp. The tungsten halogen lamp was developed to reduce lumen depreciation by eliminating bulb wall blackening.

Light-Producing Element

The tungsten halogen lamp contains a halogen such as iodine or bromine and a fill gas. The filament is made of tungsten which is heated causing the filament to incandesce producing visible light.

Enclosure

By decreasing the size of the glass envelope, the operating temperature of the lamp is increased. The enclosure or envelope is made of quartz to withstand the higher temperatures required for the tungsten iodide cycle to work. At high temperatures, the evaporated tungsten particles associates with a halogen (iodine) molecule forming tungsten iodide. Instead of being deposited on the bulb wall, the combined tungsten iodine molecule is returned to the hot filament, freeing the iodine molecule to pick up another evaporated tungsten molecule. This cleaning action minimizes the deposit of tungsten on the bulb wall, and results in an increased lumen output throughout the life of the lamp. Figure 4-4 shows the lumen output of a standard incandescent lamp and that of a tungsten halogen lamp over the rated life of each lamp.

To operate properly, tungsten halogen lamps require relatively high temperatures. To obtain these high temperatures, the filament had to be compacted and the outer envelope made smaller (see Fig. 4-7). The smaller enclosure more closely approaches the ideal point source needed for good optical control.

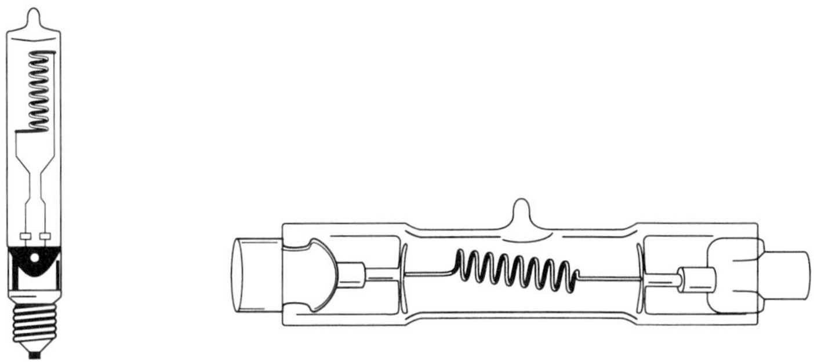

Q250CL/MC
(Mini-can screw base)

Q400T4/CL
(Recessed single contact base)

Figure 4-7. Tungsten-Halogen Lamps. (Courtesy of GTE Products Corp.)

Lamp Shape Designation:

Lamp Shapes. See Figure 4-7.

T = tubular
PAR = parabolic
MR = multisegmented reflector

Lamp Shape Designation. "T-4" = tubular—$\frac{4}{8}$th in. in diameter.

Lamp Order Code

Quartz
Finish, "Clear"
Q250CL/MC
Base, "Minican screw base"
Watts

Electrical Connection

The base provides the electrical contact and lamp positioning. Some of the more common bases are

MINICAN (miniature skirted candelabra)
RSC (recessed single contact)
PIN; MED.SKIR (medium-skirted screw base)
D.C. BAY (D.C. Bayonet)
MOGUL END PRG (Mogul end-prong base)

Operating Characteristics

Efficacy. The main objective in developing the tungsten-halogen was to maintain lumen output, but other improvements are also realized. The higher operating temperature resulting from the compact enclosure resulted in a slight improvement in the efficacy of the tungsten halogen lamp.

A Q250CL/MC lamp, which is rated at 5000 lm, has an efficacy of

$$K_{\text{system}} = \frac{5000}{250} = 20\,\frac{\text{lm}}{\text{W}}$$

Life. As result of the higher operating temperature and the random redepositing of the tungsten back on the filament, the lamp life of the tungsten halogen was increased *slightly*.

A Q250CL/MC lamp has a rated life of 2000 hr. For the same operating hours per year developed at the beginning of this chapter (3432 hr/yr), the lamp will operate in that building for

$$\text{no. yr} = \frac{\text{rated life}}{\text{oper. hr/yr}} = \frac{2000}{3432} = 0.6 \text{ yr (7 mo)}$$

Operating Variables

Voltage Variation. If the voltage to a 120 V lamp drops below 70 V the chemical bonding between the tungsten molecule and the halogen molecule stops and the lamp operates as an ordinary ''expensive'' standard incandescent. When the chemical reaction stops, bulb wall blackening will begin.

Lumen Depreciation. There is a slight improvement in lumen depreciation due to the elimination of bulb wall blackening, which prevents the loss in transmission of light through the enclosure (see Fig. 4-4).

SPD Curve. There is a slight improvement in the color rendering of a tungsten halogen due to the higher operating temperature (see Fig. 4-5). The tungsten halogen lamp still has a large amount of red or long wavelength present with a slight increase in blue or short-wavelength energy which makes the tungsten halogen slightly ''whiter'' than the ''standard'' incandescent lamp.

Energy-Saving Devices. Tungsten halogen lamps are dimmable. However, dimming will reduce the bulb temperature causing the tungsten iodine cycle to stop resulting in bulb wall blackening. Manufacturers claim that turning up the lamp to ''full on'' will clean the lamp. Extended dimming will increase lumen depreciation and reduce lamp life slightly. The low-voltage tungsten halogen lamps require a step-down transformer, which introduces potential problems in the dimming circuit.

Summary

Tungsten halogen is an expensive incandescent lamp that has a very compact envelope, which makes it an excellent lamp where **optical control** is important. It still has all the negative aspects of the "standard" incandescent, which are a short life and a low efficacy, which makes the tungsten halogen expensive to operate and maintain. Color rendition is good. The normal voltage (120 V) lamp requires no auxiliary equipment (no transformer), which results in a lower initial cost.

The low-voltage tungsten halogen lamps (MR-16) require a step-down transformer to reduce the line voltage from 120 V to 12 V. The transformer adds to the initial cost of the system and introduces a device that may require additional maintenance.

The tungsten halogen lamp operates with an internal pressure. If the lamp is improperly handled, operated at overvoltage, or operated at higher than normal temperature, the tungsten halogen lamp may fail violently. When the lamp fails, it will explode, showering the hot quartz glass into the room which may cause a fire or personal injury. It is recommended that the luminaire be enclosed with a protective lens to contain the lamp if it should experience a violent failure. The tungsten halogen lamp should not be handled with bare hands, since the oils in one's skin will act on the quartz when it is heated causing a chemical reaction that can weaken the lamp resulting in lamp failure.

GASEOUS DISCHARGE BALLASTS

Magnetic Ballasts—General Characteristics

Operating Characteristics

The remaining light sources come under the general category of gaseous discharge sources. All gaseous discharge sources require an electrical device called a **ballast.**

Gaseous discharge lamps are referred to as **zero-resistance** or **negative resistance** elements. As the additives inside the arc tube ionize, the resistance inside the arc tube decreases. This will cause the resistance to approach zero while the current approaches infinity.

$$\text{Ohm's law:}\quad I = \frac{E}{R}; \quad \text{as } R \longrightarrow 0\text{, then } I \longrightarrow \infty$$

Without a device to limit the current, the electrodes would burn up within a matter of seconds. Therefore, *all* gaseous discharge sources require a ballast.

A ballast is an electrical device that serves two primary functions; specifically,

1. It limits the current (zero-resistance element characteristics).
2. It provides starting voltage or voltage kick.

A third important feature is required for gaseous discharge lamps and systems to perform efficiently.

3. It provides power factor correction.

A ballast acts as an autotransformer to provide the voltage kick and a choke to limit the current. Most ballasts today are a *magnetic device*, which means the ballast has a **core** and **windings.** Therefore, a ballast is an inductive reactance load.

Ballast Selection

Magnetic devices such as ballasts have three characteristics the designer must be aware of to prevent problems:

1. *Noise*—humming sound due to vibration of the windings.
2. *Heat*—adds to ambient room heat, luminaire heating, and may cause premature ballast failure.
3. *Inductive reactance*—requires power factor correction.

Noise. Only *fluorescent* ballasts have a standard noise rating, for example, *A through J*. *A* is the most quiet and *F* is the noisiest. All other ballasts for the remaining gaseous discharge lamps have no standard sound rating. When dealing with gaseous discharge sources, the designer must be aware of ballast and luminaire noise.

Heat. The ballast also consumes *power*, which means that the power (watts) consumed by the ballast must be taken into consideration to evaluate system efficacy (lamp + ballast). Heat produced by an electrical device is a function of the wattage consumed by the device. Therefore, the ballast adds to the heat within the luminaire and the environment.

Power Factor (PF). Most fluorescent ballasts contain a **capacitor,** which is an integral part (inside the ballast housing) of the fluorescent ballast. The capacitor provides the power factor correction.

All other gaseous discharge ballasts require power factor correction, but the capacitor is a separate device and not integrated into the ballast.

For a resistance load such as an incandescent, the power factor, *PF*, is equal to 1.0.

For the other discharge ballasts, the uncorrected *PF* is

Low-pressure sodium:	$PF = 0.20$	Metal halide:	$PF = 0.50$
Mercury vapor:	$PF = 0.50$	High-pressure sodium:	$PF = 0.40$

If $PF = 0.50$, then the current, I, will be doubled. Most state energy codes state that if the PF is < 0.80, then the PF must be corrected to at least $PF = 0.90$.

$$\text{for ac circuits:} \quad I = \frac{P}{E \times PF}$$

The designer should always specify high-PF fluorescent ballast ($PF > 0.9$) and the addition of PF correction > 0.90 for all other ballasts.

At the end of this chapter, specific information and data are given on ballasts for each of the gaseous discharge sources described.

Electronic Ballasts

An electronic ballast transforms 60 Hz input current to 20,000 to 60,000 Hz current flow to the lamps. The high-frequency operation of fluorescent lamps can increase lamp efficacy by as much as 10% due to increased phosphor excitation. Other benefits include extended lamp life and the elimination of flicker. Electronic ballasts are available in two, three, and four lamp models. The first generation, ''hybrid,'' electronic ballast contained both magnetic devices (core and coil) and solid-state devices. Technology has been developed for a true solid-state ballast that eliminates the problems associated with the core and coil. Additional benefits of true solid-state high-frequency ballasts are reduced ballast size and weight, reduced audible noise, reduced heat, reduced losses and reduced dimming cost. Dimming of high-intensity discharge lamps has been accomplished with a solid-state high-frequency ballast.

Burning Position

Gaseous discharge lamps are sensitive to burning position. The engineer must use caution in matching the lamps selected with the final burning position of the lamp. If lamps are burned in a position other than intended, the lamp life, lumen output, and color characteristics can change. Some lamps may experience a violent failure if they are not properly installed. Manufacturer specifications should be consulted for burning position information. Typical letter designations for burning position are

BU:	base up ONLY	VBU:	vertical base up
BD:	base down ONLY	VBD:	vertical base down
HOR:	horizontal ONLY	HBU:	horizontal base up
U:	universal	HBD:	horizontal base down

Starting/Restrike

Starting Time

Starting time is the time required to bring the light output (lumens) of the lamp up to 80% of its rated lumens (cold lamp start-up).

Restrike Time

Restrike time is the time required to bring the light output (lumens) of a lamp, whose power has been momentarily interrupted (hot lamp start-up), back up to 80% of its rated lumens.

GASEOUS DISCHARGE SOURCES

Gaseous discharge sources are divided into two categories: low-pressure gaseous discharge and high-pressure gaseous discharge.

LOW-PRESSURE GASEOUS DISCHARGE SOURCES

There are two low-pressure gaseous discharge sources: fluorescent and low-pressure sodium.

Fluorescent Lamps

The first major showing of the fluorescent lamp was at the 1938–1939 New York World's Fair. The lamps were installed in vertical clusters on the flag poles along the Avenue of Flags.

Light-Producing Elements

Figure 4-8 is a schematic showing the operation of a fluorescent lamp. The fluorescent lamp requires three elements or components to produce visible light: (1) electrodes, (2) gases, and (3) phosphor.

Electrodes (Cathodes). Electrodes are the electron-emitting devices. Two types of cathodes are in use today. The **hot-cathode** lamp is a coiled coil or a triple-coiled tungsten filament coated with an alkaline earth oxide that emits electrons when heated. The electrons are boiled off the cathode at about 900°C. The **cold-cathode** lamp is a pure iron tube that also has an electron-emitting material applied inside the tube. The cold cathodes are subjected to higher voltage, releasing electrons at about 150°C. Cold-cathode lamps are used in special application such as signs and can be bent into different shapes. The hot-cathode lamp is the most common type of electrode used in fluorescent lamps for most applications. Therefore, we shall not describe cold-cathode lamps.

Gases. A small quantity of mercury droplets is placed in the fluorescent tube. During the operation of the lamp, the mercury vaporizes at a very low pressure. At this low pressure, the current flowing through the vapor causes the vapor to radiate energy principally at a single wavelength in the ultraviolet (253.7 nm) region of the spectrum. The pressure of the mercury is regulated during operation by the temperature of the bulb wall.

The lamp also contains a small amount of a highly purified rare gas. Argon and argon-neon are the most common, but krypton is also sometimes used. The gas ionizes readily when a sufficient voltage is applied to the lamp. The ionized gas decreases in resistance quickly, allowing current to flow and the mercury vaporize.

Phosphor. This the chemical coating on the inside wall of the bulb or enclosure. When the phosphor is excited by ultraviolet radiation at 253.7 nm, the phosphor produces

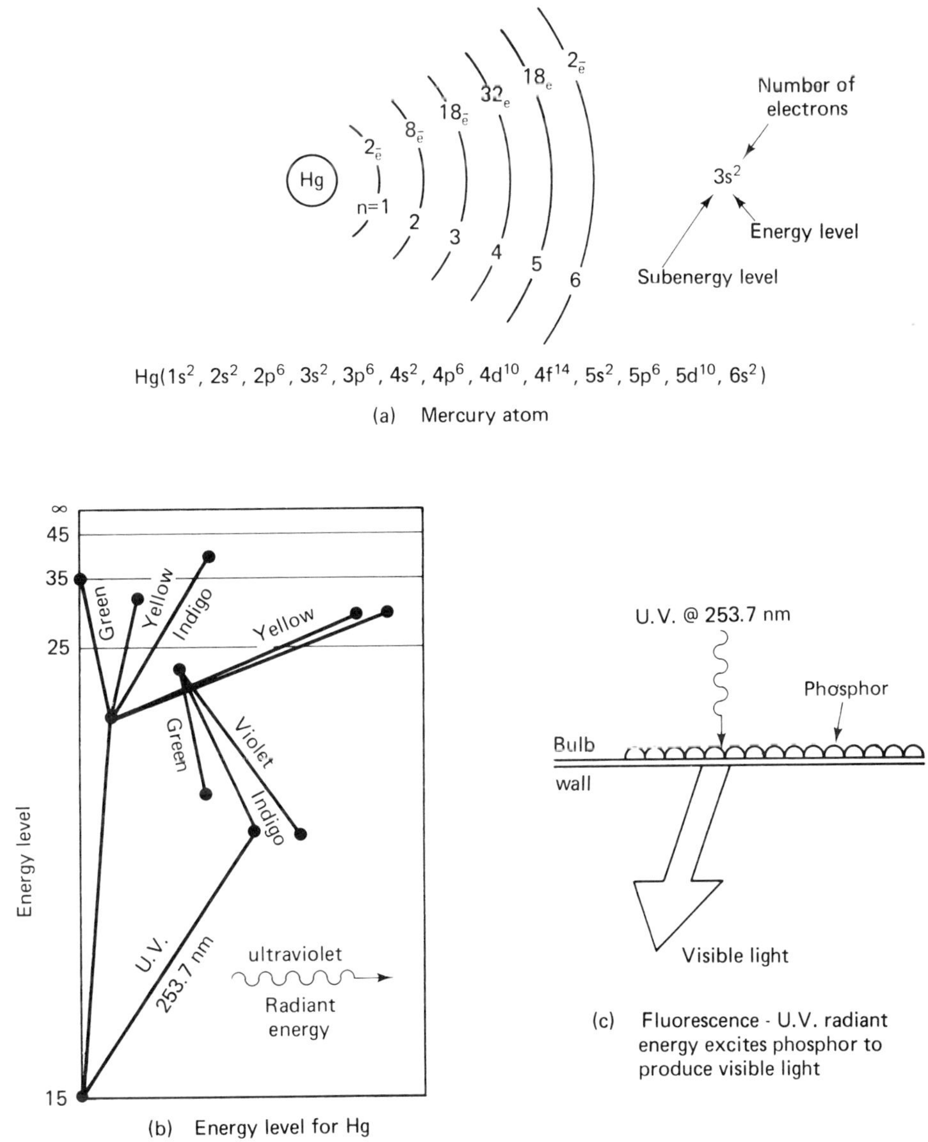

Figure 4-8. Schematic of the Operation of a Fluorescent Lamp.

visible light by fluorescence (see Fig. 4-8). That is, visible light from a fluorescent lamp is produced by the action of ultraviolet energy on the phosphor coating on the inside surface of the bulb or enclosure. The phosphor mixture can be changed to change the color of the lamp or the lamps SPD characteristics.

Enclosure

The enclosure or bulb is the glass envelope which serves two primary functions. The enclosure holds the gases in the tube *and* provides a surface to which the phosphor coating can be applied.

Lamp Shape Designation. The fluorescent lamp is designated according to its shape, diameter, and color.

Lamp Shapes. See Figure 4-9.

T = tubular	PG = power groove (high output)
C = circular	PL or TT = compact or twin tube
U = hairpin or U lamp	

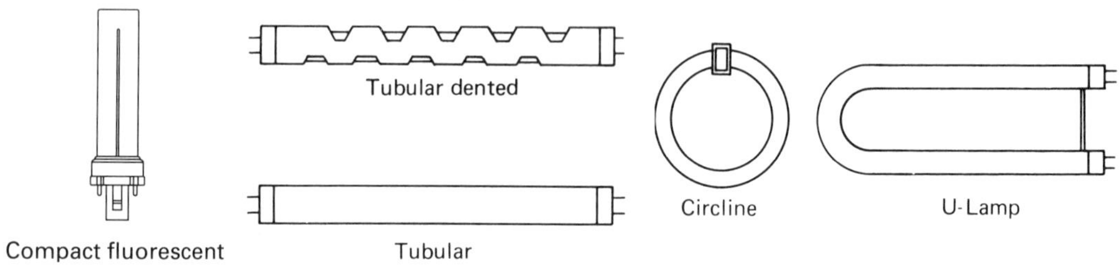

Figure 4-9. Fluorescent Lamp Shapes. (Courtesy of GE Lighting.)

Lamp Shape Designation. The lamp shape designation is a letter-number combination such as "T12" which means a tubular shape with a diameter in $\frac{1}{8}$th = $\frac{12}{8}$th or $1\frac{1}{2}$ in.

Lamp Ordering Code. The ANSI Lamp Ordering Code is a combination of letters and numbers that represent different attributes of the lamp.

The "F" which indicates a fluorescent is followed by the "nominal" watts or length of the lamp which is followed by the lamp shape designation, the lamp color, and sometimes other lamp characteristics. The preheat and rapid-start lamps use the "nominal" lamp wattage in their designation while the lamps designated high output (HO), very high output (VHO), instant-start, and power groove use the nominal length in their designation.

Fluorescent
Shape, "Tubular"
Rapid Start
F40T8/41K/RS
Color
Diameter, $\frac{8}{12}$ in.
Watts

Some examples follow:

F20T12/CW = 2′ preheat	F40CW/U/3 = 3″ (or 6″) leg spacing
F30CW = 3′ rapid start	F48T12/CW/HO = 800ma lamp
F40CW = 4′ rapid start	F96PG17/CW = power groove
F40T8/41K = T8 rapid start	F48T12/CW/1500 or VHO = 1500 ma
FC12T9/CW = circular	F5TT/27K = compact twin tube (5, 7, 9, 13, 18 W)
F96T12/CW = slimline	

Electrical Connection

The base forms the electrical connection of the lamp to the sockets and acts to support and align the lamp. Five basic types of bases are associated with the fluorescent lamp:

1. *BiPin* (miniature, medium, mogul): The most common is the *Medium BiPin*: 12″, 15″, 18″, 2′, 3′, 4′—T lamps and U lamp. *Miniature Bi-Pin*: 6″, 9″, 12″, 21″—T lamps. *Mogul BiPin*: 4′, 5′.
2. *Single-Ended BiPin*: PL or TT lamps.
3. *Single Pin*: Instant-start/slimline lamps.
4. *Recessed Double Contact*: HO and VHO—T and PG.
5. *4-Pin*: Circline.

Operating Characteristics

Efficacy. An F40CW fluorescent lamp is rated at 3150 lumens.

$$K_{\text{lamp}} = \frac{3150}{40} = 78.8\,\frac{\text{lm}}{\text{W}}$$

The system efficacy includes the lamp watts plus the ballast watts. If a 40 watt lamp is operated on a single-lamp ballast that consumes 14 watts or on a two-lamp ballast that consumes 12 watts, the system efficacy will be as follows:

$$\text{one-lamp ballast} = 54 \text{ watts}$$

$$\text{system watts} = 40 + 14 = 54 \text{ watts}$$

$$K_{\text{system}} = \frac{3150}{54} = 58.3\,\frac{\text{lm}}{\text{W}}$$

$$\text{two-lamp ballast} = 92 \text{ watts}$$

$$\text{system watts} = (2 \times 40) + 12 = 92 \text{ watts}$$

$$K_{\text{system}} = \frac{2 \times 3150}{92} = 68.5\,\frac{\text{lm}}{\text{W}}$$

A three-lamp luminaire used to require a two-lamp ballast and a one-lamp ballast. The wide use of the three-lamp luminaire during the energy crisis of the 1970s has brought about the introduction of the three-lamp ballast.

$$\text{three-lamp ballast} = 140 \text{ watts}$$

$$\text{system watts} = (3 \times 40) + 20 = 140 \text{ watts}$$

$$K_{\text{system}} = \frac{3 \times 3150}{140} = 67.5 \frac{\text{lm}}{\text{W}}$$

Life. Fluorescent lamp life is dependent on the **burning cycle per start.** The life rating for fluorescent lamps is given on the basis of 3-hour burning cycle per start. In 1973, a new collector gas was introduced into fluorescent lamps. This gas substantially reduces the boiling off of the electron emission material from the electrodes every time that the lamp is started. Therefore, the life of the lamp is no longer affected by more frequent on/off cycling of the lamp. However, the importance of burning cycle can be seen in the *extended life* of the fluorescent when the burning cycle is greater than the standard test cycle of 3 hours/start. Continuous burning of a fluorescent lamp can increase the life by a factor of 2.5 or more. Other gaseous discharge lamps use a standard test cycle of 10 hours/start. An F40CW fluorescent lamp has a rated life of 20,000 hr.

$$\text{no. yr} = \frac{\text{rated life}}{\text{oper. hr/yr}} = \frac{20{,}000}{3432} = 5.8 \text{ yr}$$

Operating Variables

Effect of Temperature. The most efficient lamp operation is achieved when the ambient temperature is between 70°F and 90°F (see Fig. 4-10) for a fluorescent lamp. Lower temperatures cause a reduction in mercury pressure, which means that less ultraviolet energy is produced; therefore, less UV energy is available to act on the phosphor

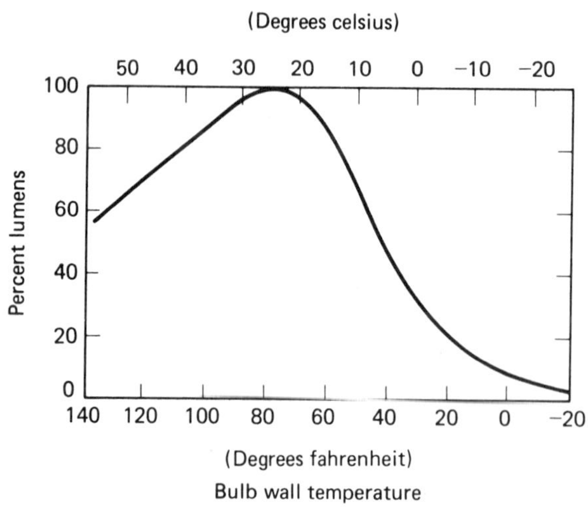

Figure 4-10. Temperature Effects on Fluorescent Lamps. (Courtesy of GE Lighting.)

and less light is produced. High temperature causes a shift in the wavelength produced that is nearer to the visual spectrum. The longer wavelengths have less effect on the phosphor and therefore less light output.

Fluorescent lamps can be operated down to a temperature of 50°F on a standard ballast. However, as Figure 4-10 indicates, the light output (lumens) will be less if the ambient temperature is outside the 70°F to 90°F range. Special low-temperature ballasts are available for starting and operating fluorescent lamps at 0°F and −20°F. These ballasts provide a higher starting voltage, and they usually contain a thermal starting switch. However, the low-temperature ballast may start the lamp, but it will *not* overcome the dramatic loss in light output (lumens).

TABLE 4-2 Temperature Conversion Equations

To obtain	Conversion equation
°C	$°C = \frac{5}{9}(°F - 32)$
°F	$°F = \frac{9}{5}(°C + 32)$
°K	$°K = °C + 273$

Effect of Humidity. Starting voltage requirements are affected by the electrostatic charge on the outside surface of a fluorescent lamp. Moist, humid air has unfavorable effects on the surface charge. This factor must be taken into account when the relative humidity exceeds 65%. A silicone coating on the outside surface of the lamp and the proper distance between the lamp and metal housing will usually solve starting problems under any conditions of humidity. However, dirt accumulation on the lamp will nullify the effects of the silicone coating and cause starting difficulties. Cleaning the lamp with an abrasive cleaner may also remove the silicone coating.

Burning Position. The fluorescent lamp should be burned in a horizontal position. Vertical operation of fluorescent lamps causes a nonuniform distribution of gases in the lamp resulting in a reduction in light output and uniformity. In a vertical position, the mercury droplets are concentrated near the lower cathode increasing deterioration of the cathode resulting in a reduction in life.

Stroboscopic Effect. Stroboscopic is a Greek word that means "to see motion." The arc stream extinguishes during each reversal (120 times per second) of the sine wave; however, the phosphor coating continues to radiate light during this brief period. Generally, this is not noticeable, but it can make high-speed machinery appear to stand still. The use of a series sequence ballast on rapid-start circuits will eliminate this problem. Another solution is to use a lead-lag ballast, which puts one lamp out of phase with the other in a two-lamp unit. This results in one lamp being at maximum light output while the other lamp is at zero output. The net effect is to eliminate the flicker.

Hot-Cathode Circuits. There are three types of hot-cathode fluorescent lamps. They are defined by the circuits for which they are designed:

1. Preheat
2. Instant start
3. Rapid start

Preheat Circuit. The preheat circuit was the first type to be developed. The preheat circuit was introduced at the 1938–1939 New York World's Fair. It requires a separate starter that preheats the electrodes, causing an emission of electrons. This causes the internal resistance to drop, which allows the arc to strike. The preheating process requires a few seconds; hence the slow, blinking start that is characteristic of the preheat circuit. Preheating may be accomplished by a *manual* button starter or an *automatic* starter. The starter applies current to the electrodes of the lamp for a sufficient time to heat them and then automatically (or manually) removes the current from the electrode, causing the voltage to be applied between the electrode and striking the arc. All tubular lamps less than 3 feet in length and the lower-wattage "compact" fluorescent lamps (7 watts to 18 watts) are preheat lamps. Power factor correction may be required for some preheat ballasts.

Instant-Start Circuit. In 1944, the instant-start circuit was introduced to overcome the slow starting characteristics of the preheat circuit. The instant-start circuit eliminates the need for a separate starter and thus simplifies the system and maintenance. A sufficiently high voltage is applied between the electrodes to break down the resistance of the lamp and strike the arc. The arc quickly heats up the filament electrodes, which then supply electrons to sustain the arc. Since no preheating is required with instant-start lamps, a single-pin contact is sufficient. This type of lamp is called a *slimline* lamp.

Rapid-Start Circuit. In 1952, the rapid-start circuit and lamp were developed. It starts quickly without the need for a separate starter. A rapid-start ballast is smaller and more efficient than the instant-start ballast for the same wattage. The rapid-start circuit utilizes low-resistance electrodes that are heated continuously with very low losses.

The rapid-start lamp is the most common lamp and is best suited for most applications. Rapid-start circuits can be flashed and dimmed quite efficiently.

Circline (circular) lamps are available for operation in rapid-start circuits. Two-foot hairpin or U-lamp fluorescents also are designed for use in rapid-start circuits. Tubular fluorescent lamps which are 3 feet and longer (430 mA, 800 mA, and 1500 mA) and the higher wattage "compact" fluorescent lamps (18 watts to 39 watts) are rapid-start lamps. Rapid-start lamps can be used in both preheat and rapid-start circuits. However, a lamp carrying the designation "preheat" cannot be used in a rapid-start circuit. Rapid-start circuits are classified according to the lamp input current rating:

RS ("standard")	430 mA
Circline	390, 420, 430 mA

HO = high output	800 mA
XHO, PG, VHO, SHO, T10 = very high output	1500 mA

Lumen Depreciation. As with most light sources, the fluorescent lamp decreases in light output (lumens) as the lamp ages throughout its rated life.

SPD Curve. The color or SPD of a fluorescent lamp depends on the phosphor coating on the inside surface of the bulb wall. The spectral power distribution curve consists of two components: (1) a continuous or smooth portion and (2) a line spectrum. The lines or bars on the SPD curve represent visible light that is generated directly by the mercury arc; the smooth portion is due to the action of the ultraviolet energy on the phosphor coating. The SPD of a fluorescent lamp can be changed by modifying the type and mix of phosphors used to coat the lamps. There are six standard ''white'' fluorescent lamps on the market (see Fig. 4-11):

CW = cool white	CWX = cool white deluxe
WW = warm white	WWX = warm white deluxe
W = white	D = daylight

This assortment of ''white'' fluorescent lamps has been developed to meet nearly all needs for white light. They are referred to as **standard white lamps** because all six are available from every major manufacturer of fluorescent lamps. In addition to these standard whites, each manufacturer markets special ''whites'' and colored fluorescent lamps.

A choice among fluorescent lamps always means a compromise between efficacy and color. Selection of the best color rendition usually means a reduction in efficacy. The CW, WW, and W lamps have higher efficacy. The CW, WW, W, and D are weak in red, resulting in poor color rendition characteristics. The CWX and WWX are designed to give the best color rendition to objects and people at a reasonable efficacy. This is achieved by the addition of red phosphors to the mixture. However, since the eye is less responsive to red energy, the luminous efficacy of the CWX and WWX is reduced by about 30% compared to the CW and WW lamps.

A number of new fluorescent lamps are using a two-digit number designation to represent the color or SPD curve. The number normally represents the first two digits of the *correlated* color temperature (see the discussion at the beginning of this chapter and Chapter 3) of the lamp. Numbers such as 27, 30, 35, or 41 represent correlated color temperatures of 2700K, 3000K, 3500K, and 4100K, respectively. The sources that designate color utilizing a correlated color temperature do not match any of the standard white fluorescent when plotted on the CIE chromaticity diagram. It is also important to realize that a given isotemperature line on the CIE chromaticity diagram (Fig. 4-1) such as 4100K, extends above and below the blackbody radiator curve. This means that source rated a 4100K could be plotted anywhere along the 4100K isotemperature line, resulting in a major shift in the color rendering characteristics between two lamps both of which are rated at 4100K.

There is no true white light source in terms of color rendition. Of the standard

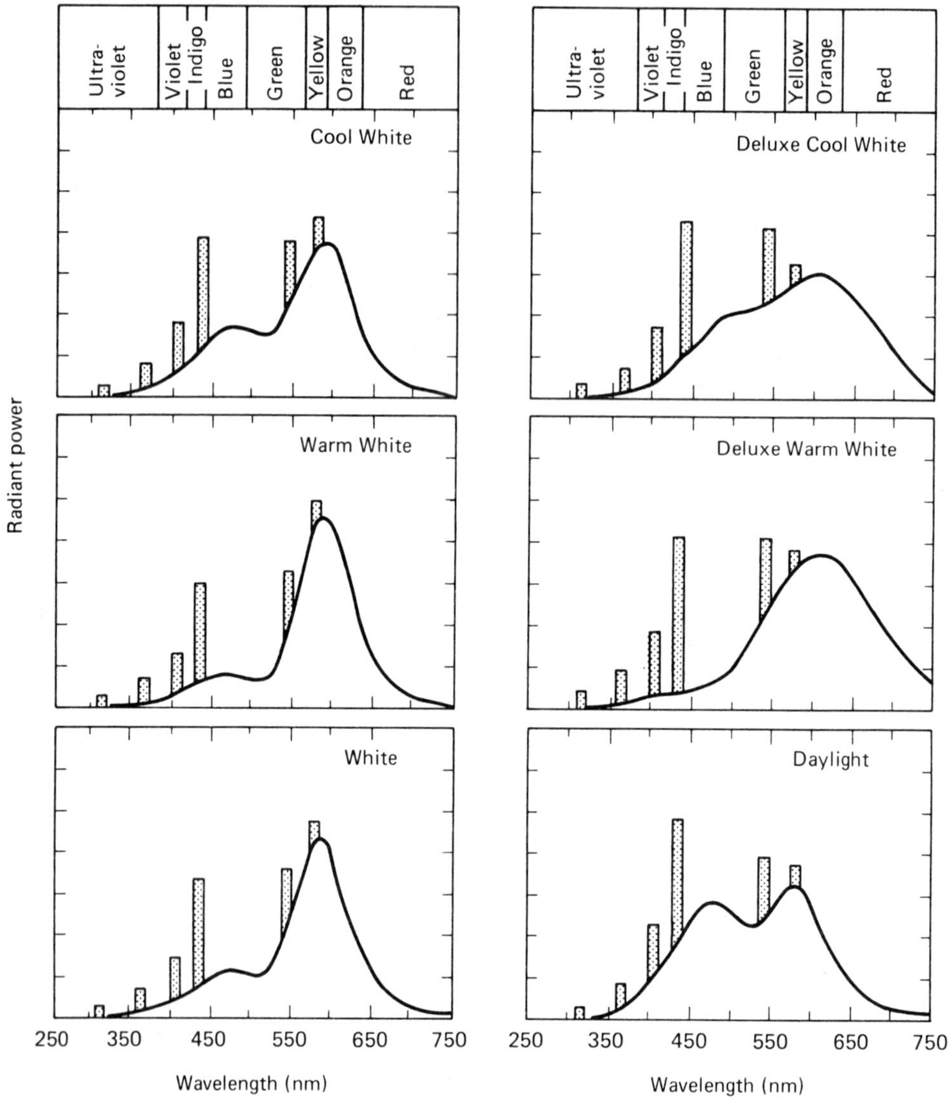

Figure 4-11. Spectral Power Distribution for the Standard Fluorescent Lamps. (Courtesy of GE Lighting.)

"white" fluorescent lamps, the CWX produces the best color rendition on all object colors, but CWX has a lower lamp efficacy. Premium lamps are available that feature better color rendering and improved efficacy, but at a substantially higher price.

Starting Voltage. The ballast for a 40 watt fluorescent lamp requires a starting voltage or voltage kick of 335 volts to start the lamp.

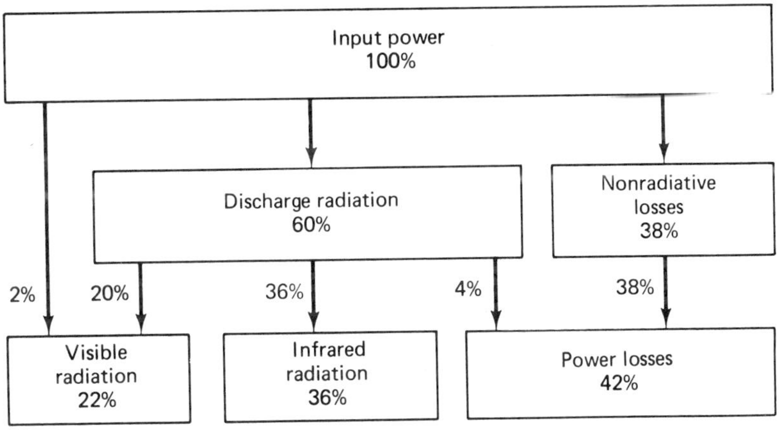

Figure 4-12. Energy Distribution for a Fluorescent Lamp 78.8 lm/W (40 W Produces 22% Visible Radiation).

Energy Distribution. Of the total input energy into the fluorescent lamp, only 22% ends up as visible light (see Fig. 4-12).

Energy-Saving Devices. The least expensive and most efficient lighting control that saves energy is the ac snap switch which provides the on/off control of the lights. The cost of fluorescent dimming is continuing to decrease. Modern SCR dimmers will save energy when the lighting level is reduced or dimmed.

Energy-saving lamps are designed to operate at a lower wattage on the same ballast as conventional fluorescent lamps and on newer energy-saving ballasts. The efficacy of some energy-saving lamps decreased; others have an increase in efficacy. Energy-saving lamps may cause premature ballast failure on old ballasts due to capacitor overloading. A high-power-factor two-lamp ballast contains a starting capacitor and a power factor correcting capacitor. On older systems that are retrofitted with energy-saving lamps, an increase of 6% in the current to the starting capacitor will cause ballast failure. New ballast designs have eliminated this problem. Energy-saving lamps should only be considered for retrofitting an existing installation that was improperly designed and overlit; they should not be used for new installations unless the new energy-saving ballast is specified. Energy-saving lamps should not be used where dimming systems are used. The lamp may flash or swirl and show end blackening due to cathode deterioration.

Energy-saving ballast must meet current energy efficiency regulations and should be specified for most applications. Standard lamps can be operated on most energy-saving ballasts and actually perform better than energy-saving lamps. The lighting designer should be cautious when specifying energy-saving systems (single manufacturer's lamp and ballast) if the ballast guarantee is valid only when using the manufacturer's energy-saving lamps. This type of a system guarantee prevents the building owner from bidding competitively on a lamp contract. The ballast section at the end of this chapter discusses the federal standards for the circle ''E'' ballast.

For economic reasons, the four-lamp fluorescent luminaire was preferred over a three-lamp luminaire prior to the energy crisis. The four-lamp luminaire may have produced lighting levels higher than required, but at the time energy cost was lower. Because of the lower energy cost, it was more economical to buy a luminaire with two two-lamp ballasts than was to purchase one two-lamp ballast and one one-lamp ballast for a three-lamp luminaire. The one-lamp ballast costs about the same as a two-lamp ballast. The energy crisis created a high demand for three-lamp luminaire to reduce lighting levels and reduce energy consumption. The result of this demand has been the introduction of the three-lamp ballast, which is a single unit that operates all three lamps.

Removing one lamp from a two-lamp or four-lamp luminaire may appear to be a simple solution for reducing energy consumption in an existing building where unnecessarily high illuminance levels exist in noncritical task areas. Removal of a lamp from a luminaire does not deenergize the ballast and can lead to overheating of the ballast and a violent ballast failure. This problem was solved by the development of a **dummy tube** that replaces one fluorescent tube to complete the series circuit of the two-lamp ballast. The dummy tube is constructed of glass and contains a capacitor that offsets the ballast inductance. The capacitance restores the system to its normal power factor while allowing the remaining fluorescent to operate without damaging the ballast. A typical two-lamp, 40 watt, rapid-start circuit will show a 62% drop in the wattage consumed when one lamp is replaced with a dummy tube. However, the light output of the two-lamp luminaire will decrease by 67% of the original light output. This will result in a slight decrease in efficacy.

The use of dummy tubes should be limited to retrofit applications where lighting levels are excessive in noncritical visual areas since the dummy tubes are quite expensive. Removal of a lamp from a luminaire also creates an ugly stripping or nonuniform appearance to the surface of the lens.

Summary

Fluorescent sources are available in a large number of wattages, lengths, shapes, and colors. The relatively high efficacy and long life make fluorescent lighting an excellent choice for a number of applications. The length and diameter of a fluorescent lamp results in a linear source which limits optimum lighting distribution control to a high-quality refractor (lens).

Low-Pressure Sodium Lamps

The low-pressure sodium lamp has been used extensively in Europe since the 1940s. A major marketing campaign for the lamp commenced in the United States in 1972. The low-pressure sodium lamp has the highest lamp efficacy of all sources, but it is monochromatic (single-wavelength) *yellow*.

Light-Producing Element

The light-producing element is an arc tube. The arc tube is U shaped and constructed of borate glass. The tube is dimpled to maintain a uniform distribution of sodium throughout the arc tube. The arc tube contains a small amount of argon and neon to aid in starting the lamp. The pressure inside the arc tube is approximately 1×10^{-3} mm; the enclosing space between the arc tube and the outer enclosure is under a vacuum.

Visible light is produced by the action of the electrons on the sodium which results in monochromatic yellow light (589 nm = 95% and 586 nm = 5%).

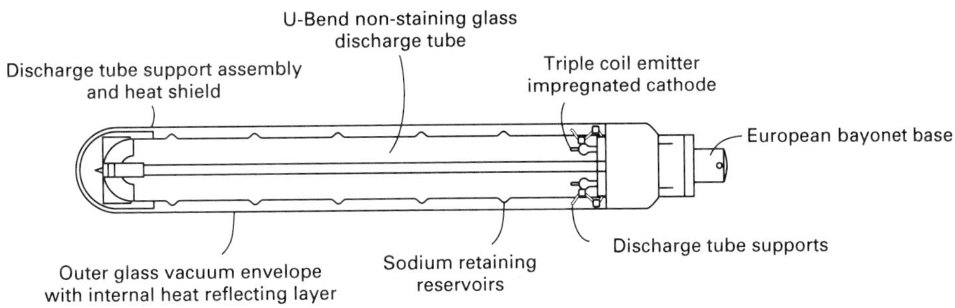

Figure 4-13. Low-Pressure Sodium Lamps. (Courtesy of Philips Lighting Company.)

Enclosure

The enclosure is made of ordinary glass and it serves one function which is to maintain the arc tube in a constant thermal environment while protecting the arc tube. The arc tube operates at a temperature of 260°C (500°F). The lamp comes in five wattages:

Lamp Shape Designation

Lamp Shape. The low-pressure sodium lamp shape (see Table 4-2) is a "T," which represent a tubular shape.

TABLE 4-2 Low-Pressure Sodium Lamps—Shape and Burning Position

Nominal Watts	MOL*	Lamp Shape	Burning Position
18	8.50	T17	UP/HOR
35	12.19	T17	UP/HOR
55	16.75	T17	UP/HOR
90	20.79	T21	HOR only
135	30.50	T21	HOR only
180	44.33	T21	HOR only

* Maximum Overall Length, in inches.

Lamp Shape Designation. The 180 W lamp ("T21") is tubular with a diameter of ($\frac{21}{8}$th in.) $2\frac{5}{8}$ in. and a length of $44\frac{1}{8}$ in.

Lamp Order Code. The ANSI Lamp Order Code for the 180 W lamp is "L74RF-180" while a manufacturer's lamp order code is "SOX180."

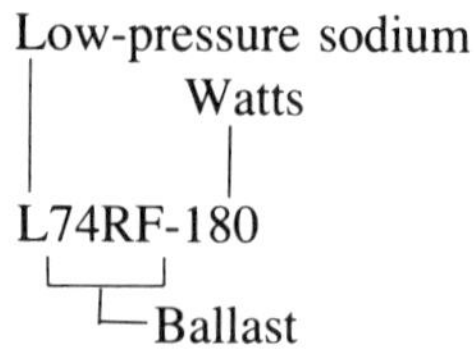

Electrical Connection

The base is a "European" bayonet base (BAY-B1) that will maintain the U-shaped arc tube in a horizontal orientation.

Operating Characteristics

Efficacy. A 180 W low-pressure sodium lamp (Tables 4-3 and 4-4) is rated at 33,000 lm. For the effects of ballast losses on efficacy, refer to the discussion of ballasts later in this chapter.

Life. The rated life for all wattages is 18,000 hours based on a burning cycle of 5 hours per start. Burning position is critical to lamp life since lamp failure is due to the migration of the sodium toward the electrodes. This migration causes an increase in the watts consumed by the lamp over its life, which results in electrode failure.

$$\text{no. yr} = \frac{18{,}000}{3432} = 5.2 \text{ yr}$$

Operating Variables. Both the lumens and watts will increase as the lamp ages, which results in a decrease in efficacy with age.

Lumen Depreciation. The lumen output increases slightly over the life of the lamp. Lumen output is said to be constant over the operating temperature range of

TABLE 4-3. Low-Pressure Sodium Lamp (Only) Efficacy

Nominal Watts	Lumens	Actual Lamp Watts (100 hr)	Lamp Efficacy (100 hr)	Actual Lamp Watts (18,000 hr)	Lamp Efficacy (18,000 hr)
18*	1,800	17	105.9	18	100.0
35	4,800	35	137.1	39	123.1
55	8,000	55	145.5	61	156.9
90	13,500	90	150.0	98	137.8
135	22,500	130	173.1	142	158.5
180	33,000	176	187.5	192	171.9
	34,000†			192	177.1

* 18 W rated at 14,000 hr.
† Lumens at end of life.

TABLE 4-4. Low-Pressure Sodium Lamp (Lamp plus Ballast) Characteristics

Nominal Watts	Lumens	Input Watts (100 hr)	System* Efficacy (100 hr)	Input Watts (18,000 hr)	System* Efficacy (18,000 hr)
18	1,800	30	60.0	33	54.5
35	4,800	60	80.0	68	70.6
55	8,000	80	100.0	90	88.9
90	13,500	125	108.0	149	90.6
135	22,500	178	126.4	226	99.5
180	33,000	220	150.0	286	115.4
	34,000†			286	118.9

* System efficacy is lamp + ballast (ballast is a high-power-factor, high-reactance type).

† Lumens at end of life.

−10°C (+14°F) to +40°C (+104°F). The effect on lumen output when the lamp is operated outside this temperature range has not been published.

SPD Curve. The light produced by a low-pressure sodium lamp is a monochromatic yellow (see Fig. 4-14). The spectral power distribution consists of two lines at 589 nm (approximately 95% of the output) and 586 nm (approximately 5% of the output). Because of the monochromatic yellow characteristics, the color rendering potential of the lamp is nonexistent. All colors appear as different shades of gray or brown except for yellow objects.

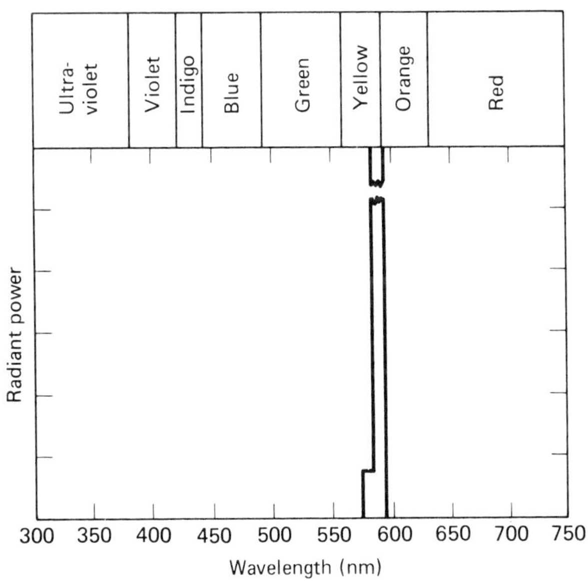

Figure 4-14. Spectral Power Distribution for a Low-Pressure Sodium Lamp.

Starting Voltage. The starting voltage or voltage kick required to start a 180 W low-pressure sodium lamp is 600 V.

Starting/Restrike Times

Low-Pressure Sodium Lamps

Starting 80% Max.	Restrike 80%
9–15 min	30 sec

Energy Distribution. Of the total input energy into a low-pressure sodium lamp, only 35.5% ends up as visible light (see Fig. 4-15).

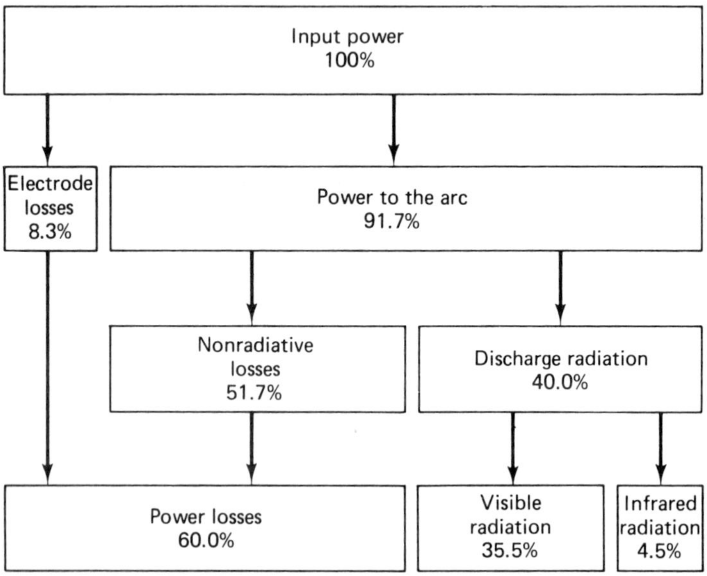

Figure 4-15. Energy Distribution of a Low-Pressure Sodium Lamp (183 lm/W (180 W) at 35.5% Visible Radiation).

Energy-Saving Devices. The least expensive and most efficient lighting control that saves energy is the ac snap switch, which provides the on/off control of the lights.

Summary

The size of a light source is critical to the optical control of a luminaire. The linear characteristic of the U-shaped arc tube limits optimum lighting distribution control to a high-quality refractor (lens).

HIGH-PRESSURE GASEOUS DISCHARGE SOURCES

There are three high-pressure gaseous discharge sources, which are referred to as **high-intensity discharge (HID) sources.** The HID sources are (1) mercury vapor, (2) metal halide, and (3) high-pressure sodium.

Mercury Vapor Lamps

Light-Producing Elements

The light-producing element is an **arc tube,** which contains two main electrodes and a **starting electrode.** The arc tube is constructed of quartz to allow ultraviolet radiation to be transmitted (see Fig. 4-16). The arc tube contains mercury and small quantities of argon, neon, and krypton. When the lamp is energized, an arc is struck between the main and starting electrode. As the mercury ionizes, resistance inside the arc tube decreases. When resistance inside the arc tube is less than external resistance, the arc jumps between the main electrodes. The mercury continues to ionize, increasing the light output. The light produced is in the typical mercury lines (404.7, 435.8, 546.1, 577.9), plus ultraviolet (UV) energy. The arc tube is operated at from 1 to 10 atmospheres of pressure.

A clear mercury vapor lamp produces a blue-green visible light. To improve the color, a phosphor coating is placed on the inside surface of the outer enclosure. The ultraviolet energy produced by the arc tube will excite the phosphor coating producing additional visible light which improves the color rendition of the mercury vapor lamp.

Enclosure

The outer enclosure serves three primary functions:

1. Ordinary glass acts as a UV filter, which prevents skin and corneal burns. All mercury vapor lamps manufactured after 1982 require a ''disconnect mechanism'' to disable the arc tube if the outer enclosure is broken to prevent dangerous UV exposure. The lamp is *not dangerous if the outer bulb is intact.*
2. It provides a constant thermal environment for the arc tube. The arc tube pressure is affected by rapid temperature change and air motion.
3. It provides a surface for a phosphor coating, which is placed on the inside of the outer enclosure to correct the color rendition. A phosphor-coated lamp will require a very large luminaire for good optical control by reflection since the coated outer enclosure becomes the light source rather than the arc tube.

Lamp Shape Designation

Lamp Shapes. See Figure 4-17.

BT = bulged tubular	T = tubular
E = elliptical	PS = pear straight
PAR = parabolic	R = reflector

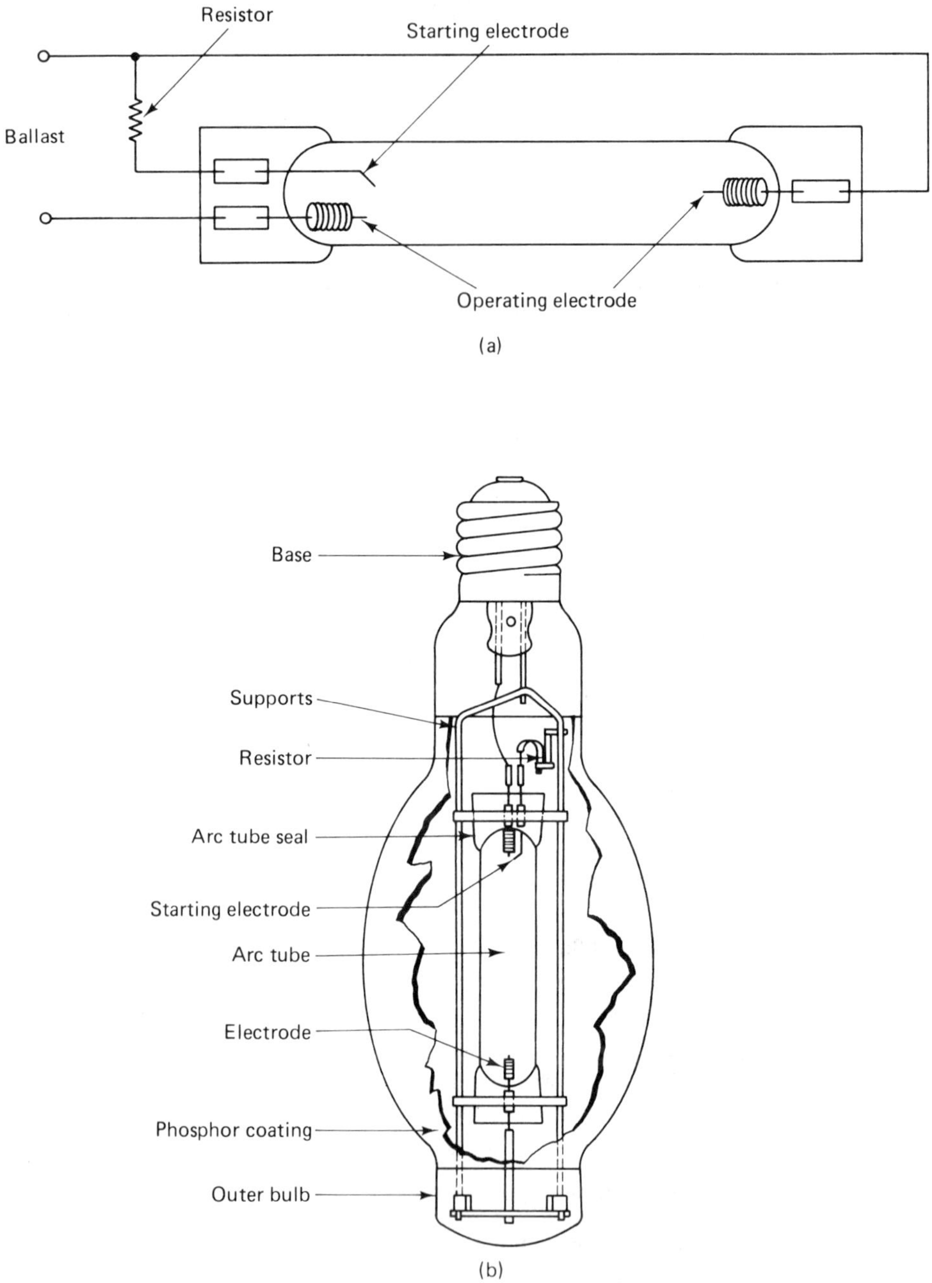

Figure 4-16. Mercury Vapor Lamp and Arc Tube. (Courtesy of Philips Lighting Company.)

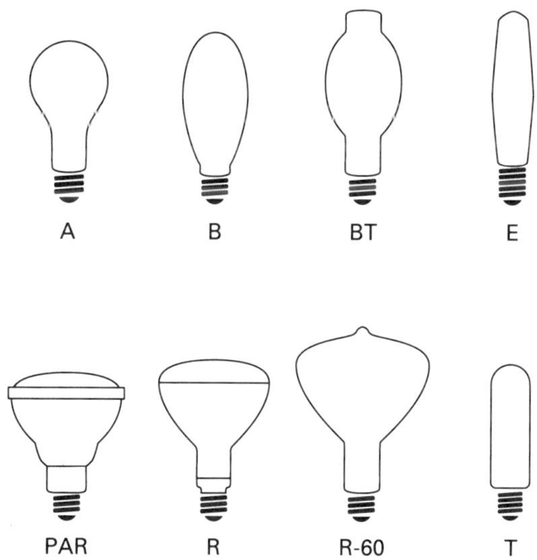

Figure 4-17. HID Lamp Shapes. (Courtesy of Philips Lighting Company.)

Lamp Shape Designation. BT-58 = bulged tubular—58/8th in. in diameter ($7\frac{1}{4}$ in.)

Lamp Finishes. Clear, DX, WDX, C (color improved).

Lamp Order Code (ANSI Code). The lamp ordering designation for mercury vapor lamps is quite different from incandescent or fluorescent lamps. The only parts that have meaning are the ''H,'' which identifies the lamp as a mercury vapor (Hg, mercury), the wattage value, and the color designation. The remaining numbers and letters are arbitrary.

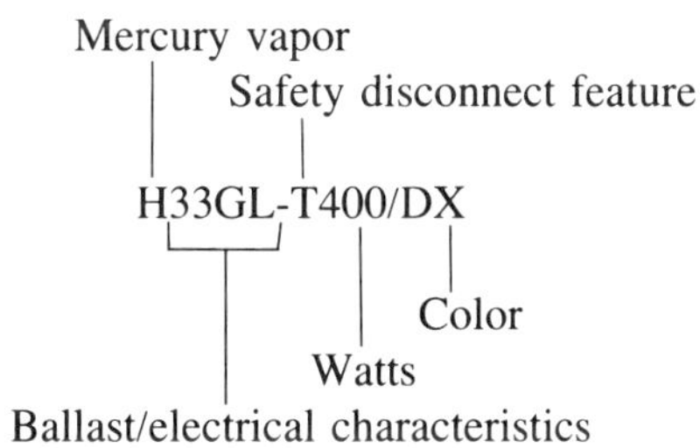

Electrical Connection

A mogul screw base is used for all higher-wattage lamps; the low-wattage lamps (40/50, 75, and 100 W) use the medium screw base. The side prong base is available for the ''PAR'' lamp.

Operating Characteristics

Efficacy. Lamp efficacy varies with the lamp wattage. The higher the lamp wattage, the higher the efficacy.

40/50 W:	25 to 30 lm/W
75, 100, 175, 250 W:	34 to 48.4 lm/W
400 W:	55 to 60 lm/W
1000 W:	57 to 63 lm/W

For an "H33GL-400/DX" lamp which is rated at 23,000 lumens, the efficacy is

$$K_{\text{lamp}} = \frac{23{,}000}{400} = 57.5\,\frac{\text{lm}}{\text{W}}$$

$$K_{\text{system}} = \frac{23{,}000}{454} = 50.7\,\frac{\text{lm}}{\text{W}}$$

Life. Life testing of all HID lamps is based on a burning cycle of 10 hr/start. The life of a mercury vapor lamp can be described in terms of its usable life or its rated life. Typically, lamps are rated on the basis of a 50% mortality curve. Because of the rapid lumen depreciation, the mercury vapor lamp is rated in terms of a higher percentage of lamps remaining to maintain a more reasonable lumen output (see Fig. 4-18).

The rated life of an H33GL-400/DX is 24,000 hr, while its usable life is 18,000 hr at 10 hr/start:

$$\text{no. yr}_{\text{rated}} = \frac{24{,}000}{3432} = 7 \text{ yr}$$

$$\text{no yr}_{\text{usable}} = \frac{18{,}000}{3432} = 5.2 \text{ yr}$$

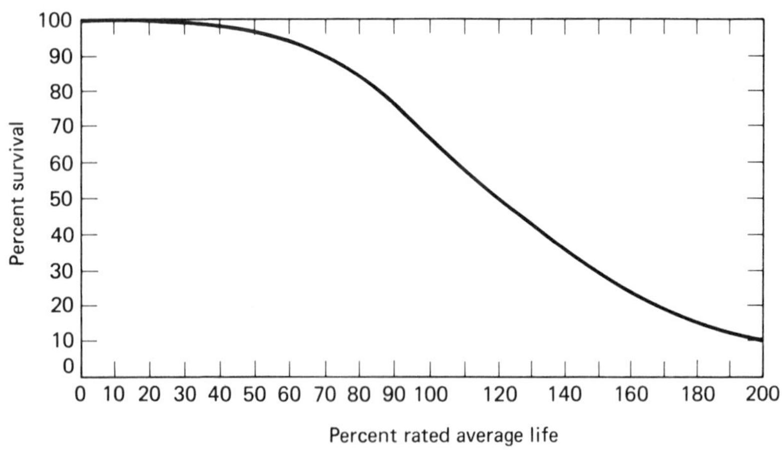

Figure 4-18. Typical Mortality Curve for Mercury Vapor Lamps. (Courtesy of Philips Lighting Company.)

Operating Variables

Self-ballasted Mercury Vapor. Self-ballasted mercury vapor lamps contain either a solid-state starting device or an incandescent filament that acts as a ballast. The solid state type of lamp should not be used in a totally enclosed luminaire because of the heat buildup associated with this lamp.

In general, self-ballasted mercury vapor lamps are 50% less efficient than standard mercury vapor lamps, but 50% more efficient than incandescent lamps. They should be limited to short-term replacement of incandescent lamps where relamping is difficult and ballast addition impractical.

Lumen Depreciation. The lumen depreciation curve for a mercury vapor lamp is quite drastic and a function of the ballast and wattage (see Fig. 4-19). The mercury vapor lamp is the *only* lamp that is listed with a rated and usable life. The usable life is well suited to the definition of "smashing point" (see Chapter 1). Light output is also a function of the supply and regulation of the voltage to the lamp.

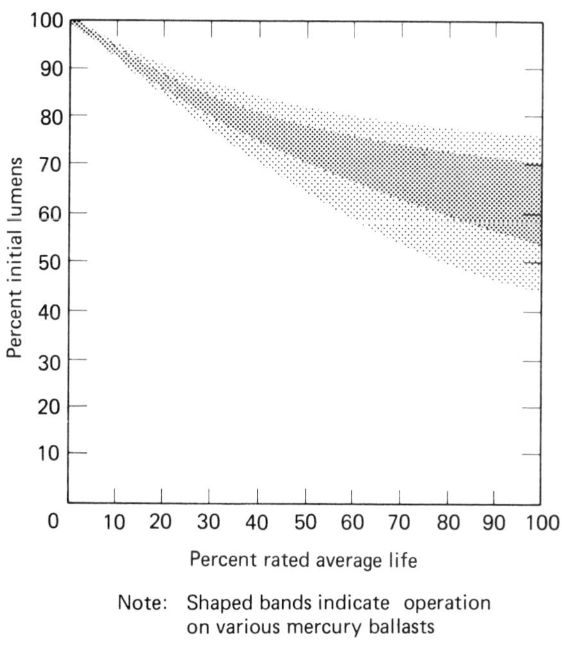

Figure 4-19. Lumen Depreciation for Mercury Vapor Lamps. (Courtesy of Philips Lighting Company.)

SPD Curve. A clear mercury vapor lamp has a predominate blue-green color characteristic associated with the mercury line spectrum. Figure 4-20 gives the SPD curve. To color correct the lamp, a phosphor coating is placed on the inside surface of the outer enclosure. The primary colors added by the phosphor are the reds and oranges. Phosphor-coated or white mercury vapor lamps are recommended for all applications where color is important. There are three standard "white" mercury vapor lamps:

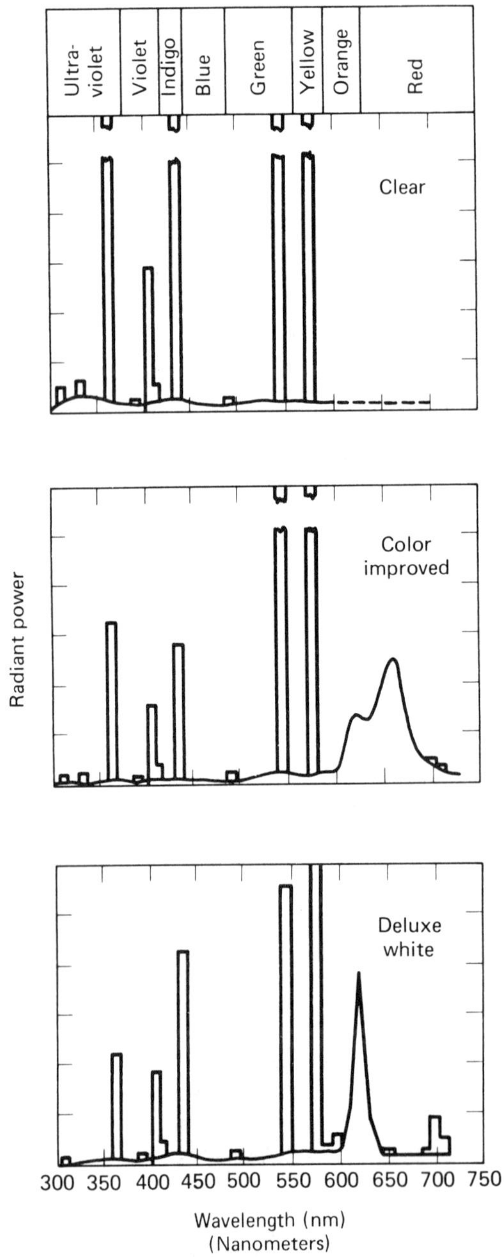

Figure 4-20. Spectral Power Distribution Curve for Mercury Vapor Lamps. (Courtesy of Philips Lighting Company.)

1. *Color improved*—very poor on reds, marginal color, not recommended.
2. *Deluxe white, DX*—increased red, good color, recommended.
3. *Warm white deluxe, WWX*—excellent reds, excellent color, highly recommended, decreased lumens.

Starting Voltage. The starting voltage is 225 V rms or 320 V peak for a 400 W mercury vapor lamp.

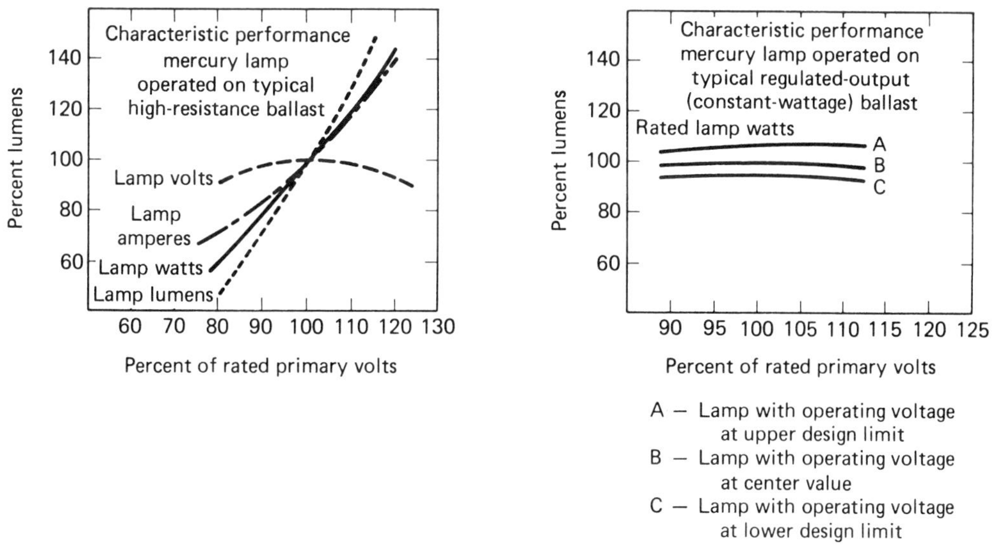

Figure 4-21. Voltage Versus Light Output. (Courtesy of Philips Lighting Company.)

Starting/Restrike Times

Mercury Vapor Lamps

Starting 80% Max.	Restrike 80%
7–10 min	7 min

Energy Distribution. Of the total input energy into the mercury vapor lamp, only 14.7% ends up as visible light (see Fig. 4-22).

Energy-Saving Devices. The development of dimming capability and the introduction of electronic ballast can result in energy savings for mercury vapor applications. Electronic ballasts have been researched since the mercury vapor lamp first appeared. A remaining problem of mercury vapor dimming is the high cost; however, the electronic

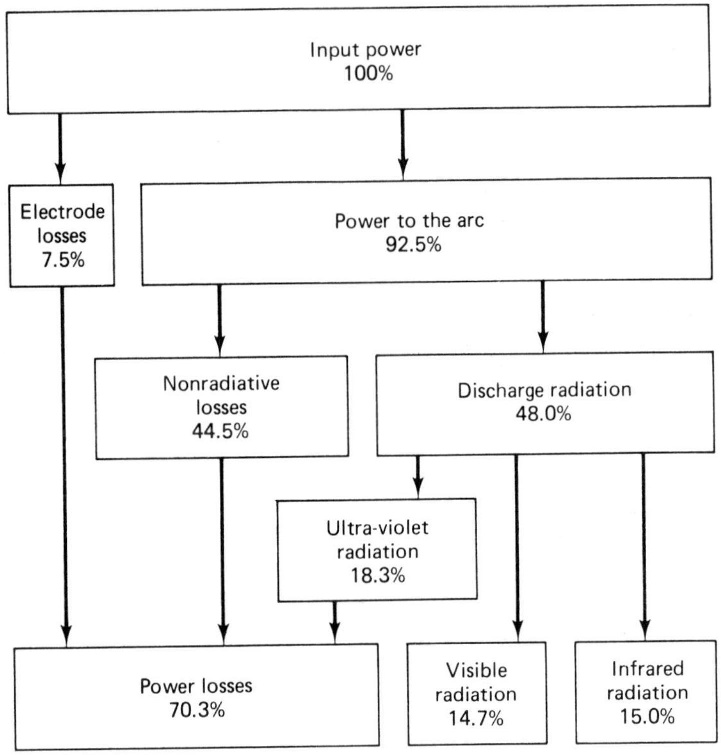

Figure 4-22. Energy Distribution for Mercury Vapor Lamps 56.3 lm/W (400 W Produces 14.7% Visible Radiation).

ballast will significantly increase lamp efficacy and total system efficiency. Other advantages of the electronic ballast are smaller size and weight, lower noise, and increased lamp life.

Summary

The primary advantage of the phosphor-coated mercury vapor lamp (DX/WWX) is excellent color rendition. Of all the standard lamps, the best possible color rendition is a toss up between the CWX, fluorescent, and the DX, mercury vapor.

The primary disadvantage of the phosphor-coated mercury vapor lamp is poor optical control (reflectors need point sources). A phosphor coated lamp is not a point source. To classify as a "point source," the distance from the source to the surface must be equal to or greater than 5× maximum dimension of the light source. For a phosphor-coated BT-56 (7 in. diameter) mercury vapor lamp to classify as a point source, the radius of a reflector would have to be 7 in. × 5 = 35 in. or almost a 6 ft diameter reflector.

Therefore, light control is achieved with a prismatic lens and reflector to control the direction of the light which result in a loss of light output due to transmission and reflection absorptions.

Another disadvantage is relatively poor efficacy compared to the other HID lamps, poor lumen maintenance, and short usable life. However, the efficacy, life, and DX color makes a low-wattage mercury vapor lamp an acceptable substitute for incandescent lamps in low-ceiling areas.

Metal Halide Lamps

Light-Producing Element

The light-producing element is called an **arc tube,** which contains two main electrodes and a **starting electrode.** The arc tube has the same construction and operating principles as the mercury lamp (see Fig. 4-23). In addition to mercury, argon, neon, and krypton, the arc tube contains halide salts (iodides) of metals. The primary additives are mercury, sodium, and scandium iodides; other additives are thallium, indium, and cesium iodides. These additives will add missing colors to the typical mercury lines, that is, red, orange, and yellow. The color of the metal halide is balanced across the spectrum. By improving the color without the need for a phospor coating, the lamp approaches a point source that will result in better potential for optical control by reflection. For horizontal burning position, one manufacturer's arc tube is arched (requires a special socket—see electrical connection) to make the temperature inside the arc tube more uniform (see Fig. 4-23).

Enclosure

The outer enclosure serves two functions:

1. It maintains the arc tube in a constant thermal environment.
2. The ordinary glass of the outer enclosure acts as an ultraviolet filter.

A phosphor coating is not necessary for good color rendition and should be avoided because of its negative effect on optical control; that is, the lamp no longer approaches a point source if it is phosphor coated.

Lamp Shape Designation

Lamp Shapes

BT = bulged tubular	T = tubular
E = elliptical	ED = ellipsoidal

Lamp Shape Designation. The enclosure is designated by a letter-number combination. Metal halide lamps are available in the shapes shown in Fig. 4-18. The number represents the maximum diameter of the outer enclosure in eighths of an inch.

E-37 = elliptical—37/8th in. ($4\frac{5}{8}$ in.)

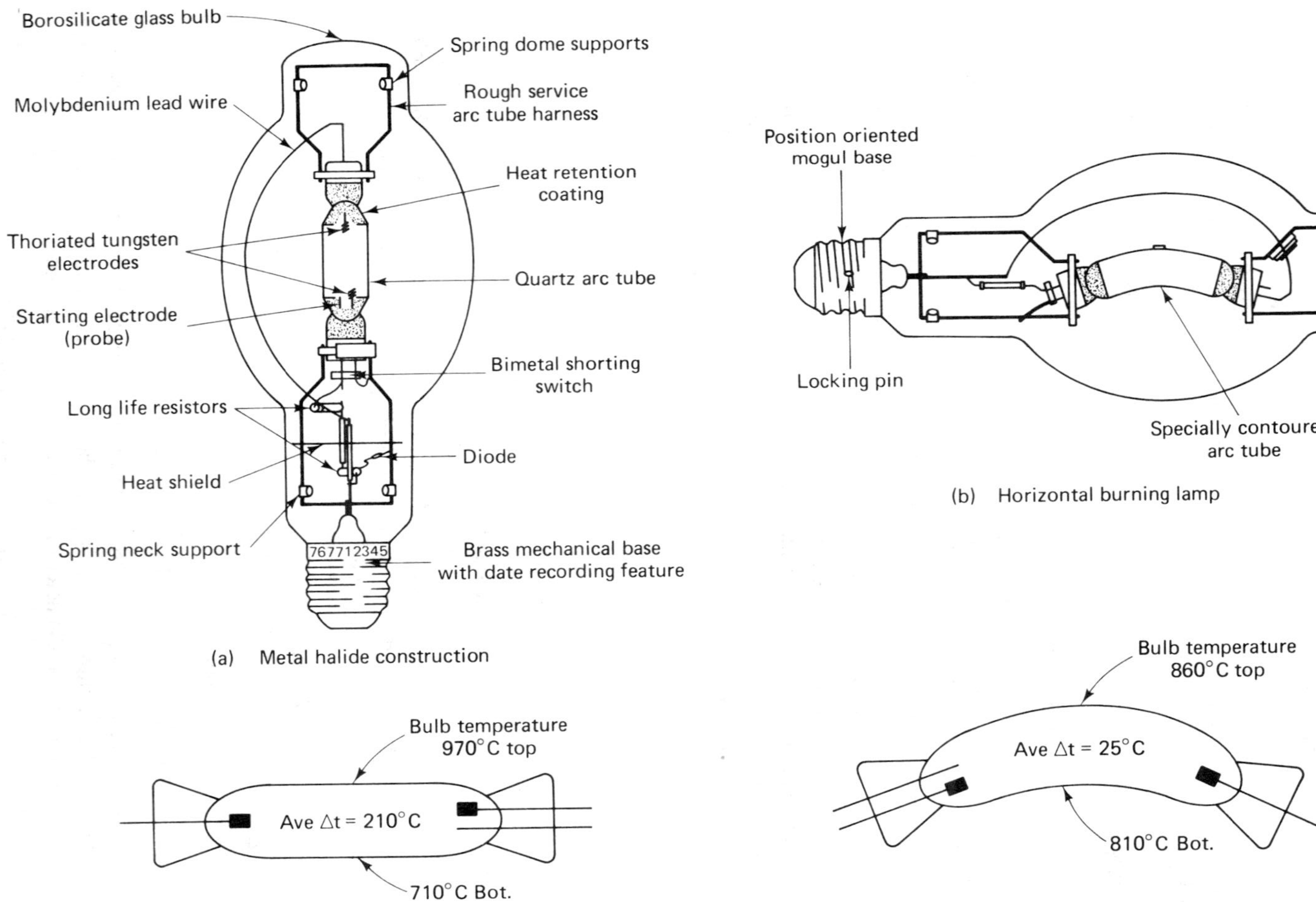

(a) Metal halide construction

(b) Horizontal burning lamp

(c) Bulb wall and internal temperature variations for horizontal operation of a straight and curved arc tube.

Figure 4-23. Metal Halide Lamp Construction—Vertical and Horizontal Burning Positions. (Courtesy of GTE Products Corp.)

Lamp Finishes. The metal halide lamp is typically a clear lamp but is also available in a C—phosphor-coated—lamp.

Lamp Order Code. The specifier should use the ANSI Lamp Order Code for metal halide lamps since each manufacturer has their own lamp order code. Use of the ANSI Code assures the specifier of an open bid on the lamp contract. The "M" or "MH" letter designation is used to identify a metal halide lamp.

```
Metal   halide
          |        No letter or T = safety
          |        |
        M59PJ-400/VBU
        └──┬──┘  |   └──Burning position
           |   Watts
        Ballast/electrical characteristics
```

Metal halide lamps are especially sensitive to burning position. Manufacturers' data should be consulted for burning position requirements.

Burning Position Codes

U = universal	HOR = horizontal only
BU = base up	BD = base down
HBU = horizontal to base up	HDB = horizontal to base down
VBU = vertical to base up	VBD = vertical to base down

Electrical Connection

The metal halide lamp uses a mogul screw base for all wattages except for the 32 W and 70 W, which use a medium screw base. Those lamps rated for horizontal *only* (HOR) burning position that contain the curved arc tube (see Fig. 4-24) use a special socket for proper positioning the lamp that has a locking pin on the lamp base. The curved arc tube should always be positioned with the curve straight up in a vertical plane. When the specifier designates an HOR lamp, the manufacturer should supply the locking pin socket. If the system does perform as expected, the designer should check to make certain that the luminaire has been shipped with the proper lamp socket.

Operating Characteristics

Efficacy. Lamp efficacies vary between manufacturers and with burning position and lamp wattage. The higher the wattage, the higher the lamp efficacy.

A horizontal burn (curved arc tube) 400 W metal halide lamp is rated at 40,000 lm.

$$K_{\text{lamp}} = \frac{40{,}000}{400} = 100\,\frac{\text{lm}}{\text{W}}$$

$$K_{\text{system}} = \frac{40{,}000}{455} = 87.9\,\frac{\text{lm}}{\text{W}}$$

Life. Life varies as a function of lamp wattage and the length of time that the lamp has been on the market. The standard practice in the lamp industry is to bring all new lamps on the market with a rating of 7500 hr. As data are developed on mortality and life, which requires long-term testing, lamp life will usually increase to 12,000 hr then 15,000 hr, 18,000 hr, and up to the lamp's maximum life. Current lamp catalogs from all manufacturers should be consulted for life ratings of lamps.

The 400 W metal halide lamp is rated at 20,000 hr.

$$\text{no. yrs of operation} = \frac{20{,}000}{3432} = 5.8 \text{ yr}$$

Operating Variables

Color Shift. The color appearance is a function of the quality control of the additives in the arc tube. Color consistency from one lamp to another is a function of the ballast, supply voltage, lamp age, and additive mixing. Color shift can cause some lamps to appear pink and some blue. Color shift from one lamp to the next lamp in the same room will exaggerate the color differences. The color shift is usually not noticeable at the work level. Where good color consistency (or match) between lamps is an important design consideration, the lamps should be *group relamped* rather than *spot relamped* because of the color change as the lamp ages.

Sensitive to Burning Position. Burning position can affect the lamp life and the lumen output. Horizontal burning of the lower-wattage lamps causes excessive temperature differentials (see Fig. 4-23) on the arc tube which can result in violent failures.

Point Source. The point source characteristics of the source allows for excellent optical control and good color rendition. However, there is a negative side to the "point source" characteristic, which is the potential for *harsh shadows*. To maximize the shadows, the overlap of lighting patterns is very important.

Lumen Depreciation. The lumen depreciation curve for a metal halide lamp is substantially better than for a mercury vapor lamp. The lumen output at the end of life for the higher-wattage lamp is 75% for a metal halide lamp (see Fig. 4-24).

SPD Curve. The metal halide lamp produces energy at all wavelengths across the visible spectrum. That is, its SPD is well balanced, which means that the lamp produces good color rendition without the need for a phosphor coating (see Fig. 4-25). The SPD is a function of the additive mixture used in the arc tube.

Starting Voltage. The starting voltage for a 400 W metal halide lamp is 383 V rms and 540 V peak.

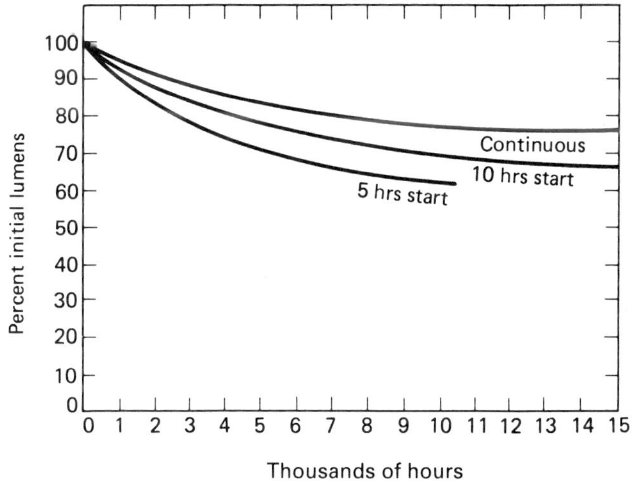

Figure 4-24. Metal Halide Lumen Depreciation. (Courtesy of GTE Products Corp.)

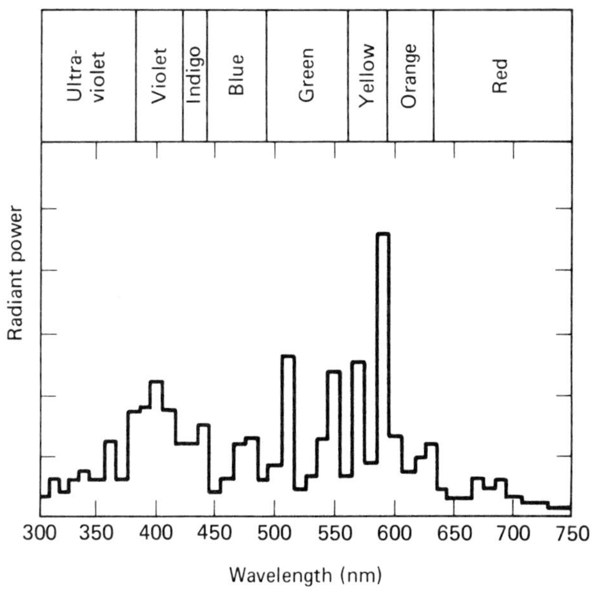

Figure 4-25. Spectral Power Distribution for Metal Halide Lamp. (Courtesy of GTE Products Corp.)

Starting/Restrike Times

Metal Halide Lamps

Starting 80% Max.	Restrike 80% Max.
5 min	10–15 min

Energy Distribution. Of the total input energy into the metal halide lamp, only 24.3% ends up as visible light (see Fig. 4-26).

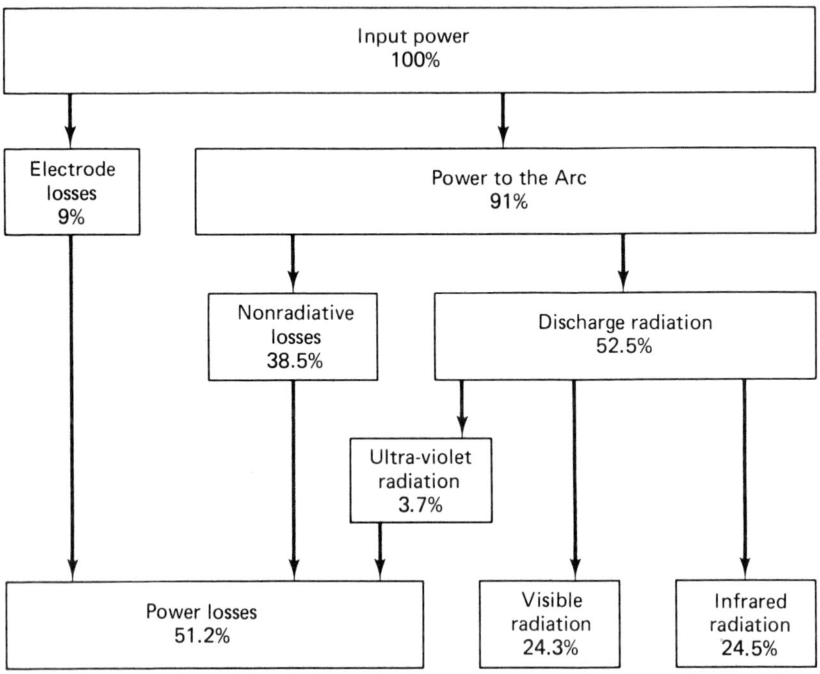

Figure 4-26. Energy Distribution for Metal Halide Lamp 100 lm/W (400 W Produces 24.3% Visible Radiation).

Energy-Saving Devices. Metal halide dimming is available but represents a very high initial cost. The 400 W lamp can be dimmed (5 min) to 47% of its total input power, which results in a 22% reduction in lumens. The 1000 W metal halide lamp can be dimmed (15 min) to 36% of its total input power, or 14% of its lumen output. As technology increases in solid-state electronic ballasts and dimming, the cost of dimming should decrease and the range of dimming should increase.

Summary (Conventional Metal Halide Lamps)

As more lower-wattage metal halide lamps are introduced into the marketplace, the application of metal halide lamps will continue to increase. The primary advantages of the metal halide lamp are its excellent optical control potential (point source), its good to excellent color rendition (without the need for a phosphor coating), its high efficacy, and its long life.

Double-Ended Tubular Metal Halide Lamps

The double-ended tubular metal halide lamp is a more compact version of the metal halide lamp. The small compact arc tube makes the source an excellent source where optical control is important. The lamp is available in wattages of 70, 150, 250, 400 in the **RCS (recessed double contact)** base configuration. The 70 watt lamp carries the ANSI Code of ''M85'' or ''M81,'' depending on the manufacturer. The 250 W and 400 W are ANSI Code designation ''M81.''

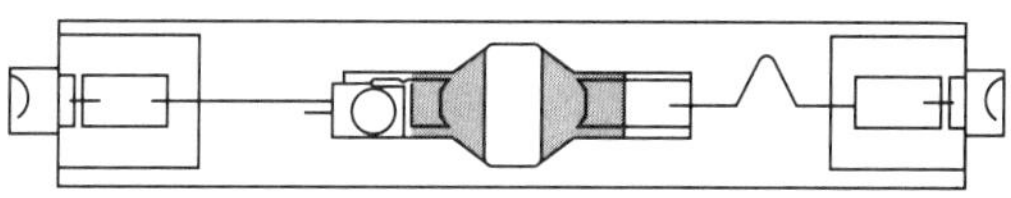

Figure 4-27. Double-Ended Tubular Metal Halide Lamps. (Courtesy of GE Lighting.)

The lighting professional needs to consult with the lamp manufacturer concerning operating restrictions and availability of higher-wattage lamps. Some lamps are also available in a ''single-ended'' base configuration. The following design considerations should be investigated before specifying this lamp.

1. The lamp should be used in luminaires with an ''interlocking,'' ultraviolet filtering glass cover lens. Since the lamp does not have an outer jacket, the lamp must be enclosed to prevent exposure to danger UV radiation levels.
2. The lamp is a ''horizontal'' burn lamp and should not be operated beyond the horizontal limits established by the manufacturer.
3. With regard to color variation versus life, as the lamp ages, the lamp will experience extreme color shifts. The ''usable'' life (see Chapter 1) should be based on a color variation of no more than a ±100°K variation in correlated color temperature. The ''rate'' life may be much longer than the lamps ''usable'' life, and the owner must be made aware of the color shift potential that may occur if the lamps are left in service to the end of their ''rated'' life.
4. Improper handling, improper burning position, and improper reflector design may cause the lamp to have a violent failure. The reflector should not redirect the light rays back through the arc stream since this will cause pressure changes and lamp failure.

High-Pressure Sodium Lamps

Light-Producing Element

The light-producing element is an arc tube. The arc tube is small in diameter to maintain a high operating temperature. Because of the small diameter, there is no starting electrode inside the arc tube. Sodium, operating at a slight pressure and high temperature, will have a corrosive effect on ordinary glass or quartz. Therefore, the arc tube is made of a ceramic material (polycrystalline, translucent alumina material). The arc tube contains xenon, an amalgam of mercury, and sodium operated at a pressure of 200 mm of Hg.

Enclosure

The enclosure helps to maintain the arc tube in a constant thermal environment and shields the arc tube from air motion.

Lamp Shape Designation

Lamp Shapes. See Figure 4-28.

T = tubular	PAR = parabolic
B = bulged	R = reflector
BT = bulged tubular	E = elliptical

Lamp Shape Designation. The lamp shape designation is a letter-number combination. The number represents the maximum diameter in $\frac{1}{8}$ in.

E-37 = elliptical with a diameter of 37/8 in. ($4\frac{5}{8}$ in.)

Lamp Finishes. The high-pressure sodium lamp produces visible light directly with no ultraviolent energy, which eliminates the need or usefulness of a phosphor coating. Therefore, the high-pressure sodium lamp is a *clear* lamp. The lamp is available in a *diffuse coating*, which adds nothing to the lumen output of the high-pressure sodium lamp. Lamps with a diffuse coating increase the optical size of the source and cause the lamp to no longer classify as a point source.

Lamp Order Code. Since each lamp manufacturer has developed their own lamp order code, it is recommended that the specifier use the ANSI Order Code. The specifier must use caution not to specify or use trade names that close acceptable lamps out of a bid.

Sodium

Watts

S51WA-400 ANSI Code

Ballast 400/D—Diffuse

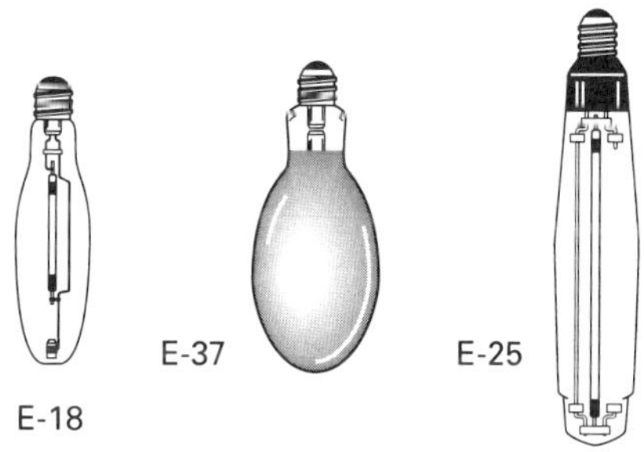

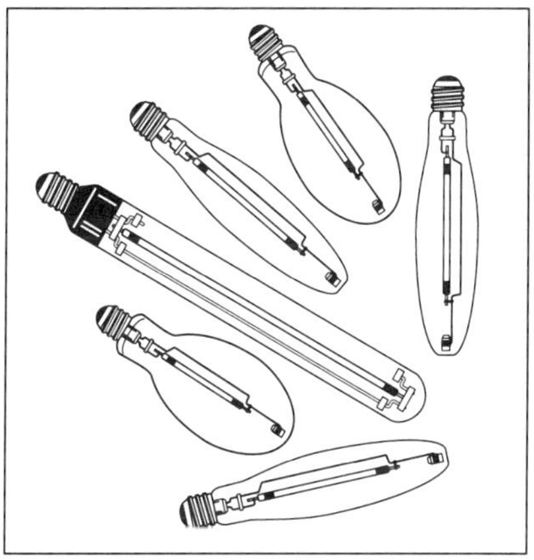

Figure 4-28. High-Pressure Sodium Lamps. (Courtesy of GE Lighting.)

Electrical Connection

The electrical connection is a screw base. The 35 W high-pressure sodium uses a medium screw base while the 50 W, 70 W, 100 W, and 150 W lamps are available in the medium or mogul screw base. Lamp wattages from 200 W to 1000 W use only the mogul base.

Operating Characteristics

Efficacy. The high-pressure sodium lamp efficacies vary as a function of the lamp manufacturer, the ballast, the burning position, and the lamp wattage.

A 400 W high-pressure sodium is rated by one manufacturer at 50,000 lm.

$$K_{\text{lamp}} = \frac{50{,}000}{400} = 125.0 \frac{\text{lm}}{\text{W}}$$

$$K_{\text{system}} = \frac{50{,}000}{465} = 107.5 \frac{\text{lm}}{\text{W}}$$

Life. Life varies as a function of wattage, ballast circuit, and manufacturer. The range is from 15,000 to 24,000 hr for the more common higher-wattage lamps.

High-pressure sodium lamps of 50 W to 1000 W are rated at 24,000 hr for a burning cycle of 10 hr/start.

$$\text{no. yrs} = \frac{24{,}000}{3432} = 7 \text{ yr}$$

Operating Variables. In addition to the ballast and a separate capacitor for power factor correction, the high-pressure sodium lamp requires a separate starting device to provide the high-voltage starting pulse.

Lumen Depreciation. The lumen depreciation curve for the high-pressure sodium lamp is the best of all high-intensity discharge lamps. The lumen output at the end of life for the higher wattage lamps is 80% (see Fig. 4-29).

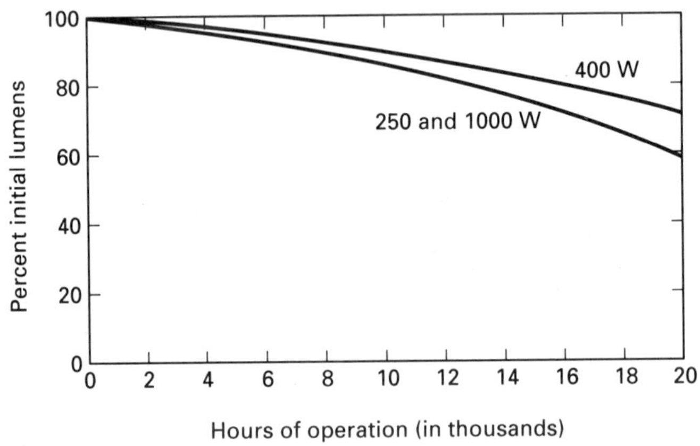

Figure 4-29. High-Pressure Sodium Lamp Lumen Depreciation. (Courtesy of GE Lighting.)

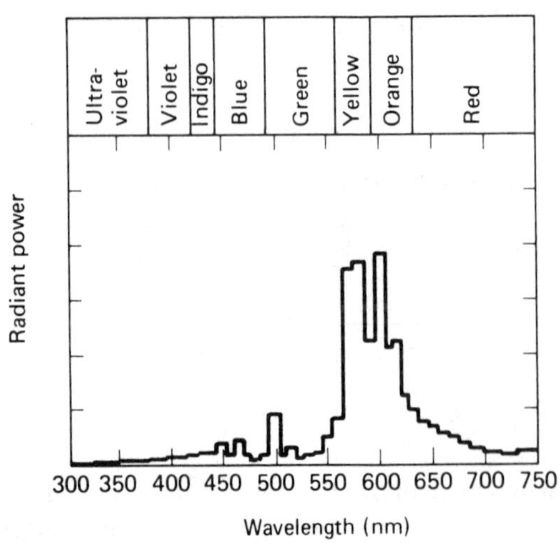

Figure 4-30. Spectral Power Distribution for High-Pressure Sodium Lamp. (Courtesy of GE Lighting.)

SPD Curve. The high-pressure sodium lamp produces energy at all wavelengths (see Fig. 4-30). However, the major portion of energy is concentrated in the yellow-orange part of the spectrum. The color characteristics of the lamp turn red objects orange and will darken the color appearance of blue and green objects. Increasing arc tube pressure seems to improve the color appearance of reds, blues, and greens. Color consistency from one lamp to another lamp is better than with metal halide lamps. However, color shifts can occur due to voltage variations and differences in ballasts.

Starting Voltage. Depending on the wattage, the high-pressure sodium lamp will require between a 2500 V to 5000 V pulse to start the lamp.

Starting/Restrike Times

High-Pressure Sodium Lamps

Starting 80% Max.	Restrike 80% Max.
3 min.	1 min.

Energy Distribution. Of the total input energy into the high-pressure sodium lamp, only 29.5% ends up as visible light (see Fig. 4-31).

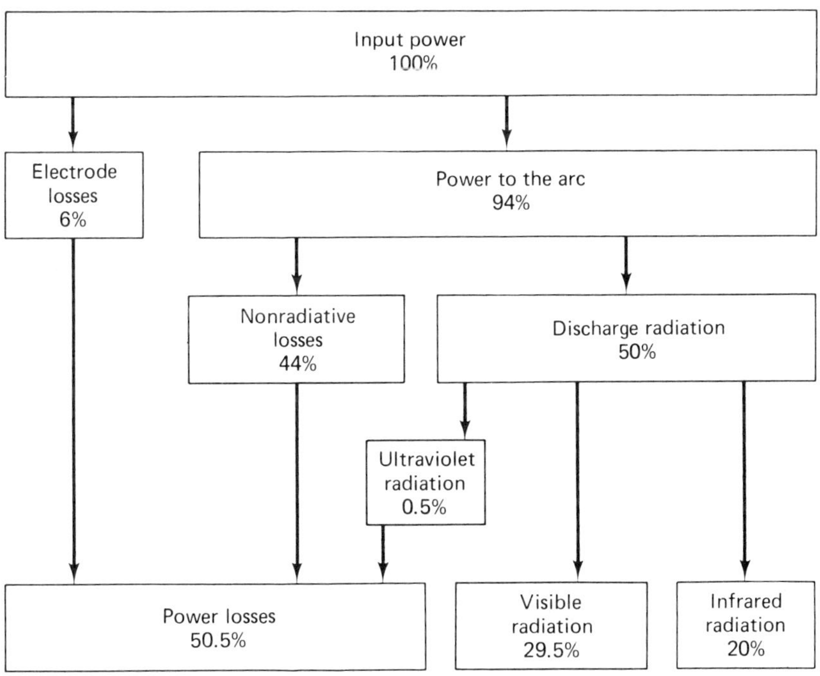

Figure 4-31. Energy Distribution for High-Pressure Sodium Lamps—125 lm/W (400 W Produces 29.5% Visible Radiation).

Energy-Saving Devices. Dimming of some wattages of high-pressure sodium lamps is possible. The 1000 W lamp can be dimmed to 38% of its total input power in approximately 15 min, with a reduction in light output to 20% of rated lumens.

Summary

The high-pressure sodium lamp has high efficacy, long life, but poor color rendition. Without a diffuse coating, the lamp approaches a point source. High-pressure sodium distorts blues and green while turning red a brilliant red-orange. By increasing the pressure inside of the arc tube, the SPD is spread across the spectrum, the super high-pressure sodium lamp has excellent color rendition with a white color appearance. The major problem associated with increasing the pressure is maintaining the arc tube seals.

BALLASTS

Low-Pressure Gaseous Discharge Ballasts

Fluorescent Ballasts

The American National Standards Institute (ANSI) sets the requirements to meet good lamp performance, such as light output, life, and reliability. Certified Ballast Manufacturers (CBM) is an organization that was established to assure the user that the ballast meets or exceeds the minimum ANSI standards. The rapid-start ballast provides a continuous heating voltage to the cathodes. The ballast supplies 4 V to the input sockets. The most common rapid start ballast operates the lamps in series. (Supply leads from ballast to lamp are red, and energy returns to the ballast from the lamp via the blue leads.) A typical fluorescent ballast will show a 1% change in lumen output for each 1% of change in rated voltage. A metal strip is required to be within 1 in. of the lamp. The metal strip acts as a capacitor that causes an effective decrease in the lamp length, which results in lower starting voltage requirements.

Typical Ballast Circuits. The primary elements of a ballast can be seen in Figure 4-32, which shows the circuits for the preheat, instant-start (slimline), and rapid-start lamps. The ballast components are

1. *Core and coil*—a reactive coil provides the voltage kick and is current limiting.
2. *Autotransformer*—boosts starting voltage to provide voltage kick.
3. *Capacitor*—a starting capacitor and a power factor correcting capacitor.

Rapid Start Ballast. Starting is a two-step process in the rapid-start ballast:

1. *Conditioning the lamp*: A continuous current preheats the cathode, causing an ionization of the argon or argon-neon. This preionization decreases the resistance of the lamp, allowing the second step.

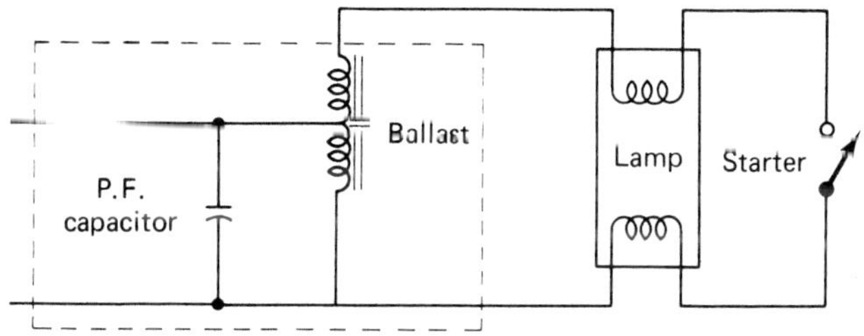

(a) Preheat circuit with an autotransformer and capacitor

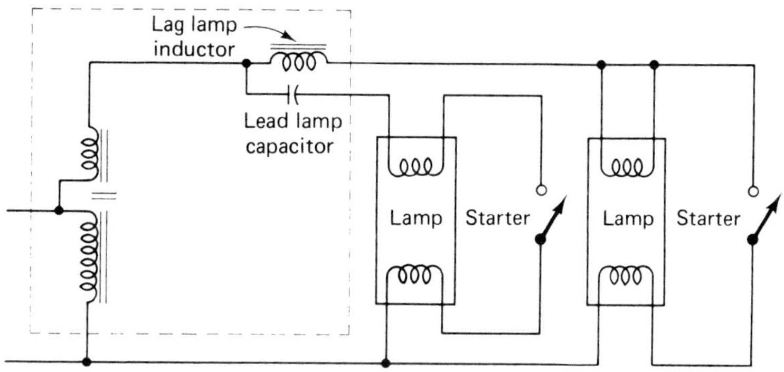

(b) Lead-lag preheat circuit

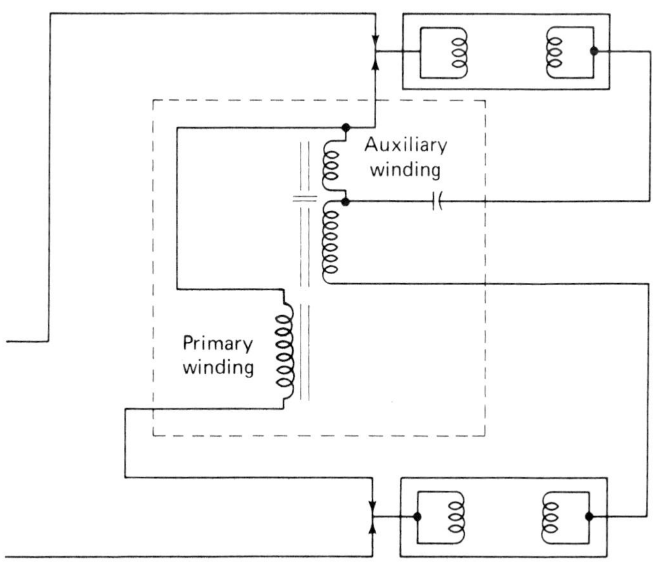

(c) Two-lamp series instant start circuit

Figure 4-32. Typical Fluorescent Ballast Circuits. (Courtesy of GE Lighting.)

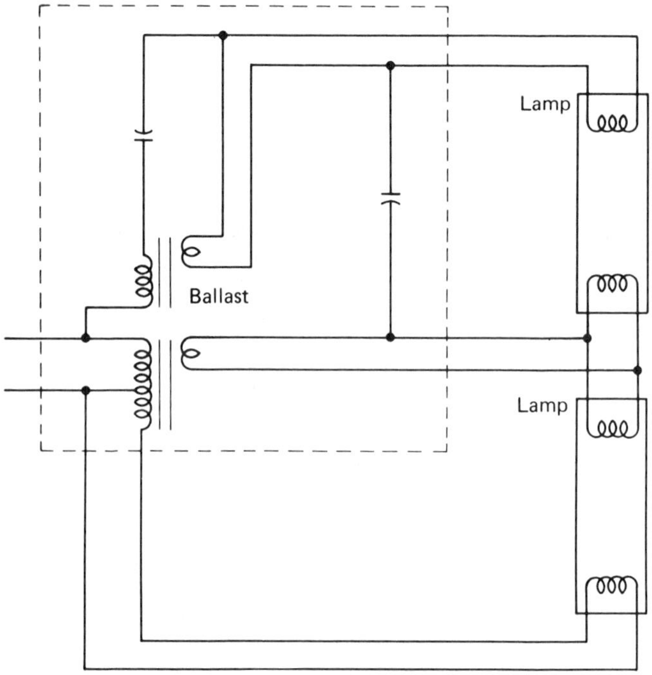

(d) Two-lamp series lead
rapid start circuit

Figure 4-32 (continued)

2. *Striking the arc*: Decreased resistance means that a lower voltage is required to strike the arc.

Trigger-Start Ballast. A trigger-start ballast is similar to the rapid-start ballast in that no separate starter is required and a continuous current is supplied to the cathodes. It is designed for use with preheat lamps of 20 W and less. It is available in a high- and low-power factor ballast.

Class "P" Ballast. The National Electric Code (NEC) states that "fluorescent fixtures for indoor installations shall incorporate ballast protection." Thermal protected ballasts that comply with Underwriter Laboratories' requirements are marked class "P." The main requirements of class P ballast are as follows:

1. That the ballast not dangerously overheat as a result of either internal faults or external abnormal conditions.
2. That the calibration of the protective device remain unchanged under the anticipated conditions of use and operation.
3. That the device perform safely when called upon to operate.

With class P protection, any overheating problem caused by misapplication becomes extremely critical. The variables that affect the ballast operating temperature are as follows:

1. *Supply voltage*: Case temperature rises 0.8°C for every 1 V above the rated supply voltage (at 120 V).
2. *Ambient temperature*: A 1°C rise in ambient temperature can raise the case temperature 0.9°C.
3. *Luminaire mounting*: Surface-mounted luminaires can have temperatures as much as 10°C higher depending on ceiling material and insulation.
4. *Luminaire design*: Variations in number of lamps, enclosure, size, material, lenses, and ballast location can account for a 20°C difference.
5. *Ballast and lamp*: Manufacturing variations can affect ballast temperature by ±5°C.

Under normal operating conditions in a 25°C ambient, class P ballasts must meet the same requirements and limitations as all listed ballasts. These are the following:

1. 90°C maximum case temperature
2. 70°C maximum capacitor temperature
3. 65°C maximum coil temperature rise

Under fault conditions, the continuous case temperature on class P ballasts is limited to 110°C case temperature under fault, and a test case temperature of 90°C means an allowable case temperature increase of only 20°C. Thus, the specifier must know the environmental conditions at each installation and specify a luminaire that will meet the design requirements.

Ballast Efficiency Standard (Circle "E" Ballast). In 1988, a federal law was passed that required the phasing out of "standard" electromagnetic ballast for four ballast types:

1. One lamp—F40 rapid start
2. Two lamp—F40 rapid start
3. Two lamp—F96T12/HO rapid start
4. Two lamp—F96T12/SL (slimline)

The "National Appliance Energy Conservation Amendments of 1988" (Public Law 100-357) requires the four ballast types listed to be designed to meet "high-efficiency" ballast design standards. The law provides for the establishment of a fluorescent lamp ballast efficiency factor, BEF. Since April 1, 1990, ballast manufacturers (foreign or domestic) could not sell a ballast that did not meet the "high-efficiency" standards established in the law. All CBM ballasts that meet the BEF requirement are marked on the ballast housing with a capital "E" printed within a circle (called a circle "E" ballast). After April 1, 1991, luminaire manufacturers cannot install ballasts in their products that do not meet the new BEF standard. That is, all luminaires bought after April 1, 1991 must contain circle "E" ballasts.

The ballast efficiency factor, is a ratio of the relative light output, RLO, to the input watts to the ballast.

$$\text{BEF} = \frac{\text{RLO}}{\text{input watts}}$$

The relative light output is a ratio of the light output (lumens) of the lamp(s) operated by a test (commercial) ballast divided by the light output of the same lamp(s) operated by a "standard reference ballast" (also called a "standard reactor"). Most ballast manufacturer's use the term ballast factor to describe the relative light output factor described in the law (BF = RLO). The RLO is specific to only one ballast/lamp combination.

$$\text{RLO} = \frac{\text{lumens (w/test ballast)}}{\text{lumens (w/standard reference ballast)}}$$

The federal ballast efficiency law applies to the four fluorescent lamp ballasts

1. That are designed to operate at nominal input voltages of 120 or 277 V.
2. That operate with an "input" frequency of 60 Hz. This would include high-frequency electronic ballast since their input side operates from a 60 Hz supply.
3. That must have a power factor, PF, of 0.90 or greater.
4. That have a BEF not less than the following:

TABLE 4-5 Minimum Ballast Efficiency Factors, BEF

Lamp Type	Nominal Input Voltage	Nominal Lamp Watts	BEF (minimum)
1-F40/RS	120	40	1.805
	277	40	1.805
2-F40/RS	120	80	1.060
	277	80	1.050
2-F96T12/SL	120	150	0.570
	277	150	0.570
2-F96T12/HO/RS	120	220	0.390
	277	220	0.390

The ballast efficiency federal standards do not apply to ballasts

1. That are designed for dimming.
2. That are used in ambient temperatures of 0°F or less (low-temperature ballasts).
3. That are low power factor (less than 0.90) ballast for use *only* in residential building applications.

Low-Pressure Sodium Ballast

The conventional low-pressure sodium ballast is an autoleakage transformer maintaining nearly constant current, with an increase in lamp voltage throughout the lamp's life, resulting in an increase in power consumption with age (see Table 4-4 and Table 4-6). Open-circuit voltage for ignition and restrike is from 390 V to 575 V. It is this high open-circuit voltage that accounts for the high ballast losses.

TABLE 4-6 Low-Pressure Sodium Ballast Characteristics (180 W Lamp)

Nominal Lamp Watts	Nominal Supply Voltages	Nominal Open-Circuit Voltage
18	120, 120/277	300
35	120/240, 208/277, 480	390
55	120/240, 208/277, 480	410
90	120/240, 208/277, 480	420
135	120/240, 208/277, 480	540
180	120/240, 208/277, 480	600

High-Pressure Gaseous Discharge Ballasts

There are no universal standards for HID ballasts. Because of this, manufacturers are not held to any regulations, which results in a large variation in lamp performance. An important consideration in the selection of HID ballasts is the lamp current crest factor. The lamp current crest factor is a ratio of the peak value of the wave to its effective (or rms) value. The maximum allowable range for the crest factor should be from 1.45 for a good reactor ballast to an upper limit of 2.0. The higher the crest factor, the greater the lumen depreciation of a lamp during its life. For best lamp performance, the preferred range should be from 1.6 to 1.8. Lamp manufacturers will provide the crest factor value necessary to produce the published lamp performance data. Most lamp manufacturer data are based on a standard reference ballast (reactor type) with a crest factor of 1.4 to 1.5.

Mercury Vapor Ballast

In general, if line voltage will vary more than ±5%, a regulating-type ballast should be used. If regulation is good, a reactor or autotransformer ballast with good crest factor should be used (see Fig. 4-33).

Reactor, Inductive Ballast. A reactor ballast has the following characteristics:

1. Nonregulating: input voltage controlled over a narrow range; maximum allowable variation of input voltage is ±5%.

Mercury Lamp Ballasts

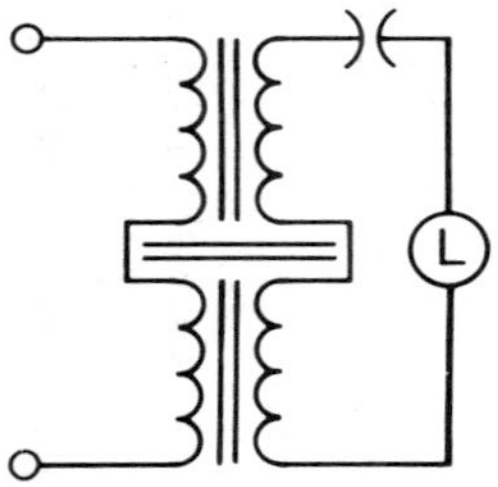

A Two-winding regulated output (constant wattage)

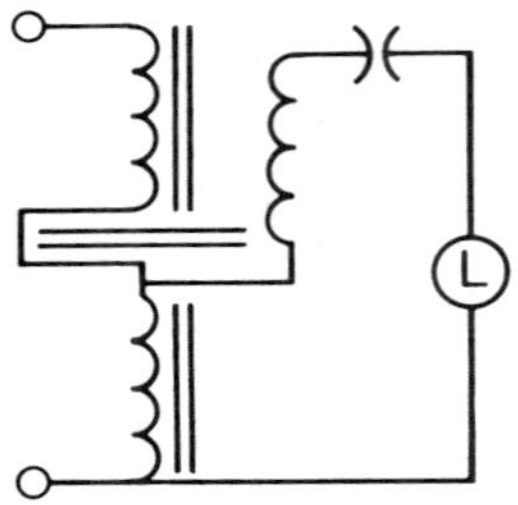

B Regulated output, auto-transformer (constant wattage)

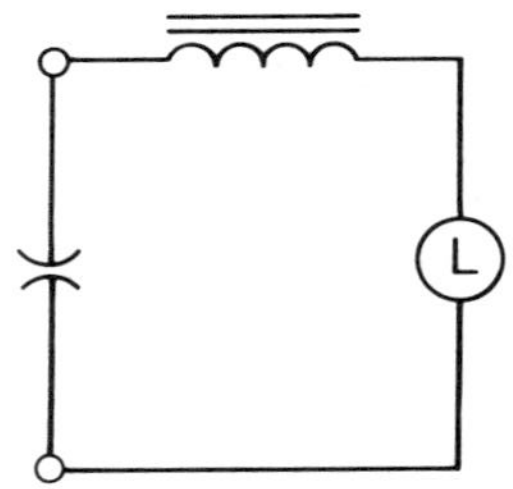

C Reactor, high power factor

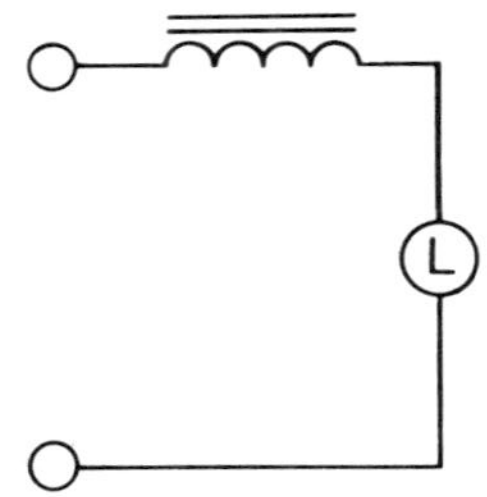

D Reactor, normal power factor

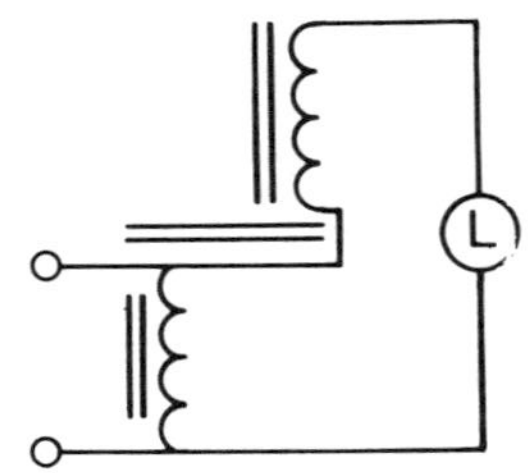

E High reactance, normal p.f. (auto-transformer)

(Not shown: two-lamp lead-lag reactor or auto-transformer; two-lamp series, isolated, regulated output; high power factor high reactance; series sequence; phase control, and others)

Table 1

Ballast Refer.	Ballast Cost	Allowable Line Voltage Range	Lamp Wattage Regulation VS. Allowable Line Volts	Lamp Wattage Regulation VS. Lamp Oper. Volts	Approx. Starting Line Volt-Amps (% of Operating)	Line Power Factor (Operating)
A	100%	±13%	±2%–3%	Proportional	20–30%	Over 90%
B	85–95%	±10%	±8%	Proportional	40–80%	Over 90%
C	50–70%	± 5%	±10%	Little change	110–165%	Over 90%
D	35–50%	± 5%	±10%	Little change	100–130%	50%
E	60–85%	± 5%	±10%	Little change	100–130%	50%
F	110–130%	±10%	±12%	Proportional	Less than 100%	Over 90%
G	150–200%	± 5%	±14%	Little change	90–100%	90%
H	200–250%	±10%	±10%	Little change	Less than 100%	90%

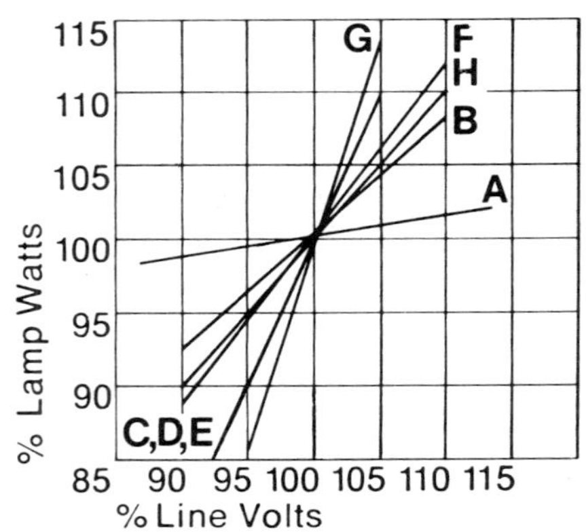

Figure 4-33. Mercury Vapor Circuit. (Courtesy of Philips Lighting Company.)

2. Available in high or low power factor.
3. Higher than normal starting current; requires heavier wire and circuit protection; primary starting current is 150% of operating current.
4. More efficient than regulating type ballast.
5. Simple, light, quiet, least expensive.
6. Current crest factor of 1.45 to 1.55.
7. Minimum line voltages: 240 and 277 for all lamps except the 1000 W and 1500 W lamps, which require 480 V supply.

Autotransformer, High-Reactance, Lag Ballast. An autotransformer ballast has the following characteristics:

1. Nonregulating: input voltage variation of ±5%.
2. High or low power factor.
3. More efficient and quieter.
4. Current crest factor of 1.45 to 1.55.
5. Line voltage of 480 V for 1000 W and 1500 W lamps, and 240 V or 277 V for all others.
6. Higher cost than reactor.
7. Used where line voltage must be stepped up to meet the starting needs of the lamp.

Premium Constant Wattage (CW), Regulator, Isolated Regulator. A premium constant wattage ballast has the following characteristics:

1. Regulation: lamp operation for large variations in line voltage of ±10%.
2. Lamp isolated from the line.
3. Most expensive form of ballast.
4. Current crest factor of 1.8 to 2.0, which is the poorest of all ballasts.
5. Can use all standard input line voltages.
6. A ±5% variation in lamp watts for line voltage range.

Constant Wattage Autotransformer (CWA), Autoregulator. A constant wattage autotransformer ballast has the following characteristics:

1. Most common compromise between regulation and cost.
2. Regulation: line voltage variation of ±10% allowable.
3. Smaller and lighter than premium constant wattage, but larger and heavier than reactor.
4. Current crest factor of 1.6 to 1.8.
5. Noise and losses lower than premium constant wattage, but worse than reactor.

Metal Halide Ballast

The metal halide lamp will not operate on a reactor ballast. Three types of ballasts can be used on metal halide lamps.

1. Regulating
2. Autoregulated
3. Lead-Peak Autotransformer
   a. Similar to CWA, except regulation is not as good
   b. Regulation: input voltage of ±10%
   c. Current crest factor of 1.6 to 1.8
   d. Higher cost than mercury ballasts

High-Pressure Sodium Ballast

The important characteristics associated with the high-pressure sodium ballast are as follows:

1. Larger and more complex; requires an **auxiliary starter.**
2. High voltage to the lamp.
3. Voltage varies over the life of the lamp. The operating characteristics are described in terms of the trapezoidal limits.
4. Current crest factor should not exceed 1.8.

Two types of ballasts are available for operating high-pressure sodium lamps:

1. Regulating HPS Ballast
   a. Regulation: ±7.5 to 10% for 1000 W and 400 W lamps; +10% for 250 W lamps.
   b. Operate on all standard voltages.

2. Reactor HPS Ballast
   a. Regulation: ±5% input voltage.
   b. Voltage: minimum of 240 V on all lamps, except 480 V for 1000 W lamps.

SUMMARY OF STARTING VOLTAGES

Starting voltage or open-circuit voltage is the voltage required to initiate the electron stream between the electrodes. Open-circuit voltage is a function of the lamp wattage. The specifier should refer to ballast manufacturers' catalogs for actual starting voltages. A representative listing of starting voltages are as follows for the light sources listed:

1. Low-pressure sodium (SOX 180): 600 V
2. Mercury vapor: 225 V to 450 V

3. Metal halide: 245 V to 450 V
4. High-pressure sodium: 2500 V to 5000 V

SUMMARY OF STARTING/RESTRIKE TIMES

TABLE 4-7 Starting/Restrike Summary

Light Source	Starting 80% Lumens (minutes)	Restrike 80% Lumens (minutes)
Low-pressure sodium (SOX180)	9–15	30 sec
Mercury vapor	7–10	7
Metal halide	5	10–15
High-pressure sodium	3	1

LIGHT SOURCE COMPARISON

Tables 4-8 through 4-10 are provided to allow comparison of the seven light sources described in this chapter. Tables 4-8 and 4-9 compare light sources in terms of the lamp only (no ballast losses), using constant wattage of 400 W and constant lumens of approximately 30,000 lumens, respectively. Table 4-10 compares the light sources with the combined effect of lamp plus ballast loss. This gives a more meaningful picture of lamp performance. However, the results can still lead to erroneous conclusions if the luminaire efficiency (distribution) is not included in the analysis (see Chapter 6). The total system (lamp + ballast + luminaire) must be analyzed completely before making design decisions on equipment selection.

REFERENCES

1. Kaufman, John E. and Jack F. Christensen, eds., *IES Lighting Handbook,* Reference Volume, 1984. New York: Illuminating Engineering Society of North America, 1984.
2. Koedam, M., R. L. C. deVaan, and T. G. Verbeek, "Further Improvements of the LPS Lamp." *Lighting Design & Application* (September 1975), pp. 39–45.
3. "Lamp Specification Guide." Somerset, NJ: Philips Lighting Company.
4. "Large Lamp Catalog." Danvers, MA: GTE Products Corporation, Sylvania Lighting Center.
5. "Selection Guide for Quality Lighting." Nela Park, Cleveland, OH: GE Lighting.

DATA SECTION

TABLE 4-8 Source Comparison Based on Equal Watts (400 W).

Lamp	Lamp designation	Quantity	Total lamp watts	System watts (each)	Initial lumens (each)	Lamp efficacy (each)	System efficacy (each)	Rated life (hrs.)	Oper. years (yrs.)	Lamp cost (each)
Incandescents										1990
A. Standard	100A	4	400	100	1,750	17.5	17.5	750	.2	$ .81
B. Tungsten-Halogen	Q400T4/CL	1	400	400	7,500	18.8	18.8	2,000	.6	$26.97
Gaseous Discharge—All Require a Ballast										
A. Low-Pressure Gaseous Discharge				2 Lamps						
1. Fluorescent	F40CW	10	400	92	3,150	78.8	68.5	20,000	5.8	$ 2.33
2. Low-Pressure Sodium	L73RE-135	3	405	178	22,500	166.7	126.4	18,000	5.2	$65.69
B. High-Pressure Gaseous Discharge										
1. Mercury Vapor	H33GL-400/DX	1	400	454	23,000	57.5	50.7	24,000	7.0	$24.12
2. Metal Halide	M59PJ-400/BU	1	400	455	40,000	100.0	87.9	20,000	5.8	$48.76
3. High-Pressure Sodium	S51WA-400	1	400	465	50,000	125.0	107.5	24,000	7.0	$52.61

Total BTU per hour = 400 W × 3.41 BTU/Hr./W = 1364 BTU/hr.

Operating Hours per Year = 3432 hrs./yr.

TABLE 4-9 Source Comparison Based on Equal Lumens (approximately 31,000 lm).

Lamp	Lamp designation	Quantity	Initial lumens (total)	Lamp watts (total)	Total BTU/Hr	System watts (each)	Lamp efficacy (each)	System efficacy (each)	Rated life (hrs.)	Oper. years (yrs.)	Lamp cost (each)
Incandescents											1990
A. Standard	100A	18	1,750	100		100	17.5	17.5	750	.2	$ .81
			(31,500)	(1,800)	6,138						
B. Tungsten Halogen	Q400T4/CL	4	7,500	400		400	18.8	18.8	2,000	.6	$26.97
			(30,000)	(1,600)	5,456						
Gaseous Discharge—All Require a Ballast											
A. Low-Pressure Gaseous Discharge						2 Lamps					
1. Fluorescent	F40CW	10	3,150	40		94	78.8	67.0	20,000	5.8	$ 2.33
			(31,500)	(400)	1,364						
2. Low-Pressure Sodium	L74RF-180	1	33,000	180		220 a	183.8	126.4	18,000	5.2	$72.38
			(33,000)	(180)	614	286 b	177.1	118.9			
B. High-Pressure Gaseous Discharge											
1. Mercury Vapor	H33GL-300/DX	2	15,700	300		345	52.3	50.7	16,000	4.7	$24.12
			(31,400)	(600)	2,046						
2. Metal Halide	M59PJ-400/BU	1	40,000	400		455	100.0	87.9	20,000	5.8	$48.76
			(40,000)	(400)	1,364						
3. High-Pressure Sodium	S55RN-150	2	16,000	150		188	106.6	107.5	24,000	7.0	$46.47
			(32,000)	(300)	1,023						

a = Initial: 33,000 lm at 100 hrs.

b = End of Life: 34,000 lm at 18,000 hrs.

TABLE 4-10. Light Source Summary.

		Operating Hours Per Year = 3432 hrs./yr.							
Lamp	Lamp designation	Initial lumens	Lamp watts	System watts (each)	Lamp efficacy (each)	System efficacy (each)	Rated life (hrs.)	Oper. years (yrs.)	Lamp cost (each)
Incandescents									1990
A. Standard	100A	1,750	100	100	17.5	17.5	750	.2	$ .81
B. Tungsten Halogen	Q250CL/MC	5,000	250	250	20.0	20.0	2,000	.6	$14.50
Gaseous Discharge—All Require a Ballast									
A. Low-Pressure Gaseous Discharge									
1. Fluorescent	F40CW	3,150	40	94 a	78.8	67.0	20,000	5.8	$ 2.33
2. Low-Pressure	L74RF-180	33,000 b	180	220 b	183.3	150.0	18,000	5.2	$72.38
		(34,000) c		(286) c		(118.9) b			
B. High-Pressure Gaseous Discharge									
1. Mercury Vapor	H33GL-400/ DX	23,000	400	454 d	57.5	50.7	24,000	7.0	$24.12
2. Metal Halide	M59PJ-400/BU	40,000	400	455 d	100.0	87.9	20,000	5.8	$48.76
3. High-Pressure Sodium	S51WA-400	50,000	400	465 d	125.0	107.5	24,000	7.0	$52.61

a = HPF-RS two-lamp ballast.

b = Initial: HPF, high reactance, 220 W at 33,000 lm (100 hr.).

c = End of Life: HPF, high reactance, 286 W at 34,000 lm (18,000 hr.).

d = CWA Ballast.

5 psychological and nonvisual concerns

Most of this book prepares the student for designing lighting systems that will most efficiently allow for the performance of a visual task. We have considered metrics such as luminance and illuminance and have presented methods for predicting them. However, as in any field involving human aspects, calculations do not represent the whole story.

Consider a playwright who uses a word processor to prepare scripts. The software assists in writing down thoughts quickly. It helps to edit, arrange, and rearrange dialog and instructions. A spelling checker points out spelling errors.

There are even grammar analyzers that will indicate any nonstandard usage. But a computer program can never tell the playwright if the play is any good. Art is subjective. Each person will have a different impression of the play. Furthermore, the execution of the play (sets, costumes, acting) may vary from what the dramatist envisioned.

Lighting is analogous. The metrics and calculation techniques we use can be extremely valuable in designing a lighting system. They can help us to avoid technical errors such as glare or insufficient visibility. They can even help to predict what a luminous environment will look like. But they cannot tell how people will respond to the space. How will the space make us feel? What kind of behavior will the space evoke? How will our bodies respond?

In this chapter we will investigate these issues. Some of them, such as the discussion of subjective impressions, will be directly applicable to the design of lighting systems. Others may serve more as background material.

The title of the chapter is somewhat of a misnomer. Many of the aspects discussed, though not "visual" in the same sense as reading or watching our playwright's production, are mediated by the visual system. The eyes are involved, but for the purposes of this chapter we will not call it "vision" when we assess the luminous environment of a space and form an emotional response to it.

Similarly, the subheadings "psychological" and "physiological" form an artificial dichotomy. The mind and body are inextricably linked. The situation is similar to the wave/particle duality of light mentioned in Chapter 1. We can only fully understand light by considering its dual nature. However, it is frequently useful to consider it as strictly a particle or strictly a wave. Likewise, when considering human feelings and behavior, it is often convenient to separate the mind from the body in our discussion.

PHYSIOLOGICAL RESPONSES

Richard Wurtman is a leading medical researcher in the field of light and human physiology. He has suggested that the influences of light on humans can be divided into direct effects and indirect effects.

Direct Effects

A direct effect of light is said to have occurred when a chemical change is observed in tissue which has absorbed light energy (37). Such effects have been identified over a wide range of wavelengths of impinging radiation extending beyond our standard definition of light (viz., visually evaluated radiant energy). We will classify these effects into three categories: those due to long-wavelength radiation (infrared), those due to medium-wavelength radiation (visible light) and those due to short-wavelength radiation (ultraviolet).

Infrared

Radiation in the range of 700 nm to 1,000,000 nm is known as infrared. We further divide the infrared range into the infrared A, infrared B, and infrared C regions as shown in Figure 5.1. In general, infrared is not visible to humans, but most people can visually detect some infrared A if it arrives at a sufficient rate. Modern illuminants radiate only small amounts of energy in the form of infrared with the notable exception of the incandescent lamp which produces nearly 90% of its radiation in this range.

Infrared passes through vacuum or clear air without appreciable loss of energy. But, when it strikes an object, it converts to heat energy. The main effects of infrared on people are mediated by this transformation into heat energy upon striking the skin. When this occurs, the local blood vessels dilate and the sweat glands are activated. In this way body temperature is regulated in much the same way as a radiator controls the temperature of an automobile engine. Sensations of warmth and cold are evoked by radiation exchange with the environment. If the skin is radiating too much to the environment, we feel cold; too much irradiation of the skin by the environment makes us feel hot.

Infrared Heating In most interior spaces, radiant heating due to light sources with high IR components (e.g., incandescent) is objectionable due to potential discomfort and costs associated with removing that heat. However, in certain applications, lighting professionals are able to capitalize on this otherwise undesirable effect through the use of infrared fixtures specifically designed for heating purposes.

IR heaters use special quartz lamps that are designed to radiate most of their

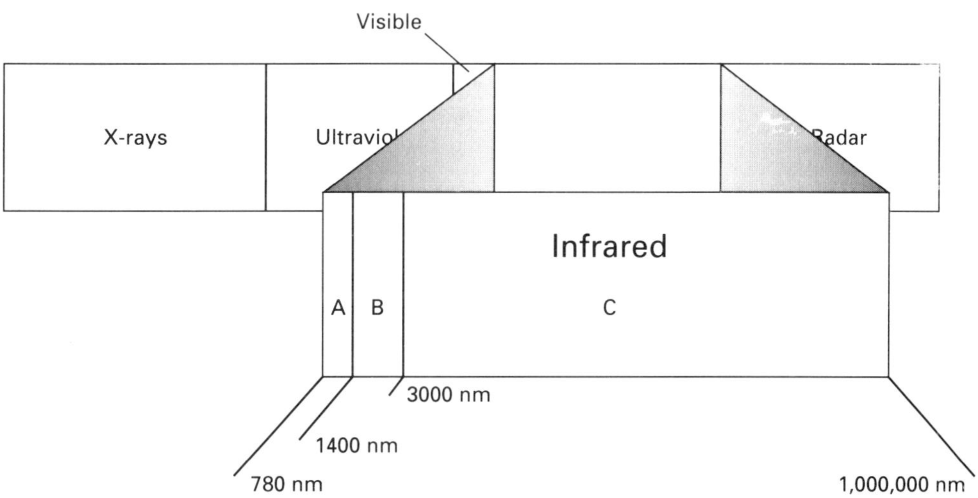

Figure 5-1. Infrared Portion of the Spectrum.

energy in the infrared A and infrared B regions of the spectrum. Thus, their radiation passes largely unattenuated through air until it strikes a person or object. The most common usage is therefore in spaces where it is impractical or undesirable to provide thermal comfort by heating the air. Exterior spaces, drafty spaces, and sparsely or infrequently inhabited rooms with large air volume are the types of spaces where spot IR heating is appropriate. Examples of such spaces include porticoes, portecocheres, or canopied entrances; outdoor walkways; certain warehouses; bus stops and transit platforms; loading docks; hangars; and some livestock and horticultural applications.

Quartz heat lamps are available in a variety of wattages from 300 to 6000 W. Rated life is on the order of 5000 hours. IR lamps are typically dual-ended T-3 lamps with flexible leads, though other configurations exist. Most IR lamps are intended to be operated horizontally. To achieve satisfactory performance in other orientations, specify lamps designed to burn in any position.

Since the basic nature of radiant flux is the same at all wavelengths, the calculation and prediction techniques used for visible light could, in theory, be adapted to radiant heating. Such rigor is rarely necessary, however, so IR heating is typically specified in terms of installed watts per unit area. Power densities for most applications will fall in the 30–70 W/ft^2 range. Lower densities (down to 10 W/ft^2) may be appropriate in still air and fairly high ambient temperatures. Densities approaching 100 W/ft^2 are necessary for snow melting and evaporation under porticoes for example. IR heating should not be attempted in windy areas as even moderate air motion will cool the skin at such a rapid rate as to overwhelm the radiant heating effect.

It is important not to confuse IR heating with other heating equipment that is frequently (somewhat erroneously) referred to as "radiant heat." Electric baseboard heating, for example, differs markedly from IR heating in that its purpose is to raise the ambient air temperature of the space, and it accomplishes this largely through convection rather than radiation. An additional substantial difference is that convection heating elements stay relatively cool (around 200°F) in comparison to the bulb wall temperature of quartz heat lamps which can exceed 600°F.

Visible Light

The main direct effect of visible light is the destruction of bilirubin, the yellow compound which occurs naturally when red blood cells die. The liver of premature infants frequently is not developed enough to remove this toxic substance. If the bilirubin level reaches excessive concentration (hyperbilirubinemia), brain damage, retardation, and even death may result. Though the mechanics are not completely understood, effective hyperbilirubinemia treatment for tens of thousands of babies annually consists of simple exposure to intense visible light (36).

A harmful direct effect of visible light is the recently described syndrome known as retinopathy of prematurity (12). Researchers believe that scar tissue observed on the retinas of premature infants may be due to environmental lighting. Designers of lighting for hospital nurseries should pay close attention to developments in this field. One possible source of excessive light on premature babies is windows. Another source is the lights used to treat hyperbilirubinemia; although the babies being treated have their eyes carefully occluded, neighboring neonates may be endangered.

During the 1960s, parapsychologists claimed to have discovered a new direct effect of visible light. The so-called "dermo-optic" sense was first reported in the Soviet Union based on the reputed ability of a subject named Rosa Kuleshova to distinguish between colors simply by touching them with her fingers (18). The popular media picked up the idea in this country and some U.S. researchers reported duplicating the Soviet results. Though largely debunked by reputable researchers at the height of the controversy (11), some were claiming that as many as one person in six had dermo-optic abilities. Episodes of sloppy science such as this and more recent examples should serve as a warning to lighting students and professionals to consider very carefully their sources of information on much heralded discoveries in lighting. Claims of manufacturers and the popular press should be verified with authoritative organizations such as the Commission Internationale de l'Eclairage, the IESNA, or other national illuminating engineering societies.

Ultraviolet

Ultraviolet radiation is that portion of the spectrum from approximately 400 nm to 100 nm. Figure 5-2 shows that radiation in this range is further classified into four zones. Ultraviolet radiation in the UV-B and UV-C zones can cause erythema (reddening of the skin) and conjunctivitis (inflammation of the conjunctiva of the eye). UV-B also causes melanogenesis (resulting in skin tanning) and contributes to skin cancer in susceptible individuals. Blacklight is also melanogenic and erythemogenic, but extreme energy densities are required (something on the order of 2000 times the levels of UV-B required for erythema). Ultraviolet of wavelength shorter than 200 nm is not a concern in architec-

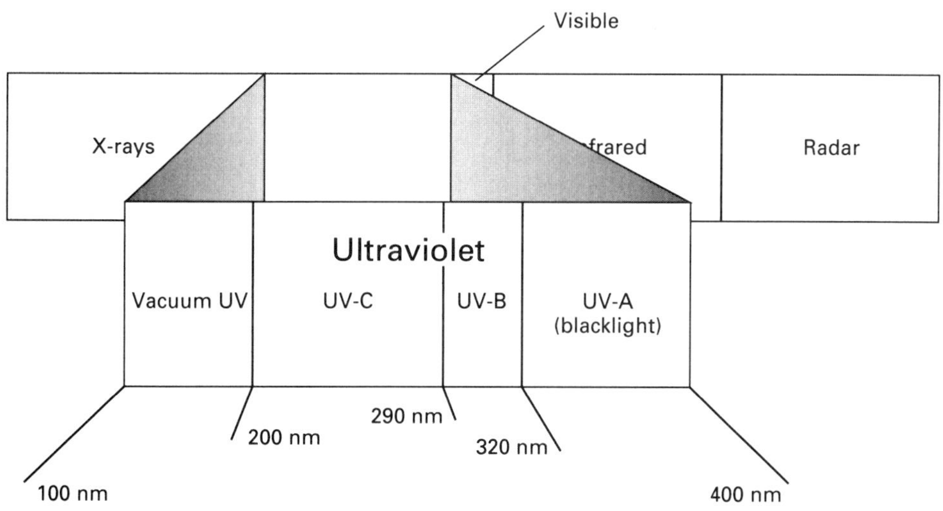

Figure 5-2. Ultraviolet Portion of the Spectrum.

tural lighting since it is strongly absorbed by air, hence the name "vacuum UV." Keep in mind that all boundaries between adjacent areas of the electromagnetic spectrum are somewhat fuzzy. It would be erroneous to think that radiation at 199 nm is totally absorbed in air whereas radiation at 201 nm is perfectly transmitted through air.

Arc discharge sources for general lighting produce some radiation in the erythemogenic range. In most cases, the majority of it is attenuated by the glass of the arc tube and the bulb or by phosphor coatings on the bulb wall. Enamel paint on the interior of luminaires are highly absorptive, as are most of the acrylic materials used for lenses. Thus, for all practical purposes, only UV-A leaves the luminaire, and the largest part of this is absorbed by building surfaces and furnishings. Only the direct component of UV-A is nonnegligible, and that is probably not dangerous.

Though lighting equipment is considered extremely safe from the standpoint of ultraviolet radiation, an exception to this safety rule occurs when the outer bulb of a mercury vapor or a metal halide lamp breaks and the arc tube continues to burn. In such cases serious and painful erythema and conjunctivitis can occur. This is why metal halide lamps on the market today have a self-extinguishing mechanism which interrupts the arc in the event of bulb breakage.

Since the body's ability to assimilate calcium needed for formation and maintenance of bones and teeth depends on the presence of vitamin D, UV's role in the formation of vitamin D is probably its most important direct effect. Though we get some from food sources, up to 90% comes from a reaction that occurs when ultraviolet light between 250 and 315 nm irradiates 7-dehydrotachysterol, a substance found in abundance in the skin. Rickets in children and softening or brittleness of the bones in the elderly are due to vitamin D insufficiencies that can be prevented or cured by exposure to small quantities of ultraviolet radiation (24). A healthy person receives all the ultraviolet necessary for vitamin D production through daily exposure of the hands and face to 15 minutes of sunlight.

As long ago as the ancient Egyptians, ultraviolet light was used for photochemotherapy. They discovered that exposure to sunlight after eating certain plants aided in treating skin afflictions. Today, derivatives of the same Egyptian plants are used as a photosensitizer in the treatment of psoriasis, a condition in which certain naturally occurring cells are present in excess. 8-MOP is given orally, followed a few hours later by ultraviolet irradiation of the skin. The result is selective destruction of the unwanted cells.

Ultraviolet light is also known to have a germicidal or bactericidal effect. The most effective range of radiation for this purpose is between 200 and 300 nm since that is the range over which the DNA molecule absorbs radiation best. When DNA absorbs radiation, it either mutates or dies.

The germicidal lamp takes advantage of this effect. These lamps are tubular, low-pressure mercury discharge lamps greatly resembling fluorescent lamps. They differ from fluorescent lamps in that they have no phosphor coating on the inner bulb wall and they are constructed of special glass that transmits short ultraviolet radiation. These lamps, which use standard fluorescent ballasts, are usually mounted in open, indirect-oriented luminaires in some hospital spaces and areas where it is desirous to kill bacteria.

Since common building finishes reflect very little UV, reflected radiation is not a problem. However, direct exposure to these lamps can cause erythema, and retinal damage could theoretically occur in about the time it takes to read the warning label. These hazards can be largely overcome by mounting the lamps in air supply ducts rather than directly in the space to be treated.

Indirect Effects

Wurtman describes an indirect response of tissue to light as resulting from neural or hormonal signals, not through direct absorption of light into the tissue itself (36). These signals result when photoreceptive cells in the retina transduce incident light energy into impulses carried to the tissue via neural, or a combination of neural and endocrine pathways. Obviously this describes the process of vision; however, nonvisual indirect effects abound.

An example of a study demonstrating the existence of nonvisual indirect effects is cited by Thorington and his colleagues. It involved raising rats under various luminous environments and noting the effects on organ development (35). That the observed strong variations across lighting treatments are mediated by the retina (as opposed to being a direct effect of light absorbed through the skin) can be demonstrated by removing the eyes of the rats, in which case no effect is observed. This leads to speculation about the influences of light on human development.

A group of researchers led by Hollwich has determined that an increase in illuminance level under fluorescent light is accompanied by an increase in hormone production in people—especially of the stress hormone cortisol (16). Their study of blind people showed that this response involves the visual pathway, and thus may be appropriately classified as an indirect effect of light.

Production of other hormones has been shown to be modified by light, especially light/dark cycles. These fundamental patterns of hormone secretion are called circadian rhythms (from Latin *circa*, meaning "about" plus *dia*, for "day"). Wurtman found that the amount of melatonin humans produce varies through a 24-hour cycle and he links this to normal daily patterns of exposure and deprivation of light (36). Melatonin is the so-called master hormone secreted by the pineal organ which affects sleep, ovulation, and the secretion of other hormones. Studies have shown that the periodicity of hormone secretion can be altered or entrained by placing subjects in an environment in which the light/dark cycle is something other than 24 hours.

What would the body do if allowed to free run in the absence of any environmental day/night or seasonal cues? This question was investigated in the celebrated 1989 research in which a young Italian woman was totally isolated in an underground test chamber for over four months. Her menstrual cycle soon stopped and she eventually developed a pattern of sleeping for 22 to 24 hours followed by up to 30 hours of activity. By the end of the study in late May, she estimated that it was only mid-March.

Results of that study, along with many others on the relationship of light and hormone secretion may eventually help us to better understand the syndrome known as seasonal affective disorder, or SAD. "Affect" is the psychologists' term for mood—

thus SAD is a mood imbalance that is related to the changing seasons. People with SAD alternate between being depressed in the fall/winter and having euthymia (normal mood) or hypomania (mild elation) in spring/summer. Jacobsen and his colleagues have given an informative discussion of the syndrome (20). They note that SAD usually begins when a person is in his or her twenties or thirties and that it is four times more common among women than men. Though some of the symptoms are the same as standard, or nonseasonal clinical depression, other symptoms are unique to SAD (see box).

SAD Winter Symptoms

Sadness

Decreased activity

Increased appetite, weight gain

Decreased libido

Increased sleep, restless sleep, daytime drowsiness

Menstrual problems

Work problems

Interpersonal problems

Milder near equator

raised items not typical of depression

Successes in treating SAD with phototherapy are widely reported. Researchers have used various methods, but most have settled on a once-daily treatment consisting of exposure to around 2500 lux for a couple of hours. One theory holds that light treatment works because people afflicted with SAD have abnormal circadian rhythms and the exposure to the bright lights causes a phase advance in melatonin secretion, thus normalizing daily rhythms. This theory is not unanimously held, and vigorous debate among medical researchers in the 1980s has left many unresolved questions about the mechanism of light treatment, the action spectrum (viz., what part of the spectrum is effective in the treatment), the optimum time of day for such treatment, and even if there is an optimum time of day.

In another area of indirect effects, Matulionis has suggested that luminance patterns may affect several physiologic events (23). He placed subjects in a small room having one of six different configurations of black, white, or gray vertical surfaces and workplane. For example, in one configuration the left wall and the left half of the front wall were black, the rest of the vertical surfaces white, and the workplane gray. Two visual tasks (drawing a row of squares and reproducing a triangle viewed for 10 seconds) were performed while measurements were made on two polygraphs. The subjects were also filmed by two video cameras. Significant differences in several variables were found in response to the different luminous environments. These included the average area and distortion of the squares drawn, the average area of the triangles, the direction of the size shift between drawn and model triangles, the number of postural shifts, heart and respiration rate, and right trapezius and right sternocleidomastoid muscle activity. The most significant differences occurred under the two split-field conditions (half of the room black, half of the room white).

Another family of indirect effects of light concerns flicker. Modern gaseous discharge light sources have the common characteristic that they flicker, or turn off and on, twice during each period of the alternating current which supplies them, thus 120 times per second in this country. The human visual system reads as continuous, or "fuses" flicker at a frequency of roughly 50 hertz (Hz). This is known as the **critical flicker frequency (CFF)** or **flicker fusion frequency (FFF).** Figure 5-3 lists many of the factors that determine FFF.

Even at suprafusion levels, flicker may affect the human nervous system according to Rey and Rey who studied this phenomenon (29). Subjects were presented with a visual task consisting of strings of random letters typed on a page and were instructed to cross off certain letters according to a prescribed set of complicated rules (e.g., cross off all consonants located within one letter from a "Z"). Performance scores and three fatigue tests were compared in an experimental design which crossed two illuminance

factors affecting

Flicker Fusion Frequency

Retinal illuminance
(directly proportional)

Location in visual field
(rods more sensitive)

Individual differences

Contrast

Color of source

Age

Figure 5-3. Determinants of Flicker Fusion.

levels (50 and 1000 lux or roughly 5 and 100 footcandles) with lighting types. The two illuminants used were both fluorescent lamps but one was driven at 100,000 hertz (i.e., flicker free) whereas the other was operated at the European standard 50 Hz. The three tests of fatigue were flicker fusion frequency (comparing before and after the visual task the specific frequency of flicker at which a subject's visual system fused the flickering light to continuous), response time to a visual stimulus, and the regularity (again both before and after performing the visual task) with which the subject could tap out 60 beats of 1 second each on a Morse key. No significant effect was measured on the illuminance level variable, but subject's fatigue as measured by each of the three tests was adversely affected by the 50 Hz source as compared to the 100,000 Hz source. In other words, working under "flickering" was associated with greater fatigue, a finding supported by Hopkinson and Collins and numerous studies which they cite (17).

Since different wavelengths of radiation produce different responses in humans, we might reasonably expect that different wavelengths within the visible range, that is, different colors of light, would have varying effects. In fact, this is the case. The color of light impinging on the retina has been reported to influence electroencephalogram readings, time estimation, reaction time, galvanic skin response, posture, and other physiological quantities (22). Goldstein claimed, in a study of subjects afflicted with Parkinson's disease, that the color red worsens the condition whereas green improves it (13).

Kearney has identified a relationship between ambient temperature and color preference (21). When subjected to a combination of one of three temperature levels and one of three light levels, volunteers' color preferences were positively correlated with wavelength under cool temperatures and negatively correlated under warm temperatures. In other words when people are cool, they prefer reds and oranges, and when warm, blues are preferred (Figure 5-4). Greene and Bell showed, however, that this phenomenon could not be "turned around," that is, that placing people in colored environments does not affect their subjective impression of ambient temperature (14). Thus it is not possible, as had been hoped for cold climates, to paint walls red, lower the thermostat, and still maintain subjective thermal comfort while saving heating energy. Also, the practice of switching from cool white fluorescents in the summer to warm white fluorescents in the winter as reportedly occurs in some offices would, according to these findings, presumably have no effect on perceived thermal comfort.

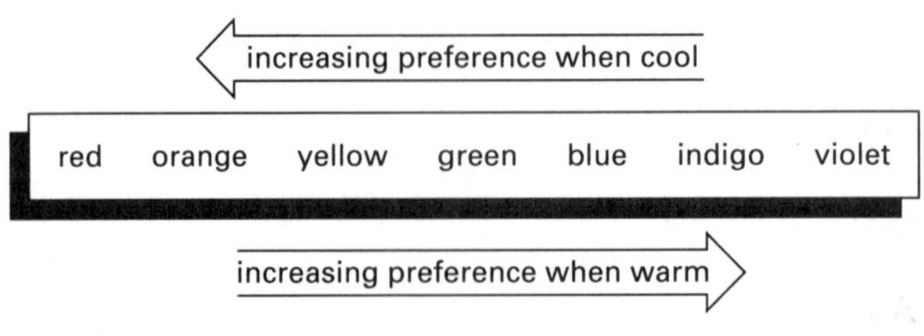

Figure 5-4. Color Preference and Ambient Temperature.

Stone and associates investigated several of the nonvisual indirect effects examined thus far (33). Their sole experimental variable was the level of illuminance, or quantity of visible radiant energy reaching the workplane. On one occasion, 1000 lux of diffuse fluorescent light was used; on another occasion, 11,000 lux (approximately 100 and 1100 footcandles, respectively). The exposure was for 4 hours in each case. The most striking finding was that pulse rate increased as soon as the higher illuminance was turned on and continued until it was extinguished. The higher level was also associated with increased sense of fatigue, decreased social affection, and reduced thermal comfort.

The chapter introduction mentioned difficulties in making a distinction between the terms ''physiological'' and ''psychological.'' If we think of the two terms as labeling adjacent portions of a continuum, then we have clearly arrived at the fuzzy boundary between the two. In the work by Stone, the question remains as to why arousal (loosely, ''alertness'') increased under the higher illuminance (11,000 lux). This is perplexing because equal or greater light levels are commonly encountered outdoors, yet the mere act of going outdoors does not evoke elevated arousal. A plausible answer is that the reaction is based on people's perception that 11,000 lux is an unusual and inappropriate illuminance level indoors. This sends the pulse rate up. However, that sounds more like what we would term a ''psychological'' response. So, for the remainder of the chapter, let's shift our thinking into the ''psychological aspects'' mode.

PSYCHOLOGICAL ASPECTS

In this section, we will label as ''psychological'' two types of phenomena: behavior and impressions. Behavior affects ways in which lighting tends to influence our actions. Impressions, sometimes called *subjective impressions*, refers to how the lighting in a space makes us feel. It also refers to our preferences about the lighted appearance and ''feel'' of a space.

Behavior

The existence of the physiological responses to colored light mentioned in the last section prompted Stark and collaborators to investigate some behavioral effects of subjects exposed to red or blue light (32). They claim a direct relationship between color stimulation and central nervous system activity but state that the exact functioning of this response is not understood.

Stark and colleagues observed 28 male or female college-age volunteers as they gambled under either red or blue light in a card game in which the odds were known in advance to the experimenters but not to the players. They claim that long wavelengths, such as red, are associated with increased frequency and intensity of responses, in short more arousing, as compared with shorter wavelengths such as green or blue. This led to the hypothesis that subjects in the red environment would bet more frequently, at higher stakes, and under greater risk than those subjected to the blue condition. And that is exactly what they found. They also determined that their subjects gambled more money, more often, and took greater risks as the experiment progressed and that this phenomenon was more markedly observed under the red condition.

The amount of noise generated by groups of people under different luminous environments was the subject of study by Sanders and his associates (30). Sound pressure level readings were taken in a darkened corridor of a university classroom building. Highly significant results indicate that groups of students congregating in the hallway made less noise under the low-light condition than with normal illuminance levels. A possible psychological interpretation is that low light levels are associated with contexts in which quietness is considered appropriate behavior, for example, church, theater, museum, or cinema. The researchers warn, however, that since the illuminance level was lowered by turning off every other luminaire, the variable ''low illuminance'' may have been confounded with ''irregular illuminance.'' In other words, the low conversational sound level may have been due to the effects of being in an unevenly lit environment rather than the effects of a dim environment. If this were the case, then these results may have to be interpreted more from a physiological point of view as in Matulionis' findings on luminance patterns cited earlier. Also, since the length of time a person spent in the environment was not controlled, the effect may, or may not be transitory.

Page and Moss studied the effects of illuminance levels on aggression (26). Their position that darkness acts as a disinhibitor was discussed in psychological terms and led to the hypothesis that subjects would deliver stronger shocks to victims in a dimly lit setting than in a brightly lit setting. Three possible explanations for such disinhibiting were (1) darkness may increase feelings of anonymity, (2) people may be conditioned to associate darkness with uninhibited behavior (bars, parties, the bedroom, etc.), and (3) darkness isolates an aggressor from his victim.

Subjects entered either a ''bright'' or ''dim'' test environment in pairs and were assigned (seemingly at random) to be either the teacher or learner in what was presented as an experiment on the effects of punishment on learning. Actually, the person chosen as the learner was a confederate of the experimenters. The subject was instructed to teach the learner pairs of nonsense syllables and, each time a mistake was made, to choose an appropriate level of electrical shock which he was told would be delivered to the learner as punishment. The learner's face was not visible to the subject and the learner showed no reaction regardless of the shock level selected (actually, no shock was received). As predicted, significantly more intense shocks were delivered in the dark than in the light environment, and thus luminance level does seem to affect aggression.

Another important behavioral aspect of lighting is its importance in the interrelated notions of orientation, attraction, and circulation. An excellent illustration of this concept is found in theater lighting. As luminance patterns change on stage, the collective gaze of the audience unconsciously shifts to exactly the spot where the lighting designer wants attention focused. In general, our attention is attracted to areas of higher luminance. This principle is put to practical use in the design of merchandise lighting. Table 5-1 gives illuminance levels as recommended by IESNA. From this it can be seen that lighting ratios of 15 to 1 are used to draw attention to emphasized merchandise. Recall that luminance ratios lower than about 3 to 1 will not usually be discernible to most observers.

TABLE 5-1. Illuminance Levels for Merchandising

Area	Volume of activity	Illuminance	
		Lux	Footcandles
Circulation	High activity	300	30
	Medium activity	200	20
	Low activity	100	10
Merchandise	High activity	1,000	100
	Medium activity	750	75
	Low activity	750	30
Feature displays	High activity	5,000	500
	Medium activity	3,000	300
	Low activity	1,500	150
Show windows			
Daytime lighting			
General		2,000	200
Feature		10,000	1,000
Nighttime lighting			
Main business districts (highly competitive)			
General		2,000	200
Feature		10,000	1,000
Secondary business districts or small towns			
General		1,000	100
Feature		5,000	500

An informal study by John Flynn gives us additional insight on this topic (10). In a restaurant-type setting, people's seat selection was observed under two lighting configurations. In one arrangement, a side wall was washed. This wall was perpendicular to the coffee bar where people pick up their beverage. In the other lighting configuration in the same space, more distant walls were washed, leaving the seating area near the coffee bar in relative obscurity. Flynn noted that, in each case, people tended to select seats that faced the lighted wall. Other factors, such as desire to sit near friends or sit at a close location, tended to override the orientation effect. Also, people were not willing to walk across the room to sit in lighting that has proven to be considered by most people as pleasant.

In other research (34), Taylor and Sucov observed people as they entered the room shown schematically in Figure 5-5. Various combinations of light levels on the left and right pathways were created. They found (in agreement with previous research) that when both paths were equally lit, about two-thirds of the subjects went to their right. When the right side illuminance was 100 times that of the left side, everyone went right. When the left was illuminated to a level 100 times higher than the right, however, a significant portion (about one quarter) still went to the right. It was concluded that people have a tendency, though far from absolute, to choose the path of higher light level.

These two studies illustrate nicely a phenomenon which we see repeatedly in behav-

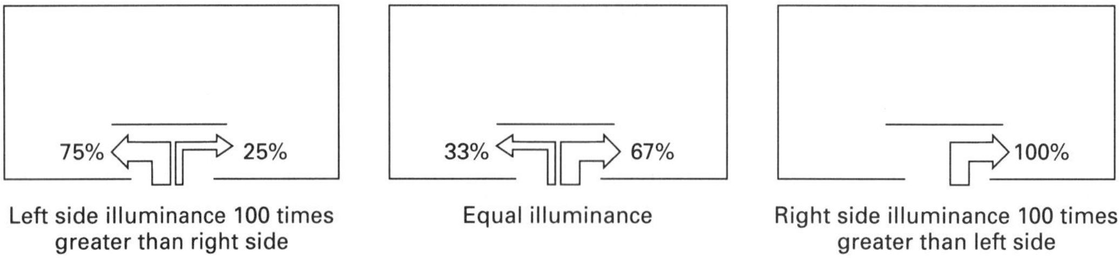

Figure 5-5. Taylor and Sucov's Study of the Effect of Illuminance on Choice of Path.

ioral studies on lighting; that is, lighting influences tend to be subtle and secondary to other considerations. The noted perception psychologist J. J. Gibson used the term ''affordances.'' The presence of a cliff ''affords'' falling off, however, just being near the edge doesn't mean you necessarily *will* fall off. In fact, given our aversion to pain, we have very strong motivation not to fall off. Similarly, we tend to be attracted to or to move toward light, however, if some contradictory influence or desire is operating simultaneously, it is quite likely that we will ignore our phototropic or mothlike behavior.

A final way in which lighting influences our behavior is seen in our response to lighting equipment and the visual problems it causes. This can be observed in almost any office where VDTs are used (which is to say, in almost any office). Many office lighting installations exist which don't take into consideration the potential for reduced VDT visibility caused by veiling reflections. In response, occupants' behavior is modified: people slouch, or put baffles on luminaires and/or shrouds on VDTs. When glare occurs on printed materials, people shift or tilt the paper to reduce glare.

In a study using an adjustable task light, Rea found that people do not always behave in ways that optimize visual task performance (28). Subjects were asked to position the desk lamp and then perform a number matching task. The results indicate that people are aware of glare and position the lamp so as to avoid glare. However, they do not optimize uniformity across the desk (which would aid in task visibility).

Impressions

Whether we like a room or not, how the room makes us feel, how we perceive objects and people within the room, all of these can be influenced by the way a room is lit.

Champness and his colleagues reported on several experiments relating social behavior and the luminous environment (7). One experiment observed people's reaction to meetings with two strangers under two types of illumination with different spectral characteristics. The subjects were asked to choose between the two strangers on a variety of scales. The lighting was found to affect peoples' perceptions about the strangers they met. The authors concluded that the spectral power distribution of illuminants is a better predictor of social effects than some commonly used measures.

Aspinall and Dewar have identified psychological responses to varying lighting orientations (1). Subjects viewed people portraying accused people on the witness stand. The accused was illuminated by one of several intensities and directions of light. After hearing one of three taped excerpts from the court proceedings (one which tended to indicate the accused's guilt, one which implied innocence, and one which was neutral), the subject was asked to determine the accused's guilt or innocence. Complex interactions between lighting and evidence were found at a statistically significant level. For example, under one lighting condition, subjects hearing the ''innocent'' tape tended to find the accused not guilty more often than those hearing the ''neutral'' tape, as expected. But under another lighting condition, those hearing the ''innocent'' story were much more likely to find the accused guilty than those subjects hearing the ''neutral'' tape. No single aspect of the lighting measured in the experiment could be isolated as being responsible for this effect. The authors conclude that lighting does affect subjective judgments of perceived guilt and that those responsible for courtroom lighting design should investigate the phenomenon more fully.

Flynn has developed a suggested methodology for investigations of subjective impressions of light (8,9). The system uses semantic differential rating scales (e.g., rate a space on a scale of 1 to 7, with 1 being ''clear'' and 7 being ''hazy'') and multidimensional scaling to identify ''modes'' of light (e.g., overhead/peripheral, uniform/nonuniform, bright/dim, and warm/cool). People were placed in a room in which several lighting systems could be switched and dimmed to achieve various appearances. They were asked to rate various lighting conditions using a variety of semantic pairs (clear/hazy, harmony/discord, large/small, etc.) Using the statistical technique known as factor analysis, Flynn was able to group the word pairs into impression categories. He then was able to match each category to the lighting modes which elicited high scores on that category. Though some investigators using this type of methodology have encountered mixed results, Flynn's own findings seem consistent with many designers' experience and form the basis of guidelines for creating certain spatial impressions. These guidelines are given in Table 5-2. For example, if a designer wants to reinforce a feeling of spaciousness, the appropriate lighting mode to use is uniform, peripheral lighting such as can be achieved by wall-wash lighting. Note that the opposite effects to those listed are also possible: to enhance a feeling of ''public'' one would use uniform, high levels of illuminance.

TABLE 5-2. Lighting Effect Guidelines Based on Flynn's Work

Effect	Enhanced by
Spaciousness	Uniform, peripheral emphasis
Privacy	Low luminance near user Higher luminances remote Nonuniform luminance patterns
Relaxation	Nonuniform peripheral emphasis Warm white light Low illuminance levels
Pleasantness	Nonuniform lighting Peripheral emphasis

Using methods similar to Flynn's, Harvey and his associates found that people have very different subjective impressions of three different fluorescent lighting systems (one troffer and two indirect systems) (15). This was true despite the fact that performance of a realistic visual task (comparing printed material to information appearing on a computer monitor) did not vary significantly among the three systems. So people's feelings about lighting will not necessarily depend on how a lighting system affects their ability to perform a visual task. We have already seen this result in Rea's study of desk lamp positioning—the lighting configuration people choose is not necessarily the optimal one for visibility.

Smith and Rea also studied subjective ratings and performance (31). They compared fluorescent, metal halide, and high-pressure sodium sources under several different luminance levels and found that neither visual performance nor subjective impressions was affected by light source differences. Lighting *level*, of course, affected performance in the normal way—increased task luminance was associated with improved visual performance, with performance improvement being asymptotic at the higher luminance levels. However, the quantity of light had varying effects on the subjective scales. Some subjective ratings (such as brightness, clarity, and stimulation) continued to increase with increased luminance, but some ratings reached a maximum and began to decline at the highest light levels (for example, comfortableness, goodness, and pleasantness). This would suggest that we can always increase people's perception of clarity and stimulation in a room by increasing the light level. However, there exists a point beyond which if we increase the light level, people will find it *less* comfortable, pleasant, and good. Judging from Smith and Rea's work, this optimum task luminance is probably on the order of magnitude we would find on white paper in a brightly lit office, or higher.

Our first response to this tends to be "How fortunate, we don't have to worry about people finding light levels unpleasant or uncomfortable unless we greatly exceed the current (North American) standards." Or we might be tempted to think that the current standards are about right since they are in the ballpark for optimizing user comfort and feeling of well-being. However, we would do well to consider what might be the cause of people's decrease in approval as luminance is increased "too far."

One possible explanation is "that's just how we're made." And this may well be so: increase brightness too much and we're uncomfortable; increase it more and it eventually becomes painful. However, there exists a clue that this is not the case, or at least not the whole story.

Consider lighting standards in Europe and most other parts of the world. Recommended light levels have traditionally been significantly lower there than in North America. This is due in part, perhaps, to the fact of historically higher energy costs. Table 5-3, for example, gives some comparative figures for the United States and France. Belcher's experiences in France led him to state that light levels considered normal in North America would elicit complaints of discomfort and unpleasantness from French observers (3). This raises again the issue of expectations about the luminous environment. It is quite plausible that what Smith and Rea detected was people's aversion for the unexpected—in this case, light levels that were "too high." Had they performed the same

TABLE 5-3 Comparison of Illuminance Standards

Location (as described in (French work code)	Illuminance (lux) French work code	U.S. practice
Interior		
Circulation zones	40	200
Stairways and storage	60	200
Work areas, locker rooms, toilets	120	300
Windowless areas designated as permanent work areas	200	500
Exterior		
Circulation zones	10	50
Permanent work areas	40	200

experiment with French subjects, they may have found the maxima of the subjective ratings shifted downwards considerably.

Rather than exposing people to a variety of lighting situations and asking their impressions (as did Smith and Rea, Harvey et al., Flynn, etc.), another possibility for investigating subjective responses is simply to ask subjects what they prefer. Butler and Biner administered a questionnaire to college students asking what light level they prefer for various activities in various locations (6). Allowable responses were A through D, with A indicating a preference for "very dark" and D signifying "very bright." Thus, sleeping in the bedroom might be rated as A, whereas working in the kitchen might elicit a D rating. Subjects were also asked questions concerning how important it was for the light level to be correct in a given situation, and how important people felt it was to be able to control the light level.

The results of this study agree well with intuition. Significant differences were found in lighting preferences for different activities and different locations. Men and women were in agreement on most preference judgments, though exceptions were noted. Importance and light level were strongly correlated: people found it more important to have the right lighting in situations where very dim or very bright levels were preferred than in situations where moderate lighting was preferred. Finally, the more important the lighting level is, the stronger is the desire to be able to control the level. Lighting designers should keep these findings in mind when deciding whether simple on/off switching is acceptable or if dimming is required for a given application.

In most of the preceding discussion on impressions, research results were presented from studies in which the participants were not "naive." In other words, subjects were aware, in at least a general way, that the investigation involved their feelings about lighting. Awareness on the part of subjects about the goals of the research can cause problems. For example, participants may try to anticipate what the researcher wants to find, and answer or behave in such a way as to confirm the theory. When researchers call attention to the lighting, it is likely that subjects will respond in a way that is different from their everyday behavior when they aren't thinking about the lighting.

It is difficult to keep participants naive when your purpose is to answer the question "how do people feel about this lighting." Some less direct approaches have been attempted, with limited success. For example, in a study by Nelson and his associates on the interactions between ambient temperature and illuminance (25), participants unaware that lighting was being studied were asked questions about their mood. In that study, temperature, illuminance, and apparent time were manipulated to investigate sedentary work fatigue. Subjects were asked to write as many short stories as possible under various combinations of room temperature, illuminance, and clock speed (faster than normal, slower than normal, or accurate). The results indicated that mood is poorer under cool/bright and warm/dim configurations, that more stories were generated in the cool environments, and that females respond better to a higher illuminance level. Belcher also assessed the mood of naive subjects (4). His work also suggests that people's level of positive affect (good mood) may vary with lighting level and that men may respond differently than women.

Productivity

Even before the advent of the electric lamp, managers pondered the question of the possible link between lighting and worker productivity. One manager made the observation that people sitting next to windows make fewer mistakes than do those in the back of the room. When productivity dropped in the winter, another manager theorized that lower winter daylight levels were the cause. Once these ideas take root, it is hard to dissuade people from them, even in light of equally plausible competing theories. For example, if window seats are assigned to the workers with greatest seniority, those sitting near the window will already be the less error-prone employees, regardless of any lighting effect. As for the observation about decreased productivity in the winter, this could be attributable to the fact that cold workers don't perform as well.

This is not to say that it is unreasonable to suggest that lighting affects productivity. On the contrary, since the effects of the luminous environment on visual task performance are so well documented, it is perfectly natural to *want* to say that an improvement in the lighting will increase worker output. The problem with making that jump, however, is the fact that, as seen in this chapter, light affects us in a plethora of ways in addition to vision. What is the *combined* effect of the vision, health, mood, and preference influences of light? It is certainly possible, at least, that the response of humans in one or more of these areas to a lighting system that optimizes visibility, could counteract the productivity gained by increasing visual performance. For example, in the findings of Smith and Rea mentioned earlier, increasing the task luminance to 1810 cd/m^2 increased visual performance, yet decreased people's comfort and assessment of pleasantness. It is certainly plausible that this decreased comfort might cause a reduction in overall performance that would offset part or all of the gain in visual performance, or possibly even swamp it.

As a hypothetical illustration, consider an average worker who, under a certain task luminance, produces 100 units of work per hour. Those units may be widgets manufactured, paragraphs typed, or any other measure of output. Let's say that at a higher luminance level, the average person can see well enough to increase output by 10 units of work per hour. We would be tempted to say that the lighting increase improves productivity. However, let's say that this higher luminance is in the range where people on average find it less comfortable and less pleasant. This may result in higher absenteeism, poorer morale, and general worker dissatisfaction. What is the cumulative effect of these negative influences? Two units' reduction? Five? If it's ten, we wasted our money in increasing the lighting. If it's greater than ten, we've paid good money for the purpose of *reducing* output. In fact, we will see in Chapter 14 on economics that, even if the net result is an increase in units per hour, if the cost of improving the lighting exceeds the *benefit* to the company of those extra units, then the "improvement" is economically undesirable.

Bitterman makes an important observation on this subject (5):

> If, for example, the introduction of new lighting fixtures into a factory is followed by an increase in production, the conclusion is drawn that the new lighting conditions are more efficient than the old. For some reason this conclusion is accepted by many people who would reject as beneath consideration the analogous proposition that the faster an automobile engine runs the more efficient is its performance. Neither in man nor machine is it safe to equate efficiency with output; increased production cannot be taken to mean increased efficiency unless it can be demonstrated that the cost of work has not increased proportionately. (p. 908)

Many research attempts have been made to increase understanding of lighting/productivity interactions. Most of the work has been plagued with problems resulting from treating a complex problem simplistically. One classic study is reported to have demonstrated that changes in illuminance level affected the production rates of women who inspected parts, assembled relays, and wound coils at Western Electric's Hawthorne plant in Chicago. Though output did improve when lighting was increased, increases in performance were also registered when lighting was decreased. Innumerable reviews of the original work indicate that variables had been unwittingly confounded (27). A likely explanation for the observed performance improvements is that the workers, who considered the experimenters as authority figures, strived to improve their output every time special attention was paid to them via a change in their working environment, regardless of whether or not the change actually represented an "improvement" in working conditions.

Jacob, in his report on several studies performed in industrial settings in Germany, claims that increased productivity definitely results from increased illumination and reduced glare (19). These studies lack rigid scientific controls, however, and in no case was the lighting returned to its prestudy condition to investigate if productivity returned to the lower level.

Though more rigorous, the research by Barnaby increases only slightly the evidence for the link between illuminance level and performance (2). This is because there are significant differences other than illuminance level between the test conditions and the environment for which inferences of increased performance are made. Barnaby claims to have compared the effect of three different illuminance levels on productivity in an insurance company's corporate office. In fact, he analyzed a laboratory number-matching task at those light levels and extrapolated the data to the real world. Simulating the task rather than observing the real task was done to "minimize psychological and physical differences" and the "build-up of fatigue, boredom, and related motivational changes resulting from working continuously for long periods of time." These are precisely the items that are likely to interact with the luminous environment if the hypothesis is correct that the visual performance aspect is not the sole influence of lighting on productivity.

Thus we see that the question of lighting and productivity is extremely complex. Unraveling the mystery is one of the greatest challenges the lighting industry faces for the remainder of the twentieth century and beyond.

REFERENCES

1. Aspinall, P. A., and J. Dewar, "Lighting and Perceived Guilt." *Lighting Research & Technology*, Vol. 12, No. 3 (1980).
2. Barnaby, J. F., "Lighting for Productivity Gains." *Lighting Design + Application*, Vol. 10, No. 2 (February 1980).
3. Belcher, M. C., "Cultural Aspects of Illuminance Levels." *Lighting Design + Application*, Vol. 15, No. 2 (February 1985), p. 49.
4. Belcher, M. C., and R. M. Kluczny, "The Possible Effects of Light on Decision Making: Some Experimental Results." Commission Internationale de l'Eclairage, *Proceedings of the 21st Session*, Venice (June 1987).
5. Bitterman, M. E., "Lighting and Visual Efficiency: The Present Status of Research." *Illuminating Engineering*, Vol. 43, No. 8 (September 1948).
6. Butler, D. L., and P. M. Biner, "Preferred Lighting Levels: Variability Among Settings, Behaviors, and Individuals." *Environment and Behavior*, Vol. 19, No. 6 (1987), p. 695.
7. Champness, B., M. Hyland, and K. Oliver, "Effects of Fluorescent Lighting on Social Behaviour and Judgements." *Architectural Psychology Newsletter*, Vol. 8, No. 1 (1978).
8. Flynn, J. E., "A Study of Subjective Responses to Low Energy and Nonuniform Lighting Systems." *Lighting Design + Application*, Vol. 7, No. 2 (February 1977).
9. Flynn, J. E., "A Guide to Methodology Procedures for Measuring Subjective Impressions in Lighting." *Journal of the Illuminating Engineering Society*, Vol. 9, No. 1 (January 1979), p. 95.
10. Flynn, J. E., T. J. Spencer, O. Martyniuk, and C. Hendrick, "Interim Study of Procedures for Investigating the Effect of Light on Impression and Behavior." *Journal of the Illuminating Engineering Society*, Vol. 3 (October 1973), p. 87.
11. Gardner, M., "Dermo-optical Perception: A Peak Down the Nose," *Science*, Vol. 151 (1966), p. 654.
12. Glass, P., G. B. Avery, K. N. S. Subramanian, M. P. Keys, A. M. Sostek, and D. S.

Friendly, "Effect of Bright Light in the Hospital Nursery on the Incidence of Retinopathy of Prematurity." *The New England Journal of Medicine*, Vol. 313, No. 7 (August 15, 1985), p. 401.

13. Goldstein, K., "Some Experimental Observations Concerning the Influence of Colors on the Function of the Organism." *Occupational Therapy and Rehabilitation*, Vol. 21 (1942), p. 147.
14. Greene, T. C., and P. A. Bell, "Additional Considerations Concerning the Effects of 'Warm' and 'Cool' Wall Colours on Energy Conservation." *Ergonomics*, Vol. 23, No. 10 (1980), p. 949.
15. Harvey, L. O., D. L. DiLaura, and R. G. Mistrick, "Quantifying Reactions of Visual Display Operators to Indirect Lighting." *Journal of the Illuminating Engineering Society*, Vol. 14, No. 1 (1984).
16. Hollwich, F., B. Dieckhues, and B. Schrameyer, "Die Wirkung des Natürlichen und Künstlichen Lichtes über das Auge auf den Hormonund Stoffwechselhaushalt des Menschen." *Klin Mbl Augenheilk*, Vol. 171 (1977), p. 98.
17. Hopkinson, R. G., and J. B. Collins, *The Ergonomics of Lighting*. London: Macdonald, 1970.
18. Ivanov, A., "Soviet Experiments in 'Eye-less Vision'." *International Journal of Parapsychology*, Vol. 6 (1964), p. 5.
19. Jacob, E., "Lighting and Productivity." *International Lighting Review* (1966), p. 103.
20. Jacobsen, F. M., T. A. Wehr, D. A. Sack, S. P. James, and N. E. Rosenthal, "Seasonal Affective Disorder: A Review of the Syndrome and Its Public Health Implications." *AJPH*, Vol. 77, No. 1 (January 1987).
21. Kearney, G. E., "Hue Preferences as a Function of Ambient Temperature." *Australian Journal of Psychology*, Vol. 18, No. 3 (1966), p. 271.
22. Küller, R., *Non-Visual Effects of Light and Colour: Annotated Bibliography*. Lund: Swedish Council for Building Research, 1981.
23. Matulionis, R. C., "Effects of Achromatic Luminance Intensities and Contrasts [sic] on Human Performance." *Man-Environment Systems*, Vol. 8, No. 4 (1978), p. 159.
24. Neer, R. M., T. R. A. Davis, A. Walcott, S. Koski, P. Schepis, L. Thorington, and R. J. Wurtman, "Stimulation by Artificial Lighting of Calcium Absorption in Elderly Human Subjects." *Nature*, Vol. 229 (1971), p. 255.
25. Nelson, T. M., T. H. Nilsson, and M. Johnson, "Interaction of Temperature, Illuminance and Apparent Time on Sedentary Work Fatigue." *Ergonomics*, Vol. 27, No. 1 (January 1984), p. 89.
26. Page, R. A., and M. K. Moss, "Environmental Influences on Aggression: The Effects of Darkness and Proximity of Victim." *Journal of Applied Social Psychology*, Vol. 6, No. 2 (1976), p. 126.
27. Parsons, H. M., "What Happened at Hawthorne?" *Science*, Vol. 183 (March 1974), p. 922.
28. Rea, M. S., "Behavioral Responses to a Flexible Desk Luminaire." *Journal of the Illuminating Engineering Society*, Vol. 13, No. 1 (October 1983), p. 174.
29. Rey, P., and J.-P. Rey, "Les Effets Comparés de Deux Eclairages Fluorescents sur une Tâche Visuelle et des Tests de 'Fatigue'." *Ergonomics*, Vol. 6, No. 4 (1963), p. 393.

30. Sanders, M., J. Gustanski, and M. Lawton, "Effect of Ambient Illumination on Noise Level of Groups." *Journal of Applied Psychology*, Vol. 59, No. 4 (1974), p. 527.

31. Smith, S. W., and M. S. Rea, "Relationships Between Office Task Performance and Ratings of Feelings and Task Evaluations Under Different Light Sources and Levels." Commission Internationale de l'Eclairage, *Proceedings of the 19th Session*, Kyoto (1979).

32. Stark, G. M., D. M. Saunders, and P. E. Wookey, "Differential Effects of Red and Blue Coloured Lighting on Gambling Behavior." *Current Psychological Research*, Vol. 2, No. 2 (April–June 1982), p. 95.

33. Stone, P. T., S. D. P. Harker, W. Z. Coley, H. R. Rutt, and G. Jones. "Light Endocrine Mechanisms and Stress," *LUTERG Report No. 156*, Department of Human Sciences, University of Technology, Loughborough, U.K.

34. Taylor, L. H., and E. W. Sucov, "The Movement of People Toward Lights." *Journal of the Illuminating Engineering Society*, Vol. 3, No. 3 (April 1974), p. 237.

35. Thorington, L., L. Parascandola, and L. Cunningham, "Visual and Biologic Aspects of an Artificial Sunlight Illuminant." *Journal of the Illuminating Engineering Society*, Vol. 1, No. 1 (October 1971), p. 33.

36. Wurtman, R. J., "The Effects of Light on the Human Body." *Scientific American*, Vol. 233, No. 1 (1975), p. 68.

37. Wurtman, R. J., "The Effects of Light on Man and Other Mammals." *Annual Review of Psychology*, Vol. 37 (1975), p. 467.

6 luminaires

INTRODUCTION

A **luminaire** is the lighting equipment that houses the light source, electrical components, and light control method. This chapter describes the various means of redirecting and controlling the light from a luminaire. The measurement of the light output and the calculations required to develop a report (photometric test report) that contains important design parameters will be discussed. The factors that affect in-place performance will be presented as well as cost considerations.

LIGHT CONTROL

The light sources discussed in Chapter 4 provide us with a source of **luminous flux.** This luminous flux is not necessarily directed where it will do the most good. For example, a bare incandescent lamp radiates in all directions. To control or redirect the light to where we want it, materials are used that have one or more of three basic properties: transparency, translucency, or opacity. Transparent materials will pass or transmit most of the light, but will refract it as it is transmitted. Opaque materials, on the other hand, will transmit no light; they reflect and absorb all the luminous energy.

When light strikes a material that has any one of the foregoing properties, it can be reflected, transmitted, refracted, absorbed, or polarized. One or more of these phenomena may occur simultaneously.

Reflection

The first method for controlling light, **reflection,** occurs in many forms: regular or specular reflection, spread reflection, diffuse reflection, mixed reflection, scattered reflection, and selective reflection (see Fig. 6-1).

1. **Regular** or **specular reflection** is the type that occurs in mirrors or highly polished metals. In such cases, the angle of reflection equals the angle of incidence which is called the **law of regular reflection.**

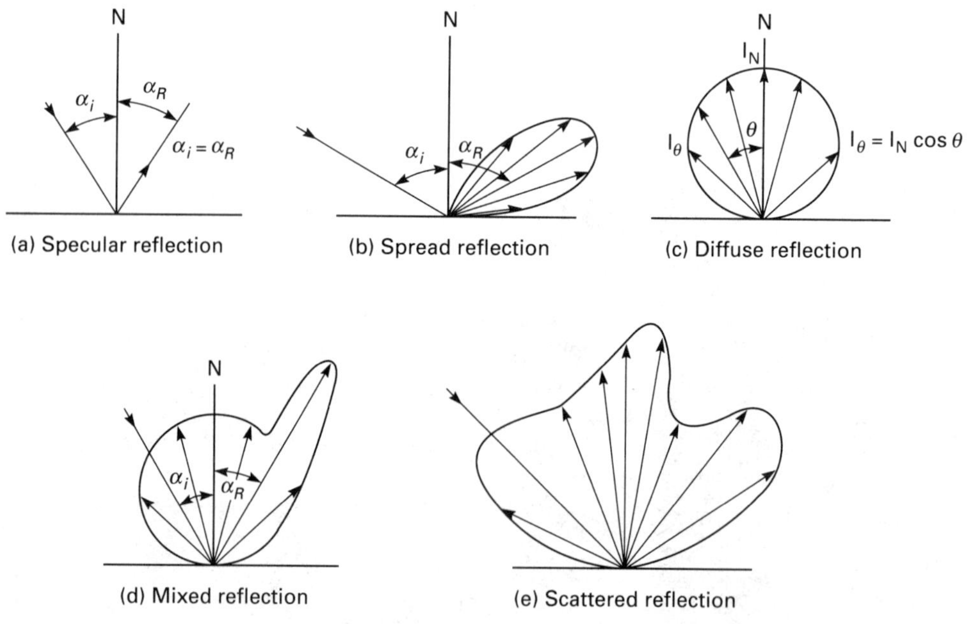

Figure 6-1. Reflection Patterns.

2. **Spread reflection** is similar to specular reflection, except the light is scattered about the ray at the angle of reflection.
3. **Diffuse reflection** is exhibited by matte white paper or flat white paint. A **perfect diffuser** would obey **Lambert's Cosine Law,** which states that the intensity reflected at angle θ is equal to the intensity reflected normal to the surface times the cosine of the angle θ between the normal ray and the reflecting ray of light.

   $$\text{Lambert's Cosine Law:} \quad I_\theta = I_N \times \cos\theta$$

   The reflected pattern from a perfect diffuser takes the form of a solid sphere called a **tangent sphere.** The reflected pattern then is not a function of the angle of incidence. This is shown in Figure 6-1.
4. **Mixed reflection** is a combination of specular, spread, and diffuse reflection. Mixed reflection probably best describes real materials.
5. **Scattered reflection** is that reflection which cannot be associated with either Lambert's cosine law or the law of regular reflection. Figure 6-1 shows a possible reflection pattern.
6. **Selective reflection** gives objects their color by the phenomena called **selective absorption** (see Chapter 3). Certain wavelengths of incident radiation are absorbed by the surface while others are reflected. The selection of which wavelength to reradiate determines the color of the object. Selective reflection can have any of the reflection patterns previously mentioned.

Control by Reflection

Generally, specular reflection is used to control light. There are four specific contours used in reflective light control: the **flat plane,** the **circle,** the **parabola,** and the **ellipse.** These basic shapes can be configured in numerous ways to direct light. For a flat plane, all incident rays are reflected at an angle equal to the angle of incidence (law of regular reflection). If the incident rays are parallel, the reflected rays will be parallel. For a circular reflector with the source at the center of the circle **(focal point),** all reflected rays return through the center. If the source is placed ahead of the center, the reflected rays will diverge as in Figure 6-2. Probably the best known reflector shape is the parabola. If the light source is placed at the focal point of a parabolic reflector, all reflected rays will be parallel (Fig. 6-2). This is the principle used in a search light to obtain a beam of parallel light rays. The final shape is the ellipse. With an ellipsoidal reflector, a source at the first focal point of the ellipse causes the reflected rays to pass through the second focal point. This principle is used in the pinhole spot from which a wide pool of light is desired from a luminaire with a small ceiling opening or aperture.

Transmission

Transmission of light through specific materials occurs in patterns similar to light reflected off materials. The types of transmission are regular, spread, diffuse, mixed, scattered, and selective. Each type produces a distribution pattern which corresponds to the reflected

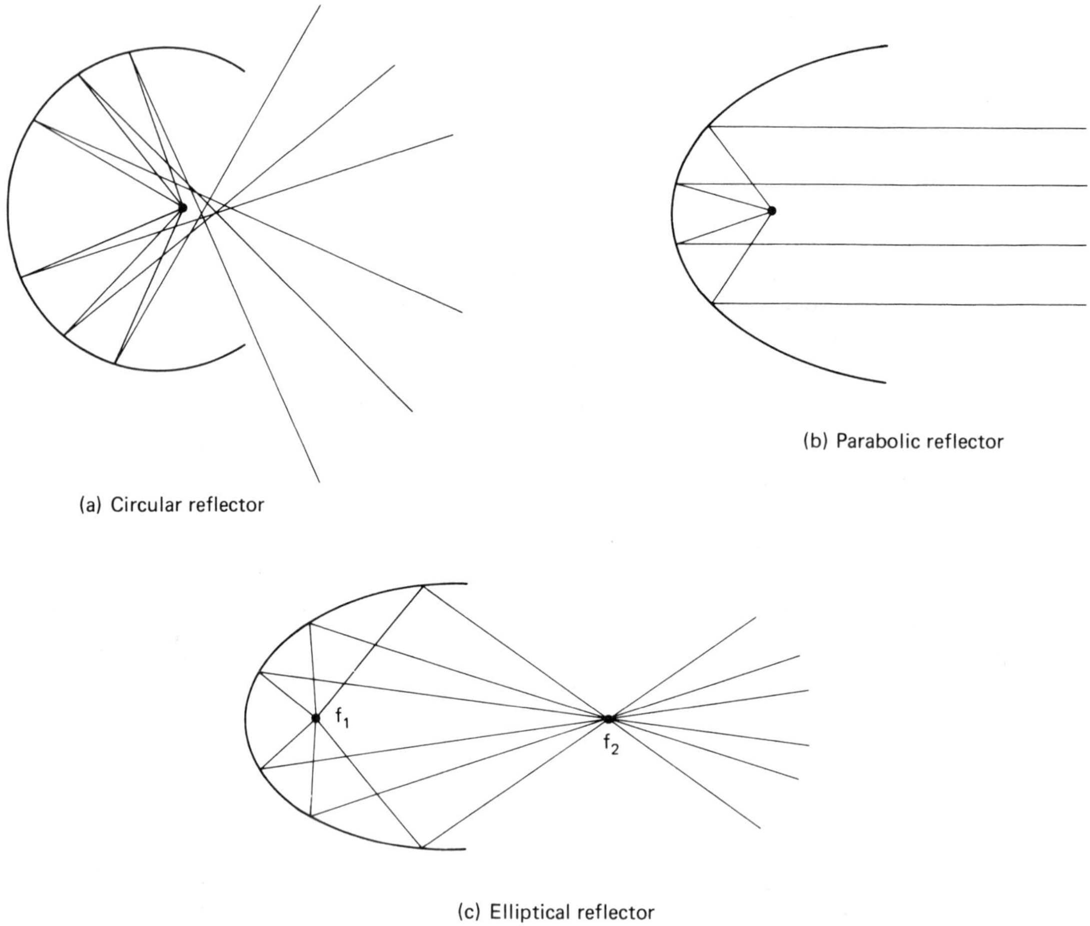

Figure 6-2. Reflector Shapes.

patterns described earlier (see Fig. 6-1), except on the side of the surface opposite the incident ray.

Refraction

Snell discovered that as light travels from one media of a given density to another media of a higher density, the light ray's speed changes slightly. As a consequence, the light ray is bent. This is known as **refraction** and follows Snell's law,

$$n_1 \times \sin \alpha_1 = n_2 \times \sin \alpha_2$$

where n is the **index of refraction** and α is the angle between the normal (perpendicular) to the surface and the light ray (Fig. 6-3). As the light ray travels from a less dense

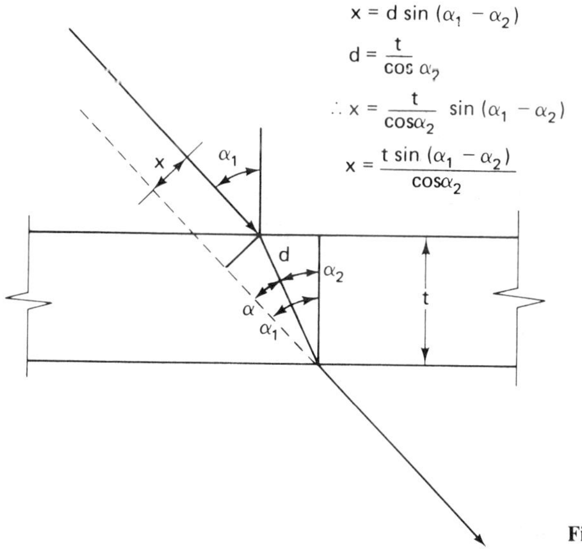

Figure 6-3. Refraction.

medium (air) through a more dense medium (glass), the light ray is bent toward the normal to the surface. As the light ray leaves the dense medium and re-enters the less dense medium (air), the ray is bent away from the normal. The emerging ray of light will be parallel to the entering ray of light but displaced a distance x. The displacement (x) can be calculated according to the formula derived in Figure 6-3. The phenomenon of refraction (2, 5) is used in flat plates, lenses, and prisms to bend light rays.

Absorption

All materials, whether they transmit or reflect light, absorb some of the energy. Absorption can be either **selective** or **nonselective.** Selective absorption, as was stated earlier, is what causes objects to have a certain color. Non-selective absorption does not affect the color appearance of the material. All wavelengths are absorbed equally in nonselective absorption, causing a change in the luminance of the material. An example of nonselective absorption in a transmitting material is a neutral density filter, which is used to change the intensity of a light ray without changing its wavelength composition.

Polarization

Light is composed of waves which are oscillating in an infinite number of planes that are perpendicular to the direction of propagation. Certain materials reflect or transmit only a portion of those oscillating waves. This light is said to be **polarized.** Figure 6-4 is a schematic representation of the polarization of light being transmitted through a polarizing material. When unpolarized light passes through a polarizing material (see Fig. 6-4), such as that found in many sunglasses, only certain waves are transmitted.

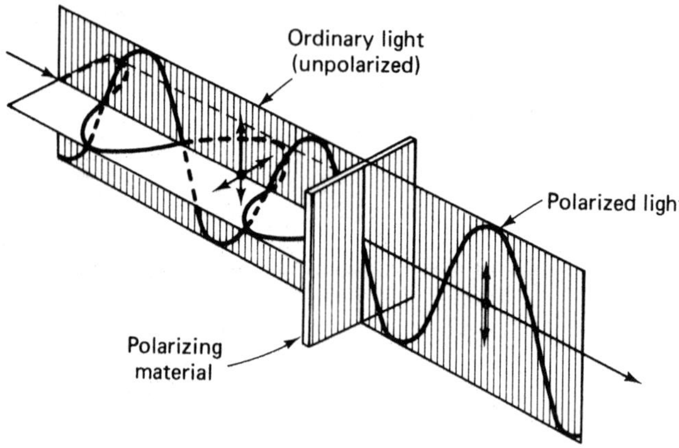

Figure 6-4. Polarization of Light.

Those that are blocked depend on the orientation of the polarizer. Most materials exhibit polarizing characteristics. Ordinary glass will block certain waves reflected from the surface. Maximum blocking for ordinary glass occurs at **Brewster's angle,** which is 57° from a normal to the glass surface (see Fig. 6-5).

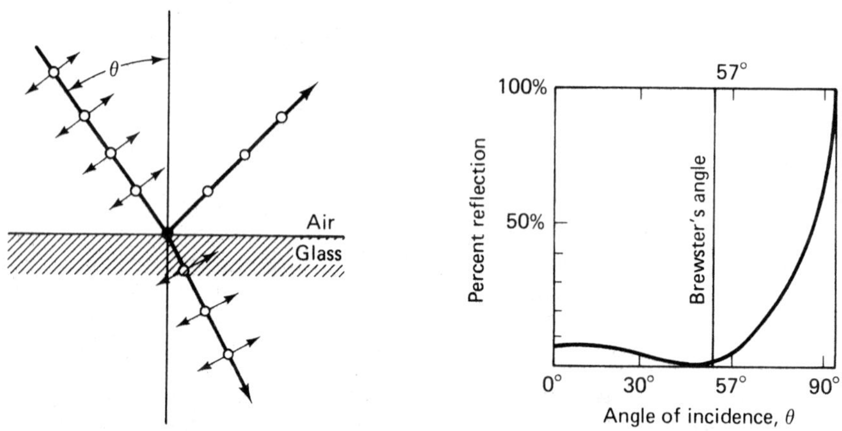

Figure 6-5. Brewster's Angle.

LUMINAIRE PHOTOMETRICS

Once a light source has been chosen and the hardware designed to direct the light, luminaire performance must be ascertained. A **photometric data sheet** is the basic technical information used to evaluate luminaire performance, to evaluate the quality of the lighting, and to determine the quantity of illumination. To create a photometric data sheet, a photometric testing laboratory measures two basic quantities which are **luminous intensity** and the **maximum luminaire luminance.** These two basic measurements are used to generate the data reported on the photometrics test. A **photometric test report** contains such information as

1. Luminaire description, which includes physical data on the luminaire
2. The luminous intensity data and distribution curves
3. Luminaire output data (zonal lumens, luminaire efficiency, and CIE/IES Luminaire Classification)
4. Luminaire luminance data (average luminance, maximum luminance, and maximum-to-average luminaire luminance ratio)
5. Coefficients of Utilization (CU)
6. Maximum spacing criterion (old terminology is spacing-to-mounting height ratio).

Luminaire Description

The **luminaire description** given on the photometric test report will usually consist of the nominal dimension, mounting or type, and control media.

Nominal Dimension

The description usually will begin with the nominal dimension of the luminaire (2 × 4, 1 × 4, 16 × 48, 2 × 2, 3 × 3) or the aperture size (8 in., 12 in.).

Mounting

1. Ceiling-mounted luminaires:
   a. Recessed unit (''troffer'' is only applied to recessed fluorescents). Recessed means that the majority of the equipment is above the ceiling plane.
   b. Surface-mounted unit:
      (1) Metal-sided or a direct unit—has metal on all four sides, and it is mounted to the ceiling.
      (2) Prismatic wraparound unit has plastic on the longitudinal sides, a prismatic bottom, and the ends may be plastic or metal.
      (3) Open bottom means the unit has louvers, cells, or no optical control under the lamps (industrial).
   c. Suspended from ceiling:
      (1) Metal sided.
      (2) Prismatic wraparound.
      (3) Open bottom.
2. Wall-mounted luminaires.
3. Floor-mounted luminaires: light columns—indirect.

Control Media

1. Acrylic lens: prismatic, translucent, patterned
2. Glass lens: fresnel, translucent, transparent, patterned
3. Louver or cells (small or large)
4. Reflectors

Test Number

The test number is assigned to a luminaire by the testing laboratory that performed the photometric test.

Shielding Angle

The shielding angle is the maximum angle, measured from **nadir,** at which the lamp or luminous element can be seen.

Bare Lamp

Bare lamp is the luminance of the bare lamp measured in candela per square foot (cd/ft^2) or in footLamberts (fL).

Luminous Intensity

The most important data used in lighting analysis and design are the **luminous intensity data.** The equipment used to measure luminous intensity from a luminaire is called a **goniophotometer.** A straight-arm goniophotometer consists of a photocell that is rotated around the light center of the test luminaire (see Fig. 6-6). As the photocell rotates

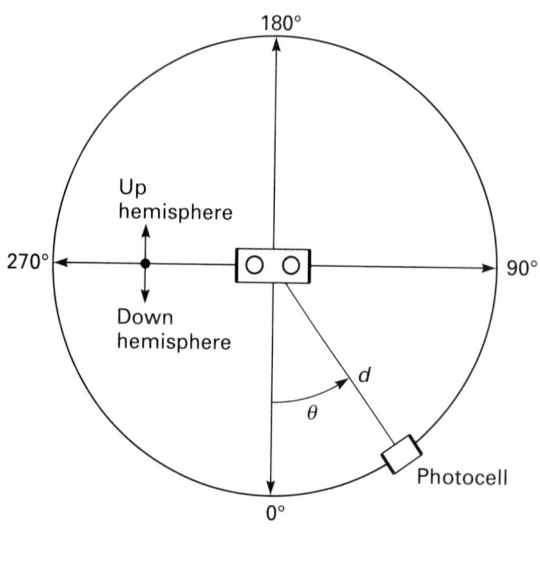

Figure 6-6. Goniophotometer Terminology.

360° around the luminaire, at a fixed test distance, the photocell sweep defines a vertical plane. The photocell may pause to make luminous intensity measurements at prescribed vertical angles called the **elevation angle,** which is denoted by the Greek letter θ. If the luminaire distribution is not symmetrical, the luminaire is rotated in the horizontal plane and the photocell is rotated 360° in the next vertical plane. The horizontal rotation is specified as a **horizontal angle** or **azimuthal angle,** which is denoted by the Greek letter α. This means that the luminous intensity, *I*, is a function of two angles. The vertical or elevation angle is theta, θ, which is measured from nadir (0°) and the horizontal or azimuthal angle is alpha, α. The vertical rotation of the photocell and the horizontal rotation of the luminaire defines a spherical space the luminaire is located inside. Calibration of the photocell is based on the assumption that the source of light (luminaire) is a **point source.** For a luminaire to classify as a point source, the test distance must be equal to or greater than five times the maximum dimension of the source of light.

The luminous intensity distribution represents the magnitude of luminous intensity from a luminaire in a particular direction. Figure 6-7 are plots of the luminous intensity distributions for the two α planes (0° and 90°) that are perpendicular to each other. The angle straight under the luminaire is the elevation angle, θ = 0°, which is also called the **nadir angle.** The center of the plot is the light center of the luminaire. A luminous intensity distribution curve connects the heads of all the intensity vectors to form a distribution solid. The radial intensity vectors are plotted on polar coordinate graph paper for each measured vertical angle, θ. A separate luminous intensity distribution curve is plotted for each vertical plane or azimuthal angle, α, which is the horizontal angle between vertical planes.

Ideally, the luminous intensity values at an infinite number of vertical angles, θ, and vertical planes, α, should be measured. Since this is impractical, a limited number of vertical angles and vertical planes are measured.

Vertical Angle, Elevation Angle, θ

For fluorescent luminaires, the *minimum* standard testing procedure requires the rotation of a photocell around a luminaire (Fig. 6-8) taking a reading at 0° (nadir), then at 5° continuing in maximum of 10° vertical steps at 15°, 25°, . . ., 75°, 85° then at 90° on to 95°, 105°, . . ., 175°, then 180°. Preferably, vertical angles should be at 2.5° steps.

Horizontal Angle (Azimuthal Angle), α

For a linear or nonsymmetrical fluorescent luminaire, the luminous intensity will vary in each vertical plane (horizontal angle), while a symmetrical luminaire, such as an incandescent downlight, will have the same luminous intensity in all vertical planes.

The vertical plane through the center of the luminaire which is parallel to the axis of the lamps (plan view, see Fig. 6-9) is called the **parallel** (or // or **along** or **lengthwise** or **0°**) **plane.** The vertical plane through the center of the luminaire which is perpendicular to the axis of the lamps (see Figures 6-7 and 6-9) is called the **perpendicular** (or ⊥ or **across** or **crosswise** or **90°**) **plane.**

For fluorescent luminaires, the old testing procedure required the testing of a

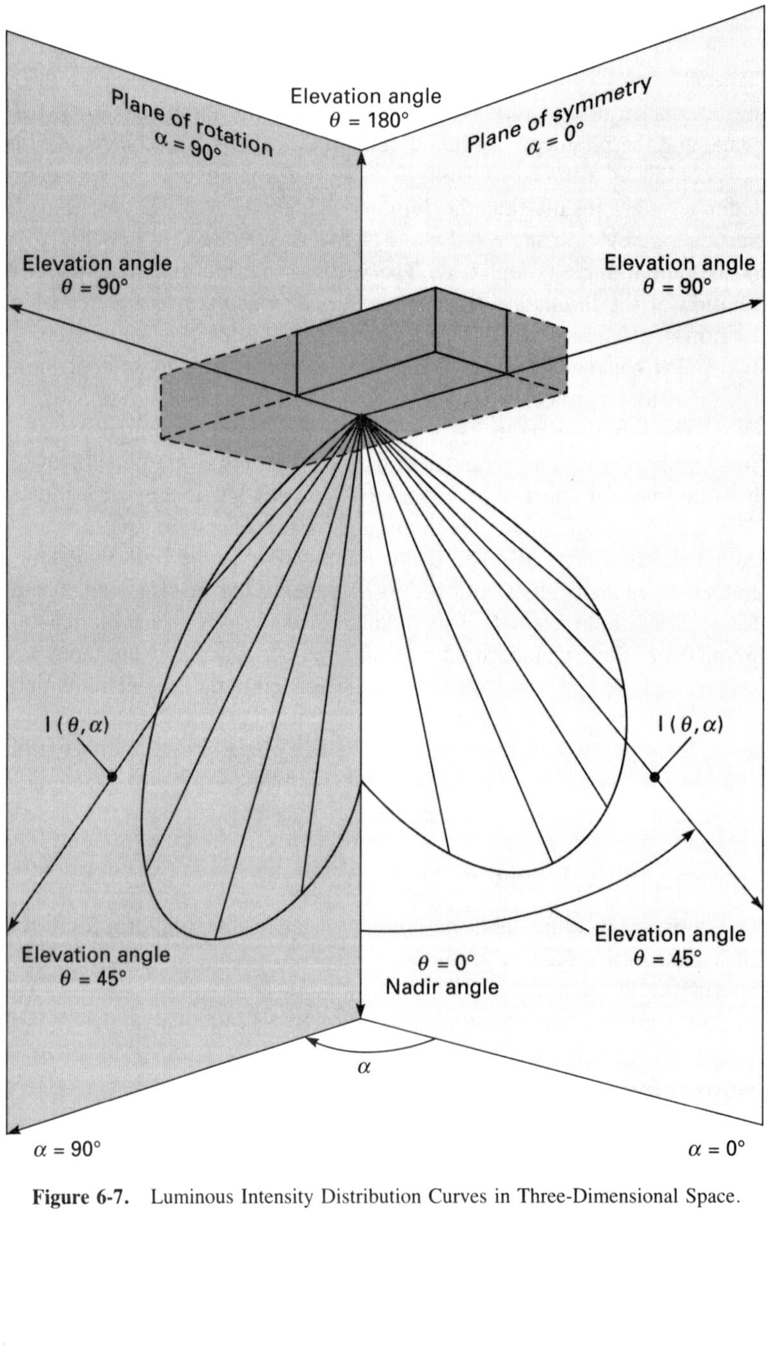

Figure 6-7. Luminous Intensity Distribution Curves in Three-Dimensional Space.

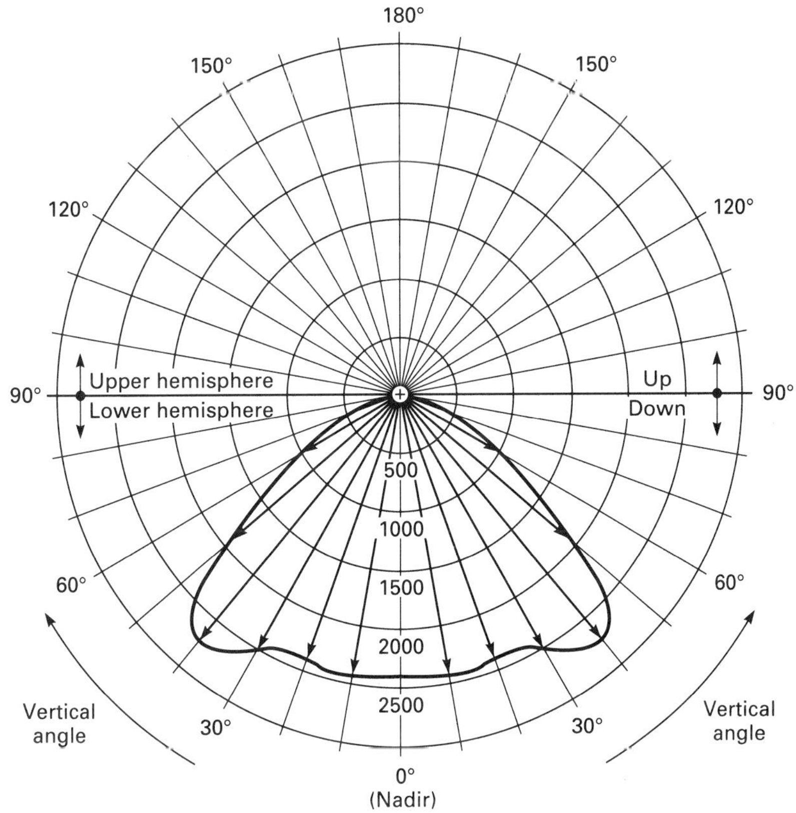

Figure 6-8. Luminous Intensity Distribution Curve.

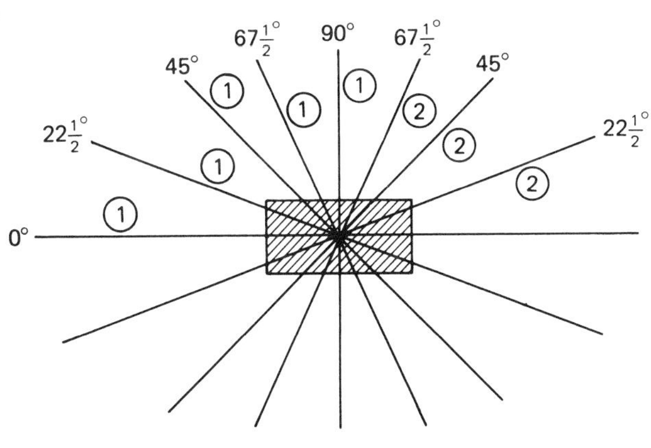

Figure 6-9. Vertical Planes Through the Luminaire (Five-Plane Photometry).

minimum of three vertical planes while today's testing procedure requires a minimum of five vertical planes.

1. Three-plane photometry:
   a. 0°, along, parallel, lengthwise
   b. 45°, two planes (left and right)
   c. 90°, across, perpendicular, crosswise

2. Five-plane photometry: (see Fig. 6-9).
   a. 0°, along, parallel, lengthwise
   b. 22½°, two planes (left and right)
   c. 45°, two planes (left and right)
   d. 67½°, two planes (left and right)
   e. 90°, across, perpendicular, crosswise

Tabulated Luminous Intensity

The luminous intensity values listed represent the tabulated values recorded during the photometer test. The tabulated values should always be used rather than picking values from the luminous intensity diagram. The angles across the top of the tabulated luminous intensity in Table 6-10 (see Data Section) are the horizontal angles, α while the angles down the side are the vertical angles, θ.

Application of Luminous Intensity Data

The measured luminous intensity data is used to

1. Evaluate the general distribution characteristics
2. Calculate the **average luminous intensity** which is used to calculate the following data:
   a. Zonal lumens
   b. Luminaire efficiency
   c. CIE/IES luminaire classification
   d. Coefficient of utilization (CU)

3. Calculate the **average luminaire luminance** which is used to calculate the following data:
   a. Maximum to average luminaire luminance ratio
   b. Disability glare

4. Point-by-point calculations of the following quantities:
   a. Illumination at a point
   b. Visual comfort probability (VCP)
   c. Equivalent sphere illumination (ESI)
   d. Luminaire spacing criterion

Output Data

Average Luminous Intensity

The luminaire is assumed to be placed in the center of a unit sphere. The unit sphere is divided into horizontal slices (Fig. 6-10) or zones that are consistent with the way in which the luminous intensity data are recorded. The angles θ_1 and θ_2 define the width of a zone. The I_θ value in each vertical plane is measured halfway between θ_1 and θ_2. I_θ is referred to as the midzonal luminous intensity (candlepower) value. The midzonal luminous intensity values in each plane for a given zone are averaged together to obtain the average luminous intensity (average midzonal candlepower). For example, the equation to calculate the average luminous intensity in the zone between 40° and 50° of a luminaire with five-plane photometry would be

$$I_{55\text{ave}} = \frac{I_0 + (2 \times I_{22.5}) + (2 \times I_{45}) + (2 \times I_{67.5}) + I_{90}}{8}$$

Note in Figure 6-9 that there is only one 0° plane and one 90° plane, while there are two planes each at 22.5°, 45°, and 67.5°. Therefore, the denominator is *8* for five-plane photometry rather than 5. For three-plane photometry, the denominator will be

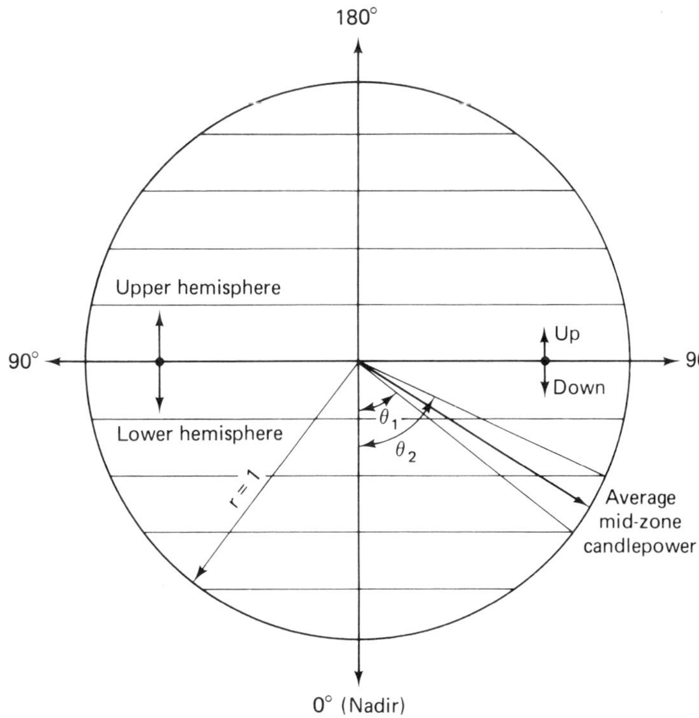

Figure 6-10. Unit Sphere for Calculating Zonal Constants.

4, that is, one 0° plane, one 90° plane, and two 45° planes. Using the luminous intensity values given in Table 6-10 in the Data Section, the average luminous intensity value for $\theta = 5°$ is calculated as follows:

$$I_{5\text{ave}} = \frac{1388 + (2 \times 1390) + (2 \times 1393) + (2 \times 1385) + 1391}{8}$$

$$= 1389$$

The average luminous intensity is calculated in a similar fashion for the remaining midzonal angles (see Table 6-1).

TABLE 6-1 Average Luminous Intensity for Luminaire Table 6-10 (Data Section)

Zone	Midzonal angle	Average intensity	Zone	Midzonal angle	Average intensity
0–10	5	1389	90–100	95	106
10–20	15	1353	100–110	105	116
20–30	25	1285	110–120	115	116
30–40	35	1159	120–130	125	115
40–50	45	937	130–140	135	96
50–60	55	615	140–150	145	75
60–70	65	365	150–160	155	63
70–80	75	221	160–170	165	49
80–90	85	135	170–180	175	39

Zonal Lumens

Zonal lumens are calculated to show the total amount and direction of the light output (lumens) radiated from a luminaire in terms of the lumens produced on each zone or horizontal strip on the surface (Fig. 6-10) of the unit sphere. The surface area of each horizontal strip can be calculated by

$$A_s = 2 \times \pi\ (\cos \theta_1 - \cos \theta_2)$$

The area, A_s, of each horizontal strip is known as the **zonal constant** or **lumen constant** for the zone between θ_1 and θ_2. The zonal constants for 5° and 10° zones can be found in Table 6-11 in the Data Section.

From the definition of luminous intensity and solid angle for a unit sphere with a radius, $r = 1$, we obtain the equation for calculating the luminous flux on a zone **(zonal lumens)** whose strip area on the surface of a unit sphere is A_s

$$I = \frac{d\Phi}{d\omega} \quad \text{and} \quad \omega = \frac{A_s}{r^2} = \frac{A_s}{(1)^2} = A_s$$

zonal lumens: $$\Phi = I \times A_s$$

Therefore, to calculate the luminous flux (lumens), Φ, on a zone, we need the area of the zone (A_s) which is the zonal constant (ZC) (Table 6-11) and the average luminous

TABLE 6-2 Zonal Lumens for the Luminaire in Table 6-10

Midzone angle	I_{ave}	Zonal constant	Zonal lumens	Midzone angle	I_{ave}	Zonal constant	Zonal lumens
5	1389	0.095	132	95	106	1.091	116
15	1353	0.283	383	105	116	1.058	123
25	1285	0.463	595	115	116	0.993	115
35	1159	0.628	728	125	115	0.897	103
45	937	0.774	725	135	96	0.774	74
55	615	0.897	552	145	75	0.628	47
65	365	0.993	362	155	63	0.463	29
75	221	1.058	234	165	49	0.283	14
85	135	1.091	146	175	39	0.095	4

intensity (I_{ave}) at the midzonal angle (Table 6-1) to calculate the zonal lumens (ZL) as follows:

$$ZL = \text{average luminous intensity} \times ZC$$
$$ZL_5 = 1389 \times 0.095 = 132 \text{ lm}$$

The light output (lumens) leaving a luminaire is divided into the lumens produced in the **down hemisphere** (0°–90°) and the **up hemisphere** (90°–180°). If light leaves a luminaire at 90°, it will be traveling parallel to the floor or ground.

An interior photometric test reports the down hemisphere data in two or three larger zones which represent the sum of the zonal lumens in the larger zones.

down hemisphere: 0–40°, indirect glare zone
40–90°, direct glare zone

All the strip zones in the up hemisphere are summed to give the total lumens in the up hemisphere.

up hemisphere: 90°–180°

The sum of all the zones from 0° to 180° is reported as the **total output.**

The zonal lumens from Table 6-2 are summed for the two larger zones in the down hemisphere and for all of the up hemisphere and given in Table 6-3.

TABLE 6-3 Lumen Summation Table for Table 6-2

Zone degrees	Sum lumens		
0–40	1838	Indirect glare zone	Down hemisphere
40–90	2019	Direct glare zone	Down hemisphere
90–180	625	Up hemisphere	
0–180	4482	Total output	

Luminaire Efficiency

The total lamp lumens represent the percentage of the total bare lamp lumens that "get out" of the luminaire which is also the **Luminaire Efficiency.** The luminaire efficiency (Table 6-4) is a ratio of the total amount of lumens leaving the luminaire to the total lumens produced by the "bare" lamps.

$$\%\text{ total lamp lumens} = \frac{\text{sum of zonal lumens}}{\text{total bare lamp lumens}} \times 100$$

TABLE 6-4 Percentage of Total Lamp Lumens for the Luminaire in Data Section (Figure 6-13)

Zone degrees	Equation	% Total lamp lumens
0–40	$\frac{1838}{2 \times 3110} \times 100$	29.5
40–90	$\frac{2019}{2 \times 3110} \times 100$	32.5
90–180	$\frac{625}{2 \times 3110} \times 100$	10.1
0–180	$\frac{4482}{2 \times 3110} \times 100$	72.1

The % total lamp lumens (0–180°) is called the luminaire efficiency, which is 72.1% for the luminaire in Figure 6-13. If 72.1% of the lumens gets out of the luminaire, then 27.9% of the lumens produced by the bare lamps is absorbed or trapped (lost) inside the luminaire.

CIE/IES Luminaire Classification

The % total distribution (Table 6-5) represents the percentage of lumens out of the luminaire in a zone to the total lumens that "get out" of the luminaire. To determine the CIE/IES classification, the percentage of total output in the up hemisphere and the down hemisphere must be determined.

$$\begin{array}{c}\%\text{ total distribution}\\ \text{in a zone}\end{array} = \frac{\text{lumens in zone}}{\text{total lumens }(0\text{–}180^\circ)} \times 100$$

The six categories (Fig. 6-11) of the CIE/IES Luminaire Classification are (1) direct, (2) semidirect, (3) direct-indirect, (4) general diffuse, (5) semi-indirect, and (6) indirect. The general diffuse category has a substantial horizontal component of luminous intensity, whereas the direct-indirect (not recognized by the CIE) has little or no horizontal component of luminous intensity.

For Table 6-5, the CIE/IES Luminaire Classification is semidirect.

TABLE 6-5 Percentage of Total Output for the Luminaire in Data Section (Figure 6-13)

Zone degrees	Equation	% Total lamp lumens	
0–40	$\frac{1838}{4482} \times 100$	41.0	86.1% Down hemisphere
40–90	$\frac{2019}{4482} \times 100$	45.1	
90–180	$\frac{625}{4482} \times 100$	13.9	(% Up hemisphere)
0–180	$\frac{4482}{4482} \times 100$	100.0	

Luminance Data

Maximum Luminaire Luminance

The maximum luminaire luminance (Figure 6-13, Data Section) is measured by scanning the surface of the lens of the luminaire searching for the maximum luminance at the **glare angles.** A luminance photometer with a circular aperture, whose area is one square inch, referred to as a ''circular square inch aperture,'' is used to search the surface of the lens. The maximum luminance is measured in the **along axis** and **across axis** at the glare angles (45°, 55°, 65°, 75°, 85°).

Average Luminaire Luminance

The average luminaire luminance both ''along'' and ''across'' is calculated for each of the glare angles utilizing the tabulated luminous intensity (''candlepower'') data (Table 6-10, Data Section) and the projected area of the luminous surface area. Examples of the different types of luminaires and what to use for the projected area are shown in Figure 6-12.

Projected Area. The projected area across (cross axis viewing) and along (lengthwise axis viewing) must be calculated. Table 6-6 shows the calculated projected areas (PA) utilizing the specific formulas for the luminaire in Table 6-10 of the Data Section.

Luminous Sides/Ends

across axis viewing: $PA = L \times [(h \times \sin\theta) + (W \times \cos\theta)]$

along axis viewing: $PA = W \times [h \times \sin\theta) + (L \times \cos\theta)]$

Metal Sides/Ends

$$PA = L \times W \times \cos\theta$$

CIE/IES luminaire classification	% Up light (90–180°)	% Down light (0–90°)	Typical distribution curve
Direct	0–10%	100–90%	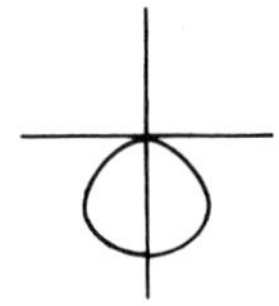
Semidirect	10–40%	90–60%	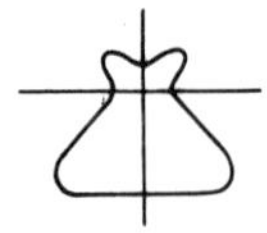
Direct-indirect*	40–60%	60–40%	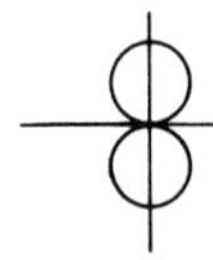 *
General diffuse	40–60%	60–40%	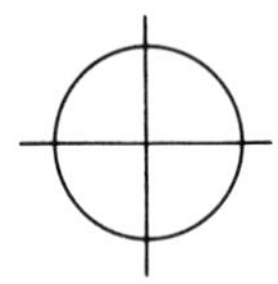
Semi-indirect	60–90%	40–10%	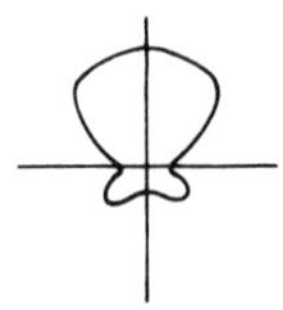
Indirect	90–100%	10–0%	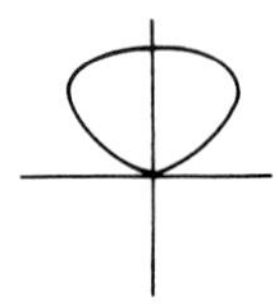

*IES classification only.

Figure 6-11. CIE/IES Classifications.

Average Luminance. The constant 144π is used to convert the average luminance in candelas per square inch (cd/in.^{2}) to footLamberts (fL) (see Chapter 1). To determine average luminaire luminance, the following equation is used:

$$\text{ave. luminance (cd/in.}^2) = \frac{\text{luminous intensity at } \theta \text{ (cd)}}{\text{projected area at } \theta \text{ (in.}^2)}$$

$$\text{ave. luminance (fL)} = \frac{\text{luminous intensity at } \theta \text{ (cd)}}{\text{projected area at } \theta \text{ (in.}^2\text{)}} \times 144\pi$$

$$L_{\text{ave}} = \frac{I_\theta}{PA_\theta} \times 144\pi$$

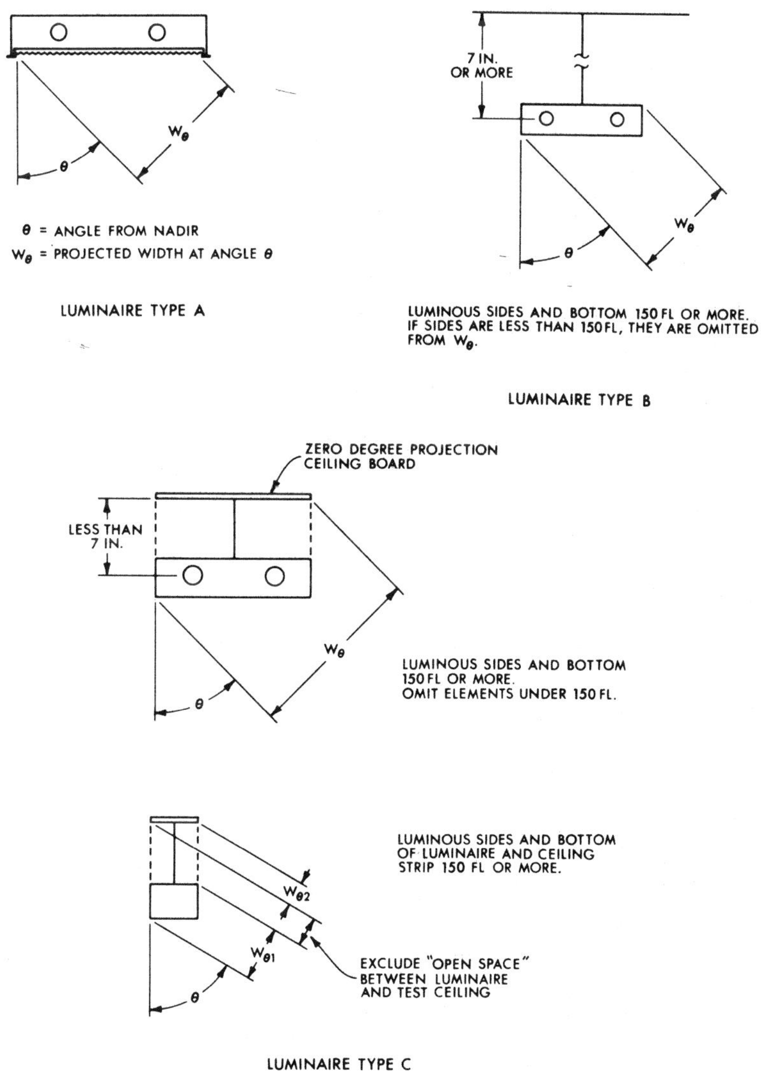

Figure 6-12. Projected Area. (Courtesy of the Illuminating Engineering Society of North America).

TABLE 6-6 Projected Area (in.²) for the Luminaire in Data Section (Table 6-10)

	Along (0°)			Across (90°)		
Angle	*PW*	*L*	*PA*	*PL*	*W*	*PA*
45	32.9	10	329	9.19	46.5	427.3
55	26.7	10	267	8.19	46.5	380.8
65	19.7	10	197	6.95	46.5	323.2
75	12.0	10	120	5.49	46.5	255.3
85	4.1	10	41	3.86	46.5	179.5

Luminance Data Table

Using the equation on page 174, the average luminance of the luminaire in Table 6-10 is tabulated in Table 6-7.

TABLE 6-7 Average Luminance, Maximum Luminance, and Maximum-to-Average Ratio (fL)

	Along (0°)					Across (90°)				
Angle	$I\theta$	*PA*	L_{ave}	L_{max}	*M/A*	$I\theta$	*PA*	L_{ave}	L_{max}	*M/A*
45	835	329	1148	2483	2.2	1018	427.3	1076	2721	2.5
55	537	267	909	1939	2.1	698	380.8	827	2211	2.7
65	295	197	678	1089	1.6	450	323.2	628	1701	2.7
75	142	120	533	850	1.6	317	255.3	560	1429	2.6
85	35	41	384	510	1.3	237	179.5	597	1463	2.5

Maximum to Average Luminaire Luminance Ratio:

The ratio of the maximum to average luminaire luminance is computed and used as an indicator of visual comfort (see Chapter 8). The maximum to average luminaire luminance values are given in Table 6-7 and shown on the photometric test report in Figure 6-13 in the Data Section.

Coefficient of Utilization

The coefficient of utilization (CU) is calculated from the average luminous intensity data. The CU value is used in the lumen method for calculating the average, uniform, horizontal illuminance. The values in a CU table (Figure 6-13, Data Section) are always listed as a whole number such as "72"; however, a CU value is *always* a decimal value such as "0.72."

Maximum Spacing Criterion

The **maximum spacing criterion** has replaced the older maximum spacing-to-mounting height ratio. Like the spacing-to-mounting height ratio (bottom of Figure 6-13, Data Section), the calculated maximum spacing criterion value is multiplied by the distance from the luminaire to the workplane to determine the maximum spacing of the luminaires. For the lumen method to be valid, the calculated maximum spacing can *not* be exceeded.

Standard Photometric Test Report

The previously calculated data are combined and printed as a standard photometric test report. An example of a complete photometer test report for the luminaire analyzed is given in Figure 6-13 in the Data Section. Additional photometric data are available in Helms (2) for working additional examples.

IES STANDARD FORMAT DATA FILES

The Illuminating Engineering Society of North America adopted a standard format for providing photometric data on computer disks. The primary data provided in the files

TABLE 6-8 IES Standard Format

```
*  <label line 1>        —each label line maximum 80 characters
*  <label line 2>
*  <label line 3>
         "
         "
         "
*  <label line n>
*  TILT = <file spec> or <INCLUDE> or <NONE>
*       <lamp to luminaire geometry>                        ┐
*       <# of pairs of angles and multiplying factors>      ├─ ONLY if
*       <angles>                                            │  TILT =
*       <multiplying factors>                               ┘  INCLUDE
*  < # lamps> <lms/lamp> <multiplier> <# vertical angles, θ>
   <# horizontal angles, α> <photometric type> <units>
   <width> <length> <height>
*  <ballast factor> <photometric factor> <input watts>
*  <listing of each vertical angle, θ separated by 2 spaces>
*  <listing of each horizontal angle, α with 2 spaces between>
*  <candela values for all θ's at 1st α angle>
*  <candela values for all θ's at 2nd α angle>
*  <candela values for all θ's at 3rd α angle>
*                  "
*                  "
*                  "
*  <candela values for all θ's at last α angle>
```

are the luminous intensity data. Other software is required to generate a "standard photometric test report" of the format shown in Figure 6-13, Data Section. The IES Standard file format is shown in Table 6-8. Each line marked with an asterisk "*" must begin a new line of data in the file. The word or phrase enclosed within the "<description>" refers to the actual data value or word (if capitalized) that is required. Data values should be separated by two blank spaces for easier viewing. All data should be saved in standard ASCII format.

LUMINAIRE SELECTION PARAMETERS

The section on luminaire photometrics was concerned with how the luminaire performs relative to its distribution of luminous energy. Equally important is how it performs mechanically. Such items as heat, noise, maintenance, and cost must be considered in the choice of a luminaire. The heat and noise produced by a luminaire are discussed in Chapter 7.

TABLE 6-9 Luminaire Selection Parameters

1. *Luminous intensity distribution*—the distribution influences the amount of light control.
2. *Amount of direct glare control*—heads-up; VCP, luminaire luminance ratios, direct glare zone (distribution).
3. *Amount of indirect glare control*—ESI: heads-down (horizontal tasks), indirect glare zone; VCP: heads-up (vertical tasks), direct glare zone.
4. *Easy maintenance*—availability of light sources and luminaire components.
5. *Mechanical construction*—strength, metal gage, thickness of lens, alignment, and rigidity.
6. *Safety*—certification: UL, CBM, NEMA, etc.
7. *Aesthetics*—appearance, appropriateness, scale, finish.
8. *Cost*—initial versus life-cycle cost; cost of lighting control system.
9. *Efficacy and luminaire efficiency.*
10. *Reliability of manufacturer*—backing up the product and reliable delivery of product.

Maintenance

Luminaires must be maintained periodically to keep them at peak performance. Maintenance includes changing lamps and cleaning the surfaces that collect dirt. To facilitate lamp changing, the luminaire should be constructed so that maintenance personnel can easily and rapidly reach the lamps. A latch and hinge arrangement is best because it avoids handling of the lens. The construction of the luminaire should be examined in terms of its rigidity, strength, and durability. These factors will determine the long-term usefulness of the luminaires. The design of the luminaire will affect the infiltration and movement of dirt inside the luminaire, which results in dirt accumulation and light loss. Any type of luminaire must be periodically cleaned. The time between cleanings can be lengthened if the luminaire is tightly constructed and designed to exclude airborne dirt; however, surface dirt accumulation will still occur and require cleaning.

A good maintenance program plays a key role in energy conservation. Lighting calculations take light losses, due to poor maintenance programs, into consideration. The net result is an increase in the number of luminaires to compensate for light loss due to inadequate maintenance. An increase in the number of luminaires means an increase in the watts for lighting, which is counterproductive to energy conservation. Poor maintenance of surfaces in the space results in the loss of luminous energy, which is absorbed by elements in the environment. Good maintenance of the luminaire and environment will reduce operating cost and conserve energy.

Cost

Good light control and sound construction are paramount for a lighting system to perform properly for many years. Like any item of quality, a luminaire that has the best operating and maintenance characteristics is usually the most expensive. Cost will be discussed in Chapter 14.

Additional Selection Factors

Numerous additional factors, such as comfort, appearance, light source, and application, are involved in the selection of a luminaire. Multiply this by the number of manufacturers and one has many thousands of luminaires from which to choose. A system proposed by Helms (1) would computerize the factors involved, and selection of luminaires could be made based on parameters input into the selection program by the designer.

It is important to realize that slight changes in the construction of the reflector, a change in lens material, a change in the physical size of the light source, or a change in the location of the light center of the source will result in a change in the luminous intensity distribution. This means that the efficiency, average luminance, maximum luminance, and the coefficient of utilization will change, even though the outward appearance of the luminaire has not changed significantly.

REFERENCES

1. Helms, R. N., ''Luminaire Selection and Design—An Alternative to Manufacturer's Catalogs.'' *Lighting Design + Application* (August 1974), p. 26.
2. Helms, R. N. *Illuminating Engineering Workbook*, 3rd ed. Lawrence: University of Kansas Publication, 1990.
3. Kaufman, John E., and Jack F. Christensen, eds. *IES Lighting Handbook*, Reference Volume, 1984. New York: Illuminating Engineering Society of North America, 1984.
4. Kingslake, R. *Applied Optics and Optical Engineering*. New York: Academic Press, 1965.
5. Williams, C. S., *Optics: A Short Course for Engineers and Scientists*. New York: Wiley-Interscience, 1972.

DATA SECTION

TABLE 6-10 Basic Test Data

Photometric data

Description:
1 × 4 surface-mounted clear prismatic acrylic wraparound
Test No.: 11417; Cat. No.: BK240
Lamps: 2F40CW; Volts: 120
Lumens/Lamp: 3110; Bare Lamp: 2360 fL
Shielding: Parl 90°, Norm. 90°

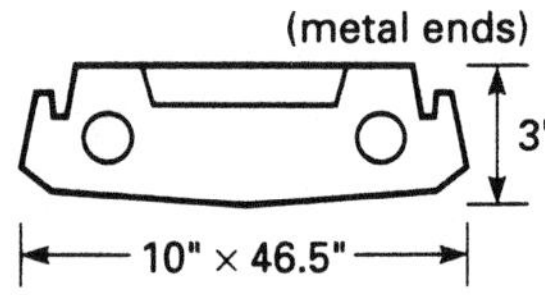

Luminous Intensity (candlepower), candelas

Angle	0°	22½°	45°	67½°	90°
5	1388	1390	1393	1385	1391
15	1336	1344	1353	1363	1370
25	1234	1260	1289	1309	1331
35	1074	1119	1167	1200	1228
45	835	890	948	982	1018
55	537	573	615	657	696
65	295	325	357	405	450
75	142	174	212	269	317
85	35	83	134	186	237
95	10	50	100	163	214
105	23	71	115	163	210
115	33	78	120	154	191
125	41	83	120	148	176
135	43	67	93	124	153
145	49	64	81	86	88
155	46	54	66	71	74
165	39	44	50	54	57
175	37	38	39	41	42

Maximum luminance, fL

Angle	0°	90°
45	2483	2721
55	1939	2211
65	1089	1701
75	850	1429
85	510	1463

S/MH = 1.4.

TABLE 6-11 Zonal Constants: $2\pi\,(\cos\theta_1 - \cos\theta_2)$

5° Zones

Zones: Lower	Zones: Upper	Zonal constant
0–5	175–180	0.0239
5–10	170–175	0.0715
10–15	165–170	0.1186
15–20	160–165	0.1649
20–25	155–160	0.2097
25–30	150–155	0.2531
30–35	145–150	0.2946
35–40	140–145	0.3337
40–45	135–140	0.3703
45–50	130–135	0.4041
50–55	125–130	0.4349
55–60	120–125	0.4623
60–65	115–120	0.4862
65–70	110–115	0.5064
70–75	105–110	0.5228
75–80	100–105	0.5351
80–85	95–100	0.5434
85–90	90–95	0.5476

10° Zones

Zones: Lower	Zones: Upper	Zonal constant
0–10	170–180	0.095
10–20	160–170	0.283
20–30	150–160	0.463
30–40	140–150	0.628
40–50	130–140	0.774
50–60	120–130	0.897
60–70	110–120	0.993
70–80	100–110	1.058
80–90	90–100	1.091

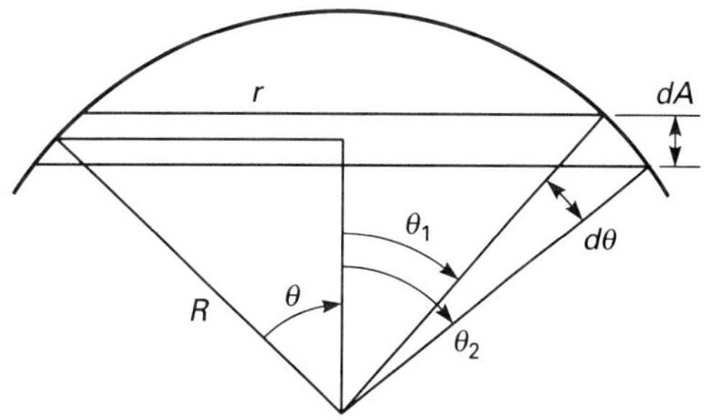

$$\omega = \frac{dA}{R^2}$$

$$d\omega = d\theta = \frac{2\pi\, r \times R d\theta}{R^2} \quad [1]$$

$$r - R_{\sin}\,\theta \quad [2]$$

$$\therefore d\theta = \frac{2\pi\,\sin\theta R^2 d\theta}{R^2}$$

$$d\theta = 2\pi\,\sin\theta\, d\theta$$

$$\int d\theta = 2\pi \int_{\theta_2}^{\theta_1} \sin\theta\, d\theta$$

$$A = 2\pi(\cos\theta_1 - \cos\theta_2)$$

TABLE 6-12 (VCP) Table for Figure 6-14

VISUAL COMFORT PROBABILITY DESCRIPTION:
2 × 4 fluorescent troffer with
prismatic batwing w/4F40CW
each rated at 3100 lumens.
100 FC Reflectances: 80/50/20

Luminaire "J"

Report no.: 10200
LUMNO = WB200

Room		Luminaires lengthwise				Luminaires Crosswise			
W	L	8.5	10.0	13.0	16.0	8.5	10.0	13.0	16.0
20	20	63	70	82	89		62	68	73
20	30	58	64	73	81	58	59	60	64
20	40	55	61	68	75	57	58	59	60
20	60	51	57	65	71	57	58	59	59
30	20	62	68	77	85	59	63	68	73
30	30	56	61	67	75	57	58	59	64
30	40	53	57	62	68	56	57	57	58
30	60	49	53	58	64	56	57	56	57
30	80	47	50	55	62	57	57	56	57
40	20	64	69	76	83	62	64	69	74
40	30	57	61	66	73	58	59	60	64
40	40	53	56	60	65	56	57	57	58
40	60	49	52	56	61	56	56	55	56
40	80	47	49	53	58	56	57	55	56
40	100	46	48	51	56	57	57	56	56
60	30	59	62	65	71	59	60	60	64
60	40	54	57	59	64	57	57	57	58
60	60	49	52	54	58	56	56	55	56
60	80	47	49	51	55	56	56	54	55
60	100	46	47	49	53	57	56	54	55
100	40	58	60	61	65	59	60	59	61
100	60	53	55	55	59	57	58	56	57
100	80	50	51	52	55	57	57	55	55
100	100	48	48	49	52	58	57	54	55

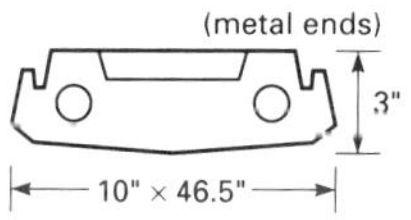

Description:
1 × 4 surface-mounted clear
prismatic acrylic wraparound
w/metal ends

Test No.: 11417 Cat. No.: BK 240
Lamps: 2F40CW Volts: 120
Lumens/Lamp: 3110 Test Dist.: 20
Shielding Angle: N: 90° L: 90°
Bare Lamp: 2360 fL Date:

LDD CATEGORY V

	Brightness Data			
	Angle	Maximum	Average	Ratio Max./Min.
Along (0°)	45°	2483	1148	2.2
	55°	1939	909	2.1
	65°	1089	678	1.6
	75°	850	533	1.6
	85°	510	384	1.3
Across (90°)	45°	2721	1076	2.5
	55°	2211	827	2.7
	65°	1701	628	2.7
	75°	1429	560	2.6
	85°	1463	597	2.5

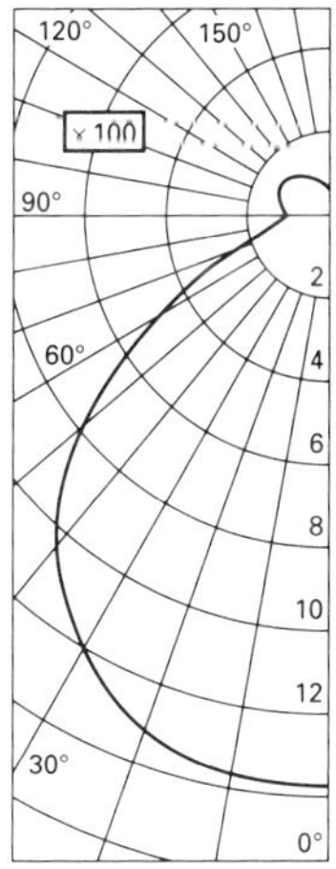

Distribution Data Average			
Angle	C.P.	Angle	C.P.
0°	1398	95°	106
5°	1389	105°	116
10°	—	115°	116
15°	1353	125°	115
20°	—	135°	96
25°	1285	145°	75
30°	—	155°	63
35°	1159	165°	49
40°	—	175°	39
45°	937	180°	40
50°	—		
55°	615		
60°	—		
65°	365		
70°	—		
75°	221		
80°	—		
85°	134		
90°	107		

Output Data			
Zone degrees	Lumens	Percent total lamp lumens	Percent total distribution
0–40°	1838	29.5	41.0
40–90°	2019	32.5	45.1
90–180°	625	10.1	13.9
0–180°	4482	72.1	100.0

Coefficient of Utilization

Floor		ρ_{fc} = 20%								
Ceiling		80%			50%			10%		
Walls		50%	30%	10%	50%	30%	10%	50%	30%	10%
Room Cavity Ratio	1	73	71	68	66	64	62	57	56	55
	2	65	60	56	58	55	52	51	49	47
	3	57	52	48	52	48	45	46	43	41
	4	51	45	41	47	42	39	41	39	35
	5	46	39	35	42	37	33	37	33	31
	6	41	35	31	37	33	29	33	30	27
	7	37	31	27	34	29	25	30	26	24
	8	33	27	23	30	25	22	27	23	21
	9	30	24	20	27	22	19	24	21	28
	10	27	21	18	25	20	17	22	18	16

Maximum recommended spacing-to-mounting height ratio above workplane is 1.4.

Figure 6-13. Standard Photometric Test Report

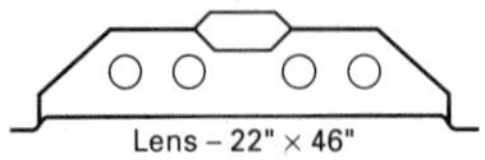

Description:
2 × 4 Troffer w/ A12 lens

Test No.: 7100 — Cat. No.: 2SP 440
Lamp: 4F40CW — Volts: 120
Lumens/Lamp: 3150 — Test Dist.: 20
Shielding Angle: N: 90° — L: 90°
Bare Lamp: 2350 fL — Date: 7/86

Luminaire: "A"
Cat. No.: WB250
Category: V
Maximum S/MH Ratio:
1.2/1.3

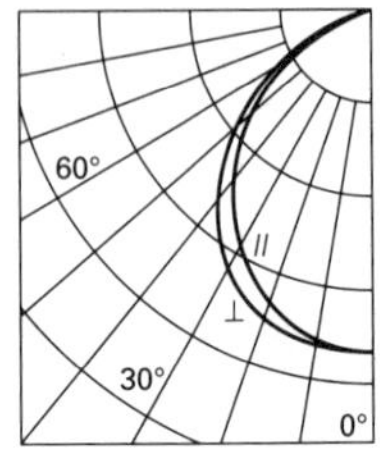

Luminance Data			
Along			
Angle	Max.	Ave.	M/A
45°	3256	1275	2.55
55°	2754	984	2.80
65°	1965	683	2.88
75°	478	461	1.04
85°	432	446	.97
Across			
45°	3015	1441	2.09
55°	1787	998	1.79
65°	1326	666	1.99
75°	519	487	1.07
85°	438	446	.98

Luminous Intensity:

Angle	Along	22½°	45°	67½°	Across
5°	3620	3629	3627	3634	3636
15°	3539	3561	3579	3612	3623
25°	3155	3195	3257	3321	3348
35°	2803	2880	2990	3080	3104
45°	2017	2140	2259	2311	2280
55°	1262	1123	1083	1136	1280
65°	646	488	414	464	630
75°	267	313	243	306	282
85°	87	131	146	135	87

Output Data			
Zone degrees	Sum lumens	Percent total lamp lumens	Percent total distribution
0–40°	4744	37.7	56
40–90°	3688	29.3	44
90–180°	0	0.0	0
0–180°	8432	66.9	100

Coefficient of Utilization

$\rho f_c = 20\%$												
ρc_c	80			70			50			30		
RCR/ρw	50	30	10	50	30	10	50	30	10	50	30	10
1	72	69	65	71	68	66	68	66	64	65	64	62
2	65	61	58	63	60	57	61	57	56	59	57	54
3	58	54	50	57	53	59	55	52	49	54	50	48
4	53	47	44	52	47	43	50	46	43	48	45	42
5	47	42	38	47	42	38	45	41	37	44	40	37
6	43	37	34	42	37	34	41	36	33	40	36	33
7	39	33	30	38	34	29	37	32	29	36	32	29
8	35	29	26	34	29	26	33	29	25	33	29	25
9	31	26	22	31	26	22	30	26	22	29	25	22
10	28	23	20	28	23	20	27	23	20	27	23	20

Figure 6-14. Photometric Test Report—2 × 4 Troffer

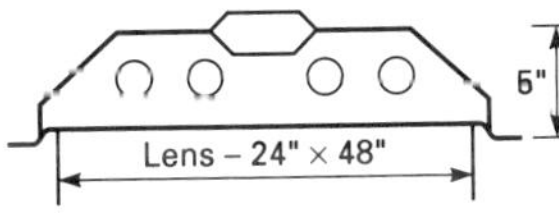

Luminaire: "J"
WB200

Category: V

Maximum S/MH
Guide: 1.7/1.2

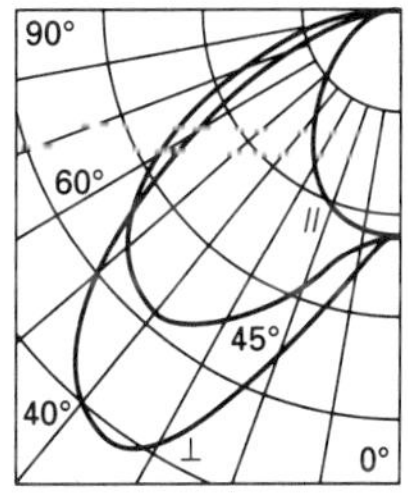

Description:
2 × 4 fluorescent troffer
with prismatic batwing

Test No.: 10200 Cat. No.: WB200
Lamps: 4F40CW Volts: 120
Lumens/Lamp: 3100 Test Dist.: 20
Shielding Angle: N: 90° L: 90°
Bare Lamp: 2360 fL Date:
By: DN Chkd. by: RH

Angle	Along	22.5	45	67.5	Across
0	1414	1414	1414	1414	1414
5	1401	1404	1413	1424	1429
15	1218	1268	1534	1768	1838
25	1103	1352	2002	2444	2580
35	903	1480	2561	3300	3464
45	539	1352	2357	2823	2857
55	272	1017	1341	1418	1426
65	245	740	876	809	703
75	186	397	485	337	211
85	146	156	140	84	54
90	0	0	0	0	0

Luminance Data			
Across			
Angle	Max	Ave	M/A
45°	3015		
55°	1787		
65°	1326		
75°	519		
85°	438		
Along			
45°	587		
55°	382		
65°	348		
75°	516		
85°	1184		

Output Data			
Zone degrees	Lumens	Percent total lamp lumens	Percent total dist.
0–40°			
40–90°			
90–180°			
0–180°			

Coefficient of Utilization

$\rho f_c = 20\%$																	
ρc_c	80				70				50			30			10		
ρ_ω	70	50	30	10	70	50	30	10	50	30	10	50	30	10	50	30	10
1	60	58	55	53	59	56	54	53	54	52	51	52	51	49	50	49	48
2	55	51	47	44	54	50	47	44	48	45	43	46	44	42	45	43	41
3	50	45	41	37	49	44	40	37	43	39	36	41	38	36	40	37	35
4	46	40	35	32	45	39	35	32	38	34	31	37	33	31	35	33	30
5	42	35	30	27	41	34	30	27	33	29	26	32	29	26	31	28	26
6	39	31	26	23	37	31	26	23	30	26	23	29	25	22	28	25	22
7	35	28	23	19	34	27	23	19	26	22	19	26	22	19	25	21	19
8	32	25	20	16	31	24	20	16	23	19	16	23	19	16	22	19	16
9	29	22	17	14	29	21	17	14	21	17	14	20	16	14	19	16	13
10	27	19	15	12	26	19	15	12	19	15	12	18	14	12	18	14	12

Figure 6-15. Photometric Test Report—2 × 4 Troffer

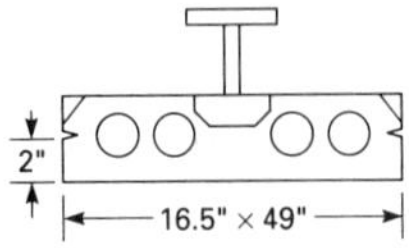

Description:
16½ × 49 Suspended prismatic
wraparound (plastic ends)

Test No.: 4380-1A
Cat. No.: 45467-4
Lamps: 4F40CW
Volts: 120
Lumens/Lamp: 3100
Test Dist.:
Shielding Angle: N: 90°
L: 90°
Bare Lamp: 2350 fL
Date: 1973
By: DB
Chkd. By: RH

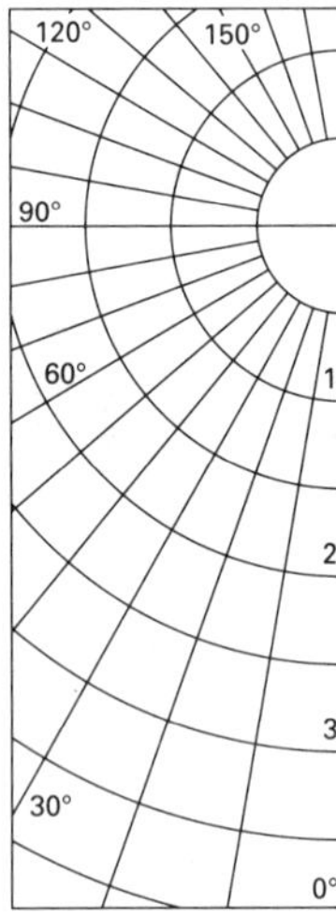

Distribution Data Average			
Angle	C.P.	Angle	C.P.
0°	—	95°	96
5°	2130	105°	240
10°	—	115°	408
15°	2362	125°	540
20°	—	135°	662
25°	2056	145°	757
30°	—	155°	830
35°	1925	165°	871
40°	—	175°	876
45°	1523	180°	—
50°	—		
55°	752		
60°	—		
65°	392		
70°	—		
75°	232		
80°	—		
85°	131		
90°	—		

Lumno = DB207

Brightness Data (Crosswise)			
Angle	Maximum	Average	Ratio Max./Ave.
45°	2230	1194	1.87
55°	1175	661	1.78
65°	763	459	1.66
75°	705	409	1.72
85°	705	457	1.54

Output Data			
Zone degrees	Lumens	Percent total lamp lumens	Percent total distribution
0-40°	3032	24.5	35.2
40-90°	2631	21.2	30.5
90-180°	2950	23.8	34.3
0-180°	8613	69.5	100.0

Coefficient of Utilization

Floor, ρf_c		For ρf_c = 20%								
Ceiling, ρc_c		80%			50%			10%		
Walls, ρw		50%	30%	10%	50%	30%	10%	50%	30%	10%
RCR	1	.67	.65	.62	.56	.55	.53	.44	.43	.42
	2	.60	.56	.52	.50	.48	.45	.39	.38	.37
	3	.53	.49	.45	.45	.42	.39	.36	.34	.32
	4	.48	.42	.38	.41	.37	.34	.32	.30	.28
	5	.43	.37	.33	.36	.32	.29	.29	.27	.25
	6	.38	.33	.29	.33	.29	.26	.26	.24	.22
	7	.35	.29	.25	.30	.26	.23	.24	.21	.19
	8	.31	.26	.22	.27	.23	.20	.22	.19	.17
	9	.28	.23	.19	.24	.20	.17	.19	.17	.15
	10	.23	.20	.17	.22	.18	.15	.15	.15	.13

Maximum recommended spacing-to-mounting height ratio above work plane is: 1.50.

Figure 6-16. Photometric Test Report—Suspended Wraparound

7 interior lighting and the luminous environment

Artificial lighting affects one's ability to perform a visual task through interaction of luminous energy, the room, and the visual display (the task itself). The source of luminous energy (lamp and luminaire) was discussed in previous chapters. This chapter investigates the role of the physical environment on the luminous environment. This discussion is included to show that the room is as important in artificial lighting as the task and luminaire.

IMPACT OF THE PHYSICAL ENVIRONMENT ON LUMINAIRE SELECTION

Seven characteristics of a room have an impact on the lighting:

1. Surface reflectances
2. Size of room
3. Shape of room
4. Windows
5. Maintenance
6. Temperature
7. Furniture or equipment in the room

Windows and their effect are discussed in Chapter 13; the effect of temperature on light sources was discussed in Chapter 4.

Surface Reflectances and Room Proportions

Reflectances of room surfaces are important for two reasons: they affect the appearances of the surfaces, and they affect the amount of light that interreflects within the space and eventually arrives at the seeing task. The former issue concerns luminance ratios, whereas the latter deals with interreflection of light.

Luminance Ratios

In Chapter 1 we saw that visibility could be influenced by differences in luminance between the visual task and its surround. If these differences are excessive (high luminance ratios), visibility is reduced. Similarly, if a wall or portion of a wall has excessive luminance, problems occur when one looks away from the task to the wall and back to the task. In this case, the visual system attempts to adapt to the higher luminance of the wall, but at that adaptation level, details of the task are hard to discern. This is known as transient adaptation. If the luminance differences are extreme enough, the environment can actually create visual discomfort.

For these reasons it is necessary for the lighting professional to ensure that luminance ratios fall within reasonable limits. Suggested luminance ratios for various conditions are found in Table 7-1.

Interreflection of Light

The light that falls on a seeing task comes from two sources, the luminaire itself (the *direct component*) and the surfaces of the room (the *interreflected component*). The room characteristics that affect the amount of light reaching the task directly from the luminaires are maintenance and temperature. The reflected component, however, is dependent on surface reflectance, location of the luminaire(s), and size and shape of the

TABLE 7-1 Recommended Luminance Ratios

	Offices/schools	Residential*	Industrial
Between task and adjacent darker surround	1 to $\frac{1}{3}$	1 to $\frac{1}{3}$ (1 to $\frac{1}{5}$)	1 to $\frac{1}{3}$ (1 to $\frac{1}{5}$)
Between task and adjacent lighter surround	1 to 3	1 to 3 (1 to 5)	1 to 3 (1 to 5)
Between task and more remote darker surround	1 to $\frac{1}{5}$	1 to $\frac{1}{5}$ (1 to $\frac{1}{10}$)	1 to $\frac{1}{10}$ (1 to $\frac{1}{20}$)
Between task and more remote lighter surround	1 to 5	1 to 5 (1 to 10)	1 to 10 (1 to 20)
Between luminaires, windows, and skylight and adjacent surround	5 to 1 (20 to 1)†	20 to 1	20 to 1
Anywhere in the visual field	5 to 1 (40 to 1)†	40 to 1	40 to 1

* Maximum allowable luminance ratio given in parenthesis ().

† Maximum allowable in spaces where critical visual work does *not* occur.

room. Each major room surface (walls, ceiling, and floor) can have a different reflectance. The reflectance of each has a different effect on the contribution of reflected light to the illumination level produced in a room.

The reflectance of the ceiling is important to the luminous environment whether the lighting system is direct or indirect. In an indirect system, light from the luminaire is cast (either exclusively or substantially) on the ceiling. In such systems, ceiling reflectance is obviously an important determinant of how much light will eventually arrive at the seeing task. But even with a totally direct lighting system (in which no light falls directly on the ceiling from the luminaire), ceiling reflectance can be critical, even though the amount of reflected light that can reach the ceiling is small. This is because the ceiling reflectance always plays a role in ceiling luminance (regardless of whether the system is direct or indirect). Ceiling luminance is important because excessive luminance ratios (between luminaires and the rest of the ceiling) can cause reduced visibility, reduced visual comfort, and may be unsightly (again, Table 7-1 will be helpful). In general, ceiling reflectances are kept as high as possible, usually around 80% to 90%. Ceilings, therefore, are almost always white, not just by tradition, but in order to achieve these high reflectances and to assure that the color characteristics of the light source are not detrimentally altered.

The coefficient of utilization (CU) is a convenient measure of the efficiency of a luminaire in getting luminous energy to the work surface in a particular room. The use of the CU in predicting illuminance levels is discussed at length in Chapter 8. The CU combines the room characteristics (geometry and reflectances) with the luminaire charac-

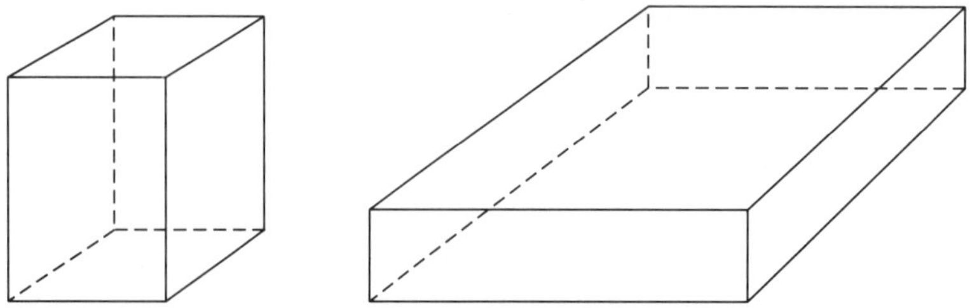

Figure 7-1. Comparison of cavity ratios.

teristics in a manner that indicates the efficiency of the interaction of the two components. In determining the CU, an intermediate metric known as the **room cavity ratio (RCR)** is used. The RCR summarizes the geometry (proportions) only of a space. Very tall spaces have high cavity ratios (on the order of 10 or greater), whereas flat spaces have low cavity ratios (on the order of 1 or less). See Figure 7-1 for comparisons of cavity ratios.

Figure 7-2 shows the effect of ceiling reflectance on the CU for direct and direct-indirect luminaires (see Chapter 6 for luminaire classifications). For the direct-indirect luminaire, the ceiling reflectance is more important because of the luminous energy that goes directly to the ceiling. This is seen by the wide gap between the 50% and 80% curves. For the direct luminaire, the difference between the 50% and 80% ceiling curves is less, which indicates the lesser role the ceiling plays in utilizing luminous energy. Note that, for both cases, the 50% and 80% lines are nearly parallel. This indicates that the ceiling effect is about the same for a given luminaire, regardless of room proportions (as summarized by the RCR).

The amount of light falling on the walls in a good lighting system should be of sufficient magnitude to provide field luminances and luminance ratios that are comfortable. Depending on reflectance properties, the wall luminance and thus the interreflected component will change. Figure 7-3 shows the effect of wall reflectance on CUs for a direct luminaire. Walls having reflectance of 70% yield substantially higher CUs than will walls with 50% reflectance. Note that in this case the lines *are not* parallel; instead the gap increases with increasing RCR. This means that reflectance of the walls has more impact on the efficiency of the lighting system in tall rooms (high RCR) than in flat rooms (low RCR).

Since greater wall reflectance makes a lighting system more efficient at delivering light to the workplane, we are tempted to think that an all-white environment would be ideal. However, the sterile feel of an all-white space can cause occupant distress, and color is usually considered necessary to provide some psychological relief. Any colored surface will reflect less light than a white surface, but with proper color choices, a pleasant, well-lighted space is possible. As a compromise, it is usually best to keep average wall reflectances in the 50% to 70% range, suggesting finishes with Munsell values of 7.5 to 8.5 (see Chapter 3).

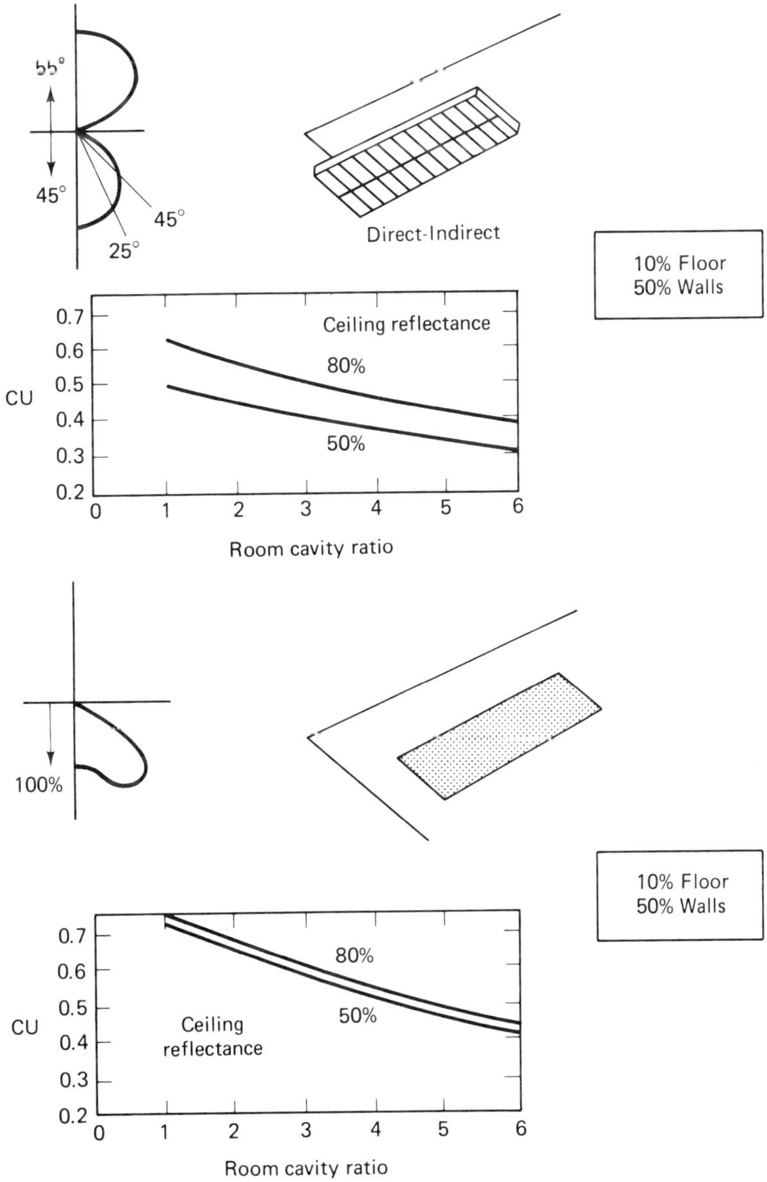

Figure 7-2. Effect of Ceiling Reflectance on the Coefficient of Utilization.

Even if the architect or interior designer provides light-colored walls, the owner may reduce the efficiency of the system by installing dark furniture, pictures, blackboards, and so on, that absorb light. If energy conservation is to be the responsibility of the design team, they must have a voice in the selection of finishes, furnishings, and both the selection and placement of furniture.

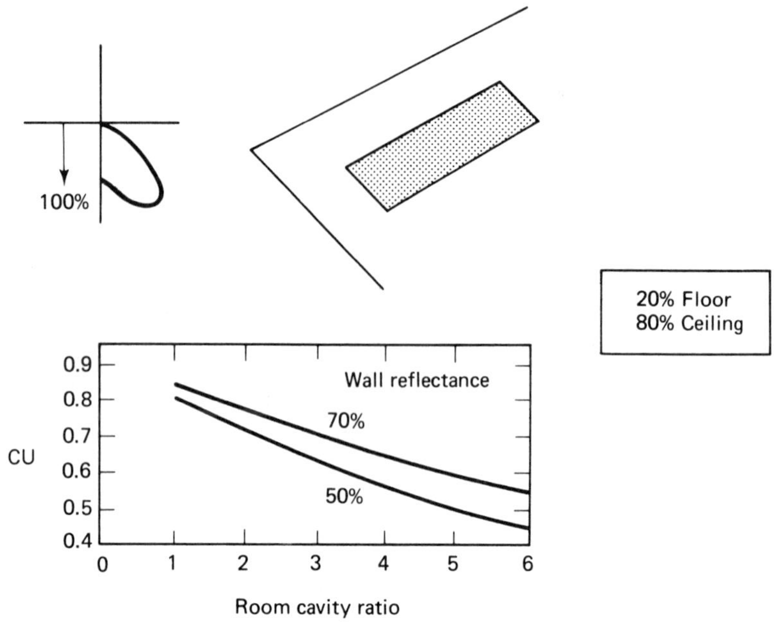

Figure 7-3. Effect of Wall Reflectance on the Coefficient of Utilization.

The importance of the floor reflectance on the visual environment varies depending on the types of visual tasks being performed. The process for determining the CU (efficiency with which the system delivers light to the top of the workplane) depends on knowing the floor reflectance of the space. However, the assumption is frequently made that the reflectance of the floor is 20%. If the floor reflectance is not 20%, then a correction factor is applied to the CU. Table 8-5 gives these correction factors for floor reflectances of 10% and 30%. We see from that table that CUs are only altered by a couple of percent over a fairly large range of floor reflectances. This indicates that the floor reflectance plays a minor role in lighting the workplane. This seems reasonable considering that the light has to be reflected off the floor, to the ceiling, and finally to the task.

If the task is oriented toward the floor, however, on the bottom of the workplane for example, floor reflectance becomes critical. An example is in the design of hangars for aerospace construction and repair. Many of the tasks are on the underside of wings or fuselages. With a low-reflectance floor, supplementary lighting is required. Bringing the floor reflectance up to 50% or 60% or higher may eliminate the need for supplementary lighting.

The size and shape of the space affect the distribution of luminous flux in the space. To see this effect, the CUs in Figures 7-2 and 7-3 should be examined. Recall that the room cavity ratio serves to summarize the proportions of a space. From the figures we see that CUs decrease as the room becomes taller (higher RCR). This is logical because in tall spaces, more of the light from the luminaires initially falls on the walls, and less falls on the workplane. A tall space depends more on interreflectance for getting light to the workplane, yet a tall space is less efficient at interreflectance. This is illustrated in Figure 7-4. The result is that for a given ceiling height, large rooms can be lit more efficiently than smaller rooms.

Maintenance

A chief factor affecting the performance of a lighting system is the deterioration of its components with age. As the light source ages, its light-generating ability declines, an effect known as **lamp lumen depreciation (LLD).** Dirt and grime buildup on luminaire surfaces reduces the amount of light which leaves the luminaire, this is called **luminaire dirt depreciation (LDD).** As paint and fabrics age and dirt accumulates on ceilings, walls, and floors, the resultant decrease in light levels is attributed to **room surface depreciation (RSD).** Other factors contributing to decreased light levels include the number of lamps burned out in a system and the condition of luminaire surfaces such as lenses. This last item was of considerable consequence in times past when most lenses and diffusers were constructed of polystyrene, which yellows and darkens with prolonged exposure to even relatively low levels of ultraviolet radiation. Today almost all these parts are made of acrylic, which has very stable transmittance properties.

The design of a lighting system should be such that the required light level is provided when the system has reached its worst state of deterioration. This ensures that occupants will receive at least the minimum illuminance requirement throughout the life of the lighting system. The extent to which the system must be "overdesigned" is dependent on how well the maintenance program is carried out once the system is

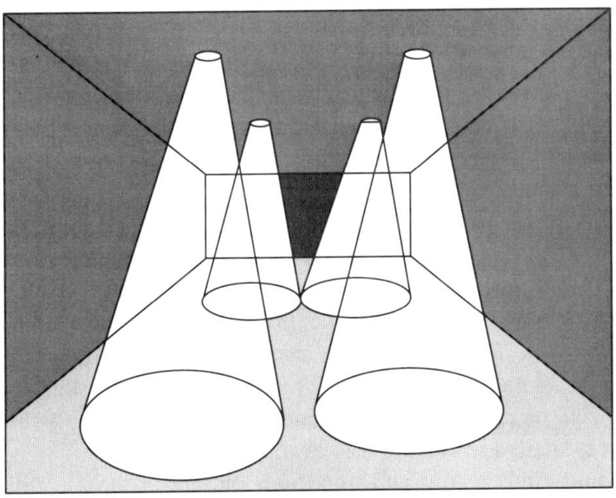

(a) Low RCR, most of light falls directly on floor or workplane

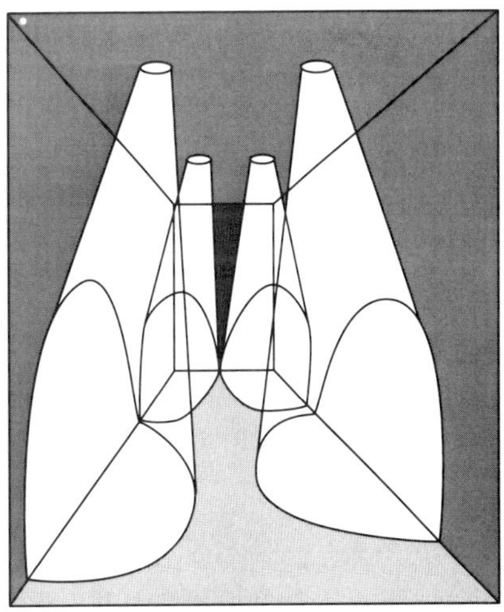

(b) High RCR, much light absorbed by walls

Figure 7-4. Effect of room cavity on utilization of light.

installed. If dirt is allowed to accumulate in luminaires, lamps are not replaced until they burn out, and room surfaces are not properly cared for, the lighting system is going to require a larger initial investment to compensate for light loss.

Lamp Lumen Depreciation

The light output of all lamps commonly used in architectural lighting decreases as they get older. Numerically it is the light emitted at 70% of rated life expressed as a percentage of initial light output. Thus if the rating of a new lamp is 1740 lumens and that lamp emits 1575 lumens after it has burned for 70% of its rated life, then its LLD is 1575/1740, or 90.5%

The effect of lumen depreciation can be reduced by planned replacement, such as group relamping where all the lamps are replaced at some interval less than 100% of rated life. Group replacement of lamps can be more economical than the spot replacement of individual burnouts. This may seem puzzling at first since group replacement implies throwing away many lamps that would otherwise continue to function well for perhaps months longer. The savings is attributable to at least two things. First, there is a savings in the cost of labor because group relamping doesn't take as much time (spot replacement requires setup and takedown time for each lamp). Second, since the light output of the system is not allowed to decline as far as it would with spot relamping, the system that will be group relamped doesn't require as many luminaires to be installed in a space initially. See Chapter 14 for the details of group versus spot relamping analysis and other economic considerations.

Luminaire Dirt Depreciation

Airborne dirt deposited on luminaire surfaces affects the amount of light leaving the luminaire and the intensity distribution of the luminaire. The amount of light leaving the luminaire decreases for the simple reason that dirty surfaces absorb more light than clean ones. The distribution is affected because specular surfaces become more diffuse when dirty, a principle that becomes obvious when you consider your image in a dusty mirror as compared with the image in a clean mirror.

The cleanliness of the environment in which a lighting system is operated is, of course, a major factor in luminaire dirt depreciation. In the cleanest of surroundings, a new, high-grade office building for example, luminaires will only lose about 10% of their output between cleanings (24 months assumed) due to dirt accumulation. In a medium environment such as would be found in light industry situations, about 20% light loss can be expected over a 24-month period. In the worst conditions such as lighting particularly dirty industrial processes, a 40% reduction could reasonably be anticipated over 24 months. More detailed information is available in Chapter 8.

In addition to the amount and kinds of dirt and impurities present in the air, dirt depreciation is also contingent on some aspects of luminaire design. Luminaire surface finishes vary in the ease with which dirt can attach itself and in the ease with which an electrostatic charge can form. Ventilated luminaires tend to collect dirt less rapidly than those with closed tops. This is due to the chimney effect by which the temperature difference between the lamp and the ambient room air cause convective air motion that helps keep dust from settling on the luminaire surfaces. On the other hand, completely closed units, if properly designed, exclude almost all dirt. These concepts are shown in Figure 7-5.

Luminaire surfaces can also lose their ability to reflect or transmit light. Anodized and enameled reflective surfaces deteriorate. Environments that contain highly caustic materials cause surfaces to deteriorate faster. Luminaire surfaces that have deteriorated typically cannot be repaired; the entire luminaire should be replaced at the next regular maintenance interval.

Cleaning luminaires at the time of lamp replacement returns the system to optimum efficiency. Cleaning at the time of relamping also allows a savings in labor cost over the cost of performing these maintenance tasks separately. Many owners balk at the cost of regular maintenance of this type. It is the responsibility of the lighting professional to educate the owner to the need for regular system maintenance and the consequences of allowing a system to deteriorate. This can be done either in person or by means of a written operating manual, or better still, both.

Room Surface Depreciation

It would seem that, if the luminaires are cleaned and relamped occasionally, the major light loss factors will have been taken care of. However, since high room surface reflectances have been shown to be a requirement for good system efficiency, any decrease in room surface reflectances will also increase light loss. Depending on the environment,

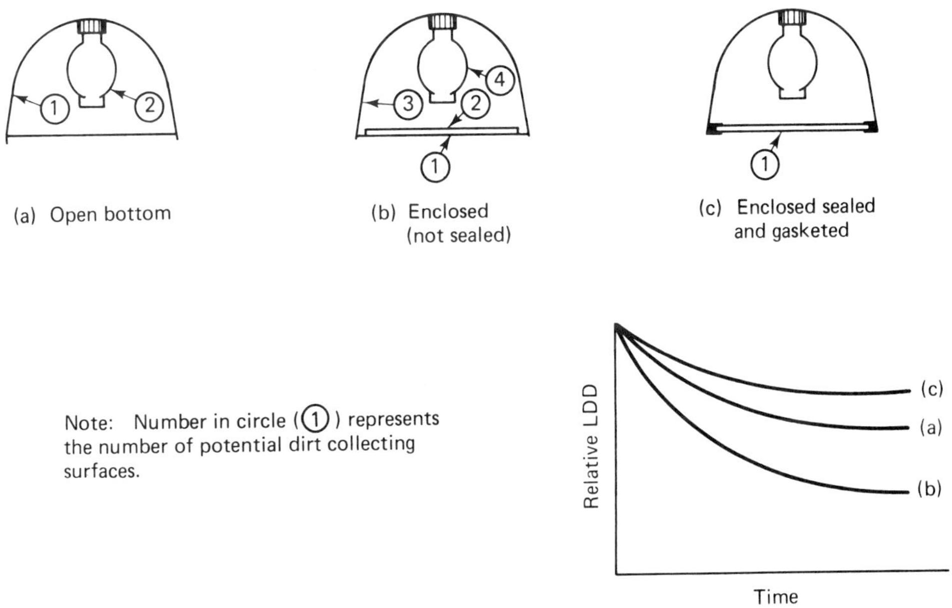

Figure 7-5. Luminaire Dirt Depreciation.

room surfaces should be washed periodically and painted when the reflectance has dropped significantly. In the cleanest environments, room surface depreciation can easily account for a 5% decline in illuminance level in the space during a 24-month cleaning cycle. For very dirty conditions the loss could be 20% or 30% or worse. Maintenance should be performed more frequently when a larger percentage of light is reflected by these surfaces (e.g., high RCRs, indirect luminaires). It should be noted that sometimes surfaces appreciate; that is, reflectance rises. Bleaching out of paint, curtains, or other materials, as well as light-colored dust, can cause the appreciation of surface reflectances.

Objects in the Environment

Most preliminary design assumes that the space is entirely empty. When furniture and equipment are placed in a room, the distribution of the light can change radically. Furniture that is light in color with a matte finish will help prevent glare and absorption of light. Shadows that occur from equipment shielding light can be prevented by properly locating equipment with respect to the lighting system. As task-lighting equipment (i.e., equipment that is located near the task and oriented properly) is improved, the problem with shadowing will be minimized.

IMPACT OF THERMAL AND ACOUSTICAL CHARACTERISTICS OF LUMINAIRES ON THE ENVIRONMENT

Besides putting light into the space, luminaires introduce thermal and acoustical changes to the space. Most of the time, lighting designers ignore these impacts of lighting equipment on the environment. This unfortunate tendency can result in spaces that, even while providing for visual tasks and aesthetics, have serious deficiencies in other areas. In order to optimize the design of the whole space, thermal and acoustical concerns must be integrated into the lighting design.

Thermal

Each watt of electrical power used in a space (lights included) is ultimately converted into heat at the rate of 3.41 British thermal units (Btu's) per hour. The transfer of this heat energy occurs by radiation, conduction, and convection. Although a welcome supplement to winter heat requirements, this extra heat must be removed when cooling is required. This load on the air conditioning system is taken care of by increasing the capacity of the cooling equipment, resulting in a substantial increase in the initial cost of the building and a substantial increase in energy costs. In fact, approximately 40% of the cooling capacity and 40% of the energy consumed by air conditioning systems of commercial office space goes toward dissipating the heat generated by lights.

It has been suggested that energy can be saved by turning off lights in existing installations, resulting in a savings in the energy used for lighting as well as the energy used for cooling. In most systems this may be true; however, in the case of systems using electric resistance heaters to maintain the design air temperature, reduction in air temperature causes the reheat coils to cycle on, increasing the air temperature to its design level. Therefore, each system must be analyzed carefully to determine the impact of lighting on the mechanical system. The coordination of lighting and mechanical design is essential to the optimization of both.

Selection of efficient sources and luminaires, proper room surface reflectances, and the use of task lighting will reduce the lighting system wattage. Any heat produced must still be considered in the design of the mechanical system.

Heat-Transfer Luminaire

If the heat from the lighting system can be utilized rather than being dumped or wasted, the energy consumed by the heating system can be reduced to the degree to which the lighting provides heat to the space. The heat from the lighting system must be redistributed to areas where it is needed. The common means of using the heat produced by the lighting is the heat-transfer luminaire. This is a type of luminaire through which a heat transfer medium (air or water) is circulated to remove unwanted heat from the unit and transfer it to a location where it can be better utilized (e.g., the perimeter of a building in winter where heat loss through the building's skin is great) or where it can more efficiently be disposed of.

Air is essential to proper ventilation of a space and can be useful as a heat-transfer medium for luminaires. The air circulation around a luminaire is essential for maximum heat transfer. However, excessive air temperature differences and air motion can affect the light output and color characteristics of fluorescent lamps due to vapor pressure variations within the lamp (see Chapter 4).

Circulating water can also be an effective heat-transfer mechanism. The primary advantage is that it requires substantially less volume for the same heat-removal capacity. The use of water to remove luminaire heat does not eliminate the need for an air system. Air is necessary for humidity control and ventilation. Some air motion may also be a desirable in a space.

The use of infrared reflecting filters in the lens of a luminaire will result in preventing infrared energy from entering the occupied space. As this traps heat in the portion of the luminaire that is recessed into the plenum, the heat removal system in the luminaires must be very efficient to remove this added heat.

The control of heat produced by lighting can lower heat gain in the surrounding space, reduce air changes required (thus reducing fan horsepower and energy consumption by fans), lower temperature differentials, reduce luminaire and ceiling temperature, and increase luminaire efficiency and life. However, heat-transfer luminaires command a cost premium over standard equipment. The increase in initial cost due to heat-transfer luminaires may be offset by the reduced cost of lamps and the reductions possible in the mechanical system.

Temperature and Lamp Flux

A secondary benefit of the heat-transfer luminaire is the potential for increased light output. Chapter 4 shows that a variation in ambient temperature can affect the light output of a fluorescent lamp. The buildup of heat can decrease light output. The heat-transfer luminaire, if properly designed, can have a reduced internal ambient temperature, resulting in more efficient operation and utilization of energy. Proper housing design and mounting details for standard luminaires will also permit some of the heat to be conducted away, but not as much as with heat-transfer luminaires.

Acoustical Considerations

Concern for environmental noise continues to increase as new technologies produce an increasingly noisy workplace. The typical ''quiet'' office will invariably have a copy machine and several personal computers whirring continually throughout the day. The clatter of typewriter keyboards has been augmented with computer keyboards, and most dot matrix printers dominate the aural environment when in service. The lighting system has also added to the noise pollution that surrounds us. Lighting equipment, especially gaseous discharge lamps (which require inductance ballasts) is a noise source.

The standard core-and-coil magnetic ballast is basically a transformer. It consists of coils of copper wire wrapped around a core of laminated metal plates. As alternating current passes through the assembly, the associated magnetic field expands and collapses at twice the input power frequency (60 Hz in North America). There is a tendency, then, for the laminae to vibrate at this frequency. This can cause an audible buzz (at 120 Hz plus harmonics), particularly if improperly mounted or used in conjunction with a poorly constructed luminaire which can act as a sounding board to amplify the noise.

There are no current industry standards for sound ratings of ballasts. However, the major fluorescent ballast manufacturers assign a letter rating of A (quietest) through F (noisiest) to their products. Table 7-2 can be used to estimate ballast noise. If many identical ballasts are in the same space, the combined sound level can be found by adding 10 times the logarithm of the number of ballasts to the intensity level of 1

TABLE 7-2 Fluorescent Ballast Sound Ratings

Sound ratings	Average noise rating (dB)
A	20–24
B	25–30
C	31–36
D	37–42
E	43–48
F	49 and up

ballast (in decibels). Therefore, 20 B-rated ballasts in a space might have an intensity level of $25 \text{ dB} + 10 \log(20) = 38 \text{ dB}$.

Fluorescent equipment manufactured for quiet indoor application (offices, shops, residences) should be equipped with an A sound-rated ballast. For most luminaire manufacturers this is standard procedure; however, the lighting design professional should always mention the A rating in the specification to assure quiet operation. Some applications (e.g., recording studios) will require a level of quiet not obtainable even with A-rated ballasts and premium luminaires. In such cases the designer may choose to locate the ballasts outside the critical area. Remote mounting of ballasts can be expensive, however, and can cause voltage drop problems which should be investigated in advance. Another alternative is to use high-frequency solid-state ballasts, a product whose importance in the market has expanded rapidly in recent years. Though currently more expensive than traditional core-and-coil types, electronic ballasts offer silent operation plus the additional benefits of enhanced dimming potential, greatly reduced energy consumption, lighter weight, increased lamp efficacy, increased lamp life, and less heat. A disadvantage of electronic ballasts is the possibility of radiofrequency interference which may be intolerable in some settings such as scientific laboratories.

Originally, HID lamps were used almost exclusively outdoors where a high level of ballast noise could usually be tolerated. With the recent proliferation of compact, low wattage, and improved color HID sources, these lamps have become commonplace in commercial interiors. Again, no standard industry sound rating exists for HID ballasts. The noise characteristics of HID ballasts used indoors should be evaluated by the designer before they are specified. As with fluorescents, luminaire construction can have important consequences for the sound transmitted to the space.

The noise generated by a luminaire is not the only effect the lighting system can have on the acoustical environment. Most lighting equipment is installed in the ceiling where acoustical tile absorbs sound. The hard surfaces of the luminaire on the other hand reflect sound. In an open office where partial-height partitions are used for privacy, an improperly located luminaire can reflect speech from one cubical to the next, thus negating the purpose of the partition. If the walls extend only up to the suspended ceiling, the plenum will be completely open. Any opening in the luminaires will allow sound to travel through one luminaire into the plenum and out another luminaire.

IMPACT OF THE DISTRIBUTION OF LUMINAIRES ON THE ENVIRONMENT

The distribution of luminaires can be described in terms of (1) the luminous output of a single luminaire, (2) the luminous output of the luminaires in the environment, and (3) the location of the luminaires in the environment. The luminous output of a single luminaire is characterized by the luminous intensity (candlepower) distribution. In Chapter 6 we discussed the calculation of zonal lumens, which are derived from luminous intensity. The distribution of zonal lumens in the upper and lower hemispheres serves as the basis for the CIE-IES luminaire classification (see Chapter 6), which describes the general characteristics of the luminous output of a single luminaire.

The distribution of luminaires (output and location) in the environment can be classified as (1) general lighting, (2) localized lighting, and (3) localized-general lighting. Each of these classifications will be described in terms of the luminous output in the environment and the location of the equipment.

General Lighting (Ambient)

General lighting produces a uniform level of illuminance throughout the environment. The uniform level is usually achieved by a symmetrical placement (layout) of luminaires in the space. Classically, this has been the approach used to solve most lighting problems. Since no one was willing to specify worker locations during the design of a project, the lighting professional illuminated the *entire* space to a level suitable for performing the most demanding visual task that would occur in the space. This approach resulted in spaces being overlit, but no one objected since energy was abundant and cheap. This method is not consistent with today's declining resources and incipient energy cost escalation. Additional pressures to reduce lighting loads come from electric utilities operating at or near capacity; extremely high peak demand charges are being implemented as a disincentive to wasted loads such as excessive general lighting.

The concept of general lighting is not dead. The general lighting approach has simply changed directions from high (task level) uniform light levels, to lower uniform light levels. The term **ambient lighting** has been coined to describe the use of lower uniform (general) lighting levels throughout the space. Ambient lighting reduces luminance ratios in the visual field to maintain visual comfort. It also provides sufficient general illumination for safe circulation within the space. The currently accepted standard is to use ambient lighting levels that produce acceptable luminance ratios within the visual field for the particular application. Ambient lighting can be provided by two modes of application: (1) direct and (2) indirect.

Direct Mode

The term ''direct'' in this sense is used to describe the location of the lighting equipment. The direct mode implies that the luminaires are attached to the ceiling and most, if not all, of the luminous intensity is directed down toward the work surfaces. Luminaires classified as direct (CIE-IES classification) are most commonly used in this mode of application because of their greater efficiency (higher CU) and lower initial cost. However, CIE-IES classifications of semidirect and direct-indirect can be used to satisfy the direct mode of application. A list of advantages and disadvantages is given in Table 7-3 to help the lighting designer evaluate the direct mode of application. The list, while not exhaustive, should encourage in-depth analysis of the lighting system put into spaces.

A lighting system placed in or on the ceiling is intended to produce illumination for good visibility. Due to their potential for excessive brightness and glare, direct-mode systems can create a focal point on the ceiling plane rather than at other more

TABLE 7-3 Analysis of Direct (CIE-IES Classification) Lighting

Advantages	Disadvantages
More energy efficient	Harsh shadows
Low initial cost	Greater potential for direct glare
Lower floor to ceiling height	Greater potential for veiling reflections and reflected glare
Easier to control and direct	Excessive luminance ratios on the ceiling
	Excessive luminance and glare causes a change in focal point to the ceiling

important areas within the environment. That is, the equipment itself becomes a focal point; the lighting system may become a visually dominant feature in the space. In most applications this is not the desired effect. The lighting equipment should be visually subordinate to other aspects of the environment. Though the light introduced into a space will have a significant impact on the character of that space, the hardware should normally be neutral.

The use of the direct mode of application, along with localized lighting, is a viable means of solving certain types of lighting problems. Through the proper selection of a direct-mode system that controls glare, the overall visibility, visual comfort, and efficiency of the lighting system can be improved.

Indirect Mode

Indirect mode implies that most, if not all, of the luminous intensity is directed toward the ceiling and is then reflected back to the workplane. The luminaires can be floor mounted or suspended from the ceiling. The luminaires can be CIE-IES classification semi-indirect or indirect. Since the ceiling becomes the primary source of light relative to the workplane, the ceiling surface should be uniform (unobstructed), highly reflective, and diffuse in character. These characteristics will improve the visual appearance of the ceiling plane and, if combined with properly designed indirect lighting equipment, will result in uniformity of the ceiling. A list of a few of the advantages and disadvantages of indirect lighting is provided in Table 7-4 for analysis of the indirect mode.

A misleading claim and common misconception relative to the indirect lighting system is that better energy utilization is possible with indirect lighting systems. Since a coefficient of utilization represents the interaction of the luminaire distribution with the room characteristics, a standard coefficient of utilization can be used as an indicator of the overall system efficiency within a given environment. Table 7-5 shows us that the indirect system can never be more energy efficient than a direct system. Table 7-5 also shows the change in coefficient of utilization for a typical fluorescent luminaire as we change the luminaire from an indirect application to a direct application.

Once the indirect mode has been selected as a means of providing ambient lighting, the question then becomes one of choosing the light source to be used. A number of systems utilize fluorescent lighting; others use HID lamps, such as phosphor-coated mercury, metal halide, or high-pressure sodium lamps.

TABLE 7-4 Analysis of Indirect (CIE-IES Classification) Lighting

Advantages	Disadvantages
Potential for maximum visual comfort	Less energy efficient
Higher ESI	Requires a uniform, unobstructed, diffuse reflecting ceiling surface
Greater direct glare control	Increase in building cost, due to increased floor-to-ceiling height for good uniformity
Blends in with the environment, reducing the potential focus point change	National Electric Code requirements and electrical distribution system may increase cost
Produces uniform, diffuse, shadow-free light	
Qualifies as tangible personal property (a bonus for depreciation and investment tax credit) if light is part of the furniture	

The use of HID lamps in an indirect application will result in better energy utilization (Table 6-5) than a fluorescent system. When the control characteristics of a fluorescent luminaire versus an HID luminaire are studied, it becomes apparent that a properly designed HID unit will always be more efficient and easier to control than a fluorescent unit since the HIDs approximate a point source. Some of the lower-mounted indirect fluorescent lighting systems (at or below eye level) utilize a low-brightness louvered system to control the direct glare. These systems are less efficient than other forms of control systems, such as lenses (see Table 7-5). A comparison of the indirect coefficients of utilization for an HID system (Table 7-5) to the direct application of the low-brightness louvered system (Table 7-6) shows the indirect application of the HID system has a higher coefficient of utilization than even the direct application of the low-brightness fluorescent system.

In addition to the selection of the light source, another critical factor in the indirect system is the distance between the luminaire and the ceiling plane. To provide both maximum spacing and good uniformity, the indirect units must produce a wide distribution of luminous intensity, known as a ''batwing'' distribution. Improperly or inadequately designed luminaires will create a spotty pattern on the ceiling surface. This will result in a change in focal point to the ceiling surface. If the ceiling is the focal point, the ceiling will again become a visually dominant feature in the space, which makes the system less attractive than the direct mode.

All of this has important implications for suspended fluorescent indirect equipment. Most equipment on the market can provide good ceiling uniformity when mounted 24 in. from the ceiling. This would require a 9 ft ceiling, an almost unheard of luxury in modern commercial office space. Some manufacturers are claiming uniform results from a 6 in. mounting distance. Since this implies radical photometric design, claims regarding good uniformity at distances closer than about 20 in. should be thoroughly investigated by the lighting professional before specifying the use of such equipment on a project.

TABLE 7-5 Typical Luminaire Coefficient of Utilization: Direct and Indirect Mode

FLUORESCENT: Coefficient of utilization prism-K12									HID: Coefficient of utilization glass cover								
	Indirect $\rho_{cc} = 0.70$				Direct $\rho_{cc} = 0.70$					Indirect $\rho_{cc} = 0.70$				Direct $\rho_{cc} = 0.70$			
RCR	0.70	0.50	0.30	0.10	0.70	0.50	0.30	0.10	RCR	0.70	0.50	0.30	0.10	0.70	0.50	0.30	0.10
1.0	0.48	0.46	0.44	0.42	0.70	0.68	0.66	0.64	1.0	0.59	0.56	0.54	0.52	0.81	0.77	0.72	0.69
2.0	0.43	0.40	0.37	0.34	0.65	0.61	0.58	0.55	2.0	0.54	0.49	0.46	0.43	0.70	0.62	0.55	0.49
3.0	0.39	0.35	0.31	0.29	0.60	0.55	0.51	0.48	3.0	0.49	0.43	0.39	0.35	0.61	0.51	0.42	0.36
4.0	0.36	0.31	0.27	0.24	0.56	0.50	0.46	0.42	4.0	0.44	0.38	0.33	0.30	0.54	0.43	0.34	0.27
5.0	0.33	0.27	0.23	0.20	0.52	0.45	0.40	0.37	5.0	0.41	0.34	0.29	0.25	0.48	0.36	0.27	0.20
6.0	0.30	0.24	0.20	0.18	0.48	0.41	0.36	0.33	6.0	0.37	0.30	0.25	0.22	0.43	0.30	0.22	0.16
7.0	0.28	0.22	0.18	0.15	0.44	0.37	0.32	0.29	7.0	0.34	0.27	0.22	0.19	0.39	0.26	0.18	0.12
8.0	0.26	0.20	0.16	0.13	0.41	0.33	0.29	0.25	8.0	0.32	0.24	0.20	0.16	0.36	0.23	0.15	0.10
9.0	0.24	0.18	0.14	0.12	0.38	0.30	0.25	0.22	9.0	0.29	0.22	0.17	0.14	0.33	0.20	0.13	0.08
10.0	0.22	0.16	0.13	0.10	0.35	0.27	0.23	0.19	10.0	0.27	0.20	0.16	0.13	0.30	0.18	0.11	0.06

TABLE 7-6 Typical Low-Brightness Fluorescent System: Direct and Indirect Modes

	Coefficient of utilization plastic louver									Coefficient of utilization low-brightness wedge							
	Indirect $\rho_{cc} = 0.70$				Direct $\rho_{cc} = 0.70$					Indirect $\rho_{cc} = 0.70$				Direct $\rho_{cc} = 0.70$			
RCR	0.70	0.50	0.30	0.10	0.70	0.50	0.30	0.10	RCR	0.70	0.50	0.30	0.10	0.70	0.50	0.30	0.10
1.0	0.37	0.35	0.34	0.33	0.54	0.52	0.51	0.49	1.0	0.33	0.31	0.30	0.29	0.49	0.48	0.47	0.46
2.0	0.34	0.31	0.29	0.27	0.50	0.47	0.44	0.42	2.0	0.30	0.27	0.25	0.24	0.46	0.44	0.42	0.41
3.0	0.31	0.27	0.24	0.22	0.47	0.42	0.39	0.37	3.0	0.27	0.24	0.21	0.20	0.44	0.41	0.39	0.37
4.0	0.28	0.24	0.21	0.19	0.43	0.38	0.35	0.32	4.0	0.25	0.21	0.18	0.16	0.41	0.38	0.35	0.33
5.0	0.25	0.21	0.18	0.16	0.40	0.35	0.31	0.28	5.0	0.22	0.19	0.16	0.14	0.38	0.35	0.32	0.30
6.0	0.23	0.19	0.16	0.14	0.37	0.32	0.28	0.25	6.0	0.21	0.17	0.14	0.12	0.36	0.32	0.29	0.27
7.0	0.21	0.17	0.14	0.12	0.34	0.29	0.25	0.22	7.0	0.19	0.15	0.12	0.10	0.34	0.30	0.27	0.25
8.0	0.20	0.15	0.12	0.10	0.32	0.26	0.22	0.20	8.0	0.17	0.13	0.11	0.09	0.31	0.27	0.24	0.22
9.0	0.18	0.14	0.11	0.09	0.30	0.24	0.20	0.17	9.0	0.16	0.12	0.10	0.08	0.29	0.24	0.21	0.19
10.0	0.17	0.13	0.10	0.08	0.28	0.22	0.18	0.16	10.0	0.15	0.11	0.09	0.07	0.27	0.22	0.19	0.17

The goal of the indirect mode should be to minimize or eliminate the spotty appearance of the ceiling surface. This requires two things in the indirect system: (1) a properly designed optical system to provide a uniform illuminance level across the surface of the ceiling and (2) proper selection of a ceiling material. Ideally, the ceiling surface should be perfectly diffuse or approach a perfect diffuser. This requires a matte, high-reflectance surface with a very slight texture to help diffuse the light on the surface. Exposed T-bar ceiling systems can create problems of reflected images off the glossy, painted surface of the T-bar. The ceiling once again becomes an undesired focal point.

The luminaire should be at a sufficient distance off the floor to prevent direct viewing of the lamp and housing. If the luminaire housing is below eye level, the more inefficient low-brightness systems must be used to control the glare.

A well-designed indirect lighting system will produce uniform luminance across the ceiling surface and uniform, diffused, shadow-free light throughout the space at the work level. Uniform diffused, shadow-free light is quite good from a visual performance standpoint. Aesthetically and psychologically, however, uniform, diffuse, shadow-free illumination can be quite boring and uninteresting. The lighting designer should take this into consideration and provide visual relief from the boredom of the uniform diffused illumination. This can be done, for example, by introducing into the space (especially around the periphery) two or three bright highlights such as by accenting a work of art.

Localized Lighting

Localized lighting is achieved by placing the lighting in the immediate vicinity of the task. Spill light from the luminaire is expected to produce some general illumination in the space. Localized lighting alone is normally not adequate to create acceptable luminance ratios in the environment, visual discomfort may result; therefore, this type of lighting system should be avoided in areas where critical visual tasks are performed.

Localized-General Lighting

Localized-general lighting, also known as task/ambient lighting, can be achieved by providing a uniform general lighting level throughout the space and localized lighting where higher task illuminance levels are required. This type of localized lighting is called **task lighting.** The general lighting requirements can be satisfied by either direct- or indirect-mode applications.

A second approach to localized-general lighting would consist of a functional arrangement of luminaires relative to the task location. The nonsymmetrical arrangement of luminaires must provide adequate general illumination. This will result in a nonuniform general illumination in the space and a nonsymmetrical luminance pattern on the ceiling plane. Nonsymmetrical patterns from direct- or indirect-mode applications can draw attention to the ceiling, possibly resulting in a visually dominant ceiling.

Task lighting can be provided by overhead luminaires mounted in the ceiling or luminaires mounted near the task area in a furniture module.

Furniture Task Lighting

Furniture task lighting has the advantage of a potential decrease in power consumption due to the decreased distance from the luminaire to the task. Because of the close proximity of the lighting equipment and the confined area of the furniture module, the possible luminaire locations are limited. The easiest location is directly in front of the task under a shelf or mounted on a backboard. This location will usually result in decreased visibility and visual discomfort, which may lead to an increase in the time required to perform visual work and thus an increase in building usage and energy consumption. Comfort control of task lighting is dependent on luminance ratios (1) within the furniture module, (2) between the luminaire and its immediate surround, and (3) between the furniture module and the far surround. Even though illuminance levels may be higher than prescribed, improperly located lighting systems relative to the task and the viewing position can result in a reduction in contrast, and hence visibility. This reduction in contrast may cause the user to sacrifice either speed or accuracy in performing work. One hopes that the user, under reduced visibility conditions, will make the proper choice between decreasing speed and lowering acceptance standards for accuracy.

Overhead Task Lighting

If task locations within the space can be defined, overhead task lighting has one major advantage: the entire ceiling is available for luminaire placement. This makes the control of visibility much easier as long as furniture locations do not change relative to luminaire positions. The relationship of luminaire location to task location is absolute and essential for the maximization of visibility. Because the luminaires are removed from the line of sight, the visual comfort potential is also quite good.

IMPACT OF THE LUMINOUS ENVIRONMENT ON THE USER

The goal of lighting design is to provide optimum visibility so that the user can perform work. Lighting is not placed in space simply to gratify the ego of a lighting designer. Lighting systems that are put into critical visual areas for purely aesthetic reasons or for purely economic reason, such as lower initial cost, may reduce visibility and create discomfort. This may result in emotional and psychological deprivation for the user, and an increase in the time spent to perform critical visual work.

REFERENCES

1. Finn, J. F., ''Efficient Application of Lighting Energy—A Luminaire Air Heat-Transfer Evaluation.'' *Lighting Design + Application* (January 1976).
2. Kaufman, John E., and Jack F. Christensen, eds., *IES Lighting Handbook,* Application Volume, 1987, New York: Illumination Engineering Society of North America, 1987.

8 quantity and quality assessment

When evaluating the design for a lighting system three questions are frequently asked. First, *How much light* (Quantity) does the system provide on the task? Second, Is that lighting *useful and comfortable* (Quality) light? Third, What does the lighting system cost? Numerical methods have been developed to assist in design and analysis of the lighting system. This chapter assists the lighting professional in evaluating two of these design considerations: **quantity** and **quality.** This chapter will briefly discuss the question of **cost,** however, a detailed coverage of cost will be found in Chapter 14 on economic analysis.

QUANTITY DETERMINATION METHODS

The most frequently calculated quantity in lighting analysis is illuminance. A great irony can be seen here since **illuminance** (a measure of light *arriving at the task*) correlates poorly with the functioning of the eye. In order to predict how the human visual system will perform, information is needed about the light *leaving the task* and arriving at the eye. We will see later that measures such as luminance and contrast yield more appropriate information.

However, it is still important for the lighting professional to be proficient in illuminance calculations. Quantity implies amount of light required to perform a visual task. The apparent ease with which illuminance levels can be verified in the field has deceived those with a less than complete understanding of lighting into believing that a single number (illuminance level) is sufficient for specifying a luminous environment. Every illumination specialist will encounter architects, engineers, interior designers, code authorities, owners, and others, who will demand to know "How many footcandles will your proposed system give me?" It is the lighting professional's duty to help educate these people and to inform them that fulfilling illuminance criteria does not necessarily ensure good design. Further, an understanding of illuminance calculations is an essential foundation for learning the more complex concepts in lighting.

ILLUMINANCE SELECTION PROCEDURE

The 1981 Application Volume of the *IES Lighting Handbook* introduced the new method of determining the recommended illuminance value, E_r, for various tasks. In previous editions of the handbook, Figure 9-80 gave single footcandle values for given tasks. In 1979, the IESNA adopted a procedure that establishes a range of three illuminance values for nine task categories. The choice of one of the three illuminance values is based on a set of weighting factors. The E value ranges will be referred to as (1) **lower** value, (2) **middle** value, and (3) **upper** value.

The illuminance selection procedure requires the lighting professional to have knowledge of

1. Type of task
2. Age of the worker
3. Importance of speed/accuracy
4. Reflectance of room surfaces or task background

The method lists nine (9) illuminance categories, designated "A" through "I" covering illuminance levels from 2 to 2000 fc (20 to 20,000 lux).

- Categories A through C—where no task activity occurs or general lighting is throughout a space.

- Categories D through F—where activities involve a visual task.
- Categories G through I—suggested a combination of general and supplementary lighting on the task.

For Categories A through C:

Two factors must be considered in determining the weighting factor:

1. Worker's age
2. Room surface reflectance which is the weighted average reflectance value.

$$\text{Weighted Average Reflectance} = \frac{(CR \times A(\text{ceiling})) + (WR \times A(\text{walls})) + (FR \times A(\text{floor}))}{A(\text{ceiling}) + A(\text{walls}) + A(\text{floor})}$$

For Categories D through I:

Three factors must be considered in determining the weighting factor.

1. Worker's age
2. Importance of speed/accuracy
3. Reflectance of the task background

Procedure:

1. Determine category from table of recommended illuminance values.
2. Determine the weighting factor by *algebraically* adding weighting factors (accounting for signs).
   a. For Categories A through C,
      (1) If the weighting factor = -2, use "lower value."
      (2) If the weighting factor = $+2$, use "upper value."
      (3) If the weighting factor is from $+1$ to -1, use "middle value."
   b. For Categories D through I,
      (1) If the weighting factor ≤ -2, use "lower value."
      (2) If the weighting factor $\geq +2$, use "upper value."
      (3) If the weighting factor is from $+1$ to -1, use "middle value."

Illuminance Calculations

In calculating illuminance, we may wish to know the illuminance at a specific point **(point-by-point method),** or we may be interested in the average uniform horizontal illuminance **(lumen method)** across the **workplane** (an imaginary plane parallel to the floor at the height of the task—frequently taken as 2.5 ft above the floor). Light arrives at a work surface in two ways: (1) directly from the luminaires, windows, and other luminous sources in the space and (2) after being reflected (and perhaps re-reflected numerous times) off walls and other room surfaces and furnishings. Appropriately, the former is called the **direct component** and the latter the **interreflected component** of illuminance at a point.

There are three methods of laying out luminaires in a space:

1. Uniform, symmetrical, or general layout: Task locations are not known (lumen method).
2. Nonuniform, nonsymmetrical, or task-oriented layout: Task locations are known (point-by-point).
3. Combination of (task + general) or supplementary lighting.

In the rest of this section we will examine these methods. First, the point-by-point method is presented as the most common way to predict the direct component of illuminance at a point under a wide variety of circumstances. Next, the lumen method of calculating average uniform horizontal illuminance is discussed.

Point-by-Point Method

The point-by-point method for interior applications is used in nonuniform lighting layout or task-oriented designs. The source of light must be classified as a point source, that is, the distance from the source to the calculation point must be equal to or greater than five times the dimension of the source of light. The point-by-point method of determining the direct component of illuminance is an application of the **Fundamental Law of Illumination,** which is defined as the **Inverse-Square Law:**

$$E_n = \frac{I_n}{D^2} \tag{8.1}$$

The Inverse-Square Law states that the illuminance, E_n, at a point on a surface normal (n) to the light ray is directly proportional to the luminous intensity, I_θ, of the source in the direction of the point and inversely proportional to the squared distance, D^2, between the source and the point of calculation. These terms are shown in Figure 1-7.

Although the Inverse-Square Law is simple, a number of assumptions limit its use. The most critical assumption is that the source must qualify as a point source. If the maximum source dimension is one-fifth the distance to the point of calculation, the error will be less than 1%, and the source can be considered a point source. Finally, **asymmetrical distributions** may cause error at particular points because the interpolation of the luminous intensity between reported photometric planes may not be accurate.

Horizontal Surface Illuminance. The Inverse-Square Law yields the illuminance on a plane perpendicular (normal) to the direction of the incident light. Usually however, the plane of interest is at an angle, θ, to the incident light (see Fig. 8-1).

In most applications, the **nadir angle** ($\theta = 0°$) of the luminaire is usually perpendicular to the ground (or workplane) which is called the **horizontal plane.** For illumination

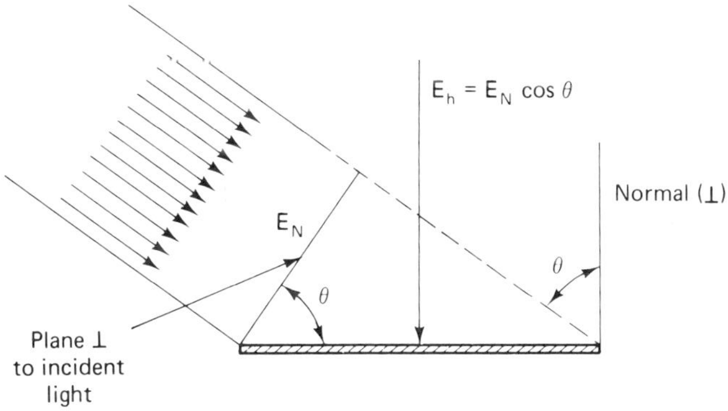

Figure 8-1. Cosine Law of Illumination.

on a horizontal plane, the **Cosine Law of Illumination** must be applied. Equation (8.1) then becomes

$$E_h = E_n \cos \theta \tag{8.2}$$

where E_h is the illuminance on a horizontal plane due to the luminous intensity incident at an angle θ to the normal. Replacing E_n in equation (8.2) with equation (8.1) gives

$$E_h = \frac{I_\theta}{D^2} \cos \theta \tag{8.3}$$

where

$$D^2 = (x)^2 + (y)^2 \quad \text{(see Fig. 8-2)}$$

Figure 8-2 shows why the cosine term is introduced. Like any vector, the illuminance vector E_n from the luminaire to the point of interest can be divided into two components, which are perpendicular to each other, using the sine and cosine functions. The "y" component of the vector, E_y, is perpendicular to the horizontal plane of interest; therefore,

$$E_h = E_y = E_n \times \cos \theta \tag{8.4}$$

When equation (8.1) is substituted into equation (8.4), the resulting equation is the same as equation (8.3). Since the E_x component is parallel to the horizontal plane, it will never strike the horizontal plane, and so E_x will make *no* contribution to the illuminance at a point on the horizontal plane. We therefore ignore E_x.

If a number of calculations are to be made of illuminance on a horizontal surface from an array of sources which are mounted at the same height, H, a useful formulation can be achieved by noting (see Fig. 8-2) that

$$D = \frac{H}{\cos \theta} \tag{8.5}$$

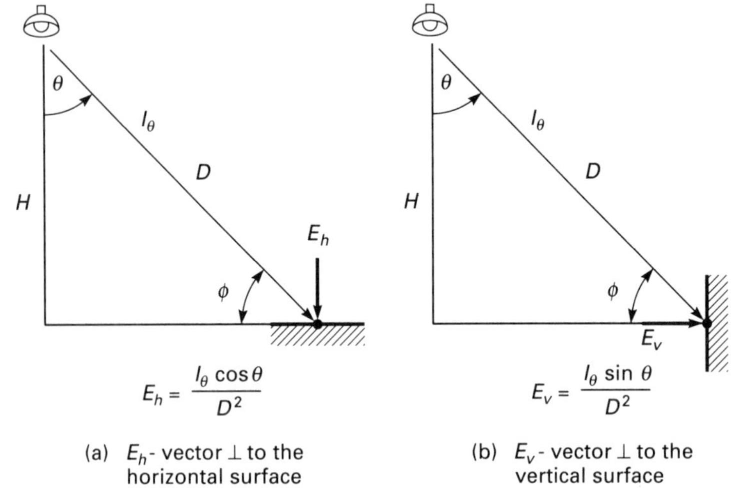

Figure 8-2. Illuminance on a Vertical or Horizontal Plane.

Substituting equation (8.5) into equation (8.3) gives the **Cosine Cubed Law of Illumination:**

$$E_h = \frac{I_\theta \cos^3 \theta}{H^2} \tag{8.6}$$

In this form (equation 8.6), only the direction of the intensity vector I_θ need be known, eliminating the calculation of the distance, D, to the point of interest.

Vertical Surface Illuminance on a Vertical Line Through the Aiming Point. The vertical line through the aiming point is formed by the intersection of a vertical photometric plane that is perpendicular to the vertical wall surface. Since (see Fig. 8-2), E_x component of the vector, E_n, is perpendicular to the vertical surface of interest, then,

$$E_v = E_x = E_n \times \sin \theta \tag{8.7}$$

When equation (8.1) is substituted into equation (8.7), the equation for the illuminance of a vertical surface (Fig. 8-2) such as a wall becomes

$$E_v = \frac{I_\theta}{D^2} \sin \theta \tag{8.8}$$

or

$$E_v = \frac{I_\theta}{D^2} \cos \Phi$$

To apply equation (8.8), the vertical plane through the point of interest and the luminaire must be perpendicular to the wall or vertical surface containing the point. Since the E_y component (see Fig. 8-2) is parallel to the vertical plane, it will never strike the vertical plane, and so E_y will make *no* contribution to the illuminance at a point on the vertical plane. We therefore ignore E_y.

Vertical Surface Illuminance on a Horizontal Line Through the Aiming Point. This type of problem is commonly encountered when calculating illumination at points from a wall wash luminaire that has photometric data given for only two planes. The two planes given are the vertical plane and the horizontal plane (which is perpendicular to the vertical plane). The intersection of the two perpendicular photometric planes defines the aiming line (see Fig. 8-3). The point where the aiming line strikes the wall is called the **aiming point.** The intersection of a vertical photometric plane which is passing through the luminaire and is perpendicular to the wall (or vertical surface), will form a vertical line on the wall. The horizontal photometric plane will form a horizontal line on the wall through the aiming point. Calculation points along the horizontal line are said to be **off-axis.** For off-axis calculation points, the E_v component of the vector, $E_{\theta h}$, is found by applying a double trigonometric correction. Since the E_x component (see Fig. 8-3) of the off-axis illumination, E_v, is perpendicular to the vertical surface at the calculation point, P, then

$$E_v = E_n \times \sin\theta_s \times \cos\alpha \tag{8.9}$$

where

$$I_\theta = I_{\theta h} \tag{8.10}$$

When equation (8.1) and equation (8.10) are substituted into equation (8.9), the formula for the illuminance off axis (point P, Fig. 8-3) on a vertical surface such as a wall becomes

$$E_v = \frac{I_{\theta h}}{D^2} \sin\theta_s \cos\alpha \tag{8.11}$$

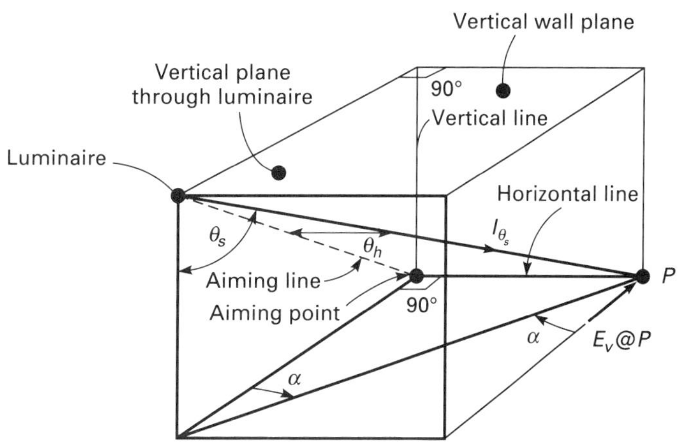

Figure 8-3. Vertical Illuminance Off Axis.

An example of calculating vertical illuminance off-axis is given at the end of this chapter in the Data Section.

Vertical Surface Illuminance on a Vertical Line Formed by an Oblique Plane: If a symmetrical downlight that has its aiming point at nadir is placed near a vertical wall, the photometric planes will intersect the wall forming vertical lines. The vertical photometric plane passing through a calculation point may be at an oblique angle to the wall (not perpendicular). The oblique plane configuration will require the application of a double trigonometric equation; however, the intensity value used will be an elevation angle, θ, in the vertical photometric plane, $I_{\theta s}$ (see Fig. 8-3).

$$E_v = \frac{I_{\theta s}}{D^2} \sin \theta_s \cos \alpha \qquad (8.12)$$

An example problem is given at the end of this chapter in which the calculation point lies on a vertical line that is formed by a vertical photometric plane (the oblique-plane case) that is *not* perpendicular to the wall.

Illuminance Calculations for Nonpoint Sources: To apply the Inverse-Square Law in a point-by-point calculation, the distance from the source of light to the calculation point must be greater than or equal to five times the maximum dimension of the source.

Example:

A 2 × 4 troffer mounted on an 8 ft ceiling will not classify as a point source for calculating illuminance at a point on a work surface that is 2.5 ft above the floor.

$$MH = 8' - 2.5' = 5.5$$

For application of the Inverse-Square Law, the distance, d, must be

$$d_{\min} = 5 \times 4' = 20' > 5.5'$$

One solution to this problem is to divide (discretize) the luminaire into smaller areas and assume that each small area has a luminous intensity distribution that has the same shape as the total 2 × 4 luminaire but the magnitude of each intensity value is scaled down.

For the example given, assume the 2 × 4 luminaire is divided into 1 × 1 segments. Each 1 ft segment would require a $d_{\min}$ of 5 ft to apply the Inverse-Square Law.

$$d_{\min} = 5 \times 1' = 5' < MH = 5.5'$$

The assumption of a proportional (fractional) equal luminous intensity from each of the 1 × 1 segments will introduce some error into the calculated value of illuminance at a point. By looking at almost any 2 × 4 lensed troffer, the lighting professional can see that the uniformity of the lens varies across the lens surface.

For the example, if the 2 × 4 troffer is divided into 1′ × 1′ segments, the troffer would consist of eight segments. Applying the assumption of proportional equal luminous intensity, all the luminous intensity values reported for the luminaire would

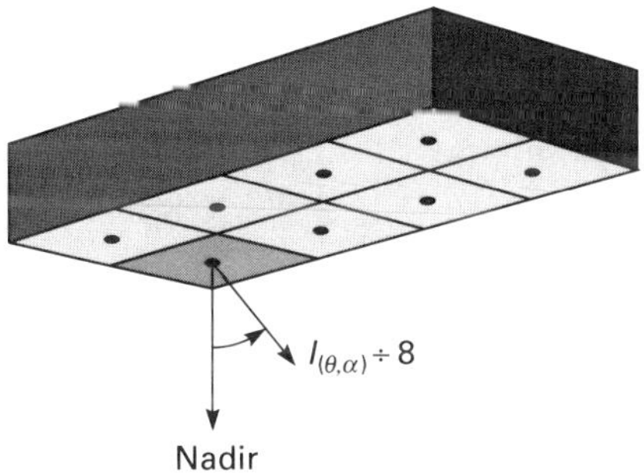

Figure 8-4. Discretization for Nonpoint Sources.

be scaled down by a factor of ⅛ and assumed to be located in the center of each 1 × 1 segment (Fig. 8-4).

$$\text{scaling factor} = \frac{\text{area of the segment}}{\text{area of the source}}$$

$$\text{scaling factor} = \frac{1 \times 1}{2 \times 4} = \frac{1}{8}$$

Summary: The Inverse-Square Law and the Cosine Law of Illumination are basic to almost every method of illumination design. Isofootcandle diagrams and tables of illuminance values are calculated by this design method. Only the direct component of illuminance can be determined using the Inverse-Square Law. Multiple interreflections from walls, ceiling, and floor may occur, making this method valid only for exterior calculations or for interior spaces where the interreflected component can be assumed to be negligible. A method of calculating the interreflected component is described in Chapter 12.

Lumen Method

Specifications often require the lighting professional to know or design for average uniform horizontal illuminance. To do this with the Inverse Square Law for a large number of points would be both tedious and expensive. In addition, a second set of calculations would have to be made to determine the interreflected components. The lumen method is used to calculate the number of luminaires required for a uniform or ''general'' lighting layout. The lumen method (8) calculates the average, ''uniform,'' horizontal maintained illuminance throughout a room. The term ''uniform'' means (1) no more than 20% variation in illuminance and (2) symmetrical lighting layout that meets the spacing criterion. The lumen method was developed to simplify calculations utilizing the following basic equations:

Analysis Formula

$$E_{ave} = \frac{(\#\ \text{lamps}/LUM)\ (\text{lm/lamp})\ (\#\ \text{of}\ LUM)(CU)\ (LLD)\ (LDD)\ (LF)\ (BF)\ (RSD)}{\text{area of the workplane}} \quad (8.13)$$

Design Formula:

$$\# \text{ of LUM} = \frac{Er \times (\text{area of the workplane})}{(\#\ \text{lamps}/LUM)\ (\text{lm/lamp})\ (CU)\ (LLD)(LDD)\ (LF)\ (BF)\ (RSD)} \tag{8.14}$$

where WP = workplane
LUM = luminaire
CU = coefficient of utilization
LLD = lamp lumen depreciation
LDD = luminaire dirt depreciation
LF = lamp factor
BF = ballast factor
RSD = room surface depreciation
LLF = light loss factor = maintenance factor, MF ($MF = LLD \times LDD \times RSD$)

Coefficient of Utilization: The coefficient of utilization (CU) is calculated by the **zonal cavity method** and is a measure of how a specific luminaire distributes light (uniform horizontal illuminance) into a given room.

The CU takes into account

1. Luminaire efficiency
2. Luminaire distribution characteristics with respect to
   a. Room proportions: room size and shape
   b. Surface reflectances
   c. Mounting height

Coefficients of utilization for a particular luminaire are found in a table format which is part of the photometric test report for that luminaire. A typical example is shown in Table 8-3. CU values are also shown in Table 8-4 in the Data Section. The CU values are typically reported as a whole number such as "63," however, the digits are actually preceded by a decimal so that the value is 0.63. The table assumes an effective floor cavity reflectance of 20% and a spacing-to-mounting height ratio of 0.40 to calculate the CU table.

Figure 8-5 shows the parameters used to determine the CU in a room. The quantity known as the **cavity ratio (CR)** is defined in general terms as

$$CR = \frac{2.5\ (\text{area of the walls})}{\text{area of the workplane}} \tag{8.15}$$

Each of the cavities formed by the luminaire plane and workplane (see Fig. 8-5) has an associated cavity ratio:

$$\text{ceiling cavity ratio } (CCR) = \frac{2.5h_{cc}\ (\text{perimeter of the walls})}{\text{area of the workplane}}$$

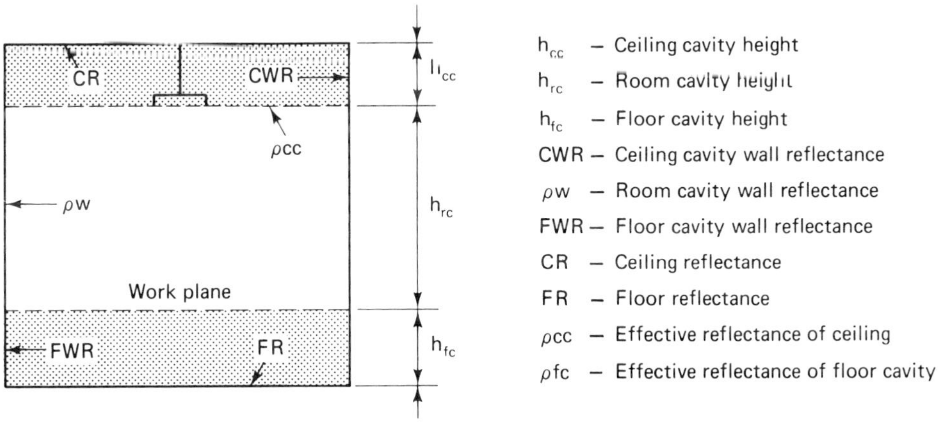

Figure 8-5. Terminology Used in the Zonal Cavity Method.

$$\text{room cavity ratio } (RCR) = \frac{2.5h_{rc} \text{ (perimeter of the walls)}}{\text{area of the workplane}}$$

$$\text{floor cavity ratio } (FCR) = \frac{2.5h_{fc} \text{ (perimeter of the walls)}}{\text{area of the workplane}}$$

For a rectangular or square room, the cavity ratio formulas are

$$CCR = \frac{5 \times h_{cc}\,(W + L)}{W \times L}$$

$$RCR = \frac{5 \times h_{rc}\,(W + L)}{W \times L}$$

$$FCR = \frac{5 \times h_{fc}\,(W + L)}{W \times L}$$

The height of the ceiling cavity, h_{cc}, is the suspension length of the luminaires; the height of the floor cavity, h_{fc} is the distance from the floor to the workplane on which the average illuminance is being calculated. The height of the room cavity, h_{rc}, is the distance between the luminaire plane and the horizontal workplane.

If the suspension length is zero, the effective ceiling cavity reflectance, ρ_{cc}, is equal to the actual ceiling reflectance, *CR*. For suspension lengths other than zero, the ceiling cavity ratio (*CCR*) is calculated. Using Table 8-5 in the Data Section with the *CWR*, *CR*, and *RCR*, we can find an effective reflectance of the ceiling, ρ_{cc}. The same procedure is used for an effective floor cavity reflectance, ρ_{fc} when the workplane is any height off the floor.

Referring to the flowchart (Fig. 8-6), the effective ceiling cavity reflectance, ρ_{cc}, the wall reflectance in the room cavity ρ_w, and the room cavity ratio must be determined. The CU value is picked from Table 8-4. This CU value is based on a ρ_{fc} equal to

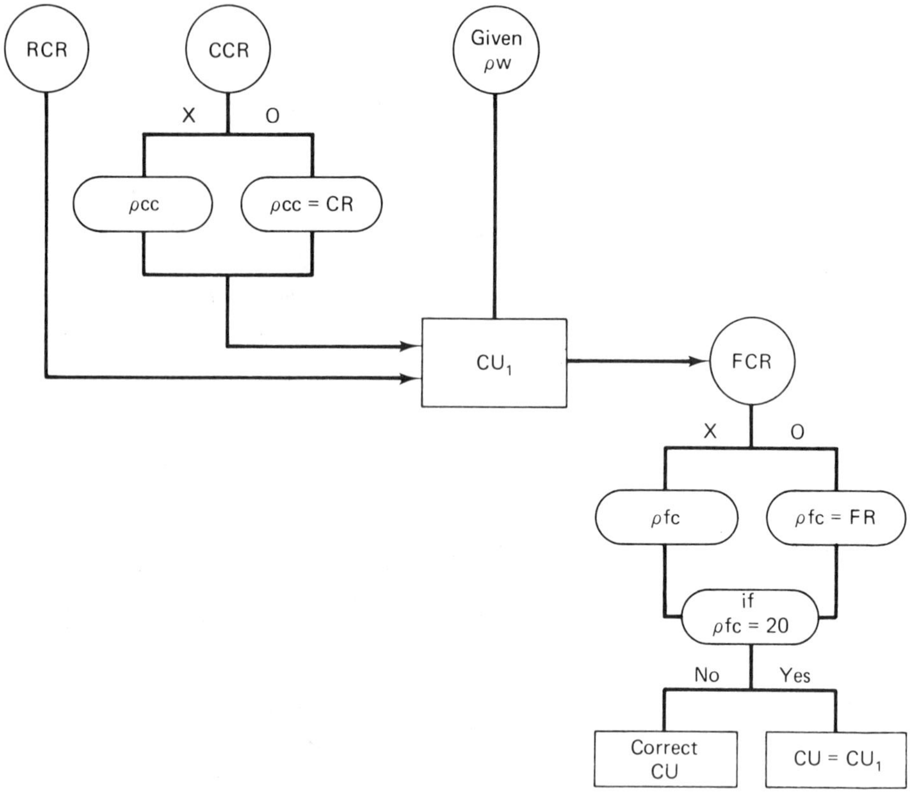

Figure 8-6. Determining the CU by the Zonal Cavity Method.

0.20. If the calculations have shown the ρ_{fc} to be something other than 20%, a correction factor must be applied to the CU. Correction factors can be found in Table 8-6 of the Data Section.

Since it would be impractical to generate tables that cover all reflectances, interpolation of the tables is required. Linear interpolation is sufficient for this type of calculation and is least time consuming. Refer to the example problems at the end of this chapter for the techniques involved. Additional examples can be found in reference 8.

Light Loss Factors: Two other quantities needed in the lumen method formula (equation 8.13 or equation 8.14) are the **lamp lumen depreciation (LLD)** and the **luminaire dirt depreciation (LDD)** factors. These factors take into account the decrease in light output of lamps over their life and the accumulation of dirt on luminaire surfaces.

The LLD factors for different lamps are found in Table 8-8. To determine the LDD factor (Table 8-7), each luminaire is categorized according to its maintenance characteristics.

The LDD category number (I, II, III, IV, V, VI) is listed on the photometric test report (Tables 8-3 and 8-4). Knowing the category, the appropriate graph (Table 8-7) can be used to find the LDD factor. Assumptions must be made as to the cleanliness of the environment and the interval between luminaire cleaning.

The **light loss factor (LLF)**, which is sometimes called the **maintenance factor (MF),** is the product of the LLD times the LDD.

Lamp Factor: The **lamp factor** is used to update a photometric test of an existing luminaire that was tested with a lamp whose lumen rating has increased since the test was first issued. The lamp factor (LF) is calculated by the formula

$$\text{lamp factor:} \quad LF = \frac{\text{actual catalog lumens}}{\text{test lumens}}$$

Ballast Factor. The lumen rating of a fluorescent lamp is based on operating the lamp on a ''standard reference ballast'' under controlled temperature and humidity. When the lamps are placed in a real luminaire with production ballasts, light output is reduced. The **ballast factor** accounts for this reduction.

The ballast factor under test conditions is 1.0. A standard off-the-shelf, high-power-factor, magnetic ballast in a well designed luminaire produces a ballast factor between 0.92 to 0.96 which was close enough to 1 that the BF was ignored. When the ''energy-efficient lamps,'' ''energy-efficient ballast,'' and the smaller-diameter lamps were introduced into the marketplace, the ballast factor became an important issue and must be included in the calculational procedure. The ballast factor for the lamp, luminaire, and ballast listed on the photometric test report applies only to that combination. If any component is substituted into the luminaire, the BF is no longer valid.

The zonal cavity method is used to calculate a CU value that is used in the lumen method to calculate the average uniform horizontal illuminance on the work-plane. To ensure that the illuminance is uniform (a necessary condition in the lumen method), the spacing criterion (spacing-to-mounting height ratio) cannot be exceeded. In addition, to prevent drop-off of illuminance near the walls, the recommended spacing procedures shown in Table 8-9 should be followed.

A number of assumptions have been made in the zonal cavity method for determining a CU value. The accuracy of the results is therefore limited. Also, it must be remembered that an average uniform horizontal illuminance is a calculated value. Use of the lumen method should be limited to noncritical, uniform (nontask-oriented) lighting and for making a preliminary determination of the approximate number of luminaires needed. Where visibility of critical tasks and nonuniform lighting layout or task-oriented lighting is important, more sophisticated techniques (see Chapter 12) should be used. The reader is referred to reference 8 for other techniques for determining the level of illuminance.

QUALITY DETERMINATION METHODS

The lighting professional needs to remember that ''footcandles'' do not necessarily mean that occupants of a space can see a task. That is, ''more light, better sight'' applies only for the limited condition of a task which is a perfect diffuser task (see Chapter 1).

An individual's sensitivity to contrast increases as background luminance increases, and background luminance increases as illuminance increases, while the task contrast is dependent on the luminance of the background, L_b, and the luminance of the object, L_o.

The factors that affect an individual's ability to see a task are

1. The location of light source
2. The location of observer (occupant of the space)
3. The reflecting properties of the task (luminance factor, β)

There are a number of important factors that will have bearing on the quality of a lighting installation. The three primary factors are **glare, luminance ratios,** and **color.** The subject of color was discussed in Chapter 3.

Glare

For a luminous environment to be visually comfortable, the occupants must not see any glare. Glare is defined as the result of any excessively bright source of light in the field of view that causes a loss in visibility and/or discomfort. Physical discomfort (discomfort glare) or loss in visual performance and visibility (disability glare) should be of primary concern to the lighting professional.

Many factors are involved when glare problems are created. One is the length of time that the excessive luminance is present. Another factor is the luminance ratio between the glare source and the surround area in the field of view. The task involved is another factor that influences glare. Important elements that impact the level of glare are source size, luminance level, position of the glare source, and the level of surround luminance. A small bright source directly on the line of sight against dark surroundings is an example of "excessive" glare.

Glare is complex and can be categorized into two major types, direct and indirect. **Direct glare** is due to any excessively bright sources of light (luminaires and/or windows) in the field of view that shines directly into the eyes, resulting in discomfort and a loss in visibility; **indirect glare** is due to any excessively bright source of light coming to the eyes indirectly (by reflection) causing a loss in visibility and/or discomfort. There are two forms of indirect glare, which are called **reflected glare** and **veiling reflections.**

Glare—Horizontal Tasks

Direct glare is associated primarily with heads-up tasks (Figure 8-7); that is, the glare produced in the visual field that affects the visual system as the individual looks around the environment. Indirect glare is associated with heads-down tasks which occurs when an individual looks down at the horizontal work surface.

The designer must evaluate each room on the basis of the tasks performed to determine which type of glare is the greater problem. In general, it is difficult to optimize the solution to both direct and indirect glare, resulting in trade-offs.

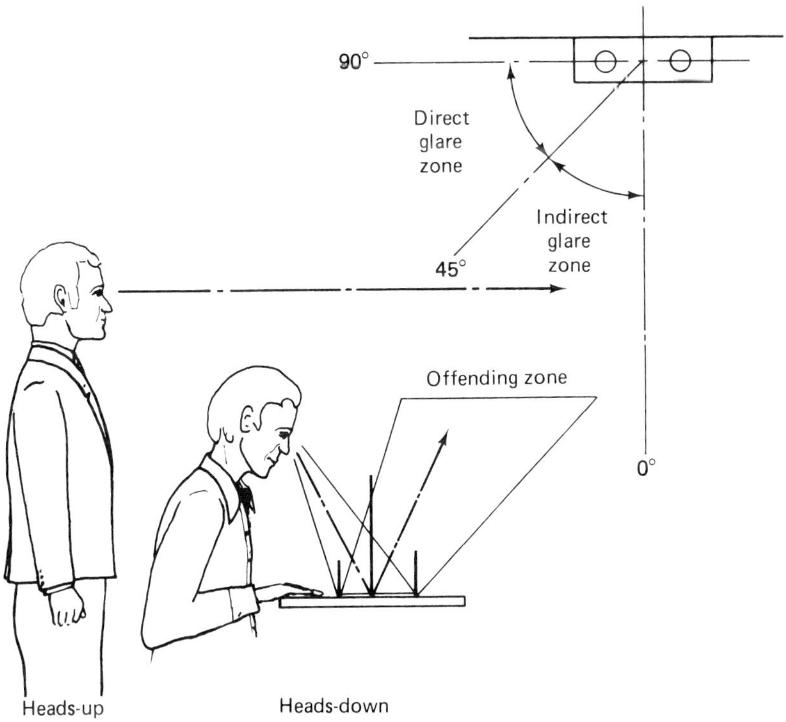

Figure 8-7. Direct and Indirect Glare Zones (Horizontal Tasks).

Direct Glare. Direct glare (heads-up tasks) is evaluated in terms of **visual comfort probability, (VCP)** (2). Relative to the luminaire, the direct glare zone (for a horizontal task) is associated with luminous intensity produced in the zone from 45° to 90° (see Figure 8-7).

Visual Comfort Probability (VCP). Direct glare is largely dependent on the characteristics of the room and the location of light sources in the field of view. The effect of direct glare on an individual in a space can be assessed by the visual comfort probability method. The Direct Glare Committee of RQQ Committee (2) of IESNA states that "direct glare may not be a problem in lighting installations if all three of the following conditions are satisfied":

1. The VCP is 70 or more.
2. The ratio of maximum-to-average luminaire luminance does not exceed 5 to 1 (preferably 3 to 1) at glare angles (45°, 55°, 65°, 75°, and 85° from nadir) in both the "crosswise" and "lengthwise" viewing directions.
3. Maximum luminaire luminance both crosswise and lengthwise should not exceed the following values:

Angle (from Nadir)	Maximum Luminance (fL)	(cd/ft^2)	(cd/m^2)
45	2250	716	7709
55	1605	511	5499
65	1125	358	3855
75	750	239	2570
85	495	158	1696

An initial judgment as to the direct glare potential of a luminaire can be made by examining the maximum and average luminance at the glare angles. A number of methods of evaluating direct glare have been devised. Visual comfort probability is the method currently accepted by the IESNA for evaluating direct glare in a room. *The VCP value represents the percentage of people that probably will not complain about the glare produced in the space.*

The VCP method involves a luminaire-by-luminaire determination of the discomfort index for each luminaire within the field of view. The individual indexes are then combined in an appropriate manner to obtain an overall rating called the **discomfort glare rating (DGR).** To obtain the numerical value of VCP, the overall DGR rating is entered on a graph. Refer to reference (4) for a more detailed description of the VCP method.

Comfort varies widely with the individual. Therefore, the VCP rating system should act only as a guide to quality design. The VCP method assesses visual comfort that is related to discomfort (discomfort glare), which is commonly associated with direct glare.

The computation of VCP for a room is dependent on the following factors:

1. Room size and shape
2. Surface reflectances
3. Illumination level
4. Luminaire type, size, and distribution
5. Number, location, and orientation of luminaires
6. Luminance of the entire visual field
7. Observer location and line of sight
8. Differences in individual glare sensitivity
9. Equipment and furniture

There are two ways to obtain a VCP value for a room: (1) by a point-by-point calculation or (2) by obtaining a manufacturers table. Naturally, the VCP at any point can be calculated, either by hand, using the method described by Guth (7) or by a computer program (4). A VCP table provided by the manufacturer utilizes a point-by-point calculation procedure, however, a number of the variables listed above must be fixed to meet

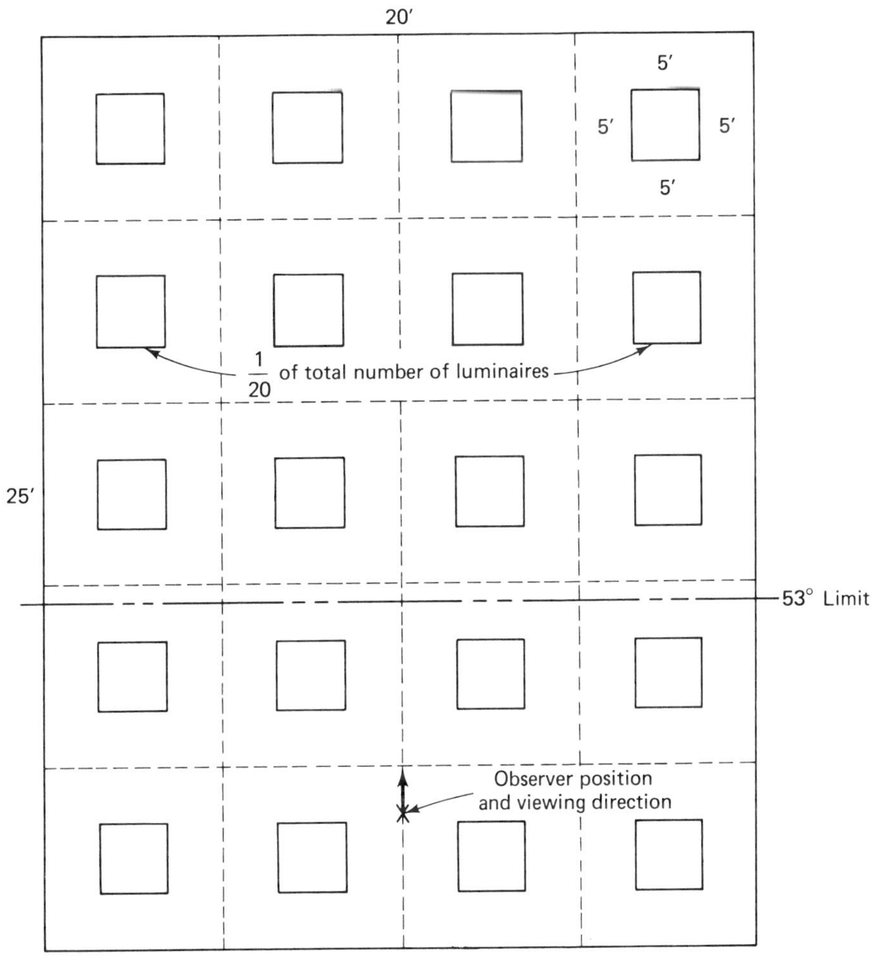

Figure 8-8. Typical Room for Calculating Visual Comfort Probability Tables.

a standard set of conditions so that results from different manufacturers will have the same basis for comparison by the lighting professional.

Figure 8-8 shows a typical rectangular room. To facilitate comparison of the VCP calculated by different manufacturers, the following standard conditions have been established:

1. Initial illuminance level of 100 fc (footcandles). (Number of luminaires determined by lumen method.)
2. Surface reflectances: 80% for ceiling, 50% for walls, and 20% for floor.
3. Mounting heights of 8.5, 10, 13, and 16 ft.
4. A standard range of room dimensions.

5. Standard layout of luminaires uniformly distributed in the room. To meet this condition, the room is divided into 5 ft-by-5 ft modules, and the total number of luminaires from step 1 is equally divided among the total number of modules (Figure 8-7).
6. An observer location, that is assumed to be the worst location, was selected to be 4 ft in front of the center of the rear wall and 4 ft above the floor.
7. A horizontal line of sight looking directly forward.
8. A limit of the visual field to 53° above the horizontal line of sight and directly forward from the observer location.

Standard tables can be used to make a preliminary evaluation of the effects of direct glare on visual comfort. However, the tables are too general to allow an accurate assessment of glare, especially if nonuniform lighting is utilized. The table requires a uniform lighting layout because of standard condition 5. The table encourages nonenergy-conserving uniform lighting layouts. Also, the standard observer location may *not* be the "worst" location or the "actual" task location.

Table 8-10 is an example of a standard VCP Table that is provided by a manufacturer. There are four different orientations possible in any given room. The width (W) column in the table is always the room dimension against which the observer's back is placed. Figure 8-9 shows the change in VCP for the luminaire in Table 8-10 as the luminaire orientation and observer location changes.

The preferred method for obtaining information on visual comfort is to analyze the actual conditions that exist in the luminous environment and obtain VCP values at specific locations within the space. The computational techniques involved are presented in reference (4). A number of microcomputer programs are available for assessing VCP values in a luminous environment for a specific layout for the lighting system.

Indirect Glare. Relative to the luminaire, the indirect glare zone is associated with luminous intensity produced in the zone from 0° to 45° (see Fig. 8-7). Indirect glare is related to *heads-down* viewing for horizontal tasks. Indirect glare involves two forms of glare: (1) **reflected glare** and (2) **veiling reflections.** One form of indirect glare (veiling reflections) can be evaluated in terms of **equivalent sphere illumination (ESI).**

Indirect glare occurs when light is reflected off a horizontal task, causing a reduction in the task contrast that may result in discomfort and a loss in visibility and visual performance.

The luminous environment interacts with the microstructure of the task, causing a change in the task contrast and/or the individual's level of sensitivity to contrast causing a change in the visibility of the task.

$$\text{seeability} = \text{visibility} = \begin{bmatrix} \text{contrast} \end{bmatrix} \times \begin{bmatrix} \text{level of} \\ \text{sensitivity} \\ \text{to contrast} \end{bmatrix} \tag{8.19}$$

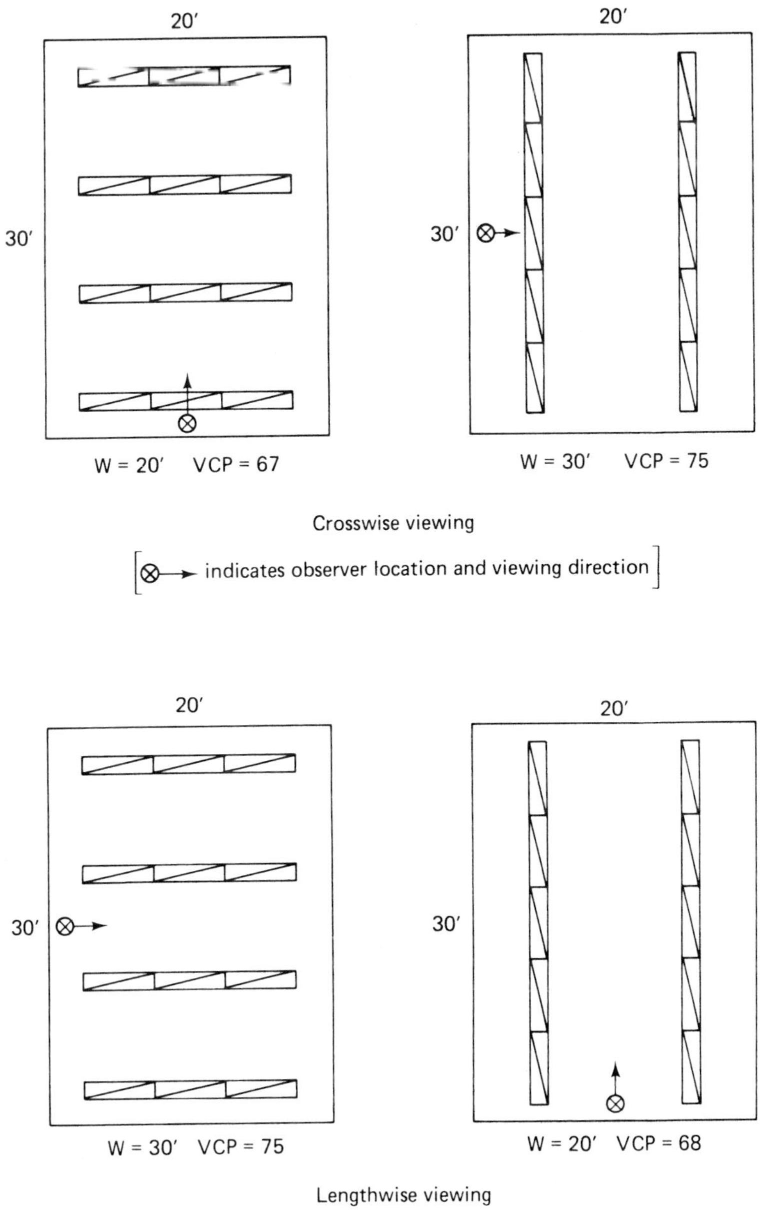

Figure 8-9. Application of Visual Comfort Probability for Table 8-10 in the Data Section.

The level of sensitivity to contrast is a function of the background luminance, L_b. The contrast (see Chapter 2) of a task is the difference in luminance between an object (print) and its background (paper).

$$C = \left| \frac{L_o - L_b}{L_b} \right| \tag{8.20}$$

where

C = contrast
L_o = luminance of the object
L_b = luminance of the background (see Fig. 2-16)

If the contrast goes to zero, the visibility (equation 8.19) will go to zero and the object (print) on the background (paper) will disappear. Adding more light under these conditions may increase an individual's level of sensitivity to contrast, but the visibility will still be zero when the contrast is zero.

The factors that affect the contrast in a real environment are the following:

1. Room dimensions
2. Room surface reflectances
3. Luminaire layout
4. Luminaire intensity distribution and polarization
5. Physical characteristics of the task
6. Viewing angle
7. Test location
8. Viewing direction

For a Perfect Diffuser

$$L = \rho \times E \tag{8.21}$$

If equation (8.21) is substituted into equation (8.20) and the illuminance is constant across the task, then

$$C = \left| \frac{\rho_b E - \rho_o E}{\rho_b E} \right| = \left| \frac{\rho_b - \rho_o}{\rho_b} \right| \tag{8.22}$$

Since a perfect diffuser has constant luminance irrespective of the direction of view, ρ_b and ρ_o will change at the same rate; therefore, the physical contrast is *constant* for a perfect diffuser. With a constant contrast, the only way to increase visibility (equation 8.19) is to increase the individual's level of sensitivity to contrast.

For a Nonperfect Diffuser. Most real visual tasks are *not* a perfect diffuser, which means the reflected light from the object (print) and the background (paper) varies as the location of the light source and the observer changes in the room. The reflecting characteristics of a nonperfect diffuser are described in terms of the luminance factor (β).

$$L = \beta \times E \tag{8.23}$$

If equation (8.22) is substituted into equation (8.20) and the illuminance is constant across the task, then

$$C = \left| \frac{\beta_b E - \beta_o E}{\beta_b E} \right| = \left| \frac{\beta_b - \beta_o}{\beta_b} \right| \tag{8.24}$$

The luminance factors β_b and β_o can be different and vary at a different rate as the viewing angle changes. Therefore, the contrast of a nonperfect diffuser varies as the viewing position and light source location change. Under certain conditions, the β_b may equal the β_o which will cause the contrast (equation 8.24) to go to zero.

Reflected Glare. Reflected glare is caused by specular or glossy surfaces reflecting images of bright sources into the eyes. Reflected glare may cause discomfort and, if reflections are excessive enough and of sufficient size, may also produce disability glare or a loss in visibility.

The bright patch of light on a specular task is an image of the light sources. The less specular or reflective the surface is, the less distinct the image will appear. The distinct image of a bright source is more annoying than the image of a diffuse source. Reflected glare is very distracting, annoying, a source of discomfort, and causes a reduction in visibility. The effects of reflected glare are more serious if the task surround is dark, owing to the luminance ratios created between the image and the surround background. Specular images in a highly polished dark walnut desk are much more harsh than they would be if reflected from a light-colored desk.

Veiling Reflections: Veiling reflections occur when light is reflected off of matte or diffuse surfaces causing a reduction in the contrast of the task. The major problem with veiling reflections is that they are not apparent; yet they reduce contrast that results in a loss in visibility and visual performance. Since most visual tasks are also nonperfect diffusers, the reduction in contrast is caused by the veil of light on the task and the differential variations between L_b and L_o.

The luminous environment interacts with the microstructure of the task (luminance factors), causing veiling reflections which can be evaluated in terms of equivalent sphere illumination. In 1968, the equivalent sphere illumination method was introduced as a numerical method for evaluating the effect of veiling reflections on task visibility. The ESI method was the first analysis tool that allowed the lighting professional to assess how a lighting system affected task contrast, hence, task visibility. It is a mathematical technique that evaluates how well an individual could see; a true quantitative evaluation of the *quality* of a lighting installation.

The ESI Method *equates* the *visibility* of a given task viewed under *reference* lighting condition *to* the *visibility* of the same task viewed under a *real* luminous environment. The reference lighting condition chosen was a *photometric sphere*. It is important to remember that a photometric sphere was *not* selected because it was the "ideal" or the "best" possible lighting. It is simply a reference lighting condition only.

It is possible and desirable to have higher ESI footcandles than "classical" footcandles. "Classical" footcandles are related directly to the watts consumed, so an energy-efficient design involves achieving the highest visibility (ESI fc) with the lowest "classical" footcandles.

$$\text{visibility} = RCS \times C \tag{8.25}$$

where

$$RCS = \text{relative contrast sensitivity}$$

$$C = \text{contrast}$$

$$\begin{bmatrix}\text{visibility under}\\ \text{reference lighting}\\ \text{conditions}\end{bmatrix} = \begin{bmatrix}\text{visibility in}\\ \text{the real}\\ \text{environment}\end{bmatrix}$$

$$\begin{bmatrix}\text{contrast}\\ \text{in a sphere}\end{bmatrix} \times \begin{bmatrix}\text{sensitivity}\\ \text{to contrast}\\ \text{in sphere}\end{bmatrix} = \begin{bmatrix}\text{contrast}\\ \text{in a real}\\ \text{environment}\end{bmatrix} \times \begin{bmatrix}\text{sensitivity to}\\ \text{contrast in a}\\ \text{real environment}\end{bmatrix}$$

Figure 8-10 gives a flow diagram of the method for determining the ESI of a task in a given room. A detailed explanation of the theory and application of the ESI method can be found in the first edition of this book (9).

CRF = contrast rendering factor
RCS = relative contrast sensitivity
RCS_e = effective relative contrast sensitivity
L_e = equivalent background luminance
ρ_o = diffuse reflectance of a given task which is a constant value for the task

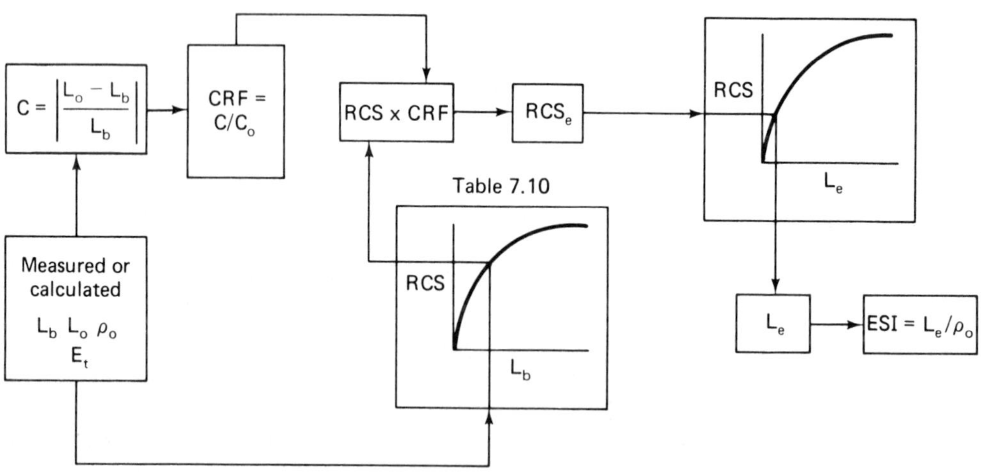

Figure 8-10. Flow Diagram for the Calculation of Equivalent Sphere Illumination.

A brief examination of the flowchart makes the calculation of ESI seem quite simple. What makes the method so complex that it requires the use of a computer? The problem is the determination of CRF, specifically the task contrast in the real environment, C, since C_o, the contrast in the sphere, is a constant for a given task.

What is needed to calculate C? The value of C is arrived at by the Blackwell contrast formula (equation 8.20). Thus it is actually the calculation of the luminance of the object and luminance of the background that makes the procedure so tedious. The equations involved in calculating L_b and L_o are

$$L_b = \Sigma[\beta(h, vs)_b \times E(h, vs)_t] \tag{8.26}$$

$$L_o = \Sigma[\beta(h, vs)_o \times E(h, vs)_t] \tag{8.27}$$

Since $\beta(h, vs)$ and $\beta(h, vs)_o$ (β = luminance factor; see Chapter 1) are constants for a given task at a calculation point, the problem actually comes down to the calculation of illuminance at a point. The illuminance at a point is made up of two components: (1) direct and (2) interreflected illuminance. As long as the luminaires can be divided into small areas, the direct component calculation simply involves the inverse-square law and the cosine law of illumination. Once the luminance of the room surfaces are determined, each surface can be treated as an emitter. That is, each surface becomes a source of light that can be divided into small areas so that the inverse-square law and the cosine law of illumination can be used to calculate the illuminance at a point. This process can be used in determining the contribution from the interreflected component. The calculation of the surface luminance involves the use of radiative luminous flux transfer theory. The computer is necessary to handle the volumes of information, data, and calculations required to determine C.

Two problems have arisen in application of the ESI method that has led to its limited use. First, research reported in the 1980s indicated a possible error in the shape of the RCS curve, which is the heart of the ESI method. Second, research to develop a table of "recommended ESI levels" was started late and stopped altogether when questions arose about the shape of the RCS curve. The first and only quantitative method of assessing the quality of lighting in terms of how well an individual can see has been shelved. For those lighting professionals who studied and understood the ESI method, the technique could still be used to compare the relative merits of two or more lighting solutions in a space where critical visual work is performed.

There are a number of microcomputer programs commercially available for predetermining ESI. The computational techniques involved are presented in reference (3). Because of the amount of data to be handled by an ESI program and the storage requirements, most programs require a hard drive and a math coprocessor.

Glare—Vertical Tasks

The increased usage of the microcomputer in the office environment has brought about the introduction of a vertical task into the luminous environment. A CRT (cathode-ray tube) screen is a very complex vertical visual task. Vertical tasks will be discussed in detail in Chapter 9—VDTs and the office environment.

Luminance Ratio

A luminance ratio represents a difference in the luminance levels between two areas within the visual field. ''Excessive'' luminance ratios are a form of glare. For example, if a bare fluorescent lamp has a luminance of 2350 fL and if a ceiling has a luminance of 50 fL, the luminance ratio between the bare lamp and the ceiling will be 47 to 1. In terms of visual comfort, the human visual system would prefer no differences in luminance levels in the visual field. Variations in the luminance levels in the visual field causes constant fluctuation in the size of the pupil and fluctuation of the individual's adaptation state. Fluctuating adaptation as the individual scans the field of view is defined as **transient adaptation factor.** This rapid iris fluctuation may be one cause of ''visual fatigue.'' Under typical interior lighting levels, there is no evidence that actual physical harm is done to the eyes by ''excessive'' luminance ratios.

To control visual comfort, the designer must minimize the luminance ratios in the visual field. The old edition of the *IES Lighting Handbook* states that ''nowhere in the visual field should the Luminance Ratio exceed 40 to 1.'' In the new edition of *IES Lighting Handbook* (10) the statement has been changed to read: ''nowhere in the visual field *where critical visual work occurs* should the luminance ratio exceed 5 to 1.'' For a south-facing window, the luminance of the window could range from 2000 fL to 8000 fL. The luminance ratio between an adjacent wall whose luminance is 40 fL and a window whose luminance is 2000 fL would be

$$\text{luminance ratio, } LR\text{: } \quad 2000:40::50:1$$
$$LR = 50 \text{ to } 1$$

A luminance ratio of 50 to 1 exceeds the old standard of 40 to 1 and greatly exceeds the new standard (5 to 1) if ''critical visual work occurs'' in the space. Buildings need not be windowless, but if the occupants of the interior space are involved in doing critical visual work, the lighting professional must think about controlling window luminance to meet acceptable luminance ratios.

Color

An understanding of color theory is important to the lighting professional because of the phenomenon known as selective absorption (see Chapter 3). The SPD of a light source (see Chapter 4) plays an important role in the color rendition of objects and people. As lighting professionals, we are responsible for selecting light sources with an SPD that enhance color rendition.

Cost Factors

Energy management and energy-efficient design have a tremendous impact on cost. As always, the final decision on which lighting system to install should be based on cost, which should not only include the first cost (*initial cost*) of the installation, but also the *operating cost* and *maintenance cost*. With the rise in energy cost and inflation in all sectors of the economy, an inexpensive system (low initial cost) could cost the owner many times more to operate and maintain.

A comparison of the cost of different lighting systems leaves out many other factors involved in the total building cost. The lighting affects the mechanical system because of heat gain from the luminaires, and heat gain and loss through windows. The floor areas, floor to ceiling heights, and floor-to-floor heights affect building volume, which again affects the mechanical and lighting systems. A discussion of the total building energy picture (6) shows that changes in lighting have little effect on the total energy used, but may greatly affect the productivity of workers.

There are a number of methods of economic analysis by which to review the different choices. The method chosen depends on the purpose and use to which the analysis is put, as well as the factors of interest in the analysis. Chapter 14 provides the lighting professional with guidance to perform an economic analysis of a lighting system.

State-of-the-Art in Quality Design

One might say that the ESI method was doomed from its beginning since its introduction predated the microcomputer. In the 1960s and early 1970s, the ESI method could only be run on ''big'' mainframe computers utilizing ''time sharing,'' which was very expensive for the user. The computer hardware and technology exists today to make the ESI method an important design tool. However, the research funds have not been made available to answer the questions raised. With Dr. H. Richard Blackwell's retirement in the late 1970s and the high cost of computer analysis of ESI, the lighting industry let over 30 years of research into a visibility-based design procedure die a silent death.

In the late 1970s, Dr. Blackwell proposed an expanded method to calculation of relative visual performance, RVP. ''Values of RVP . . . produced by a change in the level of reference illumination should prove very useful in connection with cost benefit studies of lighting applications'' (1).

This expanded procedure that expresses visibility in terms of relative performance can be combined with cost benefit analysis (6) to give a more complete indication of trade-offs in terms of energy conservation, productivity, and profits. ''You'll not use the new CIE system without a significant increase in the money expended on technological services by the engineers who design the lighting'' (5). The design engineer will not only spend additional time, but must have a greater understanding of the new system to evaluate the five variables. The VL method under study is a result of this early work of Blackwell.

As more data on the relationship of productivity to lighting are gathered and a dollar amount is attached to productivity, the economic analysis will be based on tangible figures rather than a recommended footcandle level. The relationship of productivity to the quality and quantity of illumination is the subject of current research.

The microcomputer ''explosion'' has impacted the practice of the lighting professional. There is an abundant supply of DOS based microcomputer programs available to the practitioner. The design and analysis microcomputer program can now be interfaced with computer-aided design and drawing software. With the proliferation of software in the marketplace, the lighting professional needs to determine the reliability and accuracy of the software chosen. The blind use of the "black-box" software programs by individuals not sophisticated enough to question answers that "don't look right" can lead to legal problems for the lighting professional.

There is still need for a true design microcomputer program that optimizes the luminaire layout based on the input of a specified level of illuminance at a specific point and for other design input parameters. This type of optimization procedure was first described by Ngai (11,12). Because of the complexity of optimization, progress in development of the procedure has been slow. There has been considerable discussion of optimization programs, but to date no one has made any information public.

A totally integrated synthesis program still has not been developed which would bring together all aspects of the environment by optimizing the artificial lighting, daylighting, and thermal considerations in terms of cost and benefits. This type of "super" program may become a reality in the future, but at this time the understanding of the complex interactions has limited this level of sophistication.

REFERENCES

1. Blackwell, H. R., "Energy Conservation by Selective Lighting Standards Graded in Terms of Task and Observer Characteristics." *Proceedings—The Basis for Effective Management of Lighting Energy Symposium*, Federal Energy Administration, Washington, D.C., October 29, 30, 1975.
2. Committee on Recommendations of Quality and Quantity of Illumination of the IES, "Outline of a Standard Procedure for Calculating Visual Comfort Ratings for Interior Lighting—Report No. 2." *Illuminating Engineering* (October 1966), pp. 643–666.
3. DiLaura, D. L., "On the Computation of Equivalent Sphere Illumination." *Journal of the Illuminating Engineering Society* (January 1975).
4. DiLaura, D. L., "On the Computation of Visual Comfort Probability." *Journal of the Illuminating Engineering Society* (July 1976).
5. Dorsey, D. T., and R. H. Blackwell, "A Performance-Oriented Approach to Lighting Specifications." *Lighting Design + Application* (February 1975), pp. 13–27.
6. Dorsey, R. T., "Cost-Benefit Analysis Applied to Lighting in the Energy Equation." *Lighting Design + Application* (July 1975), pp 36–38.
7. Guth, S. K., "Computing Visual Comfort Ratings for a Specific Interior Lighting Installation." *Illuminating Engineering* (October 1966).
8. Helms, R. N., "An Engineering Approach to the Zonal Cavity Technique." *Illuminating Engineering Journal*, Vol. 63, No. 5 (May 1968), pp. 287–291.
9. Helms, R. N., *Illumination Engineering for Energy Efficient Luminous Environments*, Englewood Cliffs, NJ: Prentice-Hall, 1980.
10. Kaufman, John E., and Jack F. Christensen, eds., *IES Lighting Handbook*, Application Volume, 1987. New York: Illuminating Engineering Society of North America, 1987.
11. Ngai, P. Y., "Veiling Reflections and the Design of the Optimal Intensity Distribution of a

Luminaire in Terms of Visual Performance Potential.'' *Journal of the Illuminating Engineering Society* (October 1974), pp. 53–59.

12. Ngai, P. Y., and R. N. Helms, ''Optimization—A Synthetic Approach to Lighting Design.'' *Journal of the Illuminating Engineering Society* (July 1975), pp. 286–291.

DATA SECTION

TABLE 8-1 Photometric Test Data—Wall-Wash Unit.

Description:

Incandescent wall-wash unit w/150PAR38/SP rated at 1750 lumens.

Report No. 10500

Angle	Candlepower Horz*	Candlepower Vert+
0	10000	1020
5	9730	2380
10	9300	4520
15	8700	6530
20	7910	8150
25	6790	9430
30	3150	10000
35	2400	9100
40	910	7360
45	105	5400
50	10	3180
55	0	1430
60	0	100
65	0	15
70	0	0
75	0	0
80	0	0
85	0	0
90	0	0

*Horz—horizontal plane through the aiming point.

+Vert—vertical plane aiming at 30°.

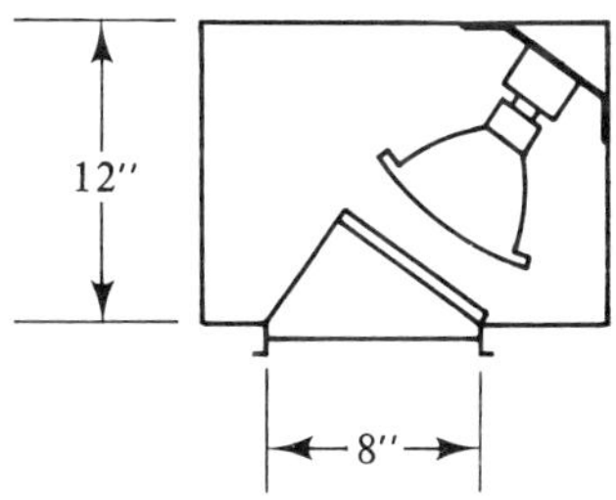

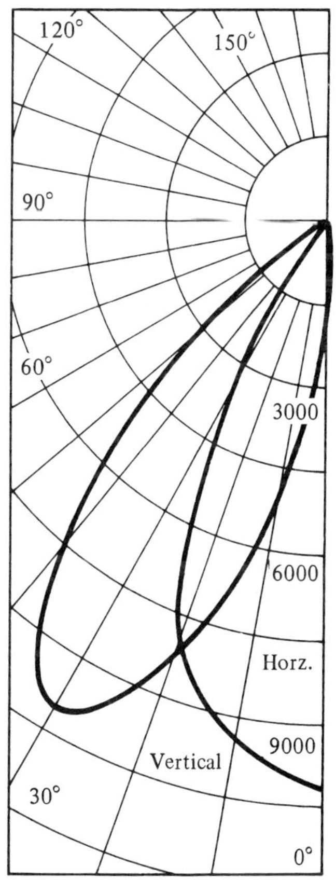

TABLE 8-2 Photometric Test Data—Downlight Unit

DESCRIPTION:

Recessed incandescent downlight with a black low-brightness cone

Report No. 10700

Angle	Candlepower	
	Black*	Clear+
0	3330	3700
5	3328	3800
10	3050	3660
15	2490	3070
20	1580	2150
25	720	980
30	230	620
35	60	100
40	15	70
45	0	0
50	0	0
55		
60		
65		
70		
75		
80		
85		
90		

*Black—150PAR38/FL with black cone.

+Clear—150PAR38/FL with clear cone.

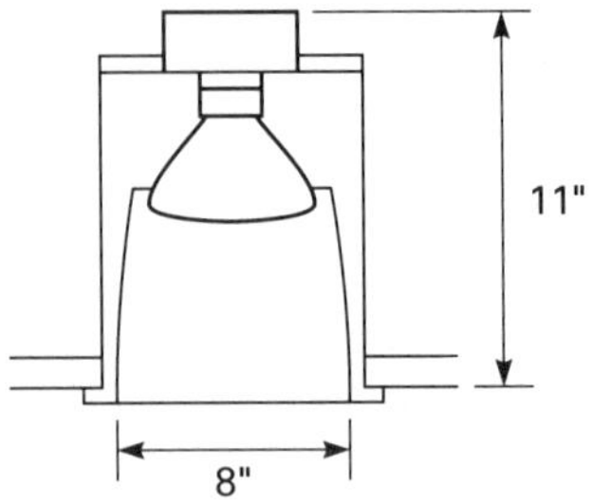

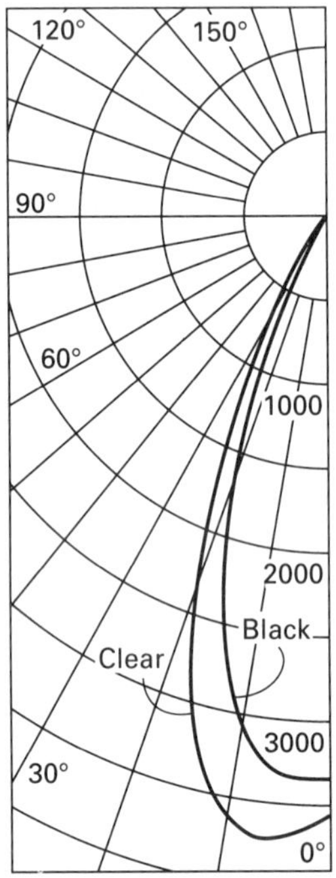

TABLE 8-3 Typical Fluorescent Coefficient of Utilization Table

Lumen method problem
Luminaire No. DB117
(w/4F40WW at 3100 lm each)
Zonal cavity coefficients of utilization: eff. floor cav. ref. = 0.20

Ceiling	80			70			50			10			0
wall RCR	50	30	10	50	30	10	50	30	10	50	30	10	0
1	0.68	0.66	0.64	0.67	0.65	0.63	0.64	0.63	0.61	0.59	0.58	0.57	0.56
2	0.62	0.58	0.55	0.60	0.57	0.54	0.58	0.56	0.53	0.54	0.52	0.51	0.49
3	0.55	0.51	0.48	0.54	0.51	0.48	0.53	0.49	0.46	0.50	0.47	0.45	0.44
4	0.50	0.45	0.42	0.49	0.45	0.41	0.48	0.44	0.41	0.45	0.42	0.40	0.39
5	0.46	0.40	0.36	0.44	0.40	0.36	0.43	0.39	0.36	0.41	0.38	0.35	0.34
6	0.41	0.36	0.32	0.40	0.35	0.32	0.39	0.35	0.32	0.37	0.34	0.31	0.30
7	0.37	0.32	0.28	0.37	0.31	0.28	0.35	0.31	0.28	0.34	0.30	0.27	0.26
8	0.33	0.28	0.25	0.33	0.28	0.24	0.32	0.27	0.24	0.30	0.27	0.24	0.23
9	0.30	0.25	0.21	0.29	0.24	0.21	0.29	0.24	0.21	0.27	0.24	0.21	0.20
10	0.27	0.22	0.19	0.27	0.22	0.19	0.26	0.22	0.19	0.25	0.21	0.19	0.17

Max S/MH = 1.43
LDD Category V.

TABLE 8-4 Typical Fluorescent Photometric Test Report

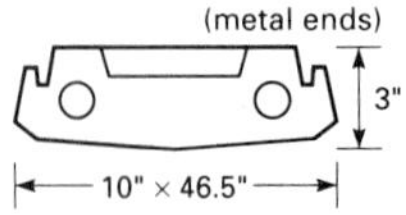

Description:
1 × 4 surface-mounted clear
prismatic acrylic wraparound
w/metal ends

Test No.: 11417	Cat. No. BK 240
Lamps: 2F40CW	Volts: 120
Lumens/Lamp: 3110	Test Dist.: 20
Shielding Angle: N: 90°	L: 90°
Bare Lamp: 2360 fL	Date:

LDD CATEGORY V

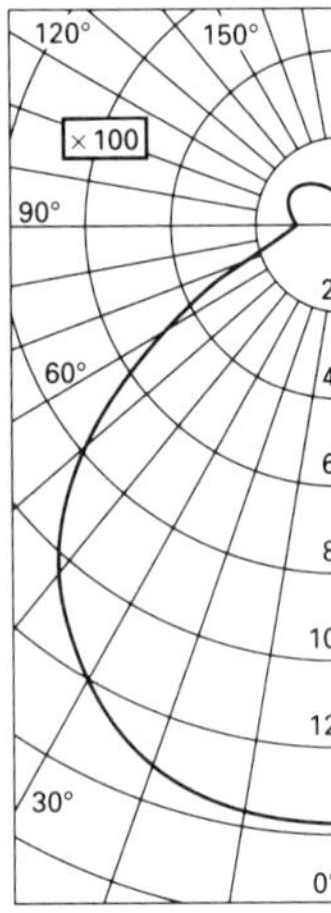

Distribution Data Average			
Angle	C.P.	Angle	C.P.
0°	1398	95°	106
5°	1389	105°	116
10°	—	115°	116
15°	1353	125°	115
20°	—	135°	96
25°	1285	145°	75
30°	—	155°	63
35°	1159	165°	49
40°	—	175°	39
45°	937	180°	40
50°	—		
55°	615		
60°	—		
65°	365		
70°	—		
75°	221		
80°	—		
85°	134		
90°	107		

	Brightness Data			
	Angle	Maximum	Average	Ratio Max./Min.
Along (0°)	45°	2483	1148	2.2
	55°	1939	909	2.1
	65°	1089	678	1.6
	75°	850	533	1.6
	85°	510	384	1.3
Across (90°)	45°	2721	1076	2.5
	55°	2211	827	2.7
	65°	1701	628	2.7
	75°	1429	560	2.6
	85°	1463	597	2.5

Output Data			
Zone degrees	Lumens	Percent total lamp lumens	Percent total distribution
0–40°	1838	29.5	41.0
40–90°	2019	32.5	45.1
90–180°	625	10.1	13.9
0–180°	4482	72.1	100.0

Coefficient of Utilization

Floor		$\rho_{fc} = 20\%$								
Ceiling		80%			50%			10%		
Walls		50%	30%	10%	50%	30%	10%	50%	30%	10%
Room Cavity Ratio	1	73	71	68	66	64	62	57	56	55
	2	65	60	56	58	55	52	51	49	47
	3	57	52	48	52	48	45	46	43	41
	4	51	45	41	47	42	39	41	39	35
	5	46	39	35	42	37	33	37	33	31
	6	41	35	31	37	33	29	33	30	27
	7	37	31	27	34	29	25	30	26	24
	8	33	27	23	30	25	22	27	23	21
	9	30	24	20	27	22	19	24	21	28
	10	27	21	18	25	20	17	22	18	16

Maximum recommended spacing-to-mounting height ratio above workplane is 1.4.

TABLE 8-5 Photometric Test Report—HID Downlight

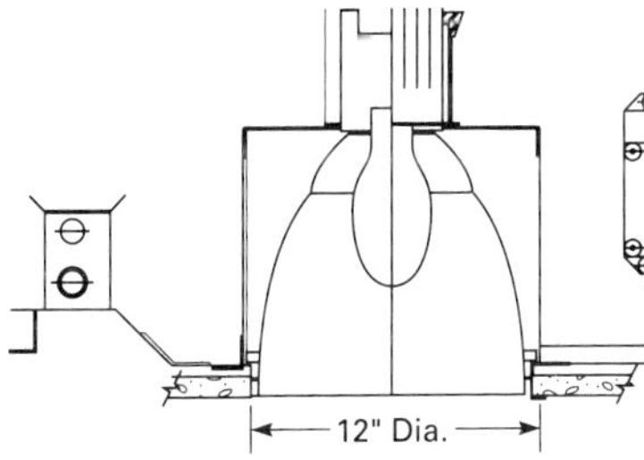

Luminaire: "H"
WB600
Category V
Max. SP = 1.03

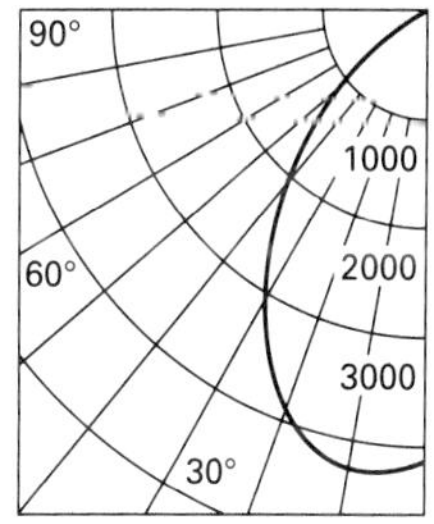

Description:
12″ open downlight w/clear
reflector cone

Test No.: 1238	Cat. No.: 1238-H9-850
Lamps: H39KC-175DX	Volts: 120
Lumens/Lamp: 8600	Test Dist.: 20
Shielding Angle: N: 53°	L: 53°
Bare Lamp:	Date: 1980
By: PR	Chkd. By: RH

Angle	I	Angle	I
0	4131	45	1599
5	4211	55	962
15	4086	65	533
25	3553	75	44
35	2865	85	0

Luminance Data			
Angle	Maximum	Average	Ratio Max./Min.
45°			
55°			
65°			
75°			
85°			

Output Data			
Zone degrees	Lumens	Percent total lamp lumens	Percent total distribution
0°–40°			
40°–90°			
90°–180°			
0°–180°			

Coefficient of Utilization

Floor		ρf_c = 20%								
Ceiling	ρc_c	80%			50%			10%		
Walls	ρ_w	50%	30%	10%	50%	30%	10%	50%	30%	10%
Room Cavity Ratio	1	82	80	79	77	76	75	72	71	71
	2	77	74	71	73	71	69	69	68	66
	3	72	69	66	69	67	64	66	64	62
	4	67	63	60	65	62	59	62	60	58
	5	63	59	56	61	58	55	59	56	54
	6	59	55	52	57	54	51	55	52	50
	7	55	51	48	54	50	47	52	49	47
	8	51	47	44	50	46	43	48	45	43
	9	48	43	40	46	42	40	45	42	39
	10	42	37	34	41	37	34	40	36	34

TABLE 8-6 Effective Ceiling or Floor Cavity Reflectance

Effective cavity reflectance

Base reflectance percent: 90

Cavity Ratio (wall reflectance percent)	90	80	70	60	50	40	30	20	10	0
0.2	89	88	88	87	86	85	85	84	84	82
0.4	88	87	86	85	84	83	81	80	79	76
0.6	87	86	84	82	80	79	77	76	74	73
0.8	87	85	82	80	77	75	73	71	69	67
1.0	86	83	80	77	75	72	69	66	64	62
1.2	85	82	78	75	72	69	66	63	60	57
1.4	85	80	77	73	69	65	62	59	57	52
1.6	84	79	75	71	67	63	59	56	53	50
1.8	83	78	73	69	64	60	56	53	50	48
2.0	83	77	72	67	62	56	53	50	47	43
2.2	82	76	70	65	59	54	50	47	44	40
2.4	82	75	69	64	58	53	48	45	41	37
2.6	81	74	67	62	56	51	46	42	38	35
2.8	81	73	66	60	54	49	44	40	36	34
3.0	80	72	64	58	52	47	42	38	34	30
3.2	79	71	63	56	50	45	40	36	32	28
3.4	79	70	62	54	48	43	38	34	30	27
3.6	78	69	61	53	47	42	36	32	28	25
3.8	78	69	60	51	45	40	35	31	27	23
4.0	77	69	58	51	44	39	33	29	25	22
4.2	77	62	57	50	43	37	32	28	24	21
4.4	76	61	56	49	42	36	31	27	23	20
4.6	76	60	55	47	40	35	30	26	22	19
4.8	75	59	54	46	39	34	28	25	21	18
5.0	75	59	53	45	38	33	28	24	20	16
6.0	73	61	49	41	34	29	24	20	16	11
7.0	70	58	45	38	30	27	21	18	14	08
8.0	68	55	42	35	27	23	18	15	12	06
9.0	66	52	38	31	25	21	16	14	11	05
10.0	65	51	36	29	22	19	15	11	09	04

Base reflectance percent: 80

Cavity Ratio (wall reflectance percent)	90	80	70	60	50	40	30	20	10	0
0.2	79	78	78	77	77	76	76	75	74	72
0.4	79	77	76	75	74	73	72	71	70	68
0.6	78	76	75	73	71	70	68	66	65	63
0.8	78	75	73	71	69	67	65	63	61	57
1.0	77	74	72	69	67	65	62	60	57	55
1.2	76	73	70	67	64	61	58	55	53	51
1.4	76	72	68	65	62	59	55	53	50	48
1.6	75	71	67	63	60	57	53	50	47	44
1.8	75	70	66	62	58	54	50	47	44	41
2.0	74	69	64	60	56	52	48	45	41	38
2.2	74	68	63	58	54	49	45	42	38	35
2.4	73	67	61	56	52	47	43	40	36	33
2.6	73	66	60	55	50	45	41	38	34	31
2.8	73	65	59	53	48	43	39	36	32	29
3.0	72	65	58	52	47	42	37	34	30	27
3.2	72	65	57	51	45	40	35	33	28	25
3.4	71	64	56	49	44	39	34	32	27	24
3.6	71	63	54	48	43	38	32	30	25	23
3.8	70	62	53	47	41	36	31	28	24	22
4.0	70	61	53	46	40	35	30	26	22	20
4.2	69	60	52	45	39	34	29	25	21	18
4.4	69	60	51	44	38	33	28	24	20	17
4.6	69	59	50	43	37	32	27	23	19	15
4.8	68	58	49	42	36	31	26	22	18	14
5.0	68	58	48	41	35	30	25	21	18	14
6.0	66	55	44	38	31	27	22	19	15	10
7.0	64	53	41	35	28	24	19	16	12	07
8.0	62	50	38	32	25	21	17	14	11	05
9.0	61	49	36	30	23	19	15	13	10	04
10.0	59	46	33	27	21	18	14	11	08	03

Base reflectance percent: 70

Cavity Ratio (wall reflectance percent)	90	80	70	60	50	40	30	20	10	0
0.2	70	69	68	68	67	67	66	66	65	64
0.4	69	68	67	66	65	64	63	62	61	58
0.6	69	67	65	64	63	61	59	58	57	54
0.8	68	66	64	62	60	58	56	55	53	50
1.0	68	65	62	60	58	55	53	52	50	47
1.2	67	64	61	59	57	54	50	48	46	44
1.4	67	63	60	58	55	51	47	45	44	41
1.6	67	62	59	56	53	47	45	43	41	38
1.8	66	61	58	54	51	46	42	40	38	35
2.0	66	60	56	52	49	45	40	38	36	33
2.2	66	60	55	51	48	43	38	36	34	32
2.4	65	60	54	50	46	41	37	35	32	30
2.6	65	59	54	49	45	40	35	33	30	28
2.8	65	59	53	48	43	38	33	30	28	26
3.0	64	58	52	47	42	37	32	29	27	24
3.2	64	58	51	46	40	36	31	28	25	23
3.4	64	57	50	45	39	35	29	27	24	22
3.6	63	56	49	44	38	33	28	25	22	20
3.8	63	56	49	43	37	32	27	24	21	19
4.0	63	55	48	42	36	31	26	23	20	17
4.2	62	55	47	41	35	30	25	22	19	16
4.4	62	54	46	40	34	29	24	21	18	15
4.6	62	53	45	39	33	28	24	21	17	14
4.8	62	53	45	38	32	27	23	20	16	13
5.0	61	52	44	36	31	26	22	19	16	12
6.0	60	51	41	35	28	24	19	16	13	09
7.0	58	48	38	32	26	22	17	14	11	06
8.0	57	46	35	29	23	19	15	13	10	05
9.0	56	45	33	27	21	18	14	12	09	04
10.0	55	43	31	25	19	16	12	10	08	03

Base reflectance percent: 60

Cavity Ratio (wall reflectance percent)	90	80	70	60	50	40	30	20	10	0
0.2	60	59	59	59	58	57	56	56	55	53
0.4	60	59	59	58	57	55	54	53	52	50
0.6	60	58	57	56	55	53	51	51	50	46
0.8	59	57	56	55	54	51	48	47	46	43
1.0	59	57	55	53	51	48	45	44	43	41
1.2	59	56	54	51	49	46	44	42	40	38
1.4	59	56	53	49	47	44	41	39	38	36
1.6	59	55	52	48	45	42	39	37	35	33
1.8	58	55	51	47	44	40	37	35	33	31
2.0	58	54	50	46	43	39	35	33	31	29
2.2	58	53	49	45	42	37	34	31	29	28
2.4	58	53	48	44	41	36	32	30	27	26
2.6	58	53	48	43	39	35	31	28	26	24
2.8	58	53	47	43	38	34	29	27	24	22
3.0	57	52	46	42	37	32	28	25	23	20
3.2	57	51	45	41	36	31	27	23	22	18
3.4	57	51	45	40	35	30	26	23	20	17
3.6	57	50	44	39	34	29	25	22	19	16
3.8	57	50	43	38	33	29	24	21	19	15
4.0	57	49	42	37	32	28	23	20	18	14
4.2	56	49	42	37	32	27	22	19	17	14
4.4	56	49	42	36	31	27	22	19	16	13
4.6	56	49	41	35	30	26	21	18	16	13
4.8	56	48	41	34	29	25	21	18	15	12
5.0	56	48	40	34	28	24	20	17	14	11
6.0	55	45	37	31	25	21	17	14	11	07
7.0	54	43	35	30	24	20	15	12	09	05
8.0	53	42	33	28	22	18	14	11	08	04
9.0	52	40	31	26	20	16	12	10	07	03
10.0	51	39	29	24	18	15	11	09	07	02

Base reflectance percent: 50

Cavity Ratio (wall reflectance percent)	90	80	70	60	50	40	30	20	10	0
0.2	50	50	49	49	48	48	47	46	46	44
0.4	50	49	48	48	47	46	45	45	44	42
0.6	50	48	47	46	45	44	43	42	41	38
0.8	50	48	47	45	44	42	40	39	38	36
1.0	50	48	46	44	43	41	38	37	36	34
1.2	50	47	45	43	41	39	36	35	34	29
1.4	50	47	45	42	40	38	35	34	32	27
1.6	50	47	44	41	39	36	33	32	30	26
1.8	50	46	43	40	38	35	31	30	28	25
2.0	50	46	43	40	37	34	30	28	26	24
2.2	50	46	42	38	36	33	29	27	24	22
2.4	50	46	42	37	35	31	27	25	23	21
2.6	50	46	41	37	34	30	26	23	21	20
2.8	50	46	41	36	33	29	25	22	20	19
3.0	50	45	40	36	32	28	24	21	19	17
3.2	50	44	39	35	31	27	23	20	18	16
3.4	50	44	39	35	30	26	22	19	17	15
3.6	50	44	39	34	29	25	21	18	16	14
3.8	50	44	38	34	29	25	21	17	15	13
4.0	50	44	38	33	28	24	20	17	15	12
4.2	50	43	37	32	28	24	20	17	14	12
4.4	50	43	37	32	27	23	19	16	13	11
4.6	50	43	36	31	26	22	18	15	13	10
4.8	50	43	36	31	26	22	18	15	12	09
5.0	50	42	35	30	25	21	17	14	12	09
6.0	50	42	34	29	23	19	15	13	10	06
7.0	49	41	32	27	21	18	14	11	08	05
8.0	49	40	30	25	19	16	12	10	07	03
9.0	48	39	29	24	18	15	11	09	07	03
10.0	47	37	27	22	17	14	10	08	06	02

Base reflectance percent	40									
wall reflectance percent	90	80	70	60	50	40	30	20	10	0
Cavity Ratio										
0.2	40	40	39	39	39	38	38	37	36	36
0.4	41	40	39	39	38	37	36	35	34	34
0.6	41	40	39	38	37	36	34	33	32	31
0.8	41	40	38	37	36	35	33	32	31	29
1.0	42	40	38	37	35	33	32	31	29	27
1.2	42	40	38	36	34	32	30	29	27	25
1.4	42	39	37	35	33	31	29	27	25	23
1.6	42	39	37	35	32	30	27	25	23	22
1.8	42	39	36	34	31	29	26	24	22	21
2.0	42	39	36	34	31	28	25	23	21	19
2.2	42	39	36	33	30	27	24	22	19	18
2.4	43	39	35	33	29	27	24	21	18	17
2.6	43	39	35	32	29	26	23	20	17	15
2.8	43	39	35	32	28	25	22	19	16	14
3.0	43	39	35	31	27	24	21	18	16	13
3.2	43	39	35	31	27	23	20	17	15	13
3.4	43	39	34	30	26	23	20	17	14	12
3.6	44	39	34	30	26	22	19	16	14	11
3.8	44	38	33	29	25	22	18	16	13	10
4.0	44	38	33	29	25	21	18	15	12	10
4.2	44	38	33	29	24	21	17	15	12	10
4.4	44	38	33	28	24	20	17	14	11	09
4.6	44	38	32	28	23	19	16	14	11	08
4.8	44	38	32	27	22	19	16	13	10	08
5.0	45	38	31	27	22	19	15	13	10	07
6.0	44	37	30	25	20	17	13	11	08	05
7.0	44	36	29	24	19	16	12	10	07	04
8.0	44	35	28	23	18	15	11	09	06	03
9.0	44	35	26	21	16	13	10	08	05	02
10.0	43	34	25	20	15	12	08	07	05	02

Base reflectance percent	30									
wall reflectance percent	90	80	70	60	50	40	30	20	10	0
Cavity Ratio										
0.2	31	31	30	30	29	29	29	28	28	27
0.4	31	31	30	30	29	28	28	27	26	25
0.6	32	31	30	29	28	27	26	26	25	23
0.8	32	31	30	29	28	26	25	25	23	22
1.0	33	32	30	29	27	25	24	23	22	20
1.2	33	32	30	28	27	25	23	22	21	19
1.4	34	32	30	28	26	24	22	21	19	18
1.6	34	33	29	27	25	23	22	20	18	17
1.8	35	33	29	27	25	23	21	19	17	16
2.0	35	33	29	26	24	22	20	18	16	14
2.2	36	32	29	26	24	22	19	17	15	13
2.4	36	32	29	26	24	22	19	16	14	12
2.6	36	32	29	25	23	21	18	16	14	12
2.8	37	33	29	25	23	21	17	15	13	11
3.0	37	33	29	25	22	20	17	15	12	10
3.2	37	33	29	25	22	19	16	14	12	10
3.4	37	33	29	25	22	19	16	14	11	09
3.6	38	33	29	24	21	18	15	13	10	09
3.8	38	33	28	24	21	18	15	13	10	08
4.0	38	33	28	24	21	18	14	12	09	07
4.2	38	33	28	24	20	17	14	12	09	07
4.4	39	33	28	24	20	17	14	11	09	06
4.6	39	33	28	24	20	17	13	10	08	06
4.8	39	33	28	24	20	17	13	10	08	05
5.0	39	33	28	24	19	16	13	10	08	05
6.0	39	33	27	23	18	15	11	09	06	04
7.0	40	33	26	22	17	14	10	08	05	03
8.0	40	33	26	21	16	13	09	07	04	02
9.0	40	33	25	20	15	12	09	07	04	02
10.0	40	32	24	19	14	11	08	06	03	01

Base reflectance percent	20									
wall reflectance percent	90	80	70	60	50	40	30	20	10	0
Cavity Ratio										
0.2	21	20	20	20	20	20	19	19	19	17
0.4	22	21	20	20	20	19	19	18	18	16
0.6	23	21	21	20	19	19	18	18	17	15
0.8	24	22	21	20	19	19	18	17	16	14
1.0	25	23	22	20	19	18	17	16	15	13
1.2	25	23	22	20	19	17	17	16	14	12
1.4	26	24	22	20	18	17	16	15	13	12
1.6	26	24	22	20	18	17	16	15	13	11
1.8	27	25	23	20	18	17	15	14	12	10
2.0	28	25	23	20	18	16	15	13	11	09
2.2	28	25	23	20	18	16	14	12	10	09
2.4	29	26	23	20	18	16	14	12	10	08
2.6	29	26	23	20	18	16	14	11	09	08
2.8	30	27	23	20	18	15	13	11	09	07
3.0	30	27	23	20	17	15	13	11	09	07
3.2	31	27	23	20	17	15	12	11	09	06
3.4	31	27	23	20	17	15	12	10	08	06
3.6	32	27	23	20	17	15	12	10	08	05
3.8	32	28	23	20	17	15	12	10	07	05
4.0	33	28	23	20	17	14	11	09	07	05
4.2	33	28	23	20	17	14	11	09	07	04
4.4	34	28	24	20	17	14	11	09	07	04
4.6	34	29	24	20	17	14	11	09	07	04
4.8	35	29	24	20	17	13	10	08	06	04
5.0	35	29	24	20	16	13	10	08	06	04
6.0	36	30	24	20	16	13	10	08	05	02
7.0	36	30	24	20	15	12	09	07	04	02
8.0	37	30	23	19	15	12	08	06	03	01
9.0	37	29	23	19	14	11	08	06	03	01
10.0	37	29	22	18	13	10	07	05	03	01

Base reflectance percent	10									
wall reflectance percent	90	80	70	60	50	40	30	20	10	0
Cavity Ratio										
0.2	11	11	11	10	10	10	10	09	09	09
0.4	12	11	11	11	11	10	10	09	09	08
0.6	13	13	12	11	11	10	10	09	08	08
0.8	15	14	13	12	11	10	10	09	08	07
1.0	16	14	13	12	12	11	10	09	08	07
1.2	17	15	14	13	12	11	10	09	07	06
1.4	18	16	14	13	12	11	10	09	07	06
1.6	19	17	15	14	12	11	09	08	07	06
1.8	19	17	15	14	13	11	09	08	06	05
2.0	20	18	16	14	13	11	09	08	06	05
2.2	21	19	16	14	13	11	09	07	06	05
2.4	22	19	17	5	13	11	09	07	06	05
2.6	23	20	17	15	13	11	09	07	06	04
2.8	23	20	18	16	13	11	09	07	05	03
3.0	24	21	18	16	13	11	09	07	05	03
3.2	25	21	18	16	13	11	09	07	05	03
3.4	26	22	18	16	13	11	09	07	05	03
3.6	26	22	19	16	13	11	09	06	04	03
3.8	27	23	19	17	14	11	09	06	04	02
4.0	27	23	20	17	14	11	09	06	04	02
4.2	28	24	20	17	14	11	09	06	04	02
4.4	28	24	20	17	14	11	08	06	04	02
4.6	29	25	20	17	14	11	08	06	04	02
4.8	29	25	20	17	14	11	08	06	04	02
5.0	30	25	20	17	14	11	08	06	04	02
6.0	31	26	21	18	14	11	08	06	03	01
7.0	32	27	21	17	13	11	08	06	03	01
8.0	33	27	21	17	13	10	07	05	03	01
9.0	34	28	21	17	13	10	07	05	02	01
10.0	34	28	21	17	12	10	07	05	02	01

Base reflectance percent	0									
wall reflectance percent	90	80	70	60	50	40	30	20	10	0
Cavity Ratio										
0.2	02	02	02	01	01	01	01	00	00	0
0.4	04	03	03	02	02	02	01	01	00	0
0.6	05	05	04	03	03	02	02	01	01	0
0.8	07	06	05	04	04	03	02	02	01	0
1.0	08	07	06	05	04	03	02	02	01	0
1.2	10	08	07	06	05	04	03	02	01	0
1.4	11	09	08	07	06	04	03	02	01	0
1.6	12	10	09	07	06	05	03	02	01	0
1.8	13	11	09	08	07	05	04	03	01	0
2.0	14	12	10	09	07	05	04	03	01	0
2.2	15	13	11	09	07	06	04	03	01	0
2.4	16	13	11	09	08	06	04	03	01	0
2.6	17	14	12	10	08	06	05	03	02	0
2.8	17	15	13	10	08	07	05	03	02	0
3.0	18	16	13	11	09	07	05	03	02	0
3.2	19	16	14	11	09	07	05	03	02	0
3.4	20	17	14	12	09	07	05	03	02	0
3.6	20	17	15	12	10	08	05	04	02	0
3.8	21	18	15	12	10	08	05	04	02	0
4.0	22	18	15	13	10	08	05	04	02	0
4.2	22	19	16	13	10	08	06	04	02	0
4.4	23	19	16	13	10	08	06	04	02	0
4.6	23	20	17	13	11	08	06	04	02	0
4.8	24	20	17	14	11	08	06	04	02	0
5.0	25	21	17	14	11	08	06	04	02	0
6.0	27	23	13	15	12	09	06	04	02	0
7.0	28	24	19	15	12	09	06	04	02	0
8.0	30	25	20	15	12	09	06	04	02	0
9.0	31	25	20	15	12	09	06	04	02	0
10.0	31	25	20	15	12	09	06	04	02	0

TABLE 8-7 Effective Floor Cavity Correction Factor. (Courtesy of the Illuminating Engineering Society of North America.)

For 30% effective floor cavity reflectance, multiply by appropriate factor below
For 10% effective floor cavity reflectance, divide by appropriate factor below

Percent effective ceiling cavity reflectance, ρ_{CC}	80			70			50			10		
Percent wall reflectance, ρ_w	50	30	10	50	30	10	50	30	10	50	30	10
Room Cavity Ratio												
1	1.08	1.08	1.07	1.07	1.06	1.06	1.05	1.04	1.04	1.01	1.01	1.01
2	1.07	1.06	1.05	1.06	1.05	1.04	1.04	1.03	1.03	1.01	1.01	1.01
3	1.05	1.04	1.03	1.05	1.04	1.03	1.03	1.03	1.02	1.01	1.01	1.01
4	1.05	1.03	1.02	1.04	1.03	1.02	1.03	1.02	1.02	1.01	1.01	1.00
5	1.04	1.03	1.02	1.03	1.02	1.02	1.02	1.02	1.01	1.01	1.01	1.00
6	1.03	1.02	1.01	1.03	1.02	1.01	1.02	1.02	1.01	1.01	1.01	1.00
7	1.03	1.02	1.01	1.03	1.02	1.01	1.02	1.01	1.01	1.01	1.01	1.00
8	1.03	1.02	1.01	1.02	1.02	1.01	1.02	1.01	1.01	1.01	1.01	1.00
9	1.02	1.01	1.01	1.02	1.01	1.01	1.02	1.01	1.01	1.01	1.01	1.00
10	1.02	1.01	1.01	1.02	1.01	1.01	1.02	1.01	1.01	1.01	1.01	1.00

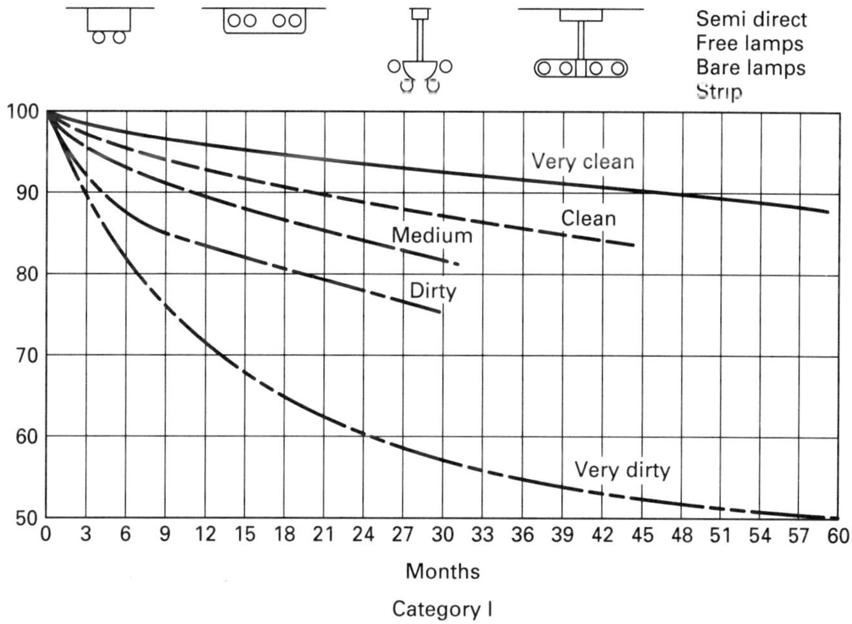

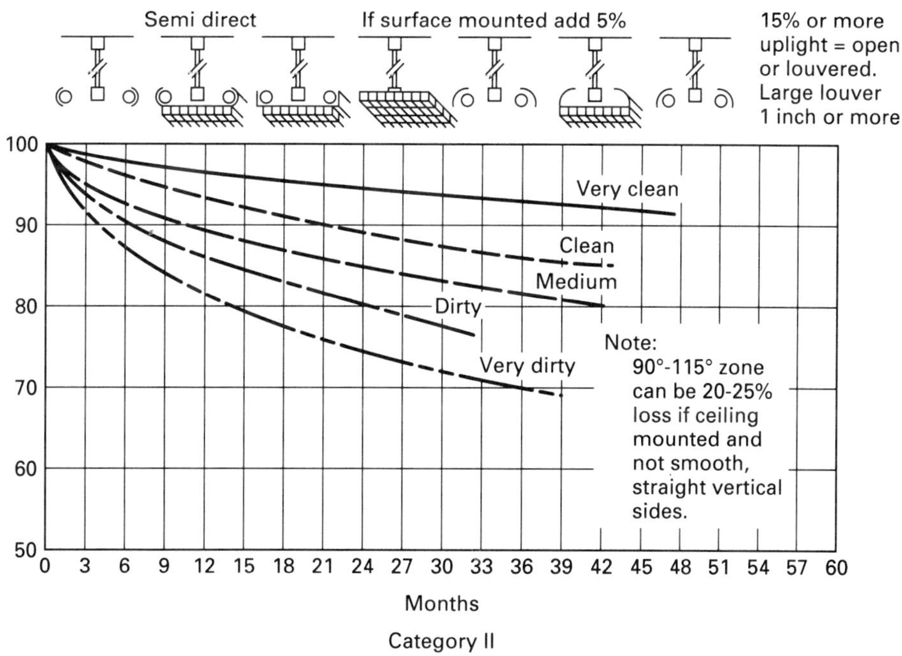

Figure 8-11. LDD Graphs. (Courtesy of the Illuminating Engineering Society of North America.)

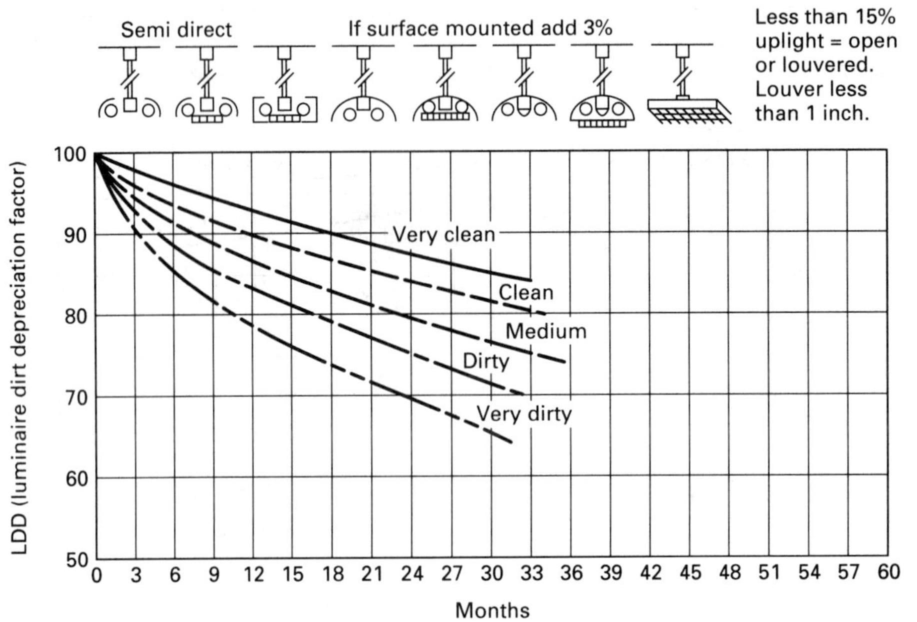

Category III

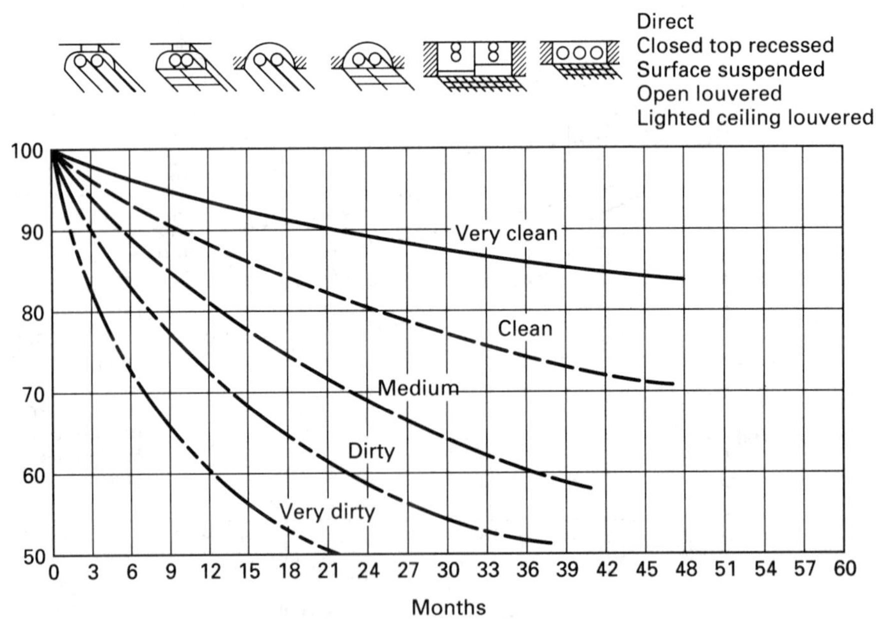

Category IV

Figure 8-11. (continued) LDD Graphs.

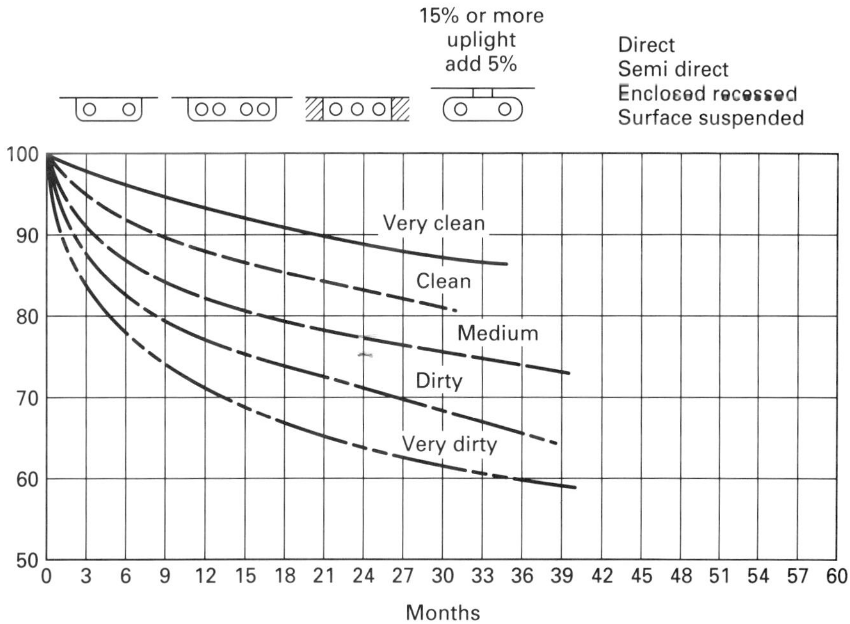

Category V

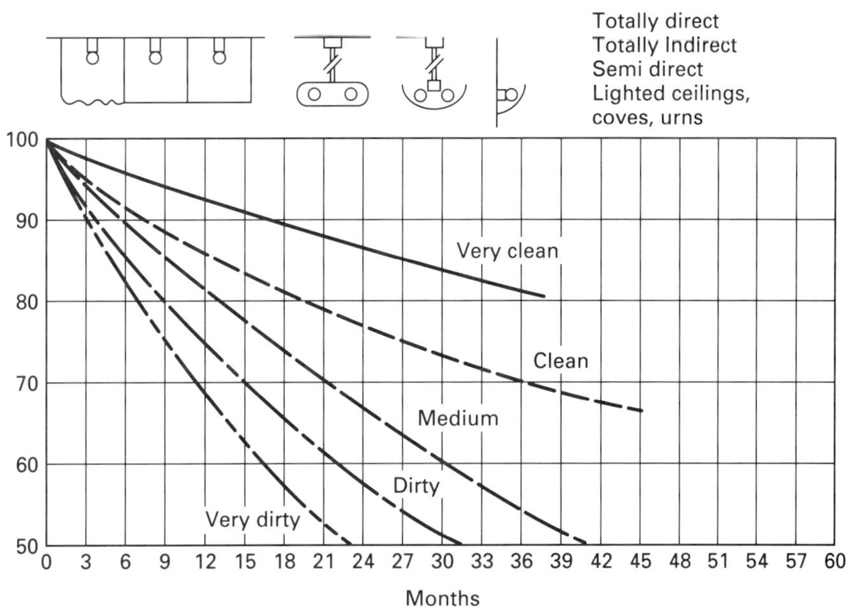

Category VI

Figure 8-11. (continued) LDD Graphs.

TABLE 8-8 Lamp Lumen Depreciation (LLD) 70 percent of life. (Courtesy of Illuminating Engineering Society of North America.)

Lamp description		LLD factor		
Incandescent				
General service	to 150 W	91		
	250 to 500 W	90		
	750 to 1500 W	86		
Silver-bowl	200 to 500 W	75		
Reflector	R40	86		
	R52 and R57	81		
Projector	PAR 38 to 64	84		
Mercury				
H39-22 KB	175 W	85		
H39-22 KC/C	175 W	83		
H39-22 KC/W	175 W	75		
H37-5 KB	250 W	85		
H37-5 KC/C	250 W	83		
H37-5 KC/W	250 W	73		
H33-1 CD	400 W	86		
H33-1 GL/C	400 W	83		
H33-1 GL/W	400 W	74		
H36-15 GV	1000 W	77		
H36-15 GW/C	1000 W	72		
H36-15 GW/W	1000 W	61		
		Hours per start		
Fluorescent		6	12	18
Instant start 425 ma				
Standard colors		88	87	85
Improved-color types		82	80	78
Rapid start 430 ma				
Standard colors		87	86	85
Improved-color types		81	80	79
Rapid start 800 ma				
Standard colors		81	79	77
Rapid start 1500 ma				
Tubular		76	74	72
Others		70	68	64

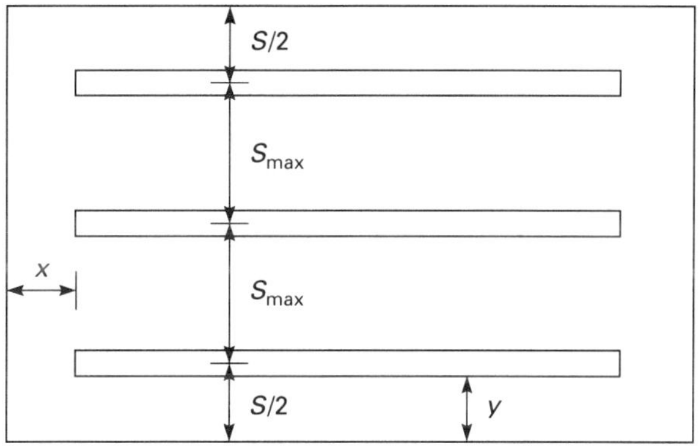

Recommended spacing for "uniformity"

Distance	Preferred	Maximum
x	6" to 12"	2'-0"
y	$2\frac{1}{2}$' to 3'	4'-0"

(a)

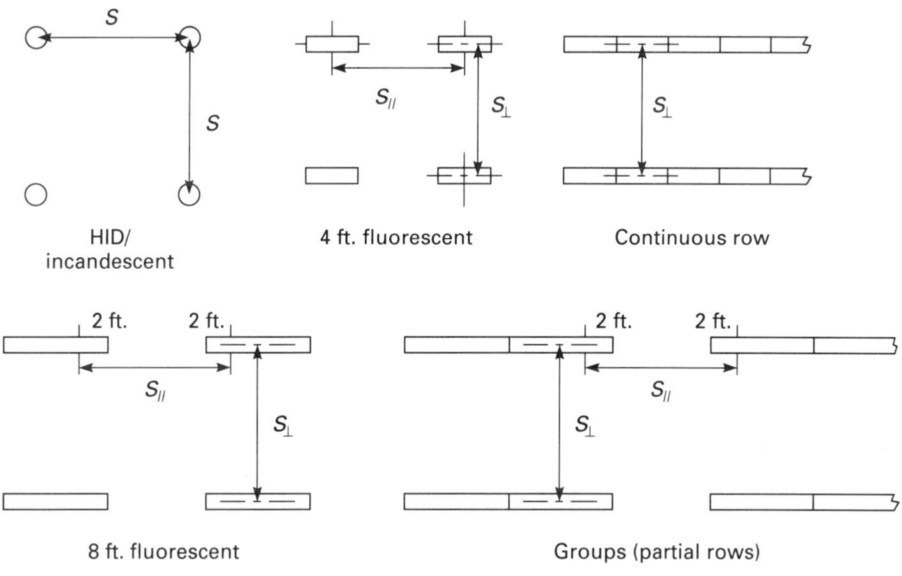

(b)

$S_{//}$ = Maximum spacing along the axis of the lamps.
$S_{\perp}$ = Maximum spacing across the axis of the lamps.

Figure 8-12. Spacing Requirements. (Courtesy of the Illuminating Engineering Society of North America.)

TABLE 8-9 Typical VCP Table

Description: 2 × 4 troffer w/6250 lens w/4F40CW, each rated at 3100 lm — LUMNO = DB117, Report No. 1857-1

100 fc room		Reflectances 80/50/20							
		Luminaires lengthwise				Luminaires crosswise			
W	L	8.5	10.0	13.0	16.0	8.5	10.0	13.0	16.0
20	20	71	74	76	77	70	73	75	75
20	30	64	68	72	73	63	67	70	71
20	40	61	64	68	69	60	63	66	67
20	60	58	60	64	66	56	59	62	64
30	20	71	75	77	76	71	75	76	75
30	30	64	68	72	72	63	67	70	70
30	40	59	63	67	68	59	62	65	66
30	60	56	59	62	64	55	58	61	62
30	80	53	57	59	61	52	55	58	59
40	20	72	76	79	78	72	76	78	76
40	30	64	68	73	73	64	68	72	71
40	40	59	63	67	69	59	62	66	67
40	60	56	58	61	64	55	57	60	62
40	80	53	56	58	60	52	55	57	59
40	100	52	54	56	58	51	53	55	57
60	30	65	69	73	74	65	68	72	73
60	40	60	63	66	69	59	62	66	68
60	60	56	58	60	63	55	57	60	62
60	80	53	55	57	60	52	54	56	58
60	100	52	54	55	57	51	53	54	56
100	40	62	65	68	71	62	65	67	70
100	60	58	60	61	64	57	59	61	63
100	80	55	57	57	60	54	56	57	59
100	100	53	54	55	57	52	54	54	56

EXAMPLE PROBLEMS

Example problem 1: Point-by-point for a wall-wash unit

Using the luminaire in Table 8-1 in the data section, calculate the initial illumination at points A, B, C, and D in the figure.

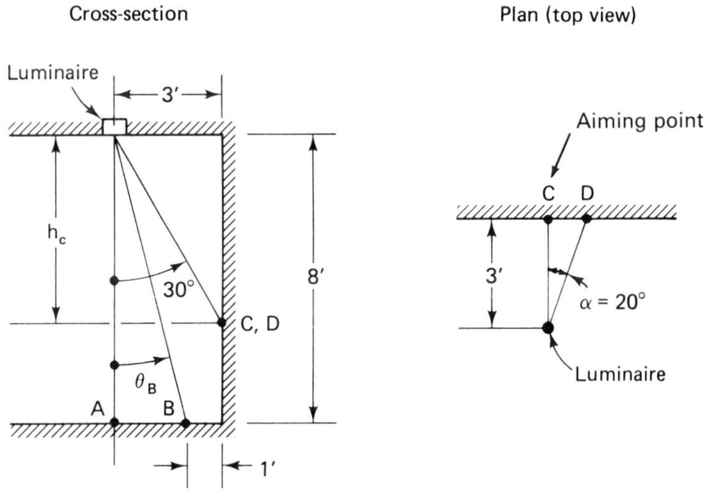

Solution:

Point A:

$$I_0 = 1020 \text{ cd}$$

$$E_A = \frac{1020}{(8)^2} = 15.94 \text{ fc}$$

Point B:

$$\theta_B = \arctan\frac{2}{8} = 14.04°$$

$$I_{\theta B} = 6128 \text{ cd} \qquad D^2 = (8.25)^2 = 68$$

$$E_B = \frac{6128}{68}\cos 14.04° = 87.4 \text{ fc}$$

Point C:

$$h_C = \frac{3}{\tan 30°} = 5.20' \qquad d = \frac{3}{\sin 30°} = 6'$$

$$I_{30} = 10{,}000 \text{ cd}$$

$$E_C = \frac{10{,}000}{(6)^2}\sin 30° = 138.9 \text{ fc}$$

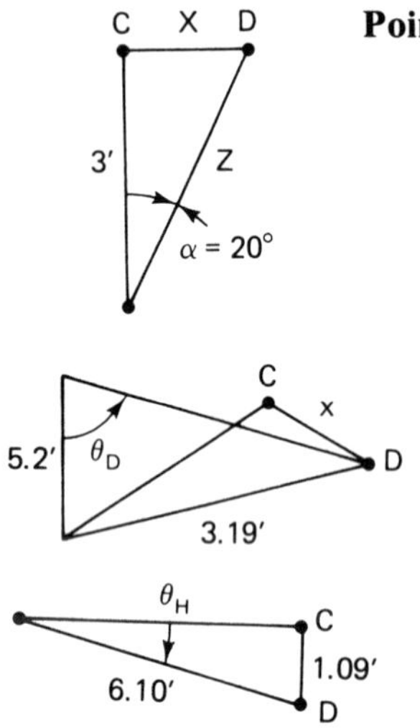

Point *D*:

$$z = \frac{3}{\cos 20^\circ} = 3.19'$$

$$x = 3 \tan 20^\circ = 1.09'$$

$$d^2 = \overline{3.19}^2 + \overline{5.2}^2 = 37.22'$$

$$\theta_D = \arctan \frac{3.19}{5.2} = 31.53^\circ \qquad \sin \theta_D = 0.523$$

$$\sin \theta_H = \frac{1.09}{6.10} \qquad \theta_H = 10.29^\circ \quad I_\theta = 9240$$

$$E_D = \frac{I_{\theta H}}{D^2} \sin \theta_D \cos \alpha$$

$$= \frac{9240}{37.22} \times 0.523 \times 0.940 = 122.05 \text{ fc}$$

Example problem 2: Point-by-point for a downlight unit.

Using the luminaire in Table 8-2 in the data section, calculate the initial illumination at points *A*, *B*, *C*, and *D* in the figure.

Cross-section

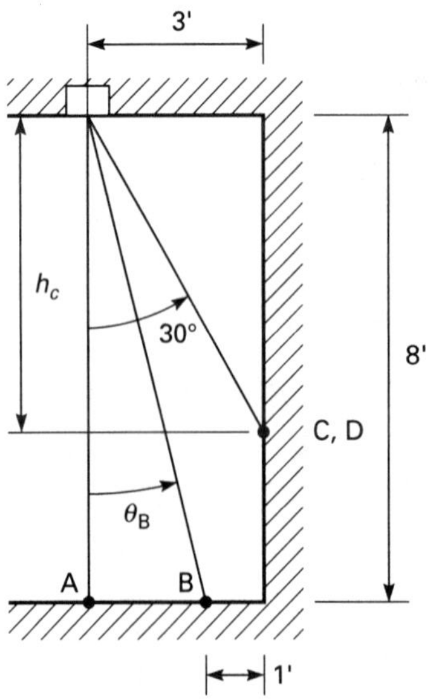

Plan (top view)

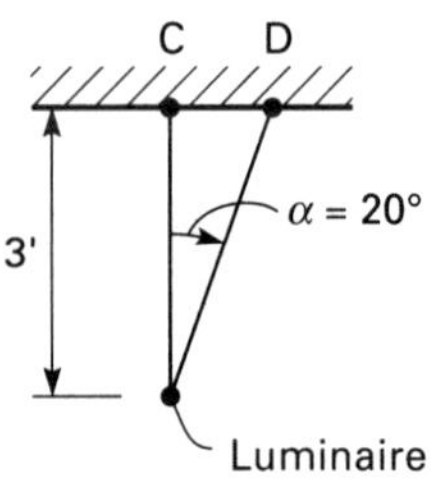

Use "clear" cone

Cross Section

Point A:
$$I_A = 3700 \text{ cd} \qquad E_A = \frac{3700}{(8)^2} = 57.8 \text{ fc}$$

Point B:
$$\theta_B = \arctan\frac{2}{8} = 14.04°$$
$$I_B = 3188 \text{ cd} \qquad D^2 = (8.25)^2 = 68.1$$
$$E_B = \frac{3188}{68.1} \times \cos 14.04 = 45.4 \text{ fc}$$

Point C:
$$h_C = \frac{3}{\tan 30°} = 5.20' \qquad d^2 = (3)^2 + (5.20)^2 = 36.04$$
$$I_C = 620$$
$$E_C = \frac{620}{36.04} \times \sin 30° = 8.6 \text{ fc}$$

Point D:
$$Z = \frac{3}{\cos 20°} = 3.19'$$

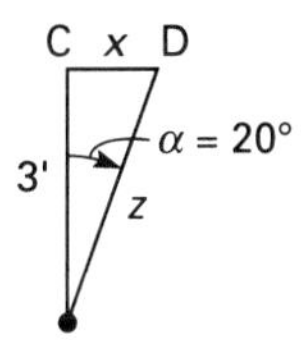

$$\theta_D = \arctan\frac{3.19}{5.20} = 31.53°$$
$$I_D = 464 \text{ cd}$$
$$d^2 = (3.19)^2 + (5.20)^2 = 37.22'$$

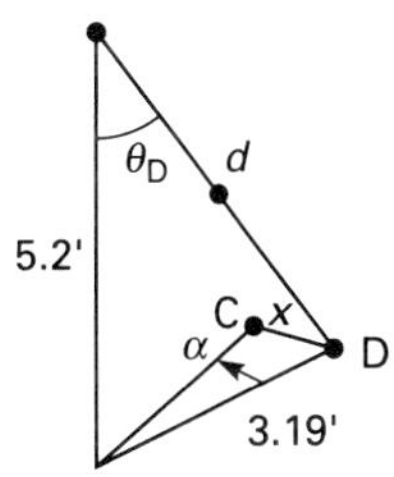

$$E_D = \frac{I\theta_D}{d^2} \times \sin\theta_D \times \cos x$$
$$E_D = \frac{464}{37.22} \times \sin 31.53 \times \cos 20 = 6.13 \text{ fc}$$

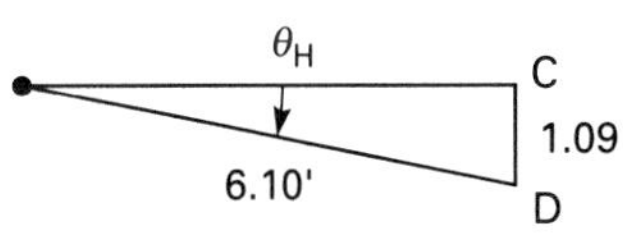

Example problem 3: Lumen method—zonal cavity

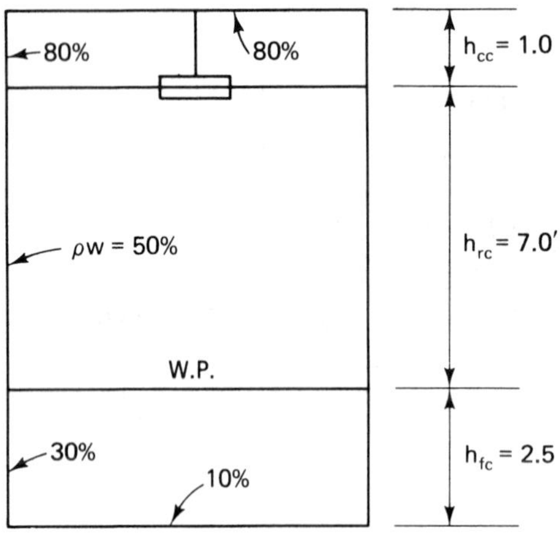

Cross-section

Room dimensions
30 x 35

$E = 50$ fc

$LDD = 0.83$ (clean and 24 months)—Figure 8-11.

$LLD = 0.87$ (6 hr/start)—Table 8-7.

Luminaire Table 8-4

2F40 CW—test lumens = 3110 ea.

actual lumens = 3150

$$LF = \frac{\text{Act. lm}}{\text{Test lm}} = \frac{3150}{3110} = 1.01$$

Solution:

$$RCR = \frac{5\,h_{rc}\,(W+L)}{W \times L} = \frac{5(7)\,(30+35)}{30 \times 35} = 2.17$$

$$CCR = \frac{5\,h_{cc}\,(W+L)}{W \times L} = \frac{5(1)(30+35)}{30 \times 35} = 0.31$$

Find ρ_{cc} by interpolation for $CR = 0.80$, $CWR = 0.80$, and $CCR = 0.31$ from Table 8-5.

$$\rho_{cc} = 78\%$$

Find CU by interpolation for $RCR = 2.17$, $\rho_{cc} = 0.78$, and $\rho w = 0.50$, using the luminaire in Table 8-4.

$$CU = 0.62 \qquad \text{for} \quad \rho_{fc} = 0.20$$

$$FCR = \frac{5\,h_{fc}(W+L)}{W \times L} = \frac{5(2.5)(30+35)}{30 \times 35} = 0.77$$

Find actual ρ_{fc} by interpolation for $FR = 0.10$, $FWR = 0.30$, and $FCR = 0.77$ from Table 8-5.

$$\rho_{fc} = 10\%$$

Find floor correction by interpolation for $RCR = 2.17$, $\rho_{cc} = 0.78$, and $\rho_w = 0.50$ from Table 8-6.

$$\text{correction factor} = \frac{1}{1.06} = 0.94$$

Corrected CU:

$$CU = 0.62 \times 0.94 = 0.58$$

$$\text{no. of luminaires} = \frac{E \times A}{\text{no. lamps} \times \text{lumens per lamp} \times CU \times LLD \times LDD \times LF}$$

$$= \frac{50 \times (30 \times 35)}{2 \times 3110 \times 0.58 \times 0.87 \times 0.83 \times 1.01}$$

$$= 20$$

9 vdt and the office environment

INTRODUCTION

The passage of legislation regulating VDT (visual display terminal) operations in the office environment has brought new urgency to the need for greater understanding of the complex nature of this newest visual task. In June 1988, Suffolk County, New York, passed the first law in the United States regulating the use of VDTs in the work environment. Since 1981, the state of Illinois and the state of Maine have debated and delayed action on various House Bills to regulate VDT operations. The state of California has also been very active in attempting

252-268

to legislate regulations for VDT operations in the office environment. In 1985, the West German law regulating VDT operations was fully implemented. With the passage of the Suffolk County law, legislation in other states, counties, and cities is more likely to follow.

The format of the law passed and the pending legislation parallels the 1978 "National Safety and Health Directive" passed into law in Sweden. The law puts requirements on the VDT equipment utilized, the operator's furniture, the lighting conditions, employee education, mandatory rest period, mandatory eye examinations, and makes the employer financially responsible for prescription glasses. The Austrian Federation of Salaried Employees has established a mandatory paid rest period of 15 minutes after every 2 hours in front of a CRT screen. A 15-minute mandatory rest period for every 2 hours of CRT operation has also shown up in a number of the laws being proposed in the United States. The longer that employers remain uneducated and uninformed about the nature of the problems facing the VDT user, the higher the ultimate cost to the employer and the more dictatorial the laws and regulations will become. If the employer becomes knowledgeable of the VDT user's problems and moves without legislation to solve the problems, the restrictive and sometimes meaningless regulations can be prevented. The problems are best solved by competent professionals working with the employer and employees rather than waiting for legislative action to force compliance that may not solve the VDT user's problems. The goal of this chapter is to provide the information necessary for the employer, the employee, and the lighting professional to become better informed about the problems facing the VDT user so that the appropriate action can be taken to eliminate most of the problems.

When one speaks of VDT usage in the office environment, one is most commonly referring to a microcomputer video screen or a computer remote terminal screen. The acronym VDT means visual display terminal. The computer screen is also called a CRT or cathode-ray tube; hence, many articles on this subject will talk about the CRT user. Other authors may use the acronym VDU for visual display unit, rather than VDT. In many cases, authors may use these three acronyms (VDT, CRT, VDU) interchangeably. Since a major problem in CRT operations centers on the images on the computer video screen, most authors write about problems associated only with the CRT. However, this chapter will define the more complex problems by going beyond the CRT and include other ergonomic problems related to the keyboard, the area surrounding the CRT screen, the VDT furniture, the user's chair, and the luminous environment in which the VDT system is located. Therefore, it is most appropriate to refer to the VDT or VDU when dealing with the total system and to refer to the CRT when dealing specifically with imaging problems on the computer video screen.

The impact of VDT usage on the office worker has resulted in negative publicity which has resulted in legislation "to provide occupational safeguards for operators of video display terminals" (Illinois House Bill 1725). During the transition period from limited VDT usage to the time when VDT usage will represent 100% of the work performed in the office environment, the design of the luminous environment in the office will be compounded by the multiple visual tasks being performed. The problems, created by the introduction of the VDT system into the office environment, have been treated as individual problems not as a total system problem. For example, the lighting engineer attempts to solve the complex interactions among VDT user/VDT system/luminous environment by treating one symptom, the lighting system. A total systems approach to the design of the VDT environment will provide "occupational safeguards" for operators of visual display terminals without the need for further legislation. The authors feel that a logical, well planned in-depth solution to the complex interactions in the VDT environment may result in solutions and guidelines for the lighting professional. A properly designed VDT environment will have a positive impact on the performance and productivity of the VDT user. Legislation will not increase performance and productivity, but it will create additional problems and cost for the employer. It is in the best interest of the employer to provide an integrated visual, thermal, and acoustical environment for the VDT user to optimize performance and productivity.

The VDT user's comfort, performance, and productivity are directly related to the physical, psychological, and emotional well-being of the VDT user. The VDT operation, working in the luminous environment, involves complex interactions among VDT user/VDT system/luminous environment. A study of the key elements must include the ergonomics of the VDT user (person), the characteristics and ergonomics of the CRT screen and the VDT furniture, and the physical aspects of the space (luminous, thermal, acoustical environment). The visual work or visual task being input into the VDT by the VDT user will be referred to as the **source document.** The specific visual work performed by the VDT user involves interactions between the source document, the human visual system, and the CRT screen. The human visual system will be affected by the images of the luminous environment in the field of view reflected into the CRT screen. The chapter will analyze these three system components which will be referred to as VDT user/VDT system/luminous environment.

APPLICATION OF THE VDT IN THE OFFICE ENVIRONMENT

VDT applications can be divided into two types (see Fig. 9-1): (1) data entry terminals and (2) conversational terminals.

Data Entry

The data entry terminal (DET) involves number key operations utilizing the right hand. The primary visual task is the source document which is viewed approximately 99% of the time while the VDT user is entering data. The user will glance at the CRT screen only 1% of the time.

Conversational

The conversational terminal (CT) involves keyboard operations with both hands. The visual task is approximately 50% viewing of the source document and 50% viewing of the CRT screen. In both the DET and CT applications, the visibility of the source document is extremely important.

(a) Data entry terminal (DET)

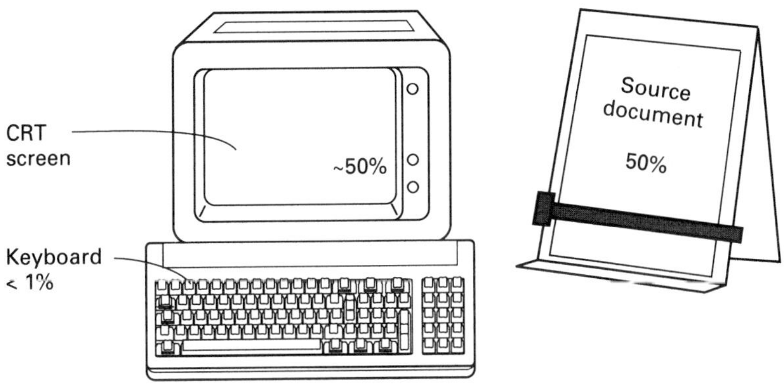

(b) Conversational terminal (CT)

Figure 9-1. Terminal Applications.

VDT USER

An understanding of the ergonomics of the VDT user (person) are crucial to the proper selection of a VDT system which includes the computer hardware and the workstation furniture. A number of studies of the ergonomics of a VDT user, the workstation, and the CRT have been conducted by human factors engineers in Sweden, Switzerland, Germany, England, and the United States since the late 1970s. The physical comfort of the VDT user can be directly related to the design of the user's chair, the furniture supporting the hardware, source document holder, and CRT screen.

Ergonomic studies establish the physical dimensions of the "standard" (percentile) anatomical measurements of the male and female body. These anatomical dimensions (see Fig. 9-2) should be used by the chair and VDT furniture designer to establish the

TABLE 9-1 Anatomical Dimensions (see Fig. 9-2).

Symbol See Fig. 9-2	Description	Male Mean Range (inches)	Female Mean Range (inches)
A	Eye height	31.8 (30.7–33.1)	29.0 (27.8–30.2)
B	Seat height	17.0 (16.0–18.0)	16.0 (15.0–17.0)
C	Elbow resting height	9.9 (8.9–10.9)	8.9 (7.9–9.0)
D	Forearm-hand length	18.7 (17.9–19.5)	17.4 (16.4–18.4)
E	Buttock-politeal length	19.8 (18.8–20.8)	18.8 (17.7–19.9)
F	Thigh clearance	6.5 (5.9–7.1)	4.9 (4.4–5.4)
G	Popliteal height	17.2 (16.3–18.1)	16.2 (15.5–16.9)

Source: NASA (1978).

Figure 9-2. Anatomical Dimension for a Seated User.

full range of adjustment required to meet all VDT users' physical attributes. Table 9-1 lists some important anatomical dimensions.

A head tilt angle of 20° forward (down) measured between the vertical axis of the spine and the head/neck axis is recommended for physical comfort of the neck and shoulders (Fig. 9-3).

Viewing distance and correction for viewing distance play an important role in VDT user comfort. The VDT user's visual task typically involves viewing a CRT screen, source document, and keyboard. These three visual tasks should be at the same viewing distance and each should be perpendicular to the line of sight. Maintaining the same viewing distances and a line of sight perpendicular to the plane of each visual task will minimize changes in accommodation and convergence. Accommodation and convergence are the visual processes involved in bringing small objects into sharp focus. Typically the viewing distance for reading is from 14 to 18 inches. Reading glasses and bifocals are normally fitted for a viewing distance of 16 inches. In theory, if the three visual tasks are at the same viewing distance, the standard 16 inch viewing distance for reading glasses and bifocals will not be correct. Why is 16 inches the wrong distance? Ergonomic studies suggest that for VDT user comfort that the forearms and hands should be horizontal and supported with an angle of 80° to 100° between the upper arm and ''horizontal'' forearm (see Fig. 9-4). This sets the vertical distance from the eyes (head tilted down 20°) to the horizontal forearm (center of the keyboard) in the range of 18 to 21 inches for a 45- to 48-inch eye height above the floor. For a 65° down viewing angle (from a horizontal line of sight) to the keyboard, the viewing distance from the eyes to the keyboard would be from 19 to 23 inches. This means that for the conditions of a horizontal forearm and the same viewing distance to the three different visual tasks, the CRT screen and source document should be approximately 21 inches from the eyes (Fig. 9-3). Therefore, reading glasses should be prescribed for a test distance of approximately 21 inches.

VDT SYSTEM

The VDT system as defined in this chapter includes the furniture that supports the microcomputer hardware, the VDT user's chair, the source document holder, and the CRT screen.

Ergonomics of the VDT Furniture

VDT Furniture

Separate or independent adjustment of the keyboard support platform, the VDT screen, and the source document holder are also critical to user comfort. There is a need for adjustable arm rests on the chair and a wrist rest on the vertically adjustable keyboard platform. A wrist rest is a padded support that should be horizontally adjustable and attached to the keyboard platform. The wrist rest reduces muscle fatigue in the forearms and wrists.

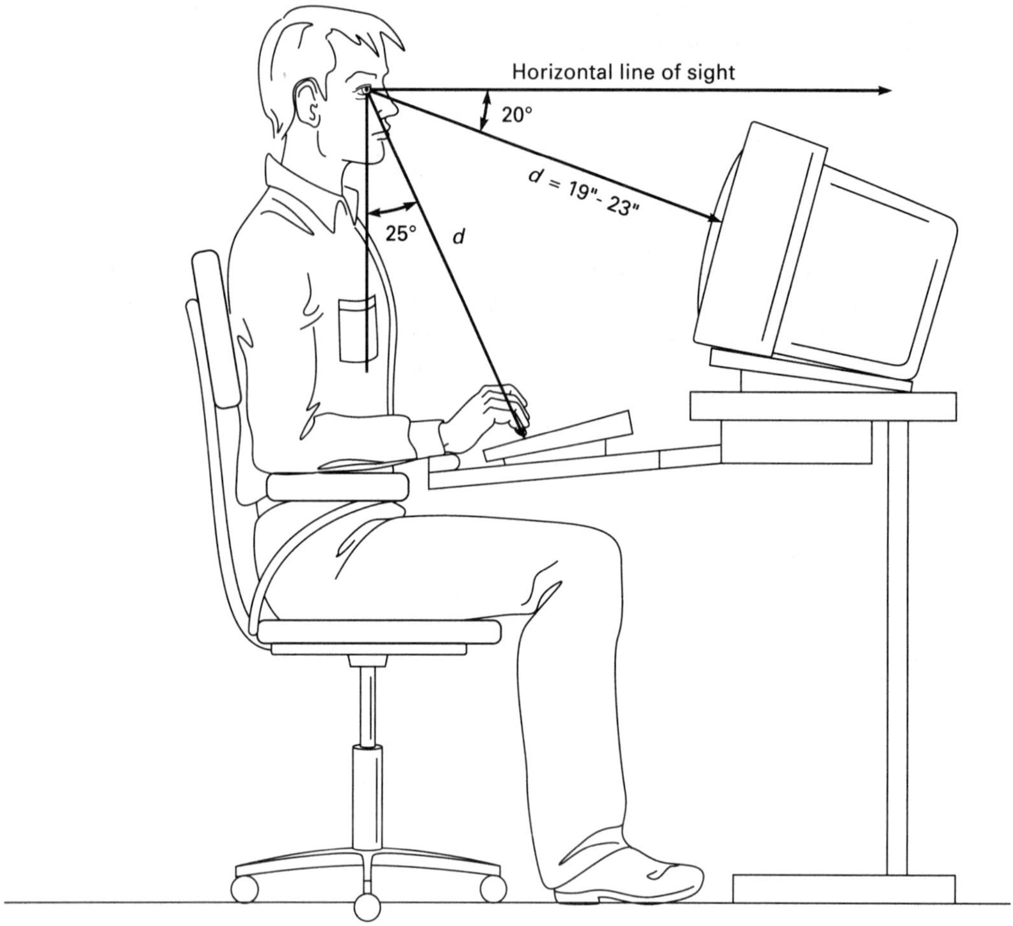

Figure 9-3. VDT System Ergonomics: Viewing Distances.

User's Chair

The adjustment of the chair relative to the keyboard and CRT screen position are critical to the skeletal/muscular comfort of the VDT user.

The VDT user's chair must meet the design conditions of adjustment and comfort. The height adjustment of the seat bottom (see Fig. 9-4) is necessary to account for anatomical differences in the length of the lower leg. A vertical and horizontal adjustment of the back support is important to establish the proper positioning in the contoured seat bottom cushion and to allow the vertical position of the support at the level of the shoulder blades of the user. The chair must have adjustable elbow ("arm") rests to maintain a horizontal forearm support. The term "elbow" rest is most appropriate since the "arm" rest should be short in length to eliminate a conflict between the "arm" rest and the VDT furniture. The function of the elbow ("arm") rest is to eliminate

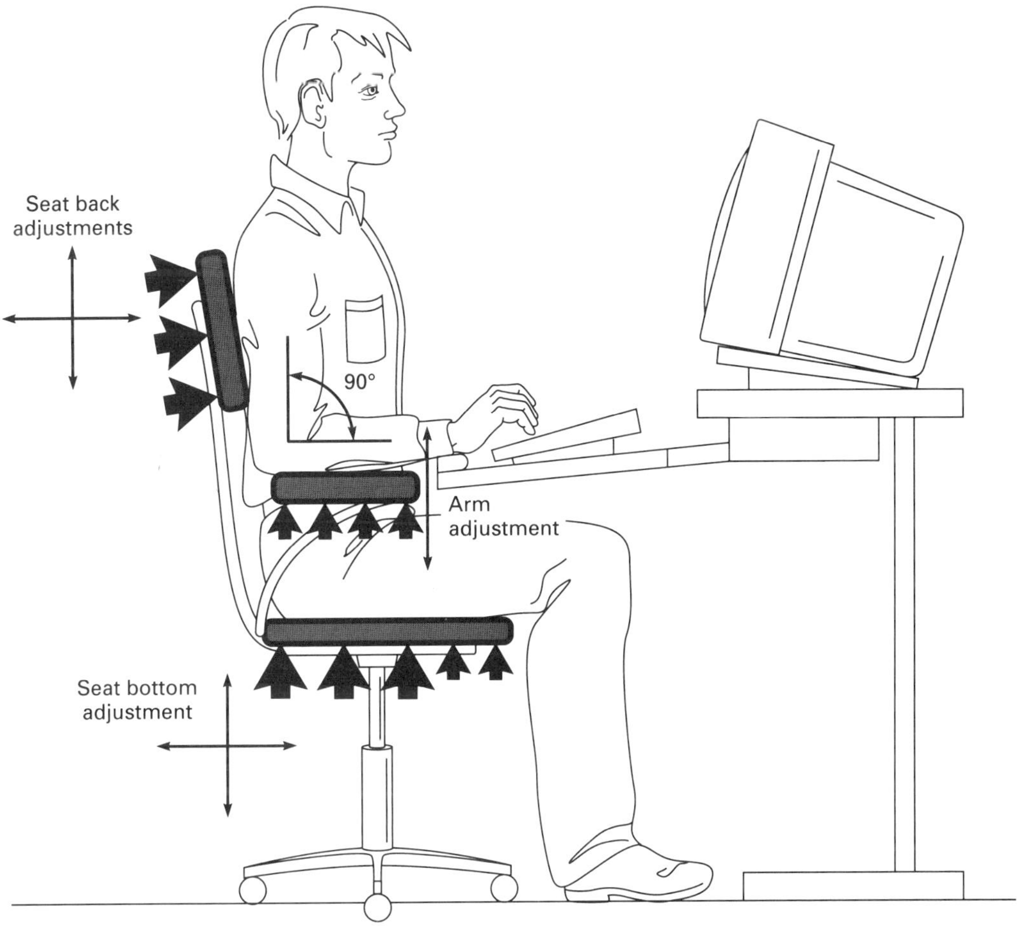

Figure 9-4. VDT System Ergonomics: The Chair.

stress in the shoulder muscles, upper back muscles, and neck by eliminating "hanging weight" of the arms by supporting the elbow. The chair must be on rollers (preferably four rollers for greater stability and easy movement) to allow the in/out positioning of the VDT user.

If a VDT user's chair is used by more than one employee, the chair should be adjusted each time for the new individual using the chair. The employee must be trained in the proper chair adjustments. The height of the chair bottom should be adjusted first to a height that allows the VDT user's feet to sit flat on the floor with a slight upward pressure on the bottom side of the upper legs (see Fig. 9-4) to minimize weight on the ankles and feet. The back rest should be moved horizontally to provide a firm back support when the buttocks is positioned comfortably in the contoured seat bottom cushion. The vertical adjustment of the back rest should place the support at the level of the shoulder blades. The vertical height of the elbow rest should be at a position so that the weight of the upper arm is not being supported by the shoulders and upper back muscles but by the elbow (''arm'') rest on the chair. It is possible to get the elbow rest too high. This should be avoided because it pushes the shoulder up in an unnatural position causing tension in the upper back muscles.

Source Document Holder

1. For Conversational Terminal Applications: Assuming the 20° downward tilt of the head and a viewing distance of 19 to 23 inches between each of the three VDT user's visual tasks, the source document (Fig. 9-5) should be (a) vertical with a slight backwards tilt of 20°, (b) at a viewing distance of 21 inches, and (c) at a vertical height where the center of the source document is at the same vertical height as the

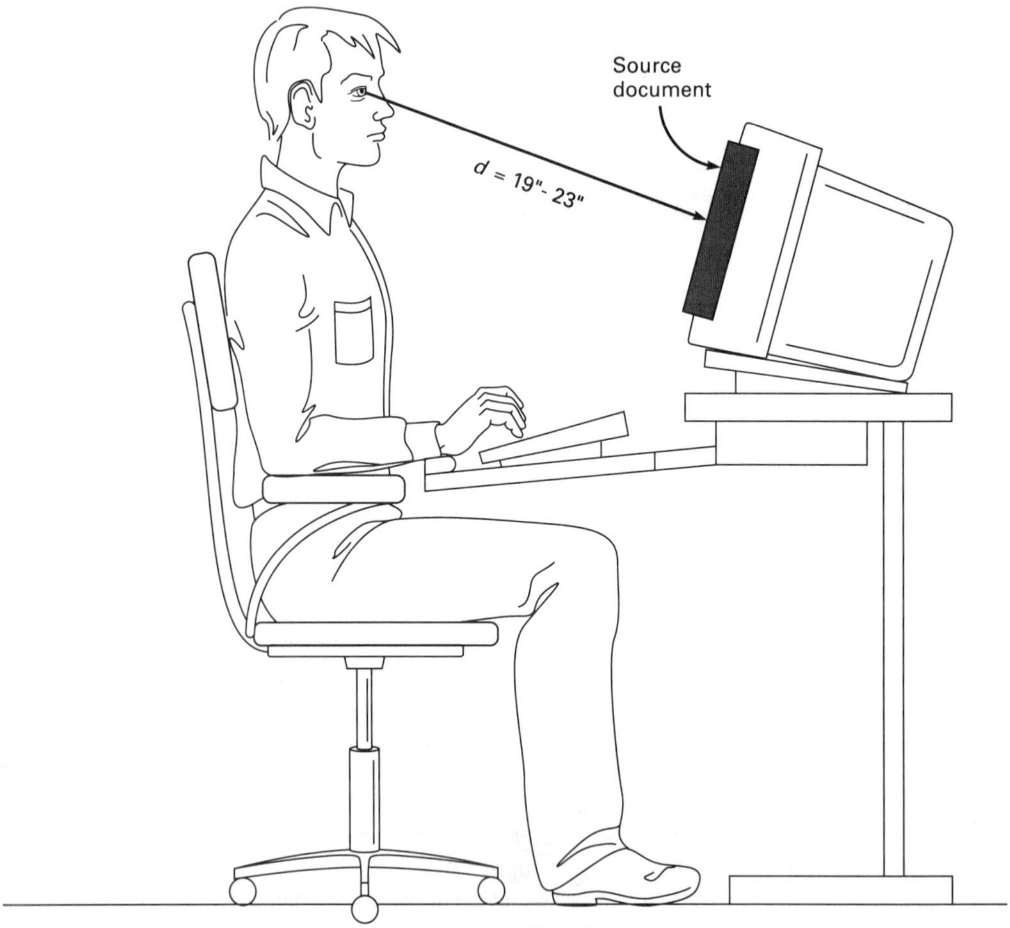

Figure 9-5. VDT System Ergonomics: The Source Document Holder.

center of the CRT screen. These conditions dictate the use of a source document holder that meets these design conditions. The position of the document holder allows for a shielded supplementary light source to be integrated into the holder. The illuminance level produced by the supplementary lighting unit should be adjustable to allow for increasing or decreasing the illuminance as the contrast and size of the detail varies from one source document to another. By fixing the shielded supplementary light unit to the document holder in a position above and slightly in front of the plane of the source document, veiling reflections and reflected glare should be eliminated because of the low angle grazing position of the light source. A properly designed source document holder (Fig. 9-5) must be strong enough to hold source documents that are the thickness of a book but adaptable enough to accept a single sheet document. To maintain the proper viewing distance and angular relationship between the supplementary light unit and the front surface of the source document, thickness adjustment for different source documents must move backward from the constant location of the front plane of the holder.

2. For Data Entry Terminal Applications: The most comfortable head-neck position is the 20° downward tilt of the head at the keyboard viewing distance which is from 19 to 23 inches. Therefore, the most appropriate conditions for a DET source document holder would be the same (Fig. 9-5) as those for conversational terminal user described earlier. However, some DET application users may prefer a more horizontal placement of the source document. If the source document is placed "horizontally," the holder should be tilted at the same angle and at the same viewing distance as the keyboard. A "horizontal" placement of the source document holder introduces indirect glare problems discussed in Chapter 8.

If a universal source document holder is to be designed that is, one for either CT or DET applications, it must have a supporting mechanism that will allow the VDT user the flexibility to position the source document in either the "vertical" or "horizontal" viewing positions.

Computer Hardware

The critical components of the computer hardware that relate directly to the VDT user are the physical characteristics of the CRT screen and the input device which is the keyboard. These two hardware components that will impact on the user's comfort and visibility will be briefly reviewed.

Keyboard

The microcomputer keyboard should be of a detachable type. A keyboard fixed to the CPU unit prevents the VDT user from making the proper keyboard position adjustments. The keys on the keyboard should be of a matte or diffuse finish to eliminate reflected glare from the keys. The keys and keyboard should be of an off-white or light-gray color to minimize luminance ratios between the keys (keyboard) and the source document.

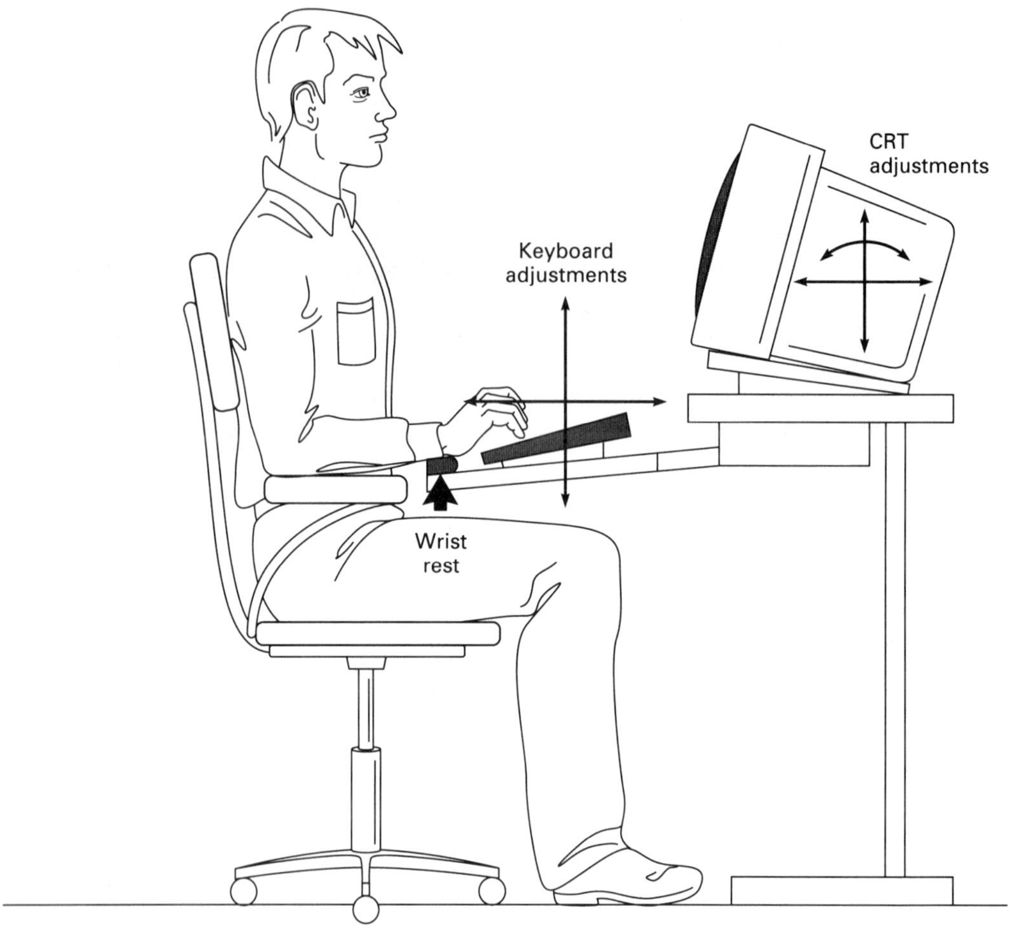

Figure 9-6. VDT System Ergonomics: The Keyboard and CRT Screen.

For the line of sight to be perpendicular to the keyboard (see Fig. 9-6), the keyboard should be tilted up toward the user approximately 25° from the horizontal plane.

CRT Screen

The CRT screen must be on a separate, adjustable platform to allow for position adjustments in a vertical, horizontal, and axial direction. The screen position (see Fig. 9-6) should be adjusted to the same viewing distance of 19 to 23 inches with the screen tilted 20° backward and the line of sight perpendicular to the plane of the screen.

CRT Screen Terminology: To understand the effects on vision and visibility, screen hardware terminology and interrelationships will be defined.

1. *Decay time*—The time it takes for the luminance of the phosphor to drop to 10% of the peak level. Decay time is characterized by the persistence of luminance.
2. *Persistance of luminance*—affects the oscillation of luminance which is measured in terms of the uniformity figure.
3. *Uniformity figure*—a ratio of the base luminance to the peak luminance. The *lower* the UF the *higher* the oscillation of luminance which is perceived as flicker.
4. *Refreshing rate*—expressed in terms of the beam movement per second. This is most commonly from 25 Hz to 80 Hz. The lower the refreshing rate the higher the oscillation of luminance hence the greater the flicker.
5. *Critical flicker frequency*—the point of transition from apparent flicker to a steady light. Flicker is dependent on the kind of phosphor coating (decay time), the refreshing time, and the luminance of the background.

CRT Screen Selection Parameters. An ideal CRT screen would utilize phosphors with a lower decay time, which would increase the persistance of luminance. Increasing the persistence of luminance will decrease the oscillation of luminance by increasing the uniformity figure. The combination of a high uniformity figure with a high refreshing rate (above 100 Hz) will assure the elimination of flicker and allow for the use of a white background screen.

The use of a white background on the CRT screen and refreshing rates of 100 Hz will reduce the luminance difference between the CRT screen and most "white" source documents. Most CRT screens on the market today use refreshing rates less than 60 Hz, which cause excessive, annoying flicker when the screen background is white.

Tilting of the CRT screen compounds the problems of reflected glare and veiling reflections (see Chapter 8). The convex surface of the glass screen increases the reflected field of view which include areas slightly in front of the screen (see Fig. 9-7) on the ceiling plane. To reduce the vertical field of view reflected in the screen, a hooding device should be placed along the top and sides of the CRT screen. A flat CRT screen will narrow the field of view reflected in the screen, but because of the need to tilt the flat screen a hooding device is still required.

The use of "glare screens" will also reduce or eliminate reflected images in the screen surface. Glare screens can be a separate device that is placed in front of the CRT screen or an optical coating that is placed directly on the CRT screen. The optical coating must withstand solutions and abrasives that are used to clean the CRT screen. The disadvantage of glare screens is that they reduce the contrast between the alphanumeric characters and the background.

Another important feature that is required for good vision and visibility is an adjustable CRT background luminance and CRT contrast. CRT contrast and background luminance adjustments are necessary to overcome losses in visibility due to veiling luminance produced by the luminous environment on the CRT screen and to overcome contrast losses due to the use of glare screens.

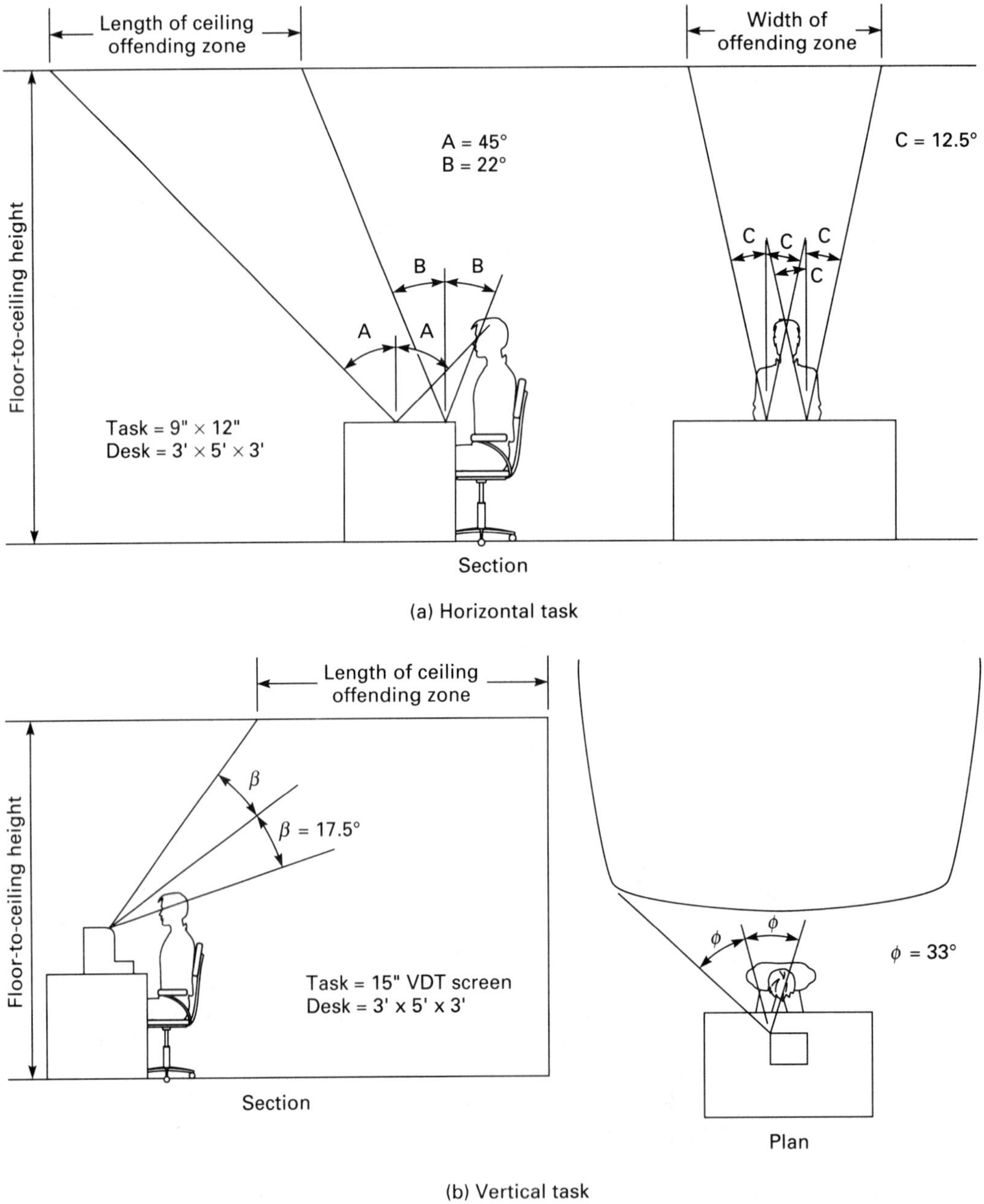

Figure 9-7. Offending Zone for Horizontal and Vertical Visual Tasks.

LUMINOUS ENVIRONMENT

The luminous environment must maximize the visibility of (1) the CRT screen, (2) the source document, and (3) the keyboard by controlling veiling reflections and reflected glare on all three visual tasks. The visibility of the alphanumeric characters on the keyboard, source document, and the CRT screen is a function of the contrast between the characters and their background. Since the VDT user's visual work for a conversational terminal primarily involves visibility of both the CRT screen and source document, adaptation to different background luminance values can affect worker visibility and visual comfort. Dark-background screens are required when the refreshing rate is low to minimize flicker. The darker screen background luminance to "white" source document background luminance ratio causes repeated adaptation changes in the visual system as the worker looks from source to screen and back. These adaptation changes will result in pupil fluctuation, accommodation change, and convergence change which affects user comfort.

The design of the luminous environment is more critical and complex because of the interaction discussed earlier. The VDT user's visual tasks (Fig. 9-8) involve head-

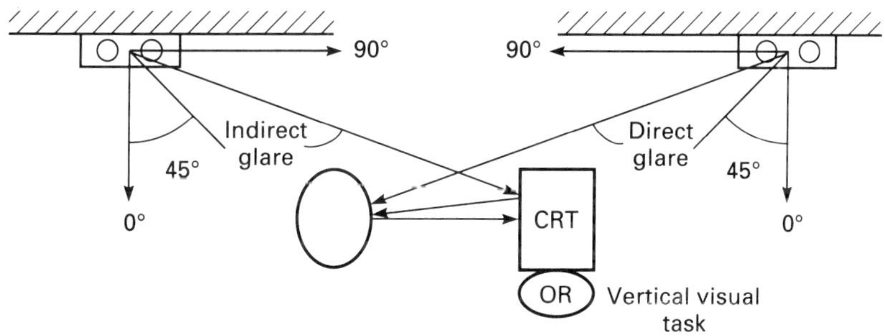

(a) CRT or vertical tasks

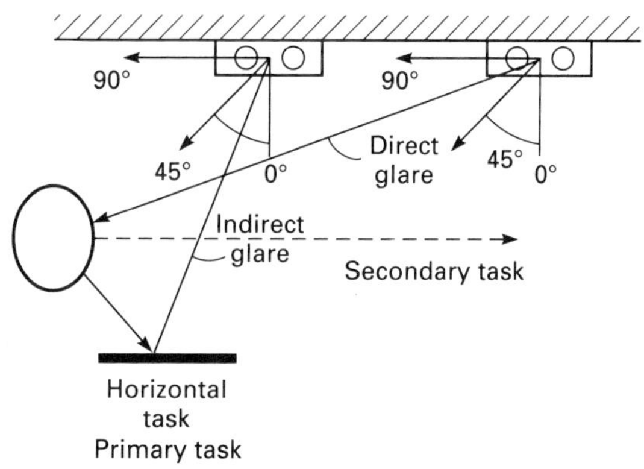

(b) Horizontal tasks

Figure 9-8. Glare for Horizontal and Vertical Visual Tasks.

up (nearly horizontal line of sight) work when viewing a ''vertically'' placed source document and the CRT screen. The glass CRT screen acts like a convex (or flat) mirror that reflects images of objects within the visual field into the CRT screen. The convex shape of the glass screen causes the visual field to be expanded as compared to a vertically placed flat mirror.

CRT SCREEN IMAGES

Since the glass acts as a mirror, the image of an object reflected in the CRT screen is a vertical, erect image, which is displaced slightly due to refraction of the light rays entering the glass. The image of an object in the glass screen appears to the viewer to be at some distance ''back inside'' the screen. The reflected image of the object in the screen is at a distance to the screen surface that is less than the true distance from the screen to the real object in the room. The appearance of a reflected image ''back inside'' the CRT screen causes the VDT user to make accommodation and convergence changes when changing gaze from the alphanumeric character on the CRT screen to the reflected image of a bright object in the CRT screen. That is, the alphanumeric characters are produced in the phosphor on the inside surface of the ''mirror'' glass surface, and the image of the object reflected in the glass surface is at a different focal distance ''back inside'' the CRT, causing rapid fluctuation in the accommodation and convergence between the alphanumeric characters and the reflected image.

TASK VISIBILITY

The design problems associated with a horizontally placed visual task (Fig. 9-8) on a desk surface are quite different from the design problems associated with a tilted, specular CRT screen and a ''vertical'' source document. The VDT user can experience losses in visibility and discomfort from both direct glare and indirect glare. A conversational terminal user experiences both types of glare while viewing the CRT screen and the source document, while a data entry terminal user experiences both types of glare while viewing the source document. Chapter 8 defines direct glare and indirect glare in terms of horizontal visual tasks. The definitions of these two terms are

1. *Direct glare*—can be caused by any excessively bright source of light coming directly to the visual system causing discomfort and/or a loss in visibility.
2. *Indirect glare*—can be caused by any excessively bright source of light coming to the visual system by reflecting off of a surface causing discomfort and/or a loss in visibility.

The VDT user of ''vertical'' visual tasks is operating in a heads-up (''horizontal'' line of sight) mode of viewing. Glare sources in the space in front of the user can become sources of direct glare into the VDT user's eyes and indirect glare on the "horizontal" source document. Glare sources in the space behind the VDT user can become sources of indirect glare on the CRT screen and "vertical" source document. The indirect glare can be reflected glare causing distinct images of the glare source in the visual task or veiling reflections causing a uniform wash of higher luminance across the surface of the visual task.

Direct glare associated with the field in front of the VDT user needs to be controlled by reducing or eliminating the high-angle luminous intensity (luminaire direct glare zone, Chap. 8) from the luminaires. High-level luminous intensity in the direct glare zone of the luminaire will increase retinal illuminance which can cause losses in visibility due to light scatter in the eye and cause discomfort due to transient adaptation changes. Luminaires with high VCP values (approaching 100) may have less potential for producing direct glare in the VDT user's eyes.

Indirect glare associated with reflected images in "vertical" visual tasks needs to be controlled by reducing luminance ratios and by reducing or eliminating high-angle luminous intensity in the direct glare zone of luminaires "behind" the VDT user. The indirect glare in the CRT screen caused by high-angle luminous intensity can be evaluated by looking at the VCP values for the luminaires behind the VDT user. To use VCP as an approximation of indirect glare potential, the VDT observer's location is assumed to be in the CRT screen looking back into the visual field behind the VDT user. Caution should be exercised in using this assessment technique because a higher VCP for backward viewing does not mean an elimination of a reflected image in the CRT screen. A higher VCP simply means a lower "brightness" to the image reflected into the CRT screen.

Luminance ratios between all objects in the room must also be controlled. Luminance ratio guidelines (see Chapter 8) are given as 3 to 1 ("preferred") with a maximum of 5 to 1 "where critical visual work occurs" (1). This luminance ratio guideline appears to be an acceptable criterion for source document visibility and visual comfort. However, when dealing with reflected images in the CRT screen a 3-to-1 luminance ratio should be considered as a "maximum" value with a preferred ratio approaching 1 to 1. If it were possible to achieve a perfect 1-to-1 luminance ratio between all objects in the environment, the level of luminance of the objects would be less critical as long as the CRT screen had a background luminance and contrast adjustment control to overcome the veiling luminance produced on the CRT screen.

Contrary to popular belief, the answer to the lighting design problem is *not* to turn off all the lights in the indirect glare zone and paint everything a 0% reflectance black. This solution would create excessive luminance ratios between the CRT screen and source document, between the dark surround of the space and the CRT screen, and between the dark surround of the space and the source document. In theory, the solution is to produce a moderate level of absolutely uniform luminance throughout the entire luminous environment. The moderate level of uniform luminance in the surround (the VDT environment) will minimize luminance ratio problems between the three visual tasks and their surrounds. The "uniform" luminance in the VDT environment will result in uniform veiling luminance on the three visual tasks. The veiling or uniform luminance of the visual tasks can be counteracted by the adjustment of the contrast and the brightness of the CRT screen and a properly placed source document holder supplementary lighting unit.

Absolute uniformity of luminance at a moderate level within the entire VDT environment is impossible to achieve. It will require that close attention be given to the luminance of the luminaire, the ceiling plane, the walls, and other luminous objects within the VDT environment. Floor-mounted, diffuse back screens (partial height partitions) and ceiling-hung, diffuse luminaire baffles may help in controlling direct glare and indirect glare for the VDT user. The reflectance of these blocking devices must be selected so that their luminance level matches the luminance of other objects in the VDT environment. The lighting professional needs to remember that the luminance is going to be a function of the level of illuminance arriving at the baffle and the reflecting characteristics of the baffle. A *luminance match* is the critical factor, *not* a reflectance match. Making all objects in the VDT environment a constant reflectance value, say, 50%, will not eliminate reflected images (see Chapter 8).

SUMMARY

We have analyzed the complex interaction among VDT user/VDT system/luminous environment that will affect a VDT user's visibility and comfort. Losses in visibility and comfort can adversely affect the VDT user's level of performance and productivity. A loss in performance and productivity can result in a loss of profits for the employer while losses in comfort will result in additional legislative action such as the National Regulation in Sweden. Legislative regulation can further reduce performance, productivity, and profit for the employer while severely restricting the lighting professional's ability to provide a comfortable, pleasant, efficient visual environment for the VDT user. The lighting professional, the architect, and the interior designer must understand the ergonomics of the VDT user as well as certain physical characteristics of the VDT system to design the visual environment properly. The designer of the VDT environment must inform the client of these complex interactions and the potential problems the client may face if a proper system approach to the design is ignored. The final design of the VDT environment must involve a total system approach that integrates VDT user/VDT system/luminous environment into the design solution.

REFERENCES

1. Kaufman, John E., and Jack F. Christensen, eds., *IES Lighting Handbook*, Application Volume, 1987. New York: Illuminating Engineering Society of North America, 1987.
2. National Aeronautics and Space Administration (NASA), ''*Anthropometric Source Book*. Vol. I, *Anthropometry for Design*, and Vol. II, *A Handbook of Anthropometric Data*. NASA Reference Publication 1024. NASA, 1978.

10 lighting and energy conservation

Lighting has been one of the prime targets of mandatory standards to reduce energy consumption. To put the role of lighting as it relates to energy conservation into perspective, we shall examine the impact of lighting on total energy resource consumption in the United States. The goal of energy conservation should be to reduce the consumption of energy resources.

Currently, almost 90% of the energy resources used in this country are fossil fuels (coal, oil, and natural gas) (7). The most critical fuels in terms of estimated reserves are oil and natural gas. Of the total resources consumed, approximately 36% are used to generate electricity. This represents an upward trend: less than 20% of energy consumption was for electricity generation in 1960, 25% in 1970, and 32% in 1980. Of that 36%, approximately 20% ends up as lighting. As Figure 10-1 shows, that means that only about 7% of U.S. energy is consumed by lighting. Figure 10-2 breaks down electricity production according to the fuel used. From this we see that only about 6% of the total energy consumption involves critical fuels for generating electricity.

Since only 6% of our energy consumption is for electricity from critical fuels, and only 7% of our energy resources are used in lighting, why is lighting a primary target for energy conservation? Because lighting is "visible." We don't often stop to think about the enormous amounts of energy used in manufacturing processes, but lighting is before us continually. "Turn off the lights" seems a simple and obvious approach to conservation. A second reason why lighting is targeted is that it represents 30 to 50% of the operating cost of a building. Thus lighting is an important concern for building owners, and therefore to the entire economy. As utility rates continue to increase, the impact of lighting on operating costs will become painfully apparent.

Make no mistake, much lighting is wasteful. Qualified lighting professionals have known for years that high levels of uniform lighting throughout a space are wasteful. For example, depending on the viewer and room finishes, many office tasks will be assigned a design target of 50 fc or higher using the IESNA illuminance selection procedure (see Chapter 8). The problem facing the designer is one of task definition and location. Since neither of these inputs is known during a typical design process (2), the designer blankets the entire space with 50 fc or more, thus wasting energy in all the parts of the space where tasks are not performed and 50 fc is not required. This uniform approach is not recommended by the IESNA, but has resulted from a lack of

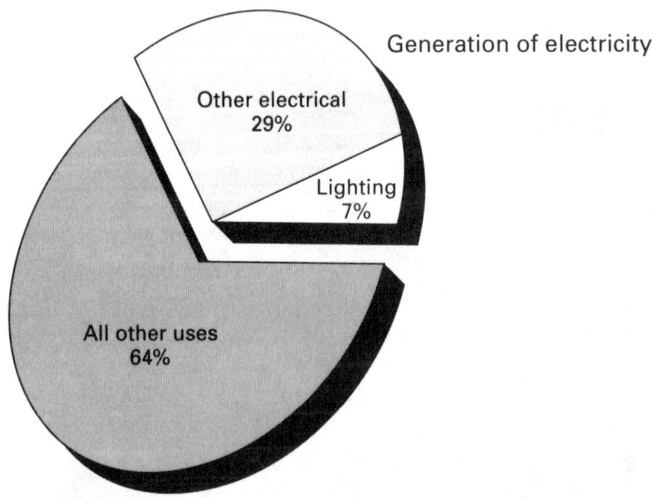

Figure 10-1. Consumption of Energy Resources in the United States by Use. Of our energy use 36% is used for generating electricity. (Source: U.S. Bureau of the Census.)

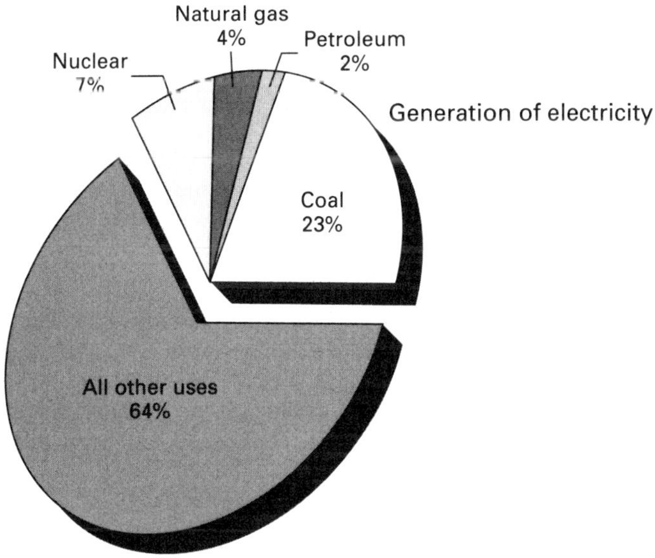

Figure 10-2. Fuels Used for Generating Electricity in the United States as a Percentage of Total Energy Consumption. (Source: U.S. Bureau of the Census.)

communication and understanding of fundamentals on the part of architects, lighting engineers, and interior designers.

LIGHTING ENERGY LEGISLATION

The OPEC oil embargo of 1973 brought the ''energy crisis'' to a head. One of the first documents that was a direct outcome of the energy crisis and the visibility of lighting was a General Services Administration/Public Building Services (GSA/PBS) publication. The GSA/PBS guidelines entitled ''Energy Conservation Guidelines for Existing Office Buildings'' was completed in early 1974. At approximately the same time a document entitled ''Design and Evaluation Criteria for Energy Conservation in New Buildings'' was prepared by the National Bureau of Standards (NBS, now National Institutes of Standards and Technology, NIST) for the National Conference of States on Building Codes and Standards (NCSBCS) on lighting for new construction. NCSBCS asked the American Society of Heating, Refrigerating, and Air Conditioning Engineers (ASHRAE) to prepare a standard for energy conservation in new buildings based on the NBS document.

In 1974, ASHRAE asked IESNA to prepare the lighting chapter, Chapter 9, of this new energy document. The importance of Chapter 9 is that it established a procedure for determining a **lighting power budget.** After being approved by all parties involved, the document was adopted by the American National Standards Institute (ANSI) and became known as ANSI/ASHRAE/IES 90-75.

TABLE 10-1 Public Law 94-163

Mandatory lighting efficiency standards
1. Be in place by 1/1/78.
2. Apply to *all* public buildings.
3. Be no less stringent than Section 9 of Standard 90-75.
4. For existing buildings contain elements deemed appropriate by the state.

In 1975, Congress passed an act entitled "Energy Policy and Conservation Act of 1975," Public Law 94-163, which was amended by Public Law 94-385, "Energy Conservation Standards for New Buildings Act of 1976." Public Law 94-163 contains mandatory lighting efficiency standards for public buildings. The major features of these laws are outlined in Tables 10-1 and 10-2.

In 1976, the Energy Research and Development Administration (ERDA) contracted with NCSBCS to codify ASHRAE 90-75. The resulting document was called "The Model Code for Energy Conservation in New Buildings." Section 9 of the model code is based on IES's lighting chapter from ASHRAE 90-75. Since its original inception, ASHRAE 90-75 has evolved. The most recent version is ANSI/ASHRAE/IES 90.1. The lighting section of that document defines the **unit power density** or UPD method of lighting power budgeting.

This digression into what may seem like dry history for some readers is extremely important. The main point for the lighting professional to keep in mind is that each state has chosen a different method for satisfying the requirements of Public Law 94-163 and 94-385. The NCSBCS model code was adopted by a number of states. Others enacted the original ASHRAE 90-75 and did not revise their statutes to keep up with the evolution of standard 90. Some states have adopted the most recent ANSI/ASHRAE/IES document. California created its own strict code in the late 1980s that was not based on any of the preceding codes. Therefore, it is important to keep abreast of the latest energy legislation for the state(s) in which you design lighting.

More recent energy documents have been published as reference guides, not intended for legislative purposes. These include the Lighting Energy Management (LEM) series from IESNA. LEM-1 established a method for lighting power limits. It was published in 1982. LEM-2 and LEM-4 appeared in 1985 and cover lighting energy limits and lighting energy analysis, respectively. LEM-3 was released in 1987 and discusses the design of energy effective lighting.

TABLE 10-2 Public Law 94-385

Energy conservation for new building requires
1. Federal action to ensure energy conservation features in new buildings.
2. Development and implementation by 1/1/80 (performance standards).
3. Performance standards to maximum practicable energy efficiency.
4. Encourage states and local governments to adopt and enforce the standards.

ILLUMINATING ENGINEERING SOCIETY ENERGY CONSERVATION RECOMMENDATIONS

In 1972, the IES prepared 12 recommendations (3, 5) for better utilization of energy in lighting design without sacrificing quality. The recommendations cover the operation, maintenance, and selection of lighting equipment. The recommendations apply to new construction as well as renovations or retrofit (upgrade). The 12 recommendations are as follows:

1. Design lighting for expected activity (light for seeing tasks, with less light in surrounding nonworking areas).
2. Design with more effective luminaires and fenestration (use systems analysis based on life cycle).
3. Use efficient light sources (higher lumen per watt output).
4. Use more efficient luminaires.
5. Use thermally controlled luminaires.
6. Use lighter finish on ceilings, walls, floor, and furnishings.
7. Use efficient incandescent lamps.
8. Turn off lights when not needed.
9. Control window brightness.
10. Utilize daylighting as practicable.
11. Keep lighting equipment clean and in good working condition.
12. Post instructions covering operation and maintenance.

To expand on these recommendations,

1. *Expected activity.* A design approach that uses lower ambient levels with higher task levels will produce an energy-efficient system. The task location must be known in order to supply the appropriate lighting level at the task location.
2. *More effective luminaires and fenestration.* Luminaires and fenestration systems should be as efficient as possible without adversely affecting comfort (VCP) and visibility (ESI).
3. *Efficient light sources.* Selection should be based not only on the efficacy (lumens/watt), but also on the life, cost, and color rendition. Color should carry as much weight as the other factors, since color has a direct bearing on the psychological behavior of people, which affects productivity and mood.
4. *More efficient luminaires.* Efficiency includes the utilization of energy in the space as well as cleaning and relamping convenience.
5. *Thermal control.* Make use of the heat produced by the lighting equipment.
6. *Reflecting surfaces.* The absorption of light in a space due to low reflectance values will reduce the efficiency of the lighting system, resulting in an increase in wattage.

7. *Efficient incandescent lamps*. If incandescent lamps must be used, the higher-wattage lamps will be slightly more efficient than will lower-wattage lamps.
8. *Lights off when not needed*. Switching off lights when not in use will result in energy savings and a reduction in operating costs. The effectiveness of the energy reduction is a function of the flexibility of the control systems.
9. *Window brightness*. Excessive glare of windows on the line of sight will affect comfort and visibility, which may reduce performance.
10. *Daylighting*. The effectiveness of daylight is dependent on the combined daylight/artificial lighting system. Unless the control system is properly designed, energy reduction is questionable.
11. *Maintenance*. Good maintenance will require fewer luminaires by increasing the utilization of the light entering the space. Maintenance should include not only spot relamping, but also room surface cleaning, painting, luminaire cleaning, and group relamping.
12. *Operating and maintenance instructions*. The design of sophisticated lighting systems and controls for energy conservation will be wasteful if the building user does not know how to properly use and maintain the system. The designer should provide instructions on how to use the system.

POWER, POWER FACTOR, AND ENERGY

The goal of good lighting design is to save energy and, at the same time, to maintain visual performance and productivity. The current emphasis in energy conservation is on the reduction of power (fewer lamps) by reducing lighting levels. This approach leads to design techniques based on watts per square foot or power design techniques.

Power

Power (true power) is measured with a wattmeter and defined as

$$P = E \times I \times \cos \theta$$

where

$$E = \text{voltage in volts}$$

$$I = \text{current in amperes}$$

$$\cos \theta = \text{power factor}$$

Power Factor

If the power factor is unity or 100%, the voltage vector will coincide with the current vector, and the resulting power will be the product of the voltage and current:

$$P = E \times I \quad \text{volt-amperes}$$

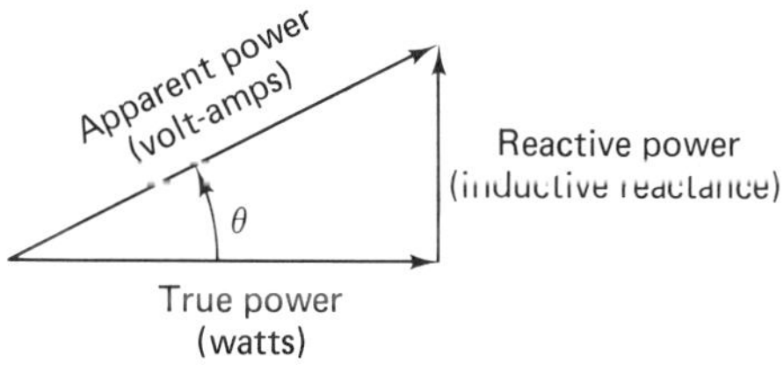

Figure 10-3. Power Triangle.

The voltage and current are said to be in phase, which means that the cosine of the angle θ is 1.0. Incandescent lamps are pure resistance loads with a unity power factor.

All gaseous discharge sources require a ballast that contains windings that produce an inductive reactance. Inductive reactance causes a phase shift between the voltage and current, resulting in a phase angle θ greater than 0 or a power factor of less than 1. Inductive reactance causes the current to lag behind the voltage or creates a lagging power factor, which results in wasted or nonproductive power. When the power factor is less than unity, nonproductive current is present, which (1) requires larger conductors (increased wire cost); (2) adds to the voltage drop (increased copper losses); (3) reduces the circuit's ability to deliver power (increased power or energy consumption); (4) overloads transformers, panels, feeders, and cables; and (5) results in power factor penalty billing. The power factor is numerically equal to the cosine of the angle θ (see Fig. 10-3).

$$\text{power factor} = \frac{\text{true power}}{\text{apparent power}}$$

$$= \frac{\text{volts} \times \text{amperes} \times \text{cosine } \theta}{\text{volts} \times \text{amperes}} = \text{cosine } \theta$$

$$= \frac{\text{wattmeter}}{\text{voltmeter} \times \text{ampmeter}}$$

The apparent power is the product of the voltage and current and is expressed in volt-amperes. The typical utility meter is a watt-hour meter that measures the true power. The generating capacity must be sufficient to deliver the apparent power, which can result in a sizable waste of power (energy) if the angle θ is large. To reduce this wasted power (energy), the power factor should be corrected to at least 0.90 (90%). Lagging power factor is usually corrected by placing a capacitor in the circuit to reduce the phase angle.

Many electric utility companies impose a penalty on customers whose power factor is too low (say, below 0.80).

Energy

The goal is energy conservation. Energy is the product of power (true power) used by a load and the amount of time that the load is in use. Energy is measured with a watt-hour meter. In other words, to be energy minded, not only the amount of electricity used (power) but also how long it is used (time) must be taken into account:

$$\text{energy} = \text{power} \times \text{time}$$

To reduce energy consumption by the lighting system, the illuminating engineer must go beyond simple lighting layouts (lighting design) and study the control of the lighting system.

Spielvogel (6) points out a number of facts collected by various agencies which led him to state that "the one factor that, more than any other, determines energy consumption of a building is how it is used." In particular, a plot of the installed lighting watts per square foot versus the total lighting energy consumption of buildings across the country shows no correlation (Fig. 10-4). If there is no correlation between designed watts per square foot and total energy used, what are the factors involved in total energy consumption? Most experts agree that one factor is the control of the system.

Electricity Rates

Since one electric utility company usually has a monopoly on providing service to a particular locality, the rates they may charge their customers are regulated by the State government. The rationale being, of course, that a company with no competition will

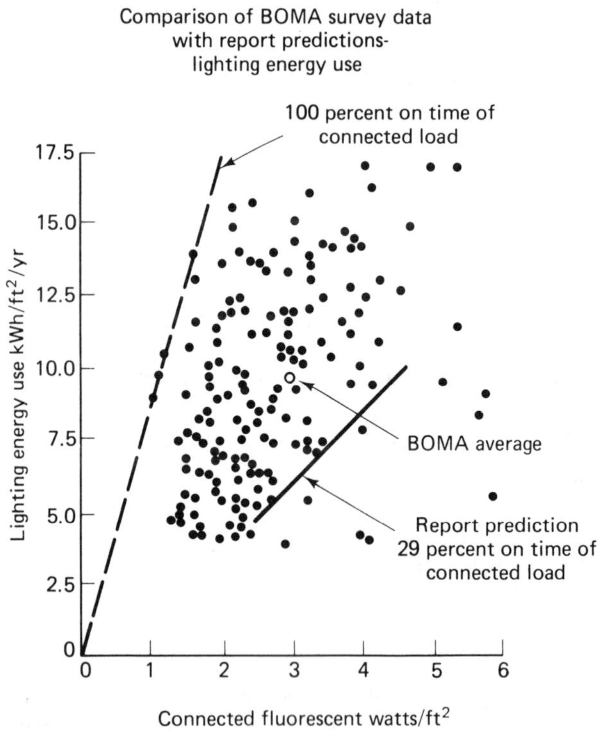

Figure 10-4. Office Building Lighting Energy Consumption Versus Installed Watts. (Courtesy of L. Spielvogel.)

charge extortionary prices. Utilities typically have different rate structures for each of several classes of customers, with the broadest division being between residential customers and nonresidential customers (commercial, industrial, and institutional).

Residential customers are charged a flat rate for the energy they consume. The national average for this rate is about 7.5 cents per kWh in 1990. Thus if a homeowner burns ten 100 W incandescent lamps for 60 hours in a certain month, they will contribute 1 kW $\times$ 60 $\times$ .075 = \$4.50 to that month's electric bill, regardless of when those lamps are on.

Commercial, industrial, and institutional (CII) customers usually buy electric energy according to a different pricing scheme. A typical rate structure for these customers is shown in Figure 10-5. The rate has two components. Part of the rate is a flat charge for energy consumed. This is similar to a residential schedule except that the national average energy charge for nonresidential customers is about 4.75 cents per kWh in 1990. In Figure 10-5 we also see that the CII energy charge may have several levels, depending on how much is consumed in a month.

In addition to a charge for the energy they consume, CII customers pay a **demand charge.** This fee is assessed monthly based on the customer's peak demand for power during the month. A meter monitors the average kilowatt load for every 30-minute period. The peak demand is then the highest of these 30-minute periods during the month. Utility companies use the peak demand as an indicator of the power that the regulatory agency requires that they have the capacity to generate. In other words, the state says that the utility must be able to provide any amount of power demanded by its customers. If a manufacturer has 100 kW of load for most of the month, but for one half-hour period requires 500 kW, then the utility's generating station must be capable of supplying the 500 kW. Excessively uneven loads on the system cause the utility to put power plants "in mothballs"; the plants must be maintained, but they are not being used and are not generating energy charge-related revenues. So the demand charge can be viewed as a penalty to encourage customers to even out their load during the month, or it can be considered the means for financing the additional generating capacity. It is not unusual, especially in areas where the electric utility is operating at or near capacity, for the demand charge to represent the largest portion of the monthly electric bill for a building. Another demand-related aspect of the CII rate structure involves seasonal rate differences. These are the result of the fact that, typically, demand is greater in the summer due to building cooling loads.

Rate structures can have an impact on the design and use of buildings. Because of the energy charge, residential and CII customers alike desire to buy energy efficient lighting and equipment, that is, devices which do the intended job with the minimum amount of power input. They are also motivated to use those devices as infrequently as possible. In addition, the demand charge provides CII users incentive to level out their loads. That is, devices that can perform their function at any time of the day should be run only in off-peak hours. This means avoiding the midafternoon peak in favor of using these loads in the middle of the night. One example of a strategy for shifting loads for large office buildings involves using chillers to make ice at night (off-peak) and storing that ice in a special receptacle under the building. During the

THE STATE UTILITIES COMMISSION

Name of Issuing Utility:	Icthus Power and Light	**Schedule #**	116A
Applicable Territory:	Zones 4 and 5	**Filing Date:**	1/1/91

GENERAL SERVICE – LARGE
Schedule GL

APPLICABILITY

This rate applies to large CII customers as defined in The State Utilities Commission regulation 23.4.6a and to the combined use of a customer at the premises where two or more classes of service to the customer at such premises are measured by separate meters, but only in the case of customers presently so billed.

TERM OF CONTRACT

Contracts under this schedule shall be for period of not less than one year from the effective date thereof, except in the case of temporary service.

RATE:

A. Demand Charge:

$10.21 per month per kW for the first 60 kW of billing demand;
$ 8.06 per month per kW for the next 340 kW of billing demand;
$ 5.14 per month per kW for all over 400 kW of billing demand.

Energy Charge:

8.012 cents per kWh for the first 120 hours use per month of the billing demand, but not more than 6,000 kWh;
6.175 cents per kWh for the next 180 hours use per month of the billing demand;
4.750 cents per kWh for the next 180 hours use per month of the billing demand;
3.014 cents per kWh for all additional kWh per month.

When a customer is taking primary electric service under this schedule, the charges under the above rate shall be reduced by 5.19 percent before application of the Tax Adjustment.

B. When a customer uses electricity as the sole means of water heating with and electric water heater of a size and design approved by the Company, the kWh used for electric water heating shall be billed at 4.750 cents per kWh.

C. Electric space heating shall be billed at 2.985 cents per kWh during the Winter Season. During the Summer Season, electric space heating shall be billed under the rates in A above.

Approval Date:	1/18/91	**For the Company:**	[signature]
Effective Date:	7/1/91	**For the Commission:**	[signature]

Figure 10-5. Typical Electric Utility Rate Schedule for Nonresidential Customers.

peak hours when air cooling is required, rather than add a load to the electrical system, the melting ice provides the necessary energy. Though daylighting strategies can be effectively used to reduce lighting demand at peak hours, no one has yet proposed a scheme by which light could be generated at night and stored until needed the next day.

LIGHTING CONTROLS

The numerous methods to control lighting energy consumption fall into two basic categories. The first type of control provides for either an **on** or **off** state; the second category provides on/off control, but in addition provides the ability to select a level of energy consumption between on and fully off. This chapter discusses on/off controls, level controls, and the state of the art of a combination of the two.

On/Off Controls

The basic on/off control is the switch. Switches are available in a number of configurations, each suited to a particular function.

Circuit-Breaker Switching

The National Electric Code requires all branch circuits to have circuit protection. Most branch circuit protection today is accomplished with circuit breakers. To save initial costs, many designs use the circuit breaker to switch the load. Although the initial cost of a switch leg is saved, circuit breakers should not be used in place of switches. State energy codes that have adopted the NCSBCS model energy codes require a minimum of one switch per circuit, which makes this practice illegal in some states. In addition, the use of circuit breakers as switches is energy inefficient. Most often the location of the panel board is inaccessible to the occupants of the space being served. In such cases, the occupants have no control over their lights, and as a consequence the lights will burn even when not needed. Also, a 20 A branch circuit loaded to maximum capacity according to code can handle 9 four-lamp fluorescent luminaires at 120 V or 22 four-lamp fluorescent luminaires at 277 V. In large areas, all the units may be needed at one time, but in offices or small areas this approach results in more than one space being served by a single circuit breaker. This will prevent energy reductions, because it keeps individual office occupants from turning off the lights as they leave the space. Although this method is often used, it is not recommended and should be avoided if energy savings are to be realized.

AC Snap Switches

The most common on/off device is the ac snap switch. The ac snap switch can be located almost anywhere within the room and can carry the full branch circuit load. For example, each private office should have at least one switch for the lights within

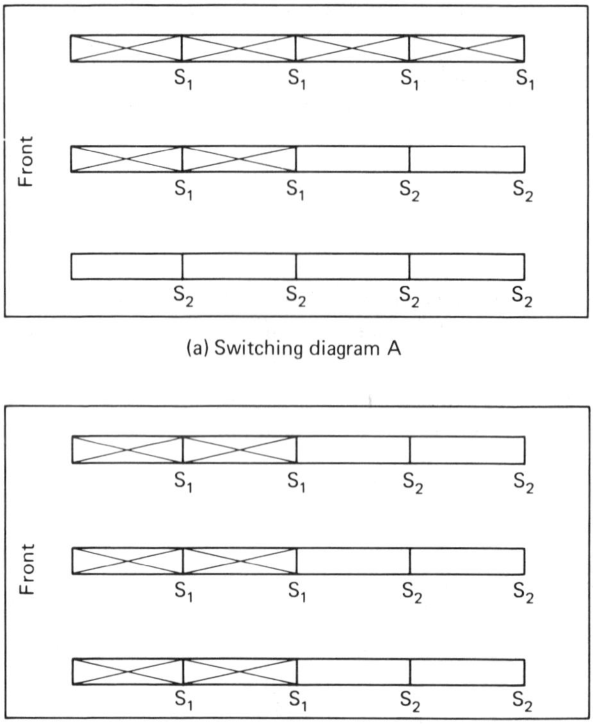

(a) Switching diagram A

(b) Switching diagram B

Figure 10-6. Switching Based on Room Function and Utilization.

that space, and larger areas can be broken up into distinct work areas with a switch in each area. For a space with more than one entrance, three-way or four-way switches should be used to provide control at each entrance.

When a number of switches are provided to control the lights in a large area, they should be circuited with the function of the space dictating the configuration. As an example, consider a lecture hall in which projection equipment is often used. One pattern of circuiting is shown in Figure 10-6(a). With switching circuited in this way, there are always lights on in the front of the room. Figure 10-6(b) is a more logical switching pattern that allows the lights in the rear of the room to remain on for taking notes while those in the front can be turned off to increase the contrast of the screen image. Open plan or office-scape designs are becoming more popular because of the increased space utilization and flexibility. Separate controls can be provided for each indirect luminaire and task light at each luminaire location. This allows for selective control of ambient lighting as well as task lighting as the space utilization changes during a day.

Although the ac snap switch is inexpensive and the most commonly installed control device, it has many disadvantages because it carries line voltage (120 or 277 V). Although its safety record is good, the possibility of death or serious injury is greater at these voltages. Also, to place the switch in its optimum functional location can be economically prohibitive because of wiring costs, as well as increased voltage drop due to long runs. The length of runs and cost become more of a factor with the use of three- or four-way switches.

As an example of both dollar savings and energy savings, consider four offices with two four-lamp luminaires per office, each office used 9 hr/day, 5 days/week. In one case, there is one switch controlling all the offices; in the other, there is one switch per office. The second alternative makes the assumption that each person is out of the office 3 hr/day and that he or she will turn off the lights upon leaving. This second assumption is optimistic, since most studies indicate that people do not turn off lights when leaving a room.

	Case 1	Case 2
Initial cost	One switch and wiring: $13	Four switches and wiring: $52
Annual electricity consumed	× 8 luminaires	× 8 luminaires
	× 200 W/luminaire	× 200 W/luminaire
	× 9 hr/day	× 9 hr/day
	× 260 days/year	× 260 days/year
	3744 kWh/year	2496 kWh/year

It costs $39 more for four switches and associated wiring, but saves 1248 kWh/year in electrical usage. The cost of electrical power would have to be 3.125¢/kWh to recover the cost of the added switches in 1 year. In many areas of the country, the cost of electricity is far above this rate; therefore, case 2 would be more economical than case 1.

Time Switches

In the preceding example the assumption was made that a person would turn off the lights in the space upon leaving the room. This is not always true, so it has been suggested that systems be used that take over the turning off of lights. One such device is the time switch, most often used with sunlamps installed in bathrooms. One manufacturer provides models with timed cycles of 0 to 5, 0 to 15, 0 to 30, and 0 to 60 min, and 0 to 6 and 0 to 12 hr. They also have 0 to 3 min, 0 to 60 min, and 0 to 12 hr times that feature a "hold" position for leaving the circuit closed continuously. The price is reasonable at about $10 each.

With a time switch, the occupant of a space must actively turn on the lights but has to take no action to turn them off. Since the behavior of individuals related to the control of lighting is not really known, the economics of the time switch should be compared to both the single switch control of four offices and individual control in each office. In the previous analysis, it was assumed that 3 hr/day would be spent out of the office. It is probable that part of that time is for short durations in which a time switch would remain on. Therefore, the time out of the office will be reduced to 2 hr with the use of time switches.

Case 3			
Initial cost		One switch and wiring: $13	
Annual		1600 W	Switches not used.
electricity	×	9 hr/day	9 hr/day
costs	×	260 days/year	
	×	4¢/kWh (assumed) = $149.76	

Case 4			
Initial cost		Four switches and wiring: $52	
Annual		1600 W	Switches not used.
electricity	×	9 hr/day	9 hr/day
costs	×	260 days/year	
	×	4¢/kWh (assumed) = $149.76	

Case 5			
Initial cost		Four time switches and wiring: $92	
Annual		1600 W	Time switch used.
electricity	×	7 hr/day	7 hr/day
costs	×	260 days/year	
	×	4¢/kWh (assumed) = $116.48	

Using a 25-year economic life and 10% cost of money, the savings investment ratio is (SIR) calculated:

$$\text{SIR} = \frac{\text{present value of future savings}}{\text{present value of incremental investment cost}}$$

$$= \frac{\text{present value factor} \times \text{yearly savings}}{\text{present value of incremental investment cost}}$$

The SIR of case 5 to case 3 is

$$\text{SIR}_{53} = \frac{9.524 \times \$33.28}{\$79} = 4.0$$

and the SIR of case 5 to case 4 is

$$\text{SIR}_{54} = \frac{9.524 \times \$33.28}{\$40} = 7.9$$

Any SIR greater than 1 economically justifies the added investment. In this situation, it is economical as well as energy efficient to install time switches when the cost of electricity is 4¢/kWh. There are other spaces, such as warehouses, bathrooms, janitor closets, or storerooms where personnel enter the space for short periods of time and neglect to turn off the lights. Time switches would prove even more advantageous in these applications. Where conventional (snap) switches are installed and used, the additional cost of installing time switches is obviously not warranted since the SIR_{51} is negative.

The major disadvantage of time switches is the audible ticking of the timing mechanism. In small areas, such as offices, this could be annoying. It would be counterproductive to improve the luminous environment and adversely affect the acoustical environment.

Low-Voltage Switching

Low-voltage remote switching is not new on the market, but is gaining in importance. All low-voltage systems consist of a magnetic relay, a transformer (putting out 24 V or less), and switches that are interconnected with low-voltage wiring. The relay switches the line voltage and current with a low-voltage command from a low-voltage switch in the space. This provides the ability to control loads from great distances, control a number of different loads from one location, and control one load from multiple locations. A pamphlet printed by a manufacturer of low-voltage equipment, ''Low Voltage Remote Control Switching'' from General Electric, Wiring Device Business Department, Warwick, Rhode Island 02886 is recommended for specific circuit suggestions.

A low-voltage system has many advantages. Since all switch legs carry 24 V or less, there is greater safety, lower copper costs (smaller gauge wires), and lower installation costs because conduit or a raceway system is not required in most states. Switches for a space can be placed in many locations, such as the supervisor's office or guard station. Flexibility can be achieved easily with low-voltage control. Each luminaire or any group of luminaires up to branch circuit capacity can be grouped on a relay. The switching pattern then can be put together by connecting the low-voltage wiring and switches as needed.

One possible configuration would be to switch off all the office lights in a building just after work hours. Then if anyone desires to stay and work overtime, they may turn on the lights again using a low-voltage switch located in the area.

Low-voltage remote-control switching becomes more attractive economically as the size and complexity of the control system increase. The initial cost of a low-voltage system may be higher than conventional methods. The initial cost is dependent on the scope and complexity of the design. The transformer, relay, wire, and switch costs may be offset by reduction in wire and installation costs. The selection of a low-voltage system based on a lower initial cost does not take into account the savings in electricity because of added control.

Time Clocks and Photocells

The controls previously discussed in this chapter require human action to initiate a change of state. Time clocks and photocells, on the other hand, do not. Time clocks and photocells are discussed together because they are conventionally used to control outdoor lighting, such as security, parking lot, roadway, and area lighting. Other than initial cost, there is no reason for not using these systems inside. There are a number of time clocks and photocells that provide different degrees of control.

The most elementary time clock has a single on time and a single off time each day of the week. It has no provision to adjust for seasonal changes in sunrise and sunset, and thus must be continually adjusted when used in outdoor applications. For indoor applications, such as offices, there is usually no need to adjust on/off times, but on weekends it is desirable not to turn on the lights. To overcome the disadvantage of continuous operation over the weekend, a time clock that is able to skip a day is available. One model provides for skipping days, but still has a single on and a single off time for the days not skipped.

If a different on or off time is needed for a certain day, and/or days need to be skipped, a 7-day time clock should be used. Again, seasonal changes in dawn and dusk must be manually compensated for when used in outdoor applications. Designed to take into account seasonal changes in the day-night cycle, astronomic clocks that have a special driving gear are available. No adjustments are needed, but they must be ordered for specific locations. Besides turning off at dawn, the clock can be set to turn off anywhere from 3:30 P.M. to 2:30 A.M. An example of this feature would be a parking lot that is never used after a certain hour at night.

Rather than using time to determine when lights should be on or off, photocells or photocontrols can be used to sense when the light level decreases (sunset) or increases (sunrise). Some photocontrols are constructed so that the level at which they are activated can be varied. This is done simply by a movable plate across the cell face. Photocells can be used alone or in conjunction with other devices. To provide lighting at sunset that is not needed all night, a photocell can be used to turn on the lights with a time control that would turn them off at a preset time. When interfaced with low-voltage switching, photocells provide another means of controlling the lighting system and its energy consumption. Photocells are an excellent choice of control because they are simple, maintenance free, independent of a source of operating power, and self-adjusting to seasonal changes.

Occupancy Sensors

The designer who plans to save energy while leaving control of the lights in the hands of the occupants risks disappointment. If everyone were energy minded, one might argue, there would be no need for devices that work independently of human interaction. The problem with this reasoning is that even the most conscientious person forgets, or has overriding motivation for leaving the lights on when not in use. Case studies have repeatedly shown that occupants cannot be relied upon to turn off the lights.

Since people won't turn off the lights, there is a growing trend toward the use of occupancy sensors to control the lights automatically. This trend is encouraged by the availability of inexpensive detectors. (In 1990 a device to control 300 ft^2 cost under \$100.) These sensors have an advantage over timers. A time clock control would leave the lights on for an entire period, say, from 7:00 A.M. to 6:00 P.M. Though a space may be nominally occupied during this period, people come and go throughout the day. The space may be vacant for 5 minutes here, 10 minutes there. These brief periods may add up to an hour or more over the course of a day. Occupancy sensors detect when someone is actually present. If the space is left unoccupied for a predetermined period (say, 3 minutes), the lights are shut off. They are automatically turned on again as soon as anyone enters the space.

Two types of occupancy sensors are common. **Motion detectors** set up a standing ultrasonic wave in the space. When the geometry of the space changes (e.g., when a person enters the room), the wave is disturbed and the detector is activated, sending a signal to a contactor to turn on the lights. **Heat sensors** detect the infrared radiation which people emit. When heat is detected, the lights come on. When no heat has been detected for a period of time, the lights are switched off. Care should be taken when using either type of sensor that a manual override switch be provided in spaces (such as projection rooms or conference rooms) where it may be desirable to have the room dark while people are present.

Level Controls

The first part of this chapter discussed a variety of means and devices for turning lights on and off. Many circumstances exist where a level of artificial illumination somewhere between fully on and off is desired.

Dimmers

The best means of controlling the level of illuminance is the dimmer. The original dimmers were resistance types that diverted some of the current through a variable resistor. Although the resistance dimmers have their advantages and disadvantages, the only characteristic of importance here is that they do not save energy. The total power used is the same whatever the light level, because the dimmer itself draws power. Most existing dimmers of this type are used exclusively in theatrical applications.

In the last 10 years, the solid-state dimmer has taken over 90% of the dimmer market. The principle of operation is simple. The current in an ac power system is a periodic function, as depicted in Figure 10-7. The power delivered to the load is a function of the shaded area under the curve. The electronic dimmer, by means of an electronic switch, turns off the current to the load for a portion of the cycle, as in Figure 10-7. Note that the shaded area under the curve is smaller, and thus less power is being delivered to the load. The electronic switch has a continuously variable operating time from fully on to fully off. Since the current cannot come on instantaneously, as the vertical line would suggest in Figure 10-7, there is a slight slope or rise time. The current will also overshoot the ac current curve a bit, after which it will oscillate a short time. The overshoot and oscillation cause radiofrequency (RF) noise and lamp filament ringing. Most manufacturers provide for RF suppression and have chokes available that reduce filament ring.

Solid-state dimmers are now made for incandescent, fluorescent, and HID lighting. Incandescent dimmers, which only require the electronic switch and RF suppression, are available in a wide range of power capacities. Wall box units are available from

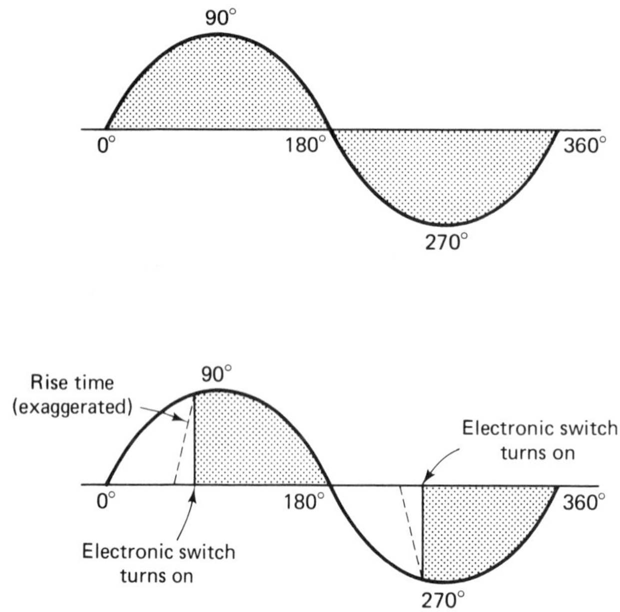

Figure 10-7 Dimmer Characteristics for Alternating Current.

600 to 2000 W. Other systems that use dimmer packages remotely located from the intensity control switch are available in sizes from 100 to 300,000 W. Remote dimmer pack systems are available with multiple remote-control stations or with a control system that allows push-button selection of a number of preset levels, full range control, and override on/off. Incandescent dimming is the only type that can be dimmed from 100% of full light output to 0%.

Fluorescent dimmers were first introduced on the market about 1962. They are not interchangeable with incandescent dimmers. They require different circuitry, a special dimming ballast, and special lamp holders. Unlike incandescent lamps, when dimmed, fluorescent lamps never go off completely. The minimum light output of any system on the market is $\frac{1}{500}$ of the maximum. This is sufficiently low that they appear to be off. Up to 30–40 W lamps may be dimmed by a single wall-box–mounted dimmer; up to 80 lamps may be controlled by a single dimming module. The dimming module system requires intensity controls to be mounted remotely from the dimming module. Because of this, dimming modules may be ganged or combined with incandescent modules. Remote-control, high-capacity systems may also be outfitted with preset-type controls. A new system introduced by one manufacturer uses an intensity control, dimming module, and electronic package that replaces the ballast. Figure 10-8 shows the increase in efficacy of this new system over the conventional dimming arrangement.

With the increased use of HID lamps, it is desirable to have dimming for them. Mercury vapor dimming systems are currently available from three manufacturers. One requires no ballast and as a consequence is much quieter than ordinary systems. Because

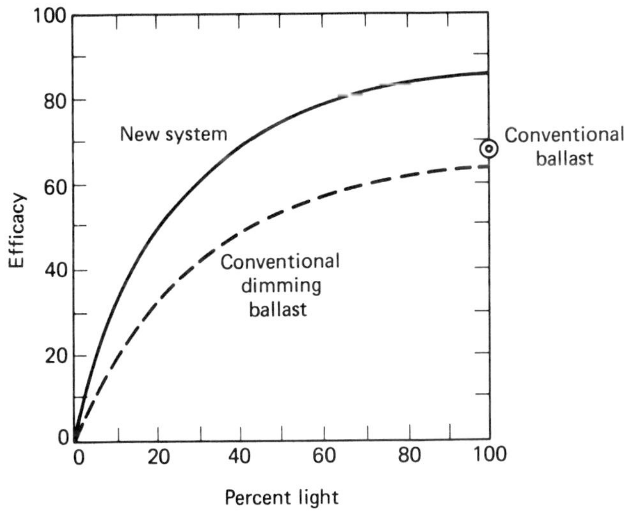

Figure 10-8 Efficacy Versus Light Output of Conventional and New Fluorescent Dimming Systems.

the arc must be maintained, minimum light output of any system is 2% of maximum. The mercury dimming system requires a dimmer module with remote intensity control and is available with a preset control package. Systems for dimming metal halide and high-pressure sodium systems are available.

The dimmer modules used in the high-power fluorescent, incandescent, and HID systems are ideal for interfacing with time clocks, photocells, or computers. Such a system has been developed by one manufacturer for the control of a mercury vapor system. It consists of a photocell, dimmers, and a control station to monitor the illuminance level and the dimmers. The selected level of illuminance is maintained by adjustments of the dimmers by the controller as changes in lumen output, daylight contribution, and dirt accumulation occur.

Dimming systems for fluorescent and HID lamps are complex and expensive. The problem lies in the inductive ballasts. Solid-state, high-frequency ballasts that make dimming easier and less costly have been developed and have recently been introduced to the market. A simple variable resistor or capacitor added to the circuit is said to allow for simple, easy dimming.

Dimmable lighting systems are expensive, and may not be applicable for every installation, but they will result in energy savings if used properly. Dimmers should be used only where it is anticipated that level control will be needed and used.

Multilevel Ballasts

A second method of level control involves the use of a multilevel ballast. With multilevel ballasts, only one type of luminaire and ballast need be specified for the job. Adjustments in levels of illumination can be made by a simple wiring change. When retrofitting of existing installations is necessary, multilevel ballasts can save money and energy in areas that were overlighted without greatly sacrificing uniformity.

Multilevel ballasts are available in two- and three-level models. The two-level ballast provides for 100% light output at 98.7 W input using two F40T12 RS lamps and 55% light output at 55.7 W. The three-level ballast has 100, 55, and 37.8% light output at 98.7, 55.7, and 37 W, respectively. A disadvantage of the multilevel ballast is that it requires an electrician to make the level change by rewiring the ballast. Local codes may not allow a switch to be placed in the level leads of a multilevel ballast. A two- or three-position switch mounted in the luminaire would make illuminance level control easier and eliminate the need for an electrician to make the modification.

Computer Controls

Computer-controlled lighting systems are becoming increasingly attractive. Such systems combine remote on/off control with two-level illuminance selection. The system consists of a controller and receiver/switch. The controller is a microcomputer and oscillator. The receiver/switch is a decoder and two relay switches. The microcomputer is programmed for different lighting patterns and the addresses of the luminaires. The address and condition codes generated are modulated by the oscillator and superimposed on the building electrical system line frequency. The message travels through the building electric distribution system. At the receiver/switch, a decoder takes the message off the line, and if the address code corresponds to the one given that switch, the condition codes are executed, turning the luminaire fully on, halfway on, or off. The computer times events, allowing for different lighting patterns to be executed at different times of the day.

The advantages of a system of this type are many. Because the commands are transmitted through the power lines, no rewiring is needed when retrofitting, no switch legs are needed, and no control wires in addition to the regular wiring is employed. By equipping each luminaire in the building with a receiver/switch, each with its own address, any lighting configuration imaginable could be programmed. In addition to preprogramming, various other devices can be used for input such as photocells and occupancy sensors. One system even allows the individual occupants of an office to change the programming by telephoning the computer and using the keypad of a touch-tone phone for numerical input.

Computer lighting controls can be stand-alone systems or part of a "smart building" system. In a smart building, feedback and control of various other building functions are all coordinated by the computer, in addition to lighting. These other functions may include heating, ventilating, cooling, fire detection and control, security, and vertical transportation. The system makes decisions regarding building performance and appropriate adjustments are executed. One such function can be monitoring of peak electrical demand. If demand exceeds a predetermined level, all building functions are checked and those not necessary are shed (turned off).

LIGHTING POWER BUDGETS

Earlier in the chapter we saw that each state has different lighting regulations. However, most states require a power budget in some form. The two most common power budgets are the original ASHRAE 90 method and the UPD (unit power density) method. Though similar, each of these methods has distinct features. We will outline each one here; an example problem for each is found in the data section.

Power budgeting is a process through which the designer arrives at a total amount of lighting power (in watts) that may be installed in a building. Therefore, two important things about power budgets must be understood. First, a power budget is not a design process. Nowhere in the procedure is the lighting professional told what equipment to select nor where to place it. Nowhere in the procedure is the question of glare addressed. Adhering to a lighting power budget in no way assures a good design. Second, limiting installed lighting power in a building does not insure energy conservation. To be convinced of this, recall that energy = power × time, and review Figure 10-4.

ASHRAE Standard 90

The procedure involves a room by room summation of lighting power. The allowable wattages are calculated based on the *intended* use of the space.

1. *Areas*. The gross area of a space (room) is called A_G. Each visual task that occurs in a space is assigned an area A_T. Typically, we assign a value of 50 ft^2 per task, the maximum allowable. However, the total of the task areas may not exceed half of the gross area. In other words, we assign each task an area which is the larger of

$$50 \text{ ft}^2 \quad \text{and} \quad \frac{A_G}{2 \text{ times the number of task locations}}$$

   Nontask area A_{NT} is equal to the sum of the task areas. If there is any space unaccounted for by the task areas and the nontask area, this is called noncritical area, A_{NC}. In other words,

$$A_{NC} = A_G - \Sigma A_T - A_{NT}$$

2. *Illuminance levels*. Illuminance levels are established for three categories of lighting systems, *task lighting* (for the task areas), *general lighting* (for the nontask area), and *noncritical general lighting* (for the noncritical area).
   a. *Task lighting*. To determine the task lighting level, E_T, follow the illuminance selection procedure of the latest edition of the IES handbook (an explanation may be found in Chapter 8 of this text).
   b. *General lighting*. This refers to the support space surrounding the task location.
      (1) $E_{gen} = E_T/3$ but not less than 20 fc.

(2) If more than one task is present (hence more than one E_T, use one-third of the weighted average of the task illuminance levels, weighted by task areas:

$$E_{gen} = \frac{1}{3} \times \frac{E_{T1}A_{T1} + E_{T2}A_{T2} + \cdots + E_{Tn}A_{Tn}}{A_{T1} + A_{T2} + \cdots + A_{Tn}}$$

c. *Noncritical general lighting*. This is typically circulation space.

$$E_{NC} = E_{gen}/3 \quad \text{but not less than 10 fc.}$$

3. *Lamp efficacy*. The following minimum values are used in determining the power budget. However, when the system is designed and installed, any lamp may be used provided that the power budget for the building is not exceeded.
   a. In areas where *moderate color rendition* is acceptable — 55 lm/W
   b. In areas where *good color rendition* is required — 40 lm/W
   c. In areas where *high color rendition* is necessary and in spaces of less than 50 ft — 25 lm/W
4. *Luminaire efficiency*. Only *approved luminaires* may be used. In order to be classified as approved for use in the procedure, the coefficient of utilization for $RCR = 1.0$ and 80/50/20 reflectances cannot be less than the following values.
   a. Tasks subject to veiling reflection *and* where visual comfort is important — $CU_{min} = 0.55$
   b. Spaces without tasks *or* tasks not subject to veiling reflections, but visual comfort *is* important — $CU_{min} = 0.63$
   c. Spaces without tasks *and* visual comfort is not important — $CU_{min} = 0.70$
5. *Assumptions*. The procedure assumes that the reflectances of the space are 80%/50%/20% for ceiling, walls, and floor, respectively. The overall light loss factor is taken as 0.70.
6. *Exceptions*. Residential space is usually excluded from this method. Specialized equipment, such as dental task lighting, is not counted toward the budget. If a space cannot meet the reflectance or light loss factor assumptions, the actual values may be substituted.

Unit Power Density (UPD) Method (4)

The UPD method is the power budget procedure currently sanctioned by the IESNA. It is similar to the old ASHRAE 90 method, but uses standardized factors to account for the coefficient of utilization, rather than finding a CU from a specific luminaire. The most recent revision of the ASHRAE standard (ASHRAE/IES 90.1-1989) also adopts this method (1). The major elements of the process are outlined here.

1. *Areas*. Each space is divided into two parts, the task area and the general area. These are expressed as percentages of the total area and are denoted $\%T$ and $\%G$, respectively. Each task is assigned an area of 50 ft^2.
2. *Illuminance levels*. The illuminance of each task, F_{et}, is determined using the IES illuminance selection procedure (see Chap. 8). The illuminance of the general area, F_{eg}, is $\frac{1}{3}$ of the average task illuminance.
3. *Base UPD*. The IES suggests that base power density should be determined from the formula:

$$P_b = \frac{(F_{et} \times \%T) + (F_{eg} + \%G)}{F_{cu} \times F_{le} \times F_{ll}}$$

   However, ASHRAE/IES 90.1-1989 has you look up the base UPD in a table listing various areas and activities. Both methods should result in similar base UPDs.
4. F_{cu}. The flux utilization factor is actually the minimum coefficient of utilization. Use the following values:

a. Tasks subject to vertical veiling reflection *and* where visual comfort is important	0.50
b. Tasks subject to horizontal veiling reflection *and* where visual comfort is important	0.65
c. Spaces without tasks *or* tasks not subject to veiling reflections, but visual comfort *is* important	0.70
d. Spaces without tasks *and* visual comfort is not a criterion	0.75

5. F_{le}. The light source factor is actually a luminous efficacy value in lumens per square foot. Use the following values:

a. Where poor color rendition is acceptable	90
b. Where fair color rendition is appropriate	80
c. Where good color rendition is appropriate	68
d. Where very good color rendition is appropriate	40
e. Where excellent color rendition, sparkle, and modeling are necessary	20

6. F_{ll}. The light loss factor is determined from task illuminance levels. For low illuminances (20 fc or lower), use 0.70. For higher illuminances, use 0.75.
7. *Weighted average UPD*. If there is more than one visual task in the space, a weighted average P_b (base UPD) is found.
8. *LPB*. Using the base UPD from (3) above, the lighting power budget for a room in watts per square foot is calculated from

$$LPB = A \times P_b \times AF$$

where A is the total area of the space and AF is the area factor. The area factor is computed from the following formula (1):

$$AF = 0.2 + 0.8(1/0.9^n)$$

where

$$n = \frac{10.21(CH - 2.5)}{\sqrt{A}} - 1$$

CH = ceiling height (use average height if ceiling is sloped)

If $AF < 1.0$, then we set $AF = 1.0$; if $AF > 1.8$, then we set $AF = 1.8$

9. *Building Interiors*. The lighting power limit for the entire building interior is simply the sum of all the individual LPBs.

REFERENCES

1. American Society of Heating, Refrigerating, and Air-Conditioning Engineers, *Energy Efficient Design of New Buildings Except New Low-Rise Residential Buildings*, ASHRAE/IES 90.1-1989. Atlanta: ASHRAE, 1990.
2. Helms, R. N., "Lighting Design Without Analysis: The Consulting Engineer's Dilemma," *Proceedings—The Basis for Effective Management of Lighting Energy Symposium*, Federal Energy Administration, October 29 and 30, 1975, Washington, D.C.
3. Kaufman, J., "Optimizing the Uses of Energy for Lighting." *Lighting Design + Application*, Vol. 3, no. 10 (October 1973).
4. Kaufman, John E., and Jack F., Christensen, eds., *IES Lighting Handbook,* Application Volume. New York: Illuminating Engineering Society of North America, 1987.
5. Ringgold, P., "In the Interest of Illumination." *Lighting Design + Application*, Vol. 2, no.11 (November 1972), pp. 1a–6a.
6. Spielvogel, L. G., "Exploding Some Myths About Building Energy Use." *Architectural Record* (February 1976), pp. 125 128.
7. U.S. Bureau of the Census, *Statistical Abstract of the United States, 109th ed.* Washington, D.C., 1989.

DATA SECTION

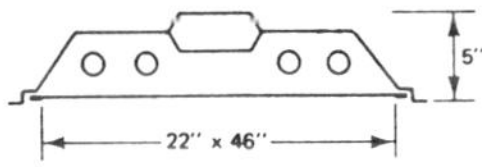

Description:
2 x 4 troffer w/6250

Test no.: 1857–1	Cat. No.: 682–4
Lamps: 4F40CW	Volts 120
Lumens/Lamp: 3100	Test Dist: 20
Shielding Angle: N: 90°	L: 90°
Bare Lamp: 2350	Date:
By: T.B.	Chkd. by:

LUMNO = DB117

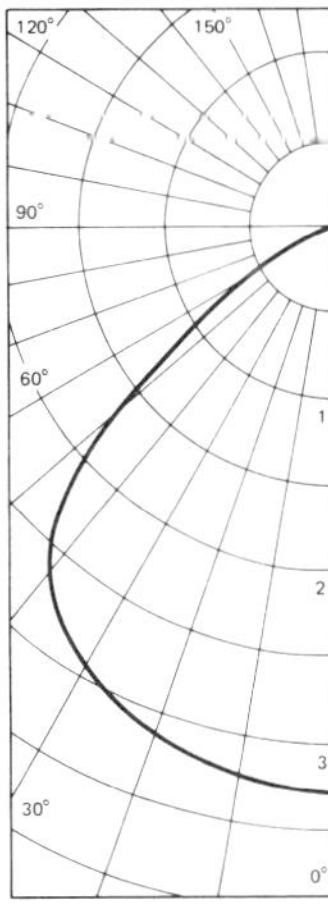

Distribution data (average)			
Angle	C.P.	Angle	C.P.
0°	3300	95°	
5°	3293	105°	
10°	–	115°	
15°	3210	125°	
20°	–	135°	
25°	3068	145°	
30°	–	155°	
35°	2858	165°	
40°	–	175°	
45°	2338	180°	
50°	–		
55°	991		
60°	–		
65°	386		
70°	–		
75°	255		
80°	–		
85°	98		
90°	0		

Luminance Data (Average)			
Angle	Maximum	Average	Ratio max./ave.
45°	2280	1478	1.5
50°	1900	1145	1.7
55°	1130	772	1.5
60°	1020	542	1.9
65°	890	408	2.2
70°	880	426	2.1
75°	860	440	2.0
80°	840	457	1.8
85°	750	503	1.5
90°	0	0	0

Output Data			
Zone Degrees	Lumens	Percent Total Lamp Lumens	Percent Total Distribution
0–40°	14437	35.8	56.2
40–90°	3458	27.9	43.8
90–180°	0	0.0	0.0
0–180°	7895	63.7	100.0

Coefficient of Utilization

Floor		ρf_c = 20%								
Ceiling		80%			50%			10%		
Walls		50%	30%	10%	50%	30%	10%	50%	30%	10%
Room Cavity Ratio	1	0.68	0.66	0.64	0.64	0.63	0.61	0.59	0.58	0.57
	2	0.62	0.58	0.55	0.58	0.56	0.53	0.54	0.52	0.51
	3	0.55	0.51	0.48	0.53	0.49	0.46	0.50	0.47	0.45
	4	0.50	0.45	0.42	0.48	0.44	0.41	0.45	0.42	0.40
	5	0.46	0.40	0.36	0.43	0.39	0.36	0.41	0.38	0.35
	6	0.41	0.36	0.32	0.39	0.35	0.32	0.37	0.34	0.31
	7	0.37	0.32	0.28	0.35	0.31	0.28	0.34	0.30	0.27
	8	0.33	0.28	0.25	0.32	0.37	0.24	0.30	0.27	0.24
	9	0.30	0.25	0.21	0.29	0.24	0.21	0.27	0.24	0.21
	10	0.27	0.22	0.19	0.26	0.22	0.19	0.25	0.21	0.19

Maximum recommended spacing-to-mounting height ratio above workplane is 1.43.

Table 10-3 Typical Photometric Test Report.

EXAMPLE PROBLEMS

Following are examples of power budget calculations for the office suite and task analysis shown.

Task Analysis*	E†	No. of tasks
General office: 100		
Total task locations, 25		
T1. Reading fair reproductions, active filing	100	5
T2. Medium pencil or ink	70	16
T3. High-contrast printed forms	30	4
Accounting: 101		
Total task locations, 15		
T1. Auditing, bookkeeping, ink or medium pencil	150	10
T2. Intermittent filing	70	5
Private office: 102		
Total task locations, 2		
T1. Ink or medium pencil	70	2

* Information obtained from architect, owner, or assumed.

† Use current IES Illuminance Selection Procedure.

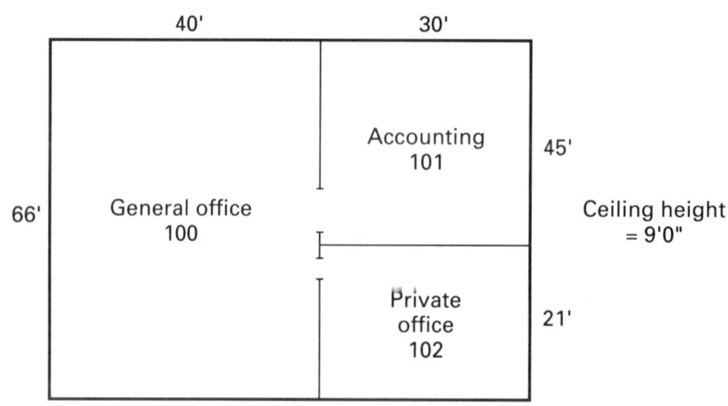

Example 1 ASHRAE 90 (Original) Method

Use luminaire LUMNO = DB117 (Figure 10-9). Luminaire meets efficiency criterion, since $CU_{act} = 0.68 > 0.55$ for tasks subject to veiling reflections.

1. General Office: 100

$$L = 66 \text{ ft}, \qquad W = 40 \text{ ft}, \qquad A_G = 2640 \text{ ft}^2$$

$$h_{rc} = 5.5, \qquad RCR = \frac{5(5.5)\,(66 + 40)}{66 \times 40} = 1.10, \qquad CU = 0.68$$

T1. Five work locations at 100 fc.

$$A_{T1} = \text{task area} = 5 \times 50\ \text{ft}^2 = 250\ \text{ft}^2$$

T2. Sixteen work locations at 70 fc.

$$A_{T2} = \text{task area} = 16 \times 50 = 800\ \text{ft}^2$$

T3. Four work locations at 30 fc.

$$A_{T3} = \text{task area} = 4 \times 50 = 200\ \text{ft}^2$$

Total task area:

$$\text{Since } \Sigma A_T = 250 + 800 + 200 = 1250\ \text{ft}^2$$
$$\Sigma A_T = 1250 < \tfrac{1}{2}\, AG = 1320\ \text{ft}^2$$

Therefore, we may use 50 ft^2 per task as assumed. Allowable power for each task is calculated from

$$\text{watts} = \frac{fc \times A}{\text{lm/W} \times CU_{act} \times 0.70}$$

Therefore,

$$\text{watts } (T_1) = \frac{100 \times 250}{55 \times 0.68 \times 0.70} = 955\ \text{W}$$

$$\text{watts } (T_2) = \frac{70 \times 800}{55 \times 0.68 \times 0.70} = 2139\ \text{W}$$

$$\text{watts } (T_3) = \frac{30 \times 200}{55 \times 0.68 \times 0.70} = 229\ \text{W}$$

General lighting level:

$$E_{\text{gen}} = \frac{1}{3}\,\frac{(E1 \times A_{\text{T1}}) + (E2 \times A_{\text{T2}}) + (E3 \times A_{\text{T3}})}{A_{\text{T1}} + A_{\text{T2}} + A_{\text{T3}}}$$

$$= \frac{1}{3}\,\frac{(100 \times 250) + (70 \times 800) + (30 \times 200)}{1250}$$

$$= 23.2\ \text{fc}$$

Nontask area, $A_{NT} = \Sigma A_T$. Therefore, $A_{NT} = 1250\ \text{ft}^2$. Allowable wattage for the nontask area is

$$\text{watts } (NT) = \frac{(23.2)\,(1250)}{55 \times 0.68 \times 0.70} = 1108\ \text{W}$$

The noncritical area, $A_{NC} = A_G - \Sigma_{AT} - A_{NT}$. Therefore $A_{NC} = 2640 - 1250 - 1250 = 140\ \text{ft}^2$

$$E_{NC} = E_{\text{gen}}/3 = 23.2/3 = 7.7\ \text{fc}$$

So we use 10 fc for E_{NC}. Allowable wattage for the noncritical area is

$$\text{watts } (NC) = \frac{(10)\ (140)}{55 \times 0.68 \times 0.70} = 53 \text{ W}$$

Total watts for room 100:

$$955 + 2139 + 229 + 1108 + 53 = 4484 \text{ W}$$

2. Accounting: 101

$$L = 45 \text{ ft}, \qquad W = 30 \text{ ft}, \qquad A_G = 1350 \text{ ft}^2$$

$$h_{rc} = 5.5, \qquad RCR = 1.53, \qquad CU = 0.65$$

T1. Ten work locations at 150 fc.

$$A_{T1} = 10 \times 50 = 500 \text{ ft}^2$$

T2. Five work locations at 70 fc.

$$A_{T2} = 5 \times 50 = 250 \text{ ft}^2$$

$$\Sigma A_T = 500 + 250 = 750 \text{ ft}^2$$

Since $\Sigma A_T = 750 > \frac{1}{2} A_G = 675$, then the allowable area for each task is

$$\frac{A_G}{2 \text{ (no. of loc.)}} = \frac{1350}{2\ (10 + 5)} = 45 \text{ ft}^2$$

Revised task areas:

$$A_{T1} = 10 \times 45 = 450 \text{ ft}^2$$

$$A_{T2} = 5 \times 45 = 225 \text{ ft}^2$$

$$\text{watts } (T_1) = \frac{150 \times 450}{55 \times 0.65 \times 0.70} = 2697 \text{ W}$$

$$\text{watts } (T_2) = \frac{70 \times 225}{55 \times 0.65 \times 0.70} = 630 \text{ W}$$

$$\text{Nontask area, } A_{NT} = \Sigma A_T = 675 \text{ ft}^2$$

$$E_{\text{gen}} = \frac{1}{3} \frac{(150 \times 450) + (70 \times 225)}{450 + 225} = 41.0 \text{ fc}$$

$$\text{watts } (NT) = \frac{41.0 \times 675}{55 \times 0.65 \times 0.70} = 1106 \text{ W}$$

$$\text{Noncritical area, } A_{NC} = A_G - \Sigma A_T - A_{NT}$$

$$A_{NC} = 1350 - 675 - 675 = 0$$

Total watts for room 101:

$$2697 + 630 + 1106 = 4433 \text{ W}$$

3. Private office: 102

$$L = 30 \text{ ft}, \qquad W = 21 \text{ ft}, \qquad A_G = 630 \text{ ft}^2$$

$$h_{rc} = 5.5, \qquad RCR = 2.23, \qquad CU = 0.60$$

T1. Two work locations at 70 fc.

$$A_{T1} = 2 \times 50 = 100 \text{ ft}^2$$

Since $\Sigma A_T = 100 < \frac{1}{2} A_G = 315$, we use 50 ft^2.

$$\text{watts } (T_1) = \frac{70 \times 100}{55 \times 0.60 \times 0.70} = 303 \text{ W}$$

Nontask area, $A_{NT} = \Sigma A_T = 100 \text{ ft}^2$

$$E_{\text{gen}} = \tfrac{1}{3}\,(70) = 23.3 \text{ fc}$$

$$\text{watts } (NT) = \frac{23.3 \times 100}{55 \times 0.60 \times 0.70} = 101 \text{ W}$$

Noncritical area, $A_{NC} = A_G - \Sigma A_T - A_{NT}$

$$A_{NC} = 630 - 100 - 100 = 430 \text{ ft}^2$$

$$E_{NC} = E_{\text{gen}}/3 = 23.3/3 = 7.8 \text{ fc (use 10 fc)}$$

$$\text{watts } (NC) = \frac{10 \times 430}{55 \times 0.60 \times 0.70} = 186 \text{ W}$$

Total watts for room 102:

$$303 + 101 + 186 = 590 \text{ W}$$

Total power budget for office suite:

$$4484 + 4433 + 590 = 9507 \text{ W}$$

Example 2 UPD Method

1. General office: 100

$$F_{eg} = 23.2 \text{ fc (from solution to example 1)}$$

$$\text{T1.} \quad \%T = \frac{250}{2640} = 0.095$$

$$\text{T2.} \quad \%T = \frac{800}{2640} = 0.303$$

$$\text{T3.} \quad \%\text{T} = \frac{200}{2640} = \frac{0.076}{0.474}$$

$$\%G = 1 - 0.474 = 0.526$$

$$P_b = \frac{\Sigma(F_{et} \times \%T) + (F_{eg} \times \%G)}{F_{cu} \times F_{le} \times F_{ll}}$$

$$= \frac{(100\,\text{fc} \times 0.095) + (70\,\text{fc} \times 0.303) + (30\,\text{fc} \times 0.076) + (23.2\,\text{fc} \times 0.526)}{0.65 \times 68 \times 0.75}$$

$$= 1.363\ \text{W/ft}^2$$

$$LPB = A \times P_b \times AF$$

$$AF = 0.2 + 0.8\ (1/0.9^n)\ n = \frac{10.21\ (CH - 2.5)}{\sqrt{A}} - 1$$

$$n = \frac{10.21\ (9 - 2.5)}{\sqrt{2640}} - 1 = 0.2916$$

$$AF = 0.2 + 0.8\ (1/0.9^{0.2916}) = 1.025$$

$$LPB = 2640\ \text{ft}^2 \times 1.363\ \text{W/ft}^2 \times 1.025 = 3688\ \text{W}$$

2. Accounting: 101

$F_{eg} = 41$ fc (from solution to example 1)

$$\text{T1.}\quad \%T = \frac{500}{1350} = 0.370$$

$$\text{T2.}\quad \%T = \frac{250}{1350} = \frac{0.185}{0.555}$$

$$\%G = 1 - 0.555 = 0.445$$

$$P_b = \frac{(150\text{fc} \times .370) + (70 \times 0.185) + (41\text{fc} \times 0.445)}{0.65 \times 68 \times 0.75}$$

$$= 2.615\ \text{W/ft}^2$$

$$n = \frac{10.21(9 - 2.5)}{\sqrt{1350}} - 1 = 0.8062$$

$$AF = 0.2 + 0.8(1/0.9^{0.8062}) = 1.071$$

$$LPB = 1350\ \text{ft}^2 \times 2.615\ \text{W/ft}^2 \times 1.071 = 3781\ \text{W}$$

3. Private office: 102. Assume VDTs present.

$F_{eg} = 70\text{fc}/3 = 23.3$ fc

$$\text{T1.}\quad \%T = \frac{100}{630} = 0.159$$

$$\%G = 1 - 0.159 = 0.841$$

$$P_b = \frac{(70\,\text{fc} \times 0.159) + (23.3\ \text{fc} \times 0.841)}{0.50 \times 68 \times 0.75}$$

$$= 1.205\ \text{W/ft}^2$$

$$n = \frac{10.21\ (9 - 2.5)}{\sqrt{630}} - 1 = 1.644$$

$$AF = 0.2 + 0.8\ (1/0.9^{1.644}) = 1.151$$

$$LPB = 630\ \text{ft}^2 \times 1.205\ \text{W/ft}^2 \times 1.151 = 874\ \text{W}$$

Total power budget for office suite:

$$3688\text{W} + 3781\text{W} + 874\text{W} = 8343\ \text{W}$$

11 exterior lighting

Exterior lighting must be well planned to optimize the use of exterior facilities while minimizing energy consumption. Security and safety are the primary functional needs associated with most exterior applications. Roadway, area, and floodlighting are the three design techniques available to satisfy these functional needs.

Once a functional need has been established, the lighting professional must select the most efficient equipment to maximize the utilization of the energy while minimizing power consumption. The quality of illumination will have a direct bearing on the safety, efficiency, and appearance of the system. Quality is dependent upon glare control, transient adaptation, color, and aesthetics.

Glare and transient adaptation will affect the safe movement of pedestrians and vehicles. Good glare control will minimize wasted spill light, which will increase the overall efficiency of the system by increasing the utilization of energy. Direct glare (Chapter 8) is the major problem encountered in most exterior lighting. It causes annoyance, discomfort, and a loss in visibility. The sensation of glare results from the luminance of a source being much higher than the surrounding luminance to which the eye is adapted. An emphasis on direct glare has been brought to the forefront in the past ten years because of the awareness on the part of people of light pollution in the environment. Transient adaptation relates to the effect created by changing the direction of view and exposing the visual system to variations in the luminance patterns in the field of view. For example, as one drives down a roadway, the visual system adapts to an ambient level of luminance. Excessively bright light sources, such as oncoming headlights and streetlights, momentarily affect the adaptation state. The greater the luminance difference (luminance ratios) between a bright source of light and a darker surrounding area, the slower the rate at which the visual system will return to its original adaptation level. It is during this readaptation process that a loss in visibility is experienced that could result in an accident. A reduction in glare, and hence luminance ratios, will also result in a reduction in the amount of clutter and confusion reaching the visual system. This decrease in clutter and confusion will create a more pleasant, safer environment for pedestrians and vehicles.

Color is also important in the exterior environment since it influences the mood and behavior of people and thus their performance. High performance is essential to the security of facilities as well as safety.

Exterior lighting can be classified into static or movable systems. A static system is one in which all the luminaires are nonadjustable on their support and each is aimed toward a single fixed point. Most roadway and pedestrian luminaires would be classified as static systems. A movable system is one in which the luminaires are aimed at different points. Floodlighting equipment with adjustable support mechanisms would be classified as movable.

Light sources are described in detail in Chapter 4. A brief summary of light sources as they apply to exterior lighting is included in this section.

1. Incandescent lamps should be avoided for exterior applications because of their low efficiency and short life. Incandescent lamps create high operating and maintenance costs.
2. Fluorescent lamps should be avoided for exterior applications because of their sensitivity to temperature and the poor quality of optical control. Enclosed and gasketed luminaires with low-temperature ballasts may maintain a sufficiently high ambient temperature to allow for operation under low temperatures; however, the enclosure that holds the heat in the unit during cold weather also holds the heat in during warm weather, which causes a drop in the lumen output (Chapter 4) of the lamps. The optimum operating ambient temperature for fluorescent lamps is

70°–90° to reach rated lumen output. Because of the physical size of the light source, optical control will be poor, resulting in lower utilization characteristics.

3. Low-pressure sodium lamps produce monochromatic (actually two yellow lines) yellow light which turns all colors gray, brown, or black, except yellow. The lamp alone has a very high efficiency. However, when the source is combined with a ballast and luminaire, the overall efficiency of the system is low. Because of the physical size of the source, optical control is poor, resulting in low utilization characteristics.
4. Mercury vapor lamps require a phosphor coating if color rendition is to be acceptable. The phosphor-coated lamp is a large source, which means that optical control by a reflector system is poor and utilization decreases. The mercury vapor lamp also has a relatively low efficacy, which makes it the third choice for exterior applications.
5. Metal halide and high-pressure sodium lamps have relatively small light-emitting elements (arc tubes), which allow for good optical control. Each of the two sources has high lamp efficacy and good system efficiency. Thus these two sources are the top choices for exterior applications. The metal halide lamp has better overall color balance and is preferred where color is important. The high-pressure sodium lamp has a higher lamp efficacy but a dominant orange appearance that may be objectionable for some applications.

ROADWAY LIGHTING

A roadway may require lighting to facilitate the safe movement of vehicles from one area to another. Roadway lighting is especially important where mixed modes of movement converge, such as vehicular/pedestrian and automobile/service vehicle. These roadway systems should use low-glare luminaires that produce very little intensity between 70° and 90° from nadir and no intensity at 90° or above. A reduction in intensity at these high angles (70° to 90°) will reduce light pollution, minimize spill light, and optimize the use of energy by placing the light where it is needed. High-angle intensity (70° and up) does little or nothing to produce illumination on a horizontal roadway surface.

The current roadway lighting standard (1) uses horizontal illuminance and uniformity of illumination as the primary design criteria for roadway lighting. Both factors are deficient in terms of what the visual system actually sees. A design based on horizontal illuminance is less than desirable, but it is currently the design procedure most commonly used. The lighting professional should be aware of the deficiencies, but may be required to use this procedure to arrive at a solution.

Safety

The purpose that roadway lighting serves is to *increase* **visibility** and **sight distance** for *increased* driver *safety*. Why should the lighting professional be interested in increasing visibility and sight distance? Adequate **stopping distance** is required to allow a vehicle to be brought to a safe stop. **Total stopping distance** is defined as

$$\text{stopping distance, } SD = \text{reaction distance} + \text{braking distance}$$

The **breaking distance** is a function of the vehicle dynamics and the pavement conditions. Breaking distance can be estimated by equating kinetic energy to work, which results in the following formula:

$$\text{breaking distance, } BD = \frac{V^2}{2 \times \mu \times g} \tag{11.1}$$

where μg represents the deceleration rate consisting of a coefficient of friction, μ, and a gravitional constant of 32 ft/sec^2.

The *BD* formula applies to vehicles of the same class with equivalent weight distribution between vehicles. The coefficient of friction, μ, represents the friction between the brake shoe and brake pad. A μ value of 0.31 when multiplied by g will produce a moderate deceleration rate without locking the brakes. **Reaction distance** is defined as

$$\text{reaction distance, } RD = \text{reaction time} \times \text{velocity (ft/sec)}$$

To convert velocity in mph to ft/sec,

$$V \text{ (ft/sec)} = X \text{ (mph)} \times \frac{5280 \text{ ft/mi}}{3600 \text{ sec/hr}} \tag{11.2}$$

$$V \text{ (ft/sec)} = X \text{ (mph)} \times 1.467$$

Figure 11-1 shows the critical relationship between the reaction time and the velocity of the vehicle to the reaction distance.

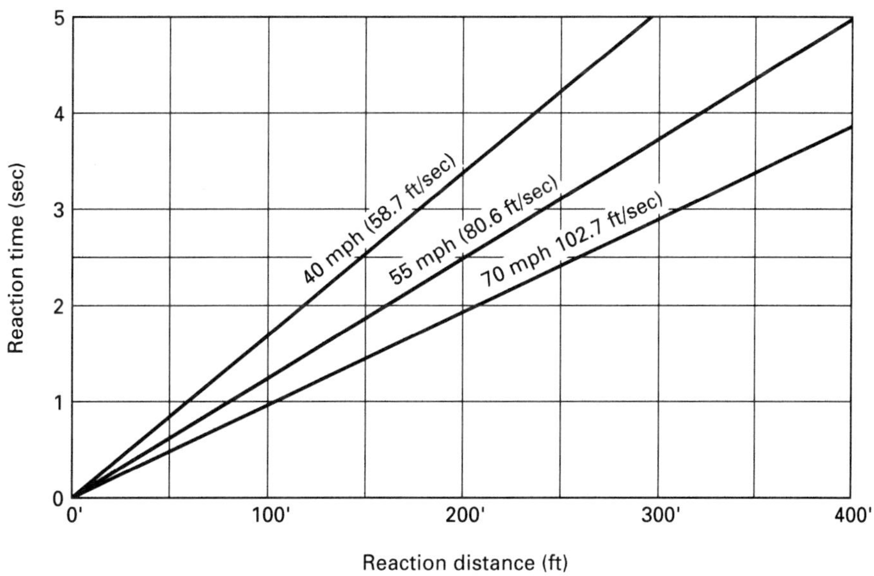

Figure 11-1. Reaction Distance.

TABLE 11-1 Reaction Distance as a Function of Reaction Time at 55 mph

Reaction Distance

Reaction Time (sec)	Reaction Distance (feet)
1	80.7
2	161.3
3	242.8
4	322.7
5	403.3

Reaction time becomes a critical component in determining the total stopping distance to bring a vehicle to a "safe" stop. Table 11-1 shows that for a given velocity (55 mph), the *longer* the reaction time, the *greater* the reaction distance; hence the *longer* the stopping distance.

Reaction time becomes one of the most critical variables involved in the complex sequence of events in safely stopping a vehicle. There are four subconscious processes involved in determining the actual driver's reaction time.

1. *Detection*. The driver must detect the hazard or object before the reaction process will begin.
2. *Perception*. Perception is triggered by the *recognition* of the physical hazard or object.
3. *Processing*. Once the driver recognizes the hazard, the brain must respond to the recognition by making a *decision* as to the potential hazard.
4. *Behavior*. The driver will then and only then begin the initiation of action.

If vehicle headlights penetrate the darkness to 250 feet on low beam and 350 feet on high beam, the driver is "overdriving" their headlights (see Table 11-2) when traveling at 55 mph on high beams (3).

Fixed roadway lighting can increase the field of view beyond the 250 ft (low beams) or 350 ft (high beams) distance the headlights penetrate the darkness. The fixed roadway lighting must be designed to increase visibility for a longer field of view and not impair or reduce visibility. Nonuniform pavement luminance and excessive glare from the roadway luminaires can reduce visibility.

A nonuniform or *spotty* roadway surface can create "clutter" and "confusion," which impairs both detection and perception (recognition) of the hazard. Excessive glare can impair visibility of an object by creating veiling glare in the eye, which causes fluctuation in the adaptation state of the driver and a reduction in contrast of the hazard. Excessive glare affects both detection and perception which will increase the reaction time, hence the total stopping distance.

TABLE 11-2. Total Stopping Distance as a Function of Velocity and Reaction Time Assuming a Constant Deceleration Rate of 0.31*g*

Reaction Time	Stopping Distance							
	30 mph	40 mph	50 mph	55 mph	60 mph	65 mph	70 mph	80 mph
1	144	237	351	418	489	566	648	830
2	188	296	425	498	577	661	751	947
3	232	354	498	579	665	756	854	1065
4	276	413	572	660	753	852	956	1182
5	320	472	645	740	841	941	1059	1299
V (ft./sec)	44	58.7	73.4	80.7	88.0	95.3	102.7	117.3
BD (feet)	100	178	278	337	401	470	546	713

[///] – low beams at 250′ [\\\\\\] – high beams at 350′

There are eight factors that will affect detection of a hazard or object in the roadway at night. They are

1. Size and shape of the object.
2. Location of the object in the field of view.
3. Stationary versus moving: If the object is moving, the direction of movement of the object relative to the driver's direction of travel becomes critical. Stationary objects are more difficult to detect.
4. Contrast between the object and its background.
5. Clutter: A complex pattern of background luminance patterns confuses the driver and reduces recognition.
6. Adaptation state of the driver.
7. Glare sources in the vicinity of the object.

The contrast of the object or hazard is a function of the luminance difference between the object and its background. If the object is dark or low reflectance against a low luminance or dark background, the contrast will approach zero making the object difficult if not impossible to see.

The roadway lighting professional has no control over the first three factors involved in detection (size/shape, location, moving/stationary). With good-quality lighting, both the contrast and the background luminance can be increased without creating glare or visual discomfort. Uniform higher background luminance will increase the adaptation state which increases the driver's level of sensitivity to contrast.

A primary factor in perception (recognition) and processing (response to recognition) is referred to as **expectancy.** That is, does the driver expect a pedestrian to be standing or walking in the middle of the roadway? Behavior or the initiation of action involves the reflex system or neurological interactions that trigger muscle actions to stop or take evasive action.

How many thousands of people are killed or permanently disabled in traffic accidents that occur at night each year? If one were to calculate the financial losses due to fatalities, bodily injury, litigation, and property damage each year attributed to night time accidents caused by poor visibility, one might ask the question—can we afford not to light our highways?

Visibility and Visual Comfort

Seeing objects on the roadway and proper guidance down the roadway are a function of driver alertness, visual comfort, and visibility. Visual comfort is affected by luminance ratios, fluctuation or luminance patterns, and field luminance. Visibility is affected by veiling luminance within the eye (reduces contrast), target (or object) luminance, pavement luminance, and the overall field luminance. Three complex interactions affect visual comfort and visibility: (1) luminaire/driver, (2) pavement/driver, and (3) luminaire/pavement.

1. *Luminaire/driver*. The luminaire affects the driver's visual comfort and visibility in terms of the luminaire's effect on adaptation, veiling luminance, sensitivity, transient adaptation, and luminance ratios.
2. *Pavement/driver*. The pavement luminance affects the driver's visual comfort and visibility. Visual comfort is a function of the luminance ratios produced on the pavement and the uniformity of their luminance patterns. Nonuniform pavement luminance can create confusion and clutter that increase discomfort. Visibility is a function of the adaptation level, which is related to pavement luminance. Uniformity of pavement luminance and the level of pavement luminance can affect the visibility of objects on the roadway in terms of the luminance difference between the pavement and object (contrast).
3. *Luminaire/pavement*. The distribution characteristic of the luminaire and the reflectance properties (luminance factor) of the pavement surface affect uniformity and level of luminance. Luminaire distribution can also increase object visibility in terms of vertical surface illuminance, which can increase depth perception and modeling of the object for easy identification.

The lighting professional normally has control over only one of the variables involved in these complex interactions and that is the luminaire. Driver/luminaire interaction relates to the control or elimination of glare from the luminaire into the drivers eyes. The pavement/driver interaction can be improved by selecting luminaire distribution patterns that create a more uniform luminance pattern. The use of uniform horizontal illuminance as the sole design criterion will not necessarily provide uniformity or prevent direct glare.

Uniformity

The human visual system does not see the luminous energy coming *to* a surface (illuminance); it sees the luminous energy coming *from* a surface (luminance) (see Fig. 11-2). A luminance design procedure is much more complex and requires information on the reflectance properties (luminance factor, β; see Chapter 1) of the roadway surface. However, with the aid of the computer, the problem is not as impossible as one might believe. Uniformity of illumination is also meaningless when talking about the uniform appearance of the pavement. The uniformity of pavement luminance is dependent upon the amount of light reflected toward the eyes (luminance) from various points on the surface. The pavement luminance is equal to the luminance factor, β, multiplied by the illuminance arriving at a point.

$$L = \beta_{(vs,h)} \times E \qquad (11.3)$$

where L = luminance of the pavement surface

β = luminance factor

E = illuminance arriving at a surface

Assume that $E = 1.0$ fc; if $\beta_{(vs,h)} = 0.0567$ for point X on the road, then,

$$L_X = 0.0567 \times 1.0 = 0.06 \text{ fL}$$

If $\beta_{(vs,h)} = 0.9409$ for point Y on the road,

$$L_Y = 0.9409 \times 1.0 = 0.94 \text{ fL}$$

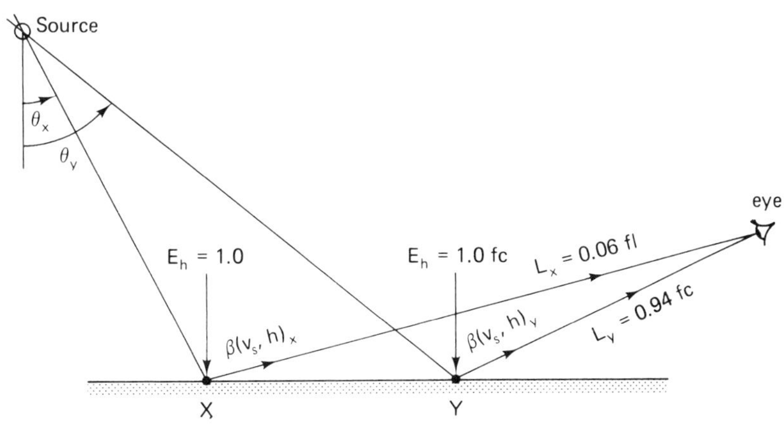

Figure 11-2. Uniformity: Illumination Versus Luminance.

This represents almost a 16-to-1 uniformity ratio in terms of luminance (what the eye sees) for a 1-to-1 ratio of illuminance.

Terminology (Fig. 11-3)

1. *Longitudinal roadway line* (*LRL*): reference lines that run parallel to the curb.
2. *Transverse roadway line* (*TRL*): reference lines that are perpendicular to the curb line or at 90° to the *LRL*.
3. *Origin*: directly under the luminaire and the point defined as the intersection of the zero mounting height (*0MH*) *LRL* and *TRL*.
4. *Reference line*: the longitudinal roadway line that passes through the origin (*0MH LRL*).
5. *Street side* (*SS*): the space located in front of the reference line.
6. *House side* (*HS*): the space located behind the reference line.
7. *Longitudinal distance* (*LD*): a distance measured in a direction parallel to the curb between the transverse roadway lines.
8. *Transverse distance* (*TD*): a distance measured perpendicular to the curb between longitudinal roadway lines.
9. *Elevation angle*, θ: the vertical angle measured from nadir (see Fig. 11-4) in a vertical plane.

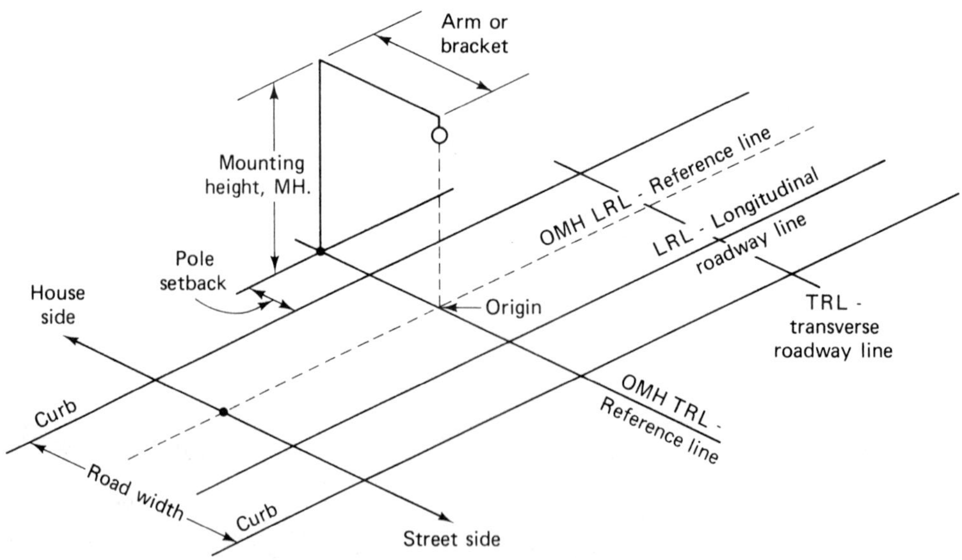

Figure 11-3. Terminology for Roadway Lighting.

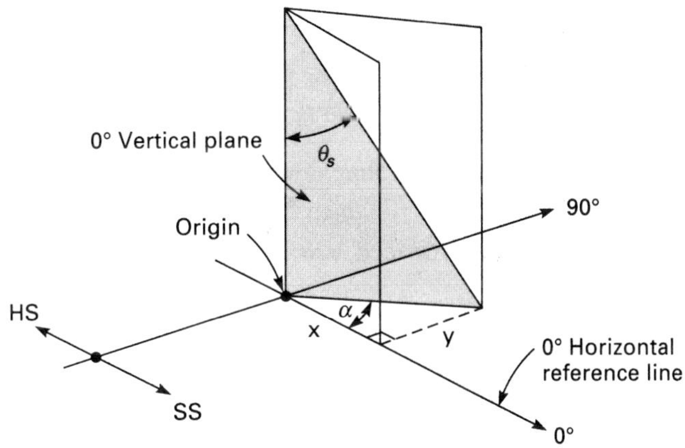

Figure 11-4. Elevation Angle, θ, and Horizontal Angle, α.

10. *Horizontal angle*, α: the horizontal or azimuth angle between vertical planes measured in the horizontal plane of the roadway. The horizontal angle is measured from the 0° vertical plane (see Fig. 11-4).
11. *Maximum cone*: defines the elevation angle, θ, at which the maximum luminous intensity (candlepower) occurs (see Fig. 11-5).
12. *Maximum vertical plane*: the horizontal angle, α, measured from the 0° horizontal angle to the vertical plane in which the maximum luminous intensity (candlepower) occurs.

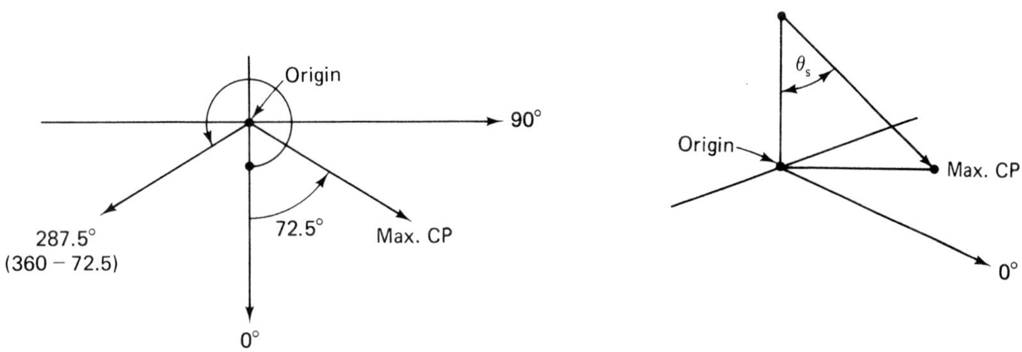

(a) Maximum vertical plane (72.5° and 287.5°)

(b) Maximum cone = 65°

Figure 11-5. Maximum Vertical Plane and Maximum Cone.

Roadway Luminaire Classification

The light distribution from roadway luminaires is classified in terms of three attributes:

1. **Spread** (short, medium, long)—which describes the vertical light distribution.
2. **Type** (I, II, III, IV, V)—which defines the lateral light distribution.
3. **Control**—which is divided into three categories (cutoff, semicutoff, noncutoff).

Spread

The classification according to spread is determined by measurements in a longitudinal direction (see Fig. 11-6). The *spread* is assigned on the basis of the location of the intersection of the maximum "candlepower" (luminous intensity) with the roadway surface.

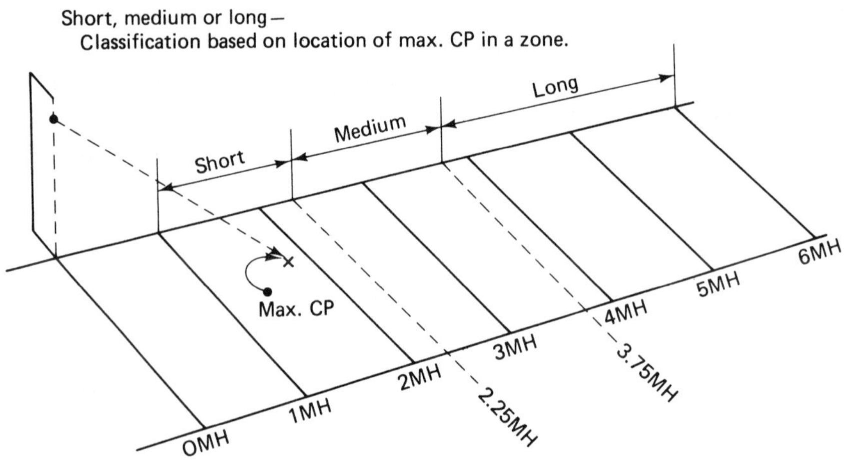

Figure 11-6. Spread Classification.

Type

The type classification is determined by measurements in a transverse direction (see Fig. 11-7). The type is assigned on the basis of the location of the half-maximum "candlepower" (luminous intensity) contour line that falls within the spread limits defined above.

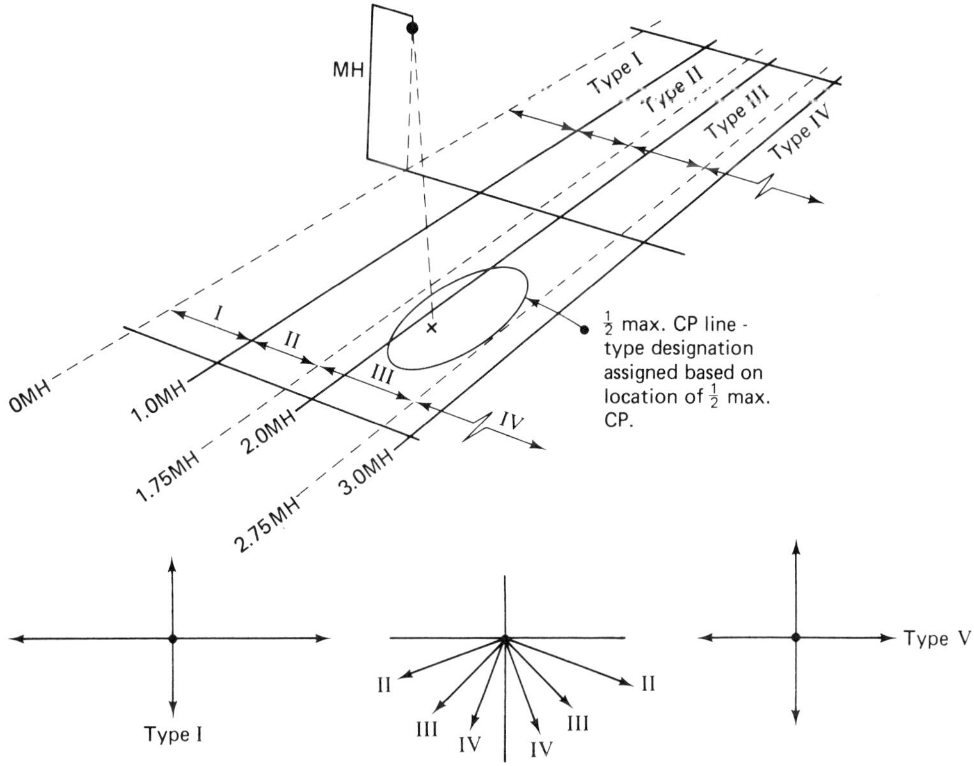

Figure 11-7. Type Classification.

Control

Three descriptive terms are used to describe control: (1) cutoff, (2) semicutoff, and (3) noncutoff (see Fig. 11-8). These terms are misleading and misrepresent the actual distribution characteristics of a luminaire (see Distribution Characteristics, this chapter).

Roadway Photometric Test Report

The standard photometric test report (Figs. 11-23 and 11-24 in the Data Section) contains an isocandela diagram, isofootcandle curve, utilization curve, descriptive information, and a summation of distribution characteristics.

Isocandela Diagram

The isocandela diagram (Fig. 11-23 in the Data Section) is usually plotted on a rectangular coordinate grid known as a **rectangular web.** The rectangular web has equally spaced vertical (longitudinal) lines and equally spaced horizontal (latitude) lines. Construction

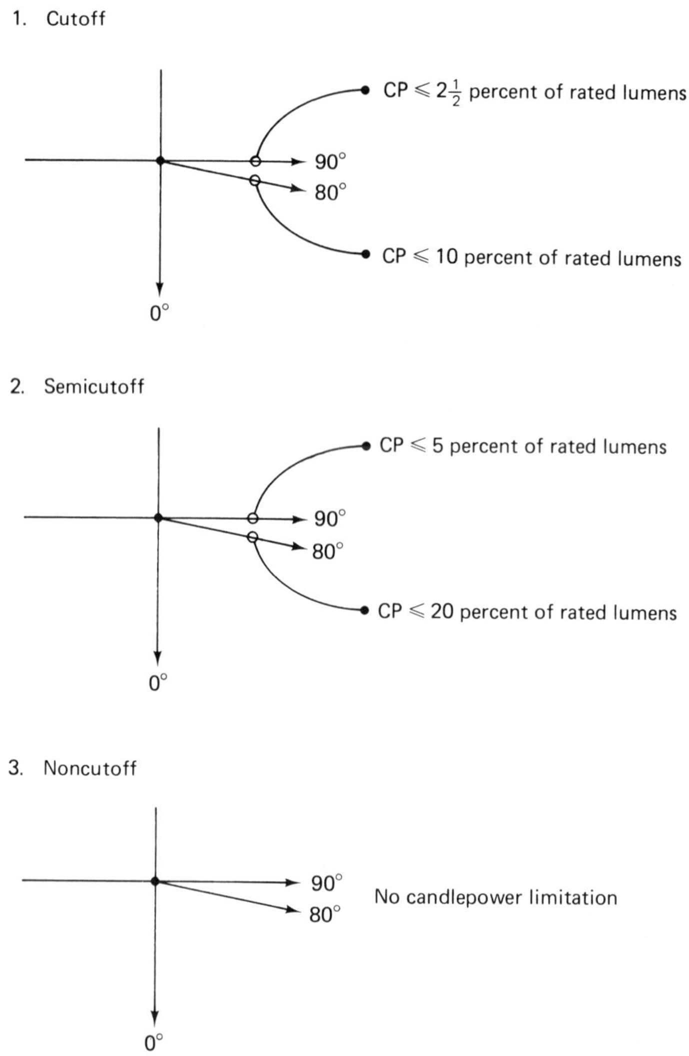

Figure 11-8. Control Classification.

of the rectangular web from a sphere (Fig. 11-9) is accomplished by straightening out the longitudinal line. The lines of latitude must increase in length to a value equal to the circumference of the equator while remaining parallel to the equator.

This forms a cylinder of equally spaced longitudinal lines and lines of latitude. If the latitude lines are cut along a constant longitude, the cylinder can be flattened out to form the rectangular web.

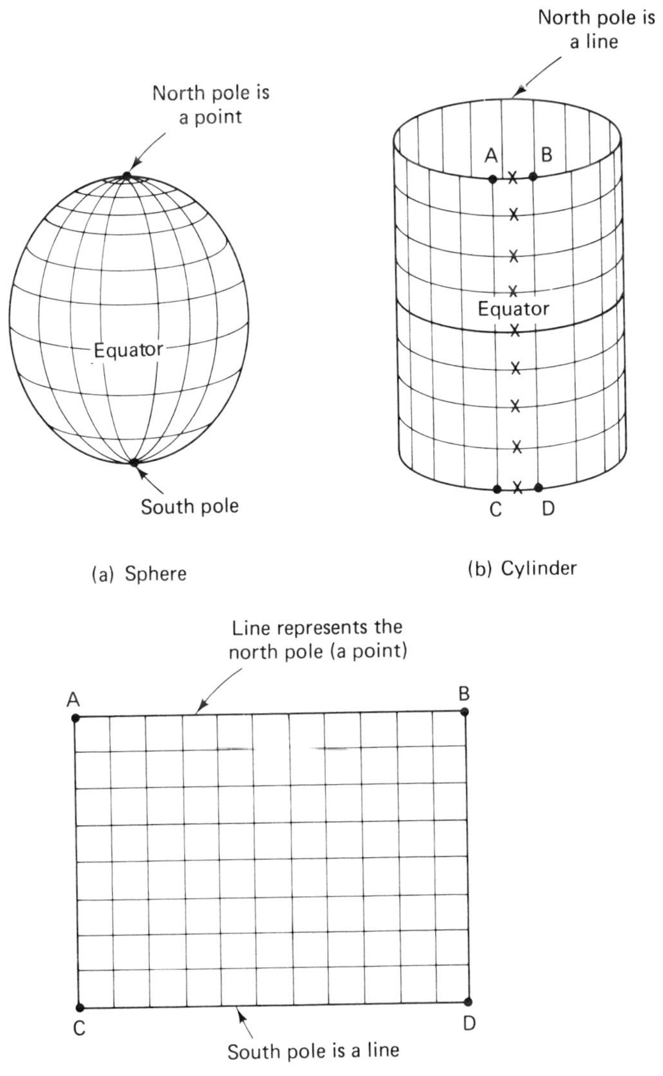

Figure 11-9. Rectangular Web Grid.

The isocandle diagram plots the luminous intensity, $I_{(\theta,\alpha)}$ values for various angles of elevation, θ, and azimuth, α, on a rectangular web. A vertical line passing through an azimuth angle represents a vertical plane of luminous intensity (candlepower, CP) that can be plotted on polar coordinate paper.

Transverse roadway lines are curved lines (Fig. 11-10) that begin at $\alpha = 0°$, $\theta = 90°$ and end at $\alpha = 180°$, $\theta = 90°$. By starting at infinity (∞) and moving along a

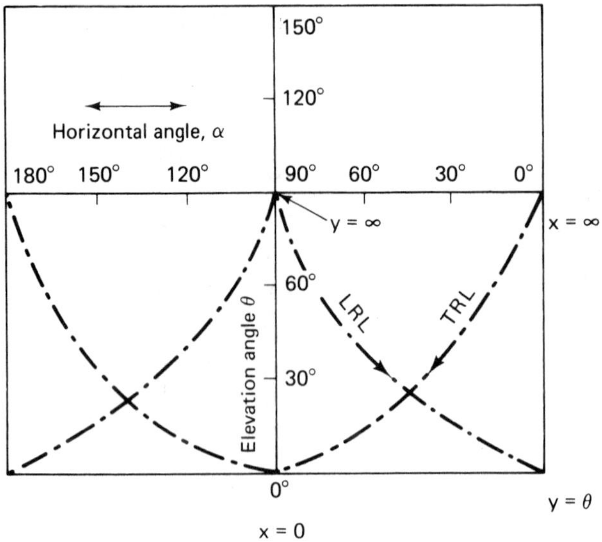

Figure 11-10. Isocandela Diagram: Transverse and Longitudinal Roadway Lines.

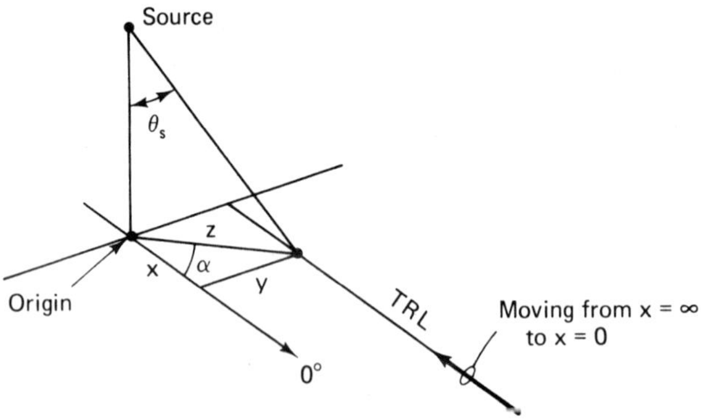

Figure 11-11. Transverse Roadway Line on Isocandela Diagram (Fig. 11-10).

transverse roadway line (*TRL*) toward the luminaire, the angular change (α and θ) of the *TRL* can be calculated and plotted on the rectangular web (see Fig. 11-11).

For Transverse Roadway Lines

For the horizontal angle, α,

$$\text{at } x = \infty, \qquad \alpha = \arctan \frac{y}{\infty}$$

Therefore, α = 0°

Moving toward the luminaire,

$$\text{as } x \rightarrow 0, \text{ then } \alpha = \arctan \frac{y}{0}$$

Therefore

$$\alpha \rightarrow 90°$$

For vertical angle, θ,

$$\text{as } x \rightarrow \infty, \text{ then } \theta = \arctan \frac{\sqrt{(\infty)^2 + (y)^2}}{h}$$

Therefore,

$$\theta = 90°$$

Moving toward the luminaire,

$$\text{as } x \rightarrow 0, \text{ then } \theta = \arctan \frac{\sqrt{(0)^2 + (y)^2}}{h}$$

Therefore, $\theta \rightarrow \theta_{min}$

Summary for Transverse Roadway Lines

$$\text{at } x = \infty, \text{ then } \alpha = 0°, \text{ and } \theta = 90°.$$

$$\text{as } x \rightarrow 0, \text{ then } \alpha \rightarrow 90°, \text{ and } \theta \rightarrow \theta_{min}$$

Longitudinal roadway lines (Fig. 11-10) are curved lines that begin at $\alpha = 90°$, $\theta = 90°$, and fan out to $\alpha = 0°$ ($\alpha = 180°$), $\theta = 0°$. Approaching the luminaire from infinity (∞) along a longitudinal roadway line, the angular change of the LRL can be calculated and plotted (see Fig. 11-12).

For Longitudinal Roadway Lines

For the horizontal angle, α,

$$\text{at } y = \infty \qquad \alpha = \arctan \frac{\infty}{x}$$

Therefore,

$$\alpha = 90°$$

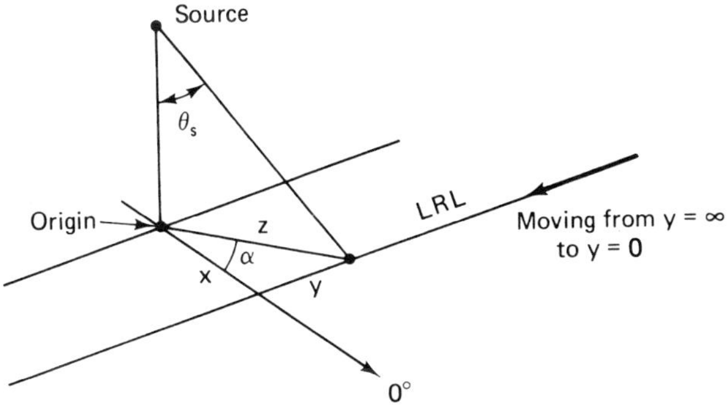

Figure 11-12 Longitudinal Roadway Line on Isocandela Diagram (Fig. 11-10).

Moving toward the luminaire,

$$\text{as } y \rightarrow 0, \text{ then } \alpha = \arctan \frac{0}{x}$$

Therefore,

$$\alpha \rightarrow 0°$$

For vertical angle, θ,

$$\text{as } y \rightarrow \infty, \text{ then } \theta = \arctan \frac{\sqrt{(x)^2 + (\infty)^2}}{h}$$

Therefore,

$$\theta = 90°$$

Moving toward the luminaire,

$$\text{as } y \rightarrow 0, \text{ then } \theta = \arctan \frac{\sqrt{(x)^2 + (0)^2}}{h}$$

Therefore,

$$\theta \rightarrow 0°$$

Summary for Longitudinal Roadway Lines

$$\text{at } y = \infty, \text{ then } \alpha = 90°, \text{ and } \theta = 90°$$

$$\text{as } y \rightarrow 0, \text{ then } \alpha \rightarrow 0°, \text{ and } \theta \rightarrow 0°$$

A simple example will demonstrate how the isocandela diagram can be used to find the luminous intensity arriving at a point on the roadway surface.

Example:

What is the luminous intensity (Fig. 11-13) arriving at a point 40 ft in front of a luminaire and 25 ft down the roadway?

$$\text{If } x = 40 \text{ ft}, \quad y = 25 \text{ ft}, \quad h = 30 \text{ ft} \qquad \text{(Fig. 11-13)}$$

$$\theta = \arctan \frac{\sqrt{(x)^2 + (y)^2}}{h} = \frac{\sqrt{(40)^2 + (25)^2}}{30} = 57.5°$$

$$\alpha = \arctan \frac{y}{x} = \arctan \frac{25}{40} = 32.0°$$

Therefore,

$$I_{(\theta,\alpha)} = 4500 \text{ cd}$$

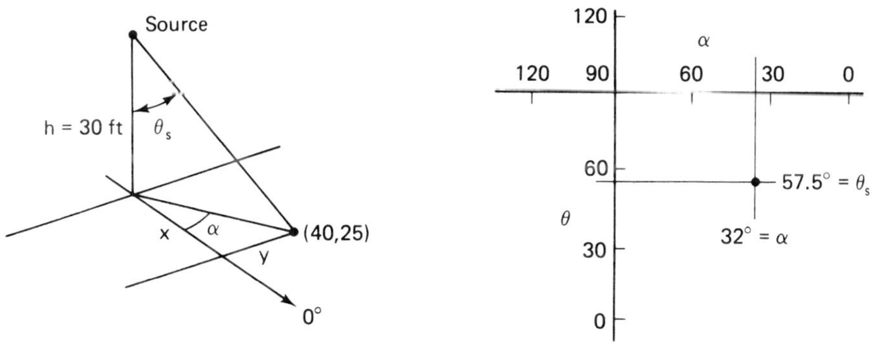

Figure 11-13. Use of Isocandela Diagram (Fig. 11-22) to Find $I_{(\theta,\alpha)}$.

Isofootcandle Diagram

The isofootcandle diagram (Figure 11-24 in the Data Section) is a contour map of illuminance values. Illuminance values of equal magnitude are connected together with lines called **isolines.** The isofootcandle diagram can be used to calculate the illuminance at a point (or points) on the roadway surface.

Two ratios must be calculated to determine the illuminance at a point from the isofootcandle diagram: (1) transverse distance (*TD*) to mounting height (*MH*) and (2) longitudinal distance (*LD*) to mounting height (*MH*).

Example:

What is the illuminance at a point 40 ft in front of the luminaire and 25 ft down the road (see Fig. 11-13) for a luminaire mounted 30 ft above the roadway?

If the transverse distance x = 40 ft and the longitudinal distance y = 25 ft, then,

$$\frac{TD}{MH} = \frac{40}{30} = 1.33$$

$$\frac{LD}{MH} = \frac{25}{30} = 0.83$$

Therefore, from Figure 11-14, the initial illuminance at the point is

$$E_h = 0.82 \text{ fc}$$

Utilization Curve

The utilization curve indicates the amount of light falling on the roadway surface in terms of a coefficient. The utilization factor (coefficient of utilization) represents the percentage of the rated bare lamp lumens that falls on two horizontal strips of infinite length one in front and one behind the luminaire. It is a measure of the efficiency of a luminaire in terms of the luminaire's ability to direct light to the roadway surface. The utilization factor (*UF*) is used in the lumen method formula to determine the average

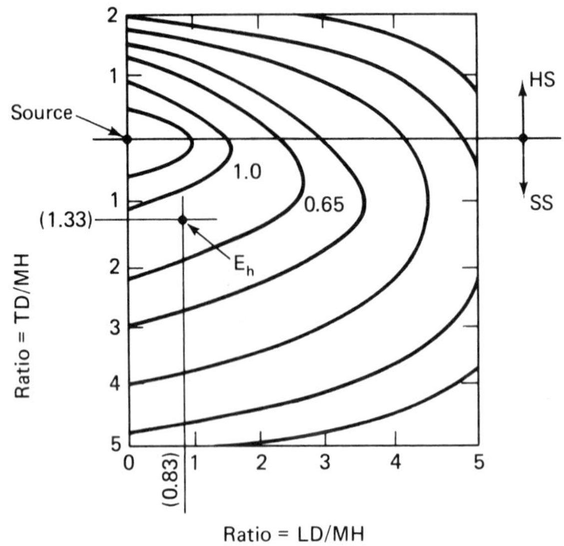

Figure 11-14. Use of Isofootcandle Diagram to Find E_h.

uniform horizontal illuminance or to determine the spacing for some recommended level of illuminance (see Fig. 11-15 and Fig. 11-24 in the Data Section).

Figure 11-16 shows utilization curves for five roadway luminaires. The differences can best be seen by studying the distribution characteristics of the luminaires (Figure 11-18). The utilization factor is found by entering the appropriate curve (street side or house side) with the corresponding ratio of transverse distance to mounting height.

The house side ratio and street side ratio are calculated:

$$HS_{\text{ratio}} = \frac{TD}{MH} = \frac{9.5}{30} = 0.32$$

$$SS_{\text{ratio}} = \frac{TD}{MH} = \frac{38.5}{30} = 1.28$$

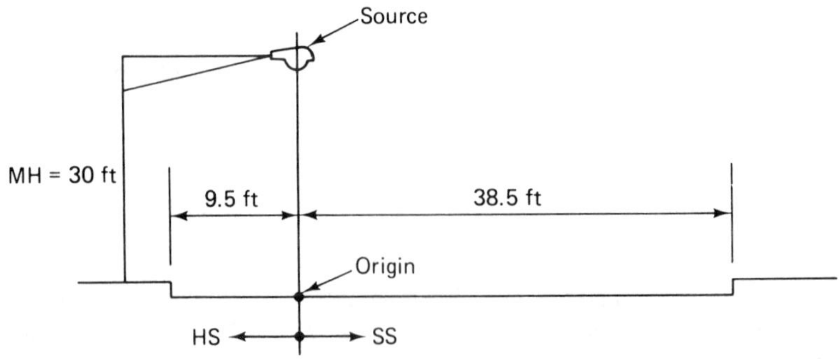

Figure 11-15. Geometry for Determining the Utilization Factor.

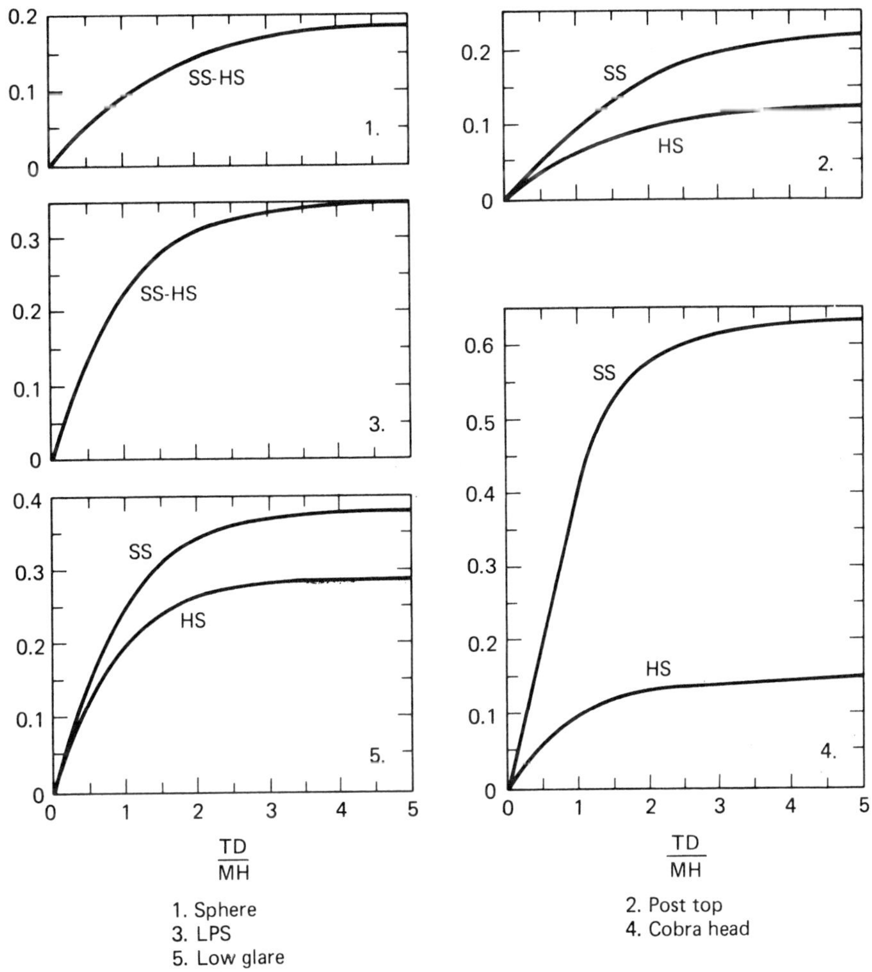

Figure 11-16. Utilization Curves for Five Luminaires.

From the utilization curves for luminaire 4 in Figure 11-16,

$$UF_{HS} = 0.05$$

$$UF_{SS} = 0.48$$

The total utilization factor is the sum of the house side and street side utilization factors:

$$UF_{\text{total}} = UF_{HS} + UF_{SS} = 0.05 + 0.48 = 0.53$$

Distribution Characteristics

The distribution of luminous flux (lumens) from a luminaire can be used to rapidly evaluate a luminaire in terms of light pollution, wasted energy, glare, and efficiency. The luminous flux distribution values represent the lumens distributed within a given zone (zonal lumens; see Fig. 11-17). The standard practice is to divide a theoretical sphere surrounding the luminaire into four zones: (1) downward hemisphere house side (*DW HS*), (2) downward hemisphere street side (*DW SS*), (3) upward hemisphere house side (*UP HS*), and (4) upward hemisphere street side (*UP SS*). Since the luminaire is assumed to be placed inside of the sphere, all the luminous flux that is directed into the lower hemisphere, 0° to 90°, is summed to represent the **downward component.** All the luminous flux that is directed into the upper hemisphere, 90° to 180°, is summed to represent the **upward component.**

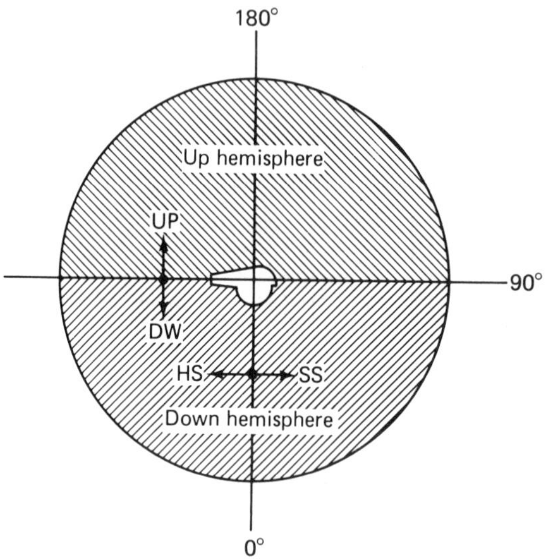

Figure 11-17 Distribution Terminology.

The flux values for a typical roadway luminaire (luminaire 4, Fig. 11-16 and Fig. 11-23, Data Section) are shown in Table 11-3 in accordance with standard practice for reporting the luminous flux distribution.

TABLE 11-3 Flux Values for Luminaire 4 (Fig. 11-22, Data Section)

Hemisphere	Zones	% Lumens	
DW SS *DW HS*	0–90	63.6 14.5	Downward = 78.1
UP SS *UP HS*	90–180	1.2 0.7	Upward = 1.9
Total	0–180	80.0	80.0

This luminaire has an efficiency of 80.0%. This means that 80.0% of the total bare lamp lumens actually gets out of the luminaire or 20.0% of the bare lamp lumens are absorbed or trapped inside the luminaire.

At first glance this may look quite good, but ask the following questions. What value is the 1.9% upward component, and will it ever reach the roadway surface? The 78.1% downward component looks good. However, it includes luminous flux in the entire downward hemisphere. What percentage of the luminous flux is at high angles (above 70°)? What effect does high-angle (70° to 90°) intensity have on glare? How effective is high-angle intensity at producing horizontal illuminance?

To answer these questions, the flux values must be reported for more zones in the downward hemisphere. Figure 11-18 shows the luminous flux distribution for the five luminaires (previously seen in Fig. 11-16) in terms of output and percentage of total.

	1. Sphere				2. Post Top			
	% Output		% Total		% Output		% Total	
	SS	HS	SS	HS	SS	HS	SS	HS
0–40	4.1	4.1	4.3	4.3	3.2	2.4	6.4	4.8
40–70	9.3	9.3	9.9	9.9	11.5	6.2	23.2	12.6
70–90	8.4	8.4	9.0	9.0	10.1	5.9	20.4	11.9
90–100	25.1	25.1	26.8	26.8	5.4	4.9	10.9	9.9
0–180	46.6	46.8	50.0	50.0	30.2	19.4	60.9	39.1
	3. LPS				4. Cobra Heat			
0–40	11.0	11.0	15.1	15.1	15.8	4.4	19.7	5.5
40–70	18.7	18.7	25.6	25.6	36.3	8.5	45.1	10.6
70–90	5.8	5.8	7.9	7.9	11.5	1.6	14.5	2.4
90–180	0.9	0.9	1.3	1.3	1.2	0.7	1.5	0.8
0–180	36.4	36.4	50.0	50.0	64.8	15.2	80.7	19.3
	5. Low glare							
0–40	16.0	15.0	20.5	19.2				
40–70	26.5	19.3	34.0	24.7				
70–90	0.9	0.3	1.1	0.4				
90–180	0.0	0.0	0.0	0.0				
0–180	43.4	34.6	55.7	44.3				

Figure 11-18 Flux Distribution for Five Luminaires.

$$\% \text{ output} = \frac{\text{lumens in a zone}}{\text{bare lamp lumens}} \times 100 \tag{11.4}$$

$$\% \text{ total} = \frac{\text{lumens in a zone}}{\text{total lumens output (0–180°)}} \times 100 \tag{11.5}$$

The sum of the % output for the house side plus street side from 0° to 180° represents the efficiency of the luminaire. The sum of the % total (*HS* + *SS*) from 0° to 180° will be 100%. A comparison of Figure 11-18 (luminaire 4) with Figure 11-23 will show the value of a more detailed breakdown of flux values.

Luminous Intensity Distribution

The optimum luminous intensity (candlepower) distribution to meet the standard design criterion (uniform horizontal illuminance) can be constructed. The construction of the optimum distribution is based on the Fundamental Law of Illumination (Inverse-Square Law) and the Cosine Law of Illumination (see Fig. 11-19).

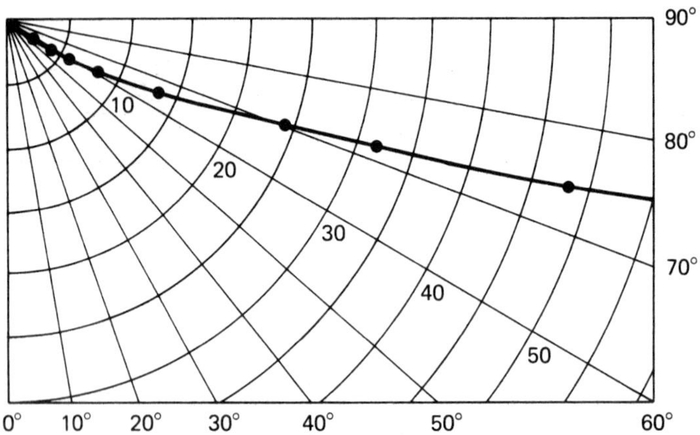

$$I_\theta = \frac{E_h H^2}{\cos^3 \theta}$$

Angle, θ	Multiplying factor	Angle, θ	Multiplying factor
0	1.0	60	8.00
10	1.05	65	13.25
20	1.21	70	24.99
30	1.54	75	57.68
40	2.22	80	190.98
50	3.77	85	1510.47
55	5.30	90	∞

$I_\theta = I_{0°}$ x Multiplying factor @ θ – (to produce uniform horizontal illumination)

Figure 11-19. Luminous Intensity Distribution for Uniform Horizontal Illuminance.

Since

$$E_h = \frac{I_\theta}{D^2} \cos \theta \qquad (11.6)$$

and

$$\cos \theta = \frac{H}{D}$$

then

$$D = \frac{H}{\cos \theta}$$

squaring both sides of the equation,

$$D^2 = \frac{H^2}{\cos^2 \theta} \qquad (11.7)$$

Therefore,

$$E_h = \frac{I_\theta}{H^2} \cos^3 \theta \qquad (11.8)$$

To find the optimum luminous intensity, I_θ, for a constant E_h (uniform horizontal illuminance), the equation can be rewritten as follows:

$$I_\theta = \frac{E_h H^2}{\cos^3 \theta}$$

If E_h = constant = 1 and H = 1 MH, then the luminous intensity, θ, must vary as the reciprocal of the cosine cubed.

$$I_\theta = \frac{1}{\cos^3 \theta} \qquad (11.9)$$

Figure 11-19 shows that the luminous intensity at high angles (70° to 90°) must be magnitudes larger than the intensity produced at nadir (0°) for uniform horizontal illuminance. At 70°, the luminous intensity must be 24.99 times the intensity at 0° (nadir), which is almost impossible to achieve. For reasons of economics, energy conservation, and glare control, the high-angle (70° to 90°) intensity should be redirected into more useful zones in the distribution.

Roadway Lighting Calculations

Illuminance Design Methods

Lumen Method: The lumen method formula is used to calculate the average uniform horizontal illuminance on the roadway or to calculate the spacing to achieve a required, maintained level of average uniform illuminance (E_r). An example of the lumen method for roadway lighting design is provided in the example problem at the end of this chapter.

$$\text{analysis formula:} \quad E_i = \frac{\text{test lumens} \times \text{utilization factor}}{\text{pole spacing} \times \text{street width}} \tag{11.10}$$

The maintained illuminance, E_m, is

$$E_m = E_i \times \text{light loss factor } (LLF) \tag{11.11}$$

$$\text{design formula:} \quad \text{spacing, } S = \frac{\text{test lumens} \times UF \times LLF \times LF}{E_r \times \text{street width}} \tag{11.12}$$

Spacing. Once the spacing, S, has been calculated to satisfy the required horizontal illuminance, the uniformity ratio specified in the standard practice (1) must be checked. The minimum-to-average uniformity of illuminance is calculated by finding the minimum illuminance level of the roadway. A layout configuration must be selected from the three standard layouts (Fig. 11-20). The minimum illuminance at one of four (or more) points must be determined. If the ratio of the minimum-to-average exceeds the recommended uniformity ratio, the design must be modified. Any one or all of the following variables can be changed to meet the uniformity ratio criterion: (1) spacing, (2) mounting height, (3) luminaire lateral placement, and/or (4) luminaire distribution. An example problem utilizing this design procedure can be found at the end of this chapter.

Light Loss Factor, LLF: The *LLF* is a number less than unity that takes into account losses in light output due to dirt accumulation and lamp lumen depreciation.

Lamp Factor, LF: The *LF* is used to increase the test lumens (lumens found on the photometric test report) to the current listed catalog lumens, and to modify isofootcandles, isocandelas (luminous intensity or "candlepower"), and lumens (luminous flux).

$$LF = \frac{\text{catalog lamp lumens}}{\text{test lumens}} \tag{11.13}$$

Mounting Height Correction, MHC: The MHC is used to modify the footcandle values for mounting heights that are different from the test mounting height.

$$MHC = \left(\frac{\text{test mounting height}}{\text{actual mounting height}}\right)^2 \tag{11.14}$$

Uniformity Ratio: The uniformity ratio is defined as a ratio of average illuminance to minimum illuminance on the roadway. The average-to-minimum ratio (3) should not exceed 3 to 1 except for residential areas, which should not exceed 6 to 1. The usefulness of this ratio was discussed previously in this chapter. (See also the example problem at the end of this chapter.)

$$\frac{E_{\text{ave}}}{E_{\text{min}}} < 3.0 \quad (6.0 \text{ for residential areas})$$

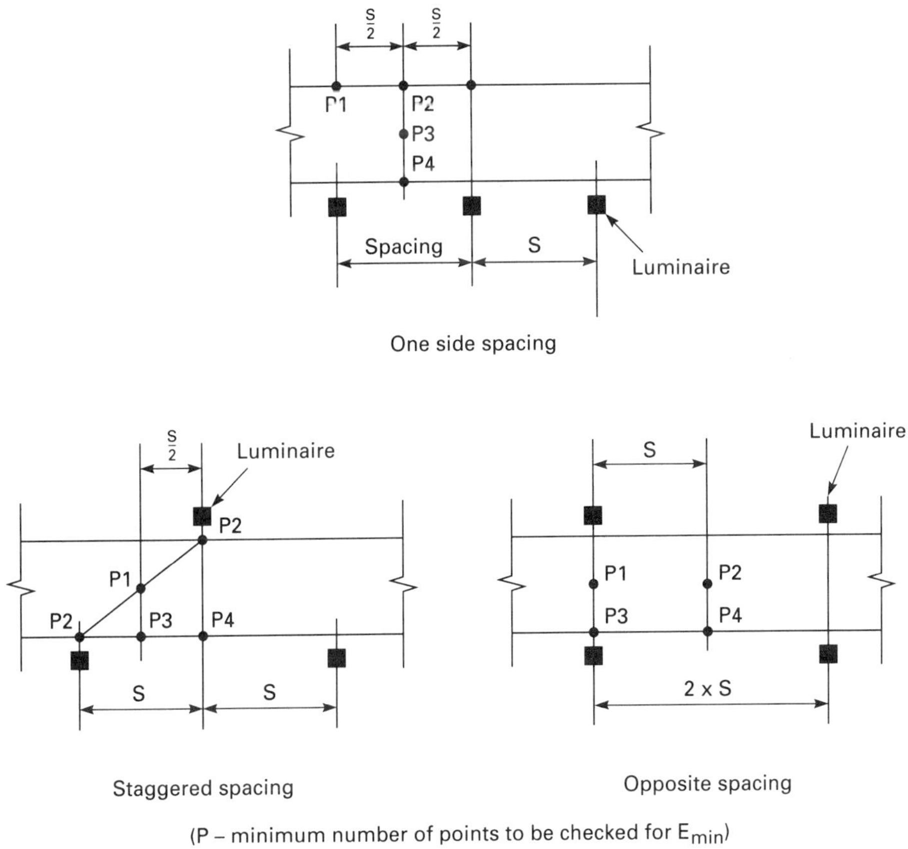

Figure 11-20. Spacing Configurations.

Point-by-Point Method. The point-by-point method (Chapter 8) utilizes the Inverse-Square Law and the Cosine Law of Illuminance to calculate the illuminance levels at various points on the roadway surface. Because of the possible rotation of the photometric plane of symmetry relative to the calculation grid and the possible "tilt" of a luminaire, the calculation of illuminance at multiple points is more easily accomplished utilizing a microcomputer, vector algebra (discussed in Chapter 12), and IES format microcomputer disks (see Chapter 6).

Luminance Design Method. In Chapter 1, Chapter 8, and in this chapter, the point was made that the human visual system does *not* see illuminance. The visual system sees objects based on the amount of light reflecting off of a surface. Uniform illuminance on a roadway surface will not necessarily result in a uniform pattern of luminance on the roadway surface. Roadway surfaces are nonperfect diffusers or semi-specular (see Chapter 6, Reflections) in their reflecting characteristics. Therefore, the reflecting characteristics of a roadway surface must be described in terms of a table of luminance factors, β (see Chapter 1). The IES Lighting Handbook (2) lists luminance factors for four different roadway finishes (R1, R2, R3, R4). Table 11-4 in the Data Section contains the luminance factors for roadway surface R3. If the actual roadway finish is not known it is common practice to assume an R3 roadway surface. From Chapter 1, the luminance factor, β, is a function of the location of the light source, the location of the observer, and the characteristics of the surface. The location of the light source is defined in terms of v_s and h_s. If the observer is assumed to be in the $h = 0°$ plane, the location of the observer is further defined by the viewing angle, v_o.

Luminance Design Procedure. The Roadway Lighting Committee of the IESNA has established the following procedure for determining roadway luminance values.

1. Use an eye height of 1.45 meters which is 4.757 feet.
2. Calculations should be made along two longitudinal lines at the transverse quarter points of each roadway lane width.
3. A minimum of 10 calculation points no more than 5 meters (16.41 ft) apart are required along each of the two longitudinal calculation lines per lane width.
4. Use one luminaire *behind* and three luminaires *ahead* of each calculation point along the calculation line.
5. Use a viewing angle, α, of 1° (v_o = 89°) measured from the roadway surface.

For a viewing angle, α = 1°, the viewing distance, *d*, from the observer to the calculation point would be

$$d = \frac{1.45}{\tan 1°} = 83.07 \text{ m} = 272.53 \text{ feet}$$

Luminance Calculational Procedure

1. Calculate transverse spacing (*TS*) of calculation lines.

$$TS = \frac{\text{lane width}}{4}$$

2. Calculate longitudinal spacing (*LS*) of calculation points.

$$\text{maximum spacing} = 5 \text{ m (16.41 ft)}$$

$$\text{minimum point} = 10$$

$$\text{staggered: luminance cycle} = 2 \times S$$

$$\text{one side: luminance cycle} = S$$

$$\text{opposite: luminance cycle} = S$$

$$\text{spacing} = \frac{\text{luminaire cycle}}{10}$$

a. If the spacing is less than 5 m (16.41 ft), use the "spacing" calculated as the distance between the calculation points.
b. If the spacing is greater than 5 m (16.41 ft), calculate the number of calculation points required by

$$\text{no. points} = \frac{\text{luminaire cycle}}{5 \text{ m (or } 16.41 \text{ ft)}}$$

Round to the next highest whole number. Then calculate the "spacing" required by

$$\text{spacing} = \frac{\text{luminaire cycle}}{\text{no. points (whole number)}}$$

3. Determine the luminaire coordinates from the origin at the luminaire #1. Include one luminaire behind and three luminaires ahead of the calculation area.

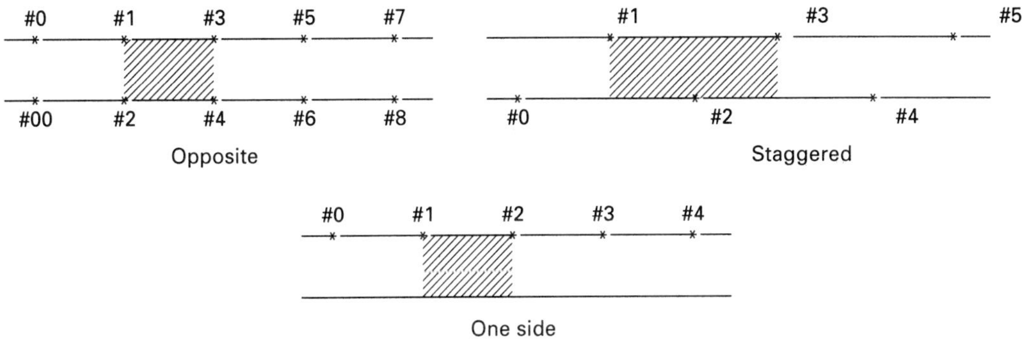

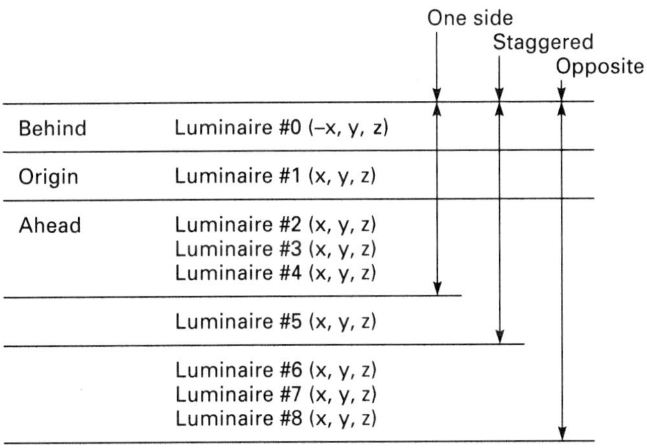

4. Determine the point coordinates from the origin along each calculation line.

$$P_{(\text{point, line})}\,(x, y)$$

5. Sum the horizontal illuminance value contributed by each luminaire at each of the calculation points, $P_{(\text{point, line})}$.

6. Sum the luminance values contributed by each luminaire at each calculation point, $P_{(\text{point, line})}$.

Table 11-4 contains the luminance factors, β, as a function of h and v_s (see Chapter 1). Most texts refer to the values in the table as the reduced luminance coefficients, r; therefore,

$$\beta_{(h,\ vs)} = \mathrm{r}_{(h,\ vs)} = q \times \cos^3(v_s) \times 10^4 \tag{11.15}$$

The $\cos^3$ expression relates to the Cosine Cubed Law of Illuminance (equation 11.8) which calculates the illuminance on a horizontal surface or finds the luminous intensity vector perpendicular to the roadway surface.

$$L_p = \sum_{1}^{n} \frac{\beta_{(h,\ vs)} \times I_{(\theta s,\ \alpha)}}{H^2} \tag{11.16}$$

7. Calculate the average luminance (L_{ave}).

$$L_{ave} = \frac{\sum_{1}^{n} L_p}{n} \tag{11.17}$$

where n is number of calculation points.

8. Calculate the longitudinal luminance uniformity (lane uniformity, L_L) per lane and choose the worst (highest) L_L for a single calculation lane as the rating.

$$L_L = \frac{L_{max}\ \text{(per lane)}}{L_{min}\ \text{(in same lane)}}\ \text{for each calculation lane}$$

9. Calculate the average luminance uniformity (LU_{ave}).

$$LU_{ave} = \frac{L_{ave}}{L_{min}}$$

where L_{min} is the minimum value of all luminance values calculated.

10. Calculate the maximum luminance uniformity (LU_{max}).

$$LU_{max} = \frac{L_{max}}{L_{min}}$$

where L_{max} is the maximum value of all luminance values calculated.

Luminance Factor Determination. The calculation of luminance values at multiple points is achieved by using the point-by-point method (Chapter 8) utilizing the Inverse-Square Law. Because of the possible "rotation" of the photometric plane of symmetry relative to the calculation grid and the possible "tilt" of a luminaire, the calculation of illumination of multiple points is more easily accomplished utilizing a microcomputer, vector algebra (discussed in Chapter 12), and IES format microcomputer files (see Chapter 6).

Disability Veiling Brightness

High disability veiling brightness (DVB) can reduce object visibility and visual acuity by reducing object contrast in the field of view as a result of a veiling luminance produced in the eye. However, increasing DVB can increase contrast sensitivity, which means that object contrast can decrease. Windshield cutoff must be taken into consideration. A driver location is selected and the DVB value (see Fig. 11-21) at a viewing position is the sum of the individual DVB values produced by each luminaire in the visual field.

$$DVB = \frac{10\pi E_v}{\theta_G^2} \tag{11.18}$$

where E_v = vertical illuminance at the eye on a plane normal (perpendicular) to the line of sight

θ_G = angle between a line from the luminaire to the eye and the line of sight

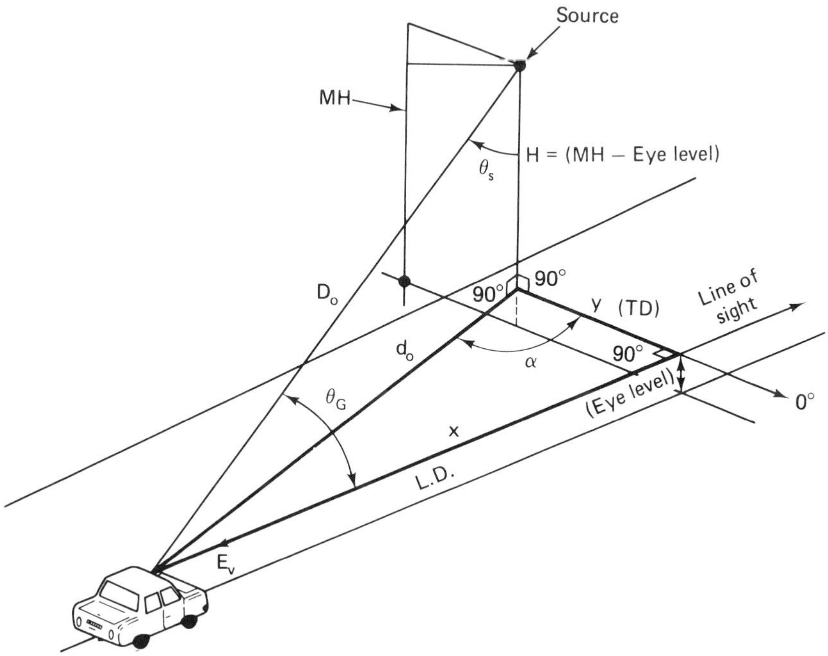

Figure 11-21. Disability Veiling Brightness Terminology.

If the windshield cutoff angle is 20°, the luminous intensity values at an elevation angle θ_s less than 70° will not contribute to the *DVB* value.

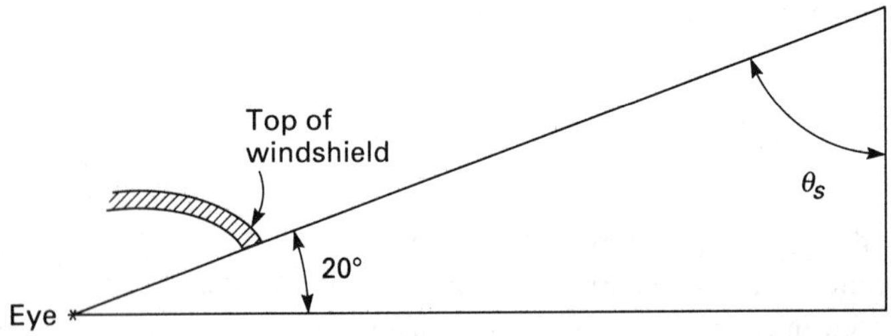

Since the approved luminance method uses an eye height of 1.45 m (4.575 ft), the procedure discussed in this section will use the 4.575 ft (1.45 m) eye height rather than the 4.3 ft eye height value used in the original work done by Rex (4, 5, 6, 7, 8).

$$E_v = \frac{I_{(\theta,\ \alpha)} \cos \theta_G}{D_o^2} \tag{11.19}$$

$$\cos \theta_G = \frac{x}{D_o} = \frac{LD}{D_o}$$

Therefore,

$$E_v = \frac{I_{(\theta,\ \alpha)} \times LD}{D_o^3} \tag{11.20}$$

then for a *DVB* per 1000 cd (assuming $I(\theta, \alpha) = 1000$ cd)

$$DVB/1000 = \frac{10\pi(1000) \times LD}{D_o^3 \theta_G^2} \tag{11.21}$$

where $\qquad D_o = \sqrt{(d_o)^2 + (H)^2}$

and $\qquad d_o = \sqrt{(x)^2 + (y)^2} = \sqrt{(LD)^2 + (TD)^2}$

Therefore, $\qquad D_o = \sqrt{(LD)^2 + (TD)^2 + (H)^2}$

H is the vertical distance from eye level (4.757 ft.) to the luminaire mounting height.

$$\theta_G = \text{arccosine} \frac{x}{D_o} = \text{arccosine} \frac{LD}{D_o} = \cos^{-1} \frac{LD}{[(LD)^2 + (TD)^2 + H^2]^{1/2}} \tag{11.22}$$

To find $I_{(\theta,\alpha)}$, calculate the angle, θ (θ_s), which is the elevation angle from nadir and the angle, α, which is the horizontal angle measured from the photometric plane of symmetry by the formula

$$\theta_s = \arctan \frac{d_o}{H} = \tan^{-1} \frac{[(TD)^2 + (LD)^2]^{1/2}}{H} \tag{11.23}$$

$$\alpha = \arctan \frac{x}{y} = \tan^{-1} \frac{LD}{TD} \tag{11.24}$$

The luminous intensity value is selected from the tabulated photometric data or IES format photometric data disk for the elevation angle, θ_s and horizontal angle, α.

Summary

The higher the DBV value at a point, the lower the target detection and the lower the visual acuity of the driver. The higher the illuminance level arriving at the driver's eye, the higher the sensitivity to object contrast in the visual field. However, the increased illuminance arriving at the eye will cause an increase in the veiling luminance produced in the eye. Veiling luminance produced in the eye reduces contrast of an object against the background as seen by the driver. That is, increased illuminance arriving at the driver's eye may increase the driver's sensitivity to contrast, but it will also reduce the apparent contrast of the object due to veiling luminance in the eye. The net effect may be a decrease in object visibility.

AREA LIGHTING

Area lighting may be required to provide security lighting and to provide for the safe movement of people. Area lighting can increase security by increasing visibility, thus reducing crimes against people and property. The safety aspects of area lighting not only include the prevention of crimes but also a reduction in accidents.

The use of movable or aimable systems for area lighting is quite common. The approach has been to place floodlighting units on poles and aim them out to flood an area with light. One must remember that, as a unit is tilted up and out, the internal parts of the luminaire and light source are exposed to the direct line of sight. This approach may be initially less expensive, but it creates glare, excessive spill, and wasteful energy utilization. Because of the effect on transient adaptation and veiling luminance produced in the eye, the visibility may be reduced to a point that the lighting is detrimental to both security and safety.

The trend in area lighting has been to move away from movable or aimable systems to the static systems used in roadway lighting. The use of low-glare (sharp cutoff), high-efficiency systems is recommended for area light. The same principles and procedures outlined in the roadway lighting section of this chapter apply to area lighting.

BUILDING FLOODLIGHTING

Building floodlighting can increase the aesthetic appearance of a selected structure as well as act as security lighting. Lighting of the exterior surface of the building can reduce crimes such as burglary and theft by making the exterior surfaces more visible to security personnel.

The primary concern in building floodlighting is to minimize glare and maximize the distribution of illumination on the building surface. The required level of illuminance is a function of the reflective properties of the building, the ambient surrounding luminance, and the aesthetic appearance desired. Building floodlighting is the terminology used to describe ground-mounted or pole-mounted equipment that is set back some distance from the building. Building lighting can also be achieved by equipment mounted directly on the structure. The type of area to be lighted, the location of equipment, and the variation in surrounding conditions impose design problems that make standardization difficult if not impossible.

The lighting of commercial buildings can serve as a means of advertising. In other words, let the architecture sell the product by attracting attention to the building and by creating a favorable impression on the passer-by. If the building is lighted properly, there should be no need for additional advertising to distract from the architecture.

Floodlighting is dependent primarily on the ability of the designer to manipulate brightness, texture, shadow, and color. The designer must establish the architectural effect that is desired with the floodlighting. It is helpful to study the daylight effects on the building. This allows the designer to select the strong daylight effects that are desirable and to attempt to reproduce these effects under floodlighting.

Daylighting effects are difficult to produce because of the directional characteristics of the sun and the diffuse characteristics of the sky. The sun is warm in color, while the diffused sky light is cool. Shadows are never black but less bright and bluer. Daylight is not constant but continually varying from day to day and hour to hour.

Illuminance Levels

The recommended illuminance levels, E_r, given in the *IES Lighting Handbook* (2) are recommended values that may vary depending on the design effects being created. The three factors that influence the level of illuminance required are

1. Reflectance of the building surface material.
2. The ambient illuminance level of the area surrounding the building.
3. The setback distance from the building and the specularity of the surface finish. If the floodlight is set very close to a building and aimed up the surface at a large tilt angle, a specular (see Chapter 6) or mirror surface will reflect most of its light up into the sky rather than back toward the observer on the ground.

   A building exterior finish of glass exhibits this characteristic. When a ray of light strikes glass at a large incident angle, the reflected ray of light approaches 100% as can be seen in Chapter 13, Figure 13-3.

Type of Equipment

Floodlighting equipment can be supplied as either open or enclosed equipment. Enclosed equipment has a higher initial cost and usually more accurate beam control. Open units usually require special bulbs with hard glass to prevent breakage caused by the cold elements (weather) touching the hot lamp enclosure. Enclosed equipment is initially more expensive; however, over a long period of time the savings due to the protection provided for the lamp and the use of standard lamps will usually pay for the higher initial cost.

Terminology (Fig. 11-25)

1. *Elevation angle or tilt angle*—the elevation angle is measured up or down from a horizontal reference axis parallel with the ground.
2. *Average maximum "candlepower"*— *old* terminology found on outdated photometric test reports. The average maximum "candlepower" is the average of the 10 highest luminous intensity ("candlepower") values recorded in the prescribed exploration of the beam pattern (10 points).

$$I_{\text{ave.max}} = \frac{\sum_{1}^{10} I_{\text{max}}}{10}$$

Example:

$$I_{\text{ave.max}} = 59{,}600 \text{ cd}$$

3. *Maximum "candlepower" (luminous intensity)*—the maximum "candlepower" value recorded in the beam pattern.
Example:

$$I_{\text{max}} = 93{,}360 \text{ cd}$$

4. *Beam limit*—In *old* terminology, the beam limit is that part of the beam in which the luminous intensity values are 10% or more of the average maximum "candlepower."
Example:

$$BL = 10\% \times 59{,}600 = 5960 \text{ cd}$$

In *new* terminology, the beam limit is that part of the beam in which the luminous intensity values are 10% or more of the maximum intensity.
Example:

$$BL = 10\% \times 93{,}360 = 9336 \text{ cd}$$

5. *Beam lumens (BLM)*—the beam lumens are the summed lumens that fall within the beam limit (see Fig. 11-25).
Example: OLD

with the beam limit at 5960 cd, the beam lumen = 21,512 lm

Example: NEW

with the beam limit at 9336 cd, the beam lumens = 19,062 lm

6. *Coefficient of beam utilization* (*CBU*) or utilization factor (*UF*)—*CBU* is a calculated ratio of the lumens effectively lighting an area to the total available Beam Lumens expressed as a percent.

$$CBU = UF = \frac{\Sigma \text{ lumens within beam limit falling on bldg}}{\text{beam lumens}}$$

The *CBU* is dependent on the following:

a. Equipment location
b. Aiming point
c. Lumens within the beam limit

The average *CBU* usually will fall within the range of 0.60 to 0.90. If less than 60% (0.60) of the beam lumens are utilized, the solution will be uneconomical and a different location or narrower beam pattern should be selected. If the CBU is over 90% (0.90), the beam spread is too narrow and the result will be spotty or nonuniform pattern of illumination.

7. *Beam spread*—the beam spread is the angle, measured in degrees, between the beam limits both horizontally and vertically (see Fig. 11-22).

horizontal angle × vertical angle

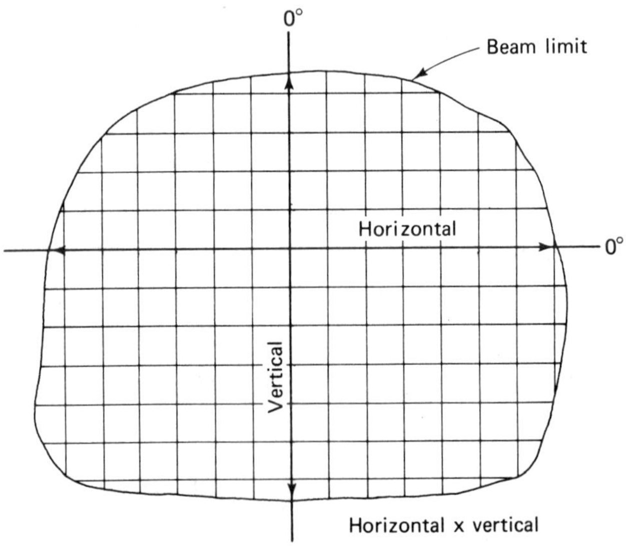

Figure 11-22. Definition of Floodlighting Beam Angles.

8. *Beam spread classifications—*

Beam Spread (degrees)	NEMA Type
10–17°	1
18–28°	2
29–45°	3
46–69°	4
70–99°	5
100–129°	6
130–up°	7

Example:

$$\text{horizontal} \times \text{vertical}$$

$$76° \times 40°$$

$$\text{NEMA type: } 5 \times 3$$

9. *Beam efficiency*—the beam efficiency is the ratio of total lumens in the beam limit to lumens generated by the "bare" lamp.

$$\text{beam efficiency} = \frac{\text{total lumens in beam limit}}{\text{total bare lamp lumens}}$$

10. *Total efficiency*—the total efficiency is the ratio of the total lumens output by the luminaire to the total lumens generated by the "bare" lamp. This value is not typically provided by the manufacturer.

$$\text{total efficiency} = \frac{\text{total lumens out of the luminaire}}{\text{total bare lamp lumens}}$$

11. *Beam footcandles*—this terminology is used by some manufacturers to produce an isofootcandle diagram that shows the footcandles on a surface normal (perpendicular) to the aiming line rather than the isofootcandles on a horizontal surface.

Floodlight Photometric Test Report

Typically, floodlighting photometric data has been reported in **Type B** photometric format. Most Type B format is actually measured in Type A (see Chapter 6) and mathematically transformed to Type B for the report. The mathematical transformation requires interpolation that introduces errors in the reported data. The IES Computer File Format (see Chapter 6) should be given in Type A format which represents the actual test data.

The standard floodlight photometric test report (Fig. 11-25 in the Data Section) consists of three major components: (1) equipment description, (2) photometric test data, (3) photometric output diagram that shows on one half of the diagram the isocandela curves in Type B format and zonal lumens within the beam limit in the other half of the output diagram. The isocandela curves and zonal lumens are presented on a rectangular coordinate grid.

Floodlighting Design Procedure

Floodlight calculations consist of analyzing the coverage and uniformity of illuminance produced on the vertical surface of a building by (1) the lumen method or (2) point-by-point method. An example of the lumen method for floodlighting design is provided in the example at the end of this chapter. Both procedures require knowledge of the following design parameters which the lighting professional must determine or assume:

1. Recommended level of illuminance, E_r.
2. A beam spread or pattern.
3. The location of the floodlighting equipment.
4. An estimate of the light loss factor.
5. An estimate of the number of floodlights required.
6. The uniformity of coverage with a check for compliance with recommended uniformity ratios. If the uniformity ratio is not met, the lighting professional must select a different beam spread and/or relocate the equipment until the uniformity ratio meets the recommended values.

Lumen Method

The lumen method is used to find the average uniform maintained illuminance on the vertical surface of the building. The lumen method formula was introduced in Chapter 8. The following formula shows the modification to the lumen method formula (Chapter 8) resulting from a difference in floodlighting terminology and reported data.

Analysis Formula

$$E_{\text{ave}} = \frac{N \times BL \times CBU \times LLF \times OV \times TF \times LF \times LP}{A} \tag{11.25}$$

where N = total number of luminaires.

BL = beam lumens of each luminaire.

$CBU\ (UF)$ = coefficient of beam utilization.

LLF = light loss factor.

OV = overvoltage factor. Operating at overvoltage is common in sports lighting; it results in increased light output and decreased life (see Chapter 4).

TF = temperature factor; usually only consider when using fluorescent lamps (see Chapter 4).

$$LF = \text{lamp factor} = \frac{\text{catalog lamp lumens}}{\text{test lumens}}$$

LP = lamp position factor; a function of the actual burning position; HID lamps burned in any position other than the position for which they were rated will have a change in output (see Chapter 4).

Design Formula

$$N = \frac{E_r \times A}{BL \times CBU \times LLF \times OV \times TF \times LF \times LP} \tag{11.26}$$

The coefficient of beam utilization has been defined as a ratio of the lumens striking the building to the total beam lumens. The *CBU* is calculated (see the example at the end of this chapter) for each floodlighting location by superimposing the building projection (Fig. 11-26) onto the rectangular coordinate system. The lumens falling on the projection of the building can be summed, and the *CBU* calculated for each luminaire and building geometry. The individual *CBU*'s for each floodlight location are summed and an **average *CBU*** is calculated and checked against the assumed *CBU* used in the design formula. With this type of coordinate system, the top and bottom of the building will be horizontal lines, while the sides will be slightly curved (see Fig. 11-26).

Lumen Method Procedure

1. Assume a design value of $CBU = 0.70$.
2. Calculate the number of floodlights, N, required to produce the recommended level of illuminance, E_r.
3. Lay out the floodlight locations or positions and select the aiming point for each luminaire.
   a. If setback is not restricted, a beginning assumption for setback distance, d, should be 2/3 to 3/4 of the overall height of the vertical surface to be illuminated.
   b. If the setback is more than 1/2 the vertical surface height, a beginning assumption for the aiming point on the surface should be 2/3 the height of the surface. As the setback moves closer than 1/2 the surface height, the aiming point will move up the surface beyond the 2/3 the height position.
   c. For tall buildings, two or more floodlights may have to be aimed at successively higher aiming points up the face of the surface. As floodlight aiming points move up the surface the beam spread will usually have to be decreased with the narrowest pattern aimed near the top of the building.
4. Plot the projection of the building on the photometric data sheet (see Fig. 11-26).
5. Calculate the *CBU* for each luminaire position and aiming point.
6. Find the weighted average of the individual *CBU* values.
7. Compare the CBU_{act} to the $CBU_{assumed}$; if the difference is great, cycle back to step 1.

8. Calculate the E_{ave} using the CBU_{act} utilizing the design formula.
9. Calculate the illuminance at a minimum of three points (top, aiming point, and bottom of the surface) on a vertical line formed by the intersection of the luminaire vertical plane with the vertical plane of the surface. Also check a minimum of three points on the edge of the building. The illuminance values for a uniformity check are calculated using the point-by-point method. Calculate the uniformity ratio as

$$\text{uniformity ratio}_{\text{average}} = \frac{E_{\text{ave}}}{E_{\text{min}}}$$

$$\text{uniformity ratio}_{\text{max}} = \frac{E_{\text{max}}}{E_{\text{min}}}$$

The literature speaks to checking uniformity but does not seem to define acceptable uniformity ratios for building floodlighting. The final decision under these conditions falls on the lighting professional. Some known data may be of assistance in making a decision on an acceptable uniformity ratio:

a. Uniformity only has meaning when making reference to the luminance of a surface not the illuminance (see Chapter 1 and Chapter 8).
b. For interior lighting design, where ambient lighting is relatively high, the following data are in the literature.
   (1) A luminance uniformity ratio (ave:min) of 3 to 1 will normally result in a visually "just perceptible" difference in luminance between two adjacent areas.
   (2) A luminance uniformity ratio (ave:min) of 5 to 1 will result in a visually "perceptible" difference in luminance between two adjacent areas; however, the difference is usually considered to be "acceptable."
c. A maximum to minimum uniformity ratio of 10 to 1 has been proposed for exterior lighting applications.

9. If the uniformity ratio is acceptable the design is complete. If the uniformity ratio is not acceptable, one must cycle back to step 1 and repeat the process.

Point-by-Point Method

The point-by-point method (Chapter 8) utilizes the Inverse-Square Law and the Cosine Law of Illumination to calculate the illuminance levels at various points on the building surface. Because of "tilt" (backward or forward) and possible "rotation" (laterally and axially) of floodlighting units, the calculation of illuminance at multiple points is more easily accomplished utilizing a microcomputer, vector algebra (discussed in Chapter 12), and IES format microcomputer files (see Chapter 6).

REFERENCES

1. Illuminating Engineering Society of North America, *American National Standard Practice for Roadway Lighting*, ANSI/IES RP-883. New York: ANSPRL, 1983.
2. Kaufman, John E., and Jack F. Christensen, eds., *IES Lighting Handbook*, Application Volume, 1987. New York: Illuminating Engineering Society of North America, 1987.
3. Ketvirtis, Antanas, *Highway Lighting Engineering*. Toronto: Foundation of Canada Engineering Corporation Limited, 1967.
4. Rex, Charles H., "Improving Seeing Efficiency with Roadway Lighting." *Traffic Engineering* (August 1956).
5. Rex, Charles H., "Comfort and Visibility Ratings for Roadway Lighting." *Illuminating Engineering* (May 1959).
6. Rex, Charles H., "Relative Visual Comfort Evaluations of Roadway Lighting." *Illuminating Engineering*, Vol. 60, No. 3 (March 1960).
7. Rex, Charles H., "Effectiveness Ratings for Roadway Lighting." *Illuminating Engineering* (July 1963).
8. Rex, Charles H., "Roadway Lighting for the Motorist." *Illuminating Engineering*, Vol. 62, No. 2, (February 1967).

EXAMPLE PROBLEMS

Example Problem: Roadway Lighting

Given the following roadway geometry, determine the spacing required to produce 0.8 fc average maintained. Find the minimum footcandles and check the minimum-to-average illuminance uniformity ratio. Use the luminaire shown in Figures 11-23 and 11-24 with staggered spacing.

Assume the following:

$$\text{catalog lumens} = 32{,}000 \text{ lm}$$

$$MH = 45 \text{ ft}$$

$$LLF = 0.70$$

$$E_r = 0.8 \text{ fc}$$

Cross section:

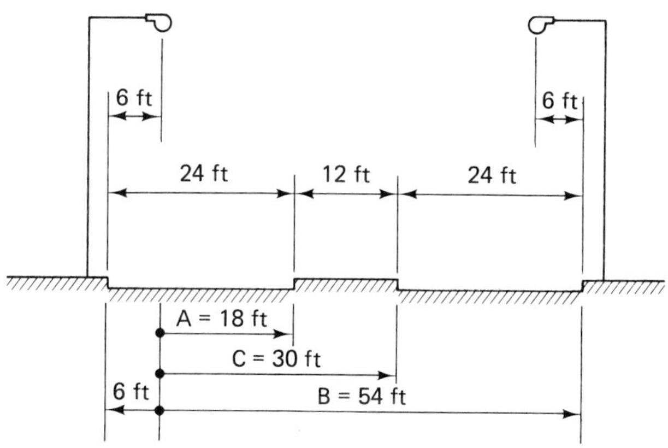

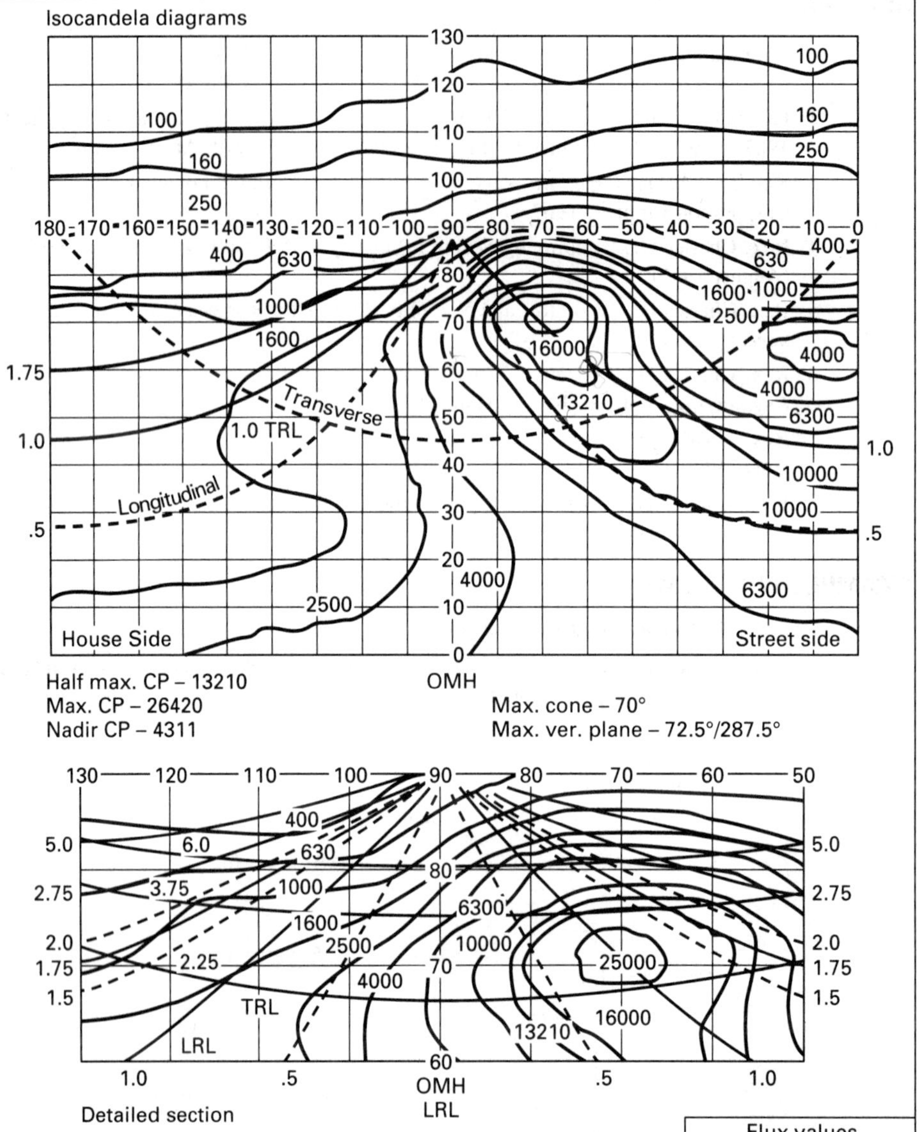

Flux values		
	Lms	%
Dw SS	20034	63.6
Up SS	378	1.2
Dw HS	4568	14.5
Up HS	220	.7
Total	25200	80.0

Test lamp rated at 31500 lumens

Isocandela, lumen, & footcandle values corrected by a ratio of act. lms ÷ 31500.

Fig. 11-23. Typical Roadway Photometric Test Report.

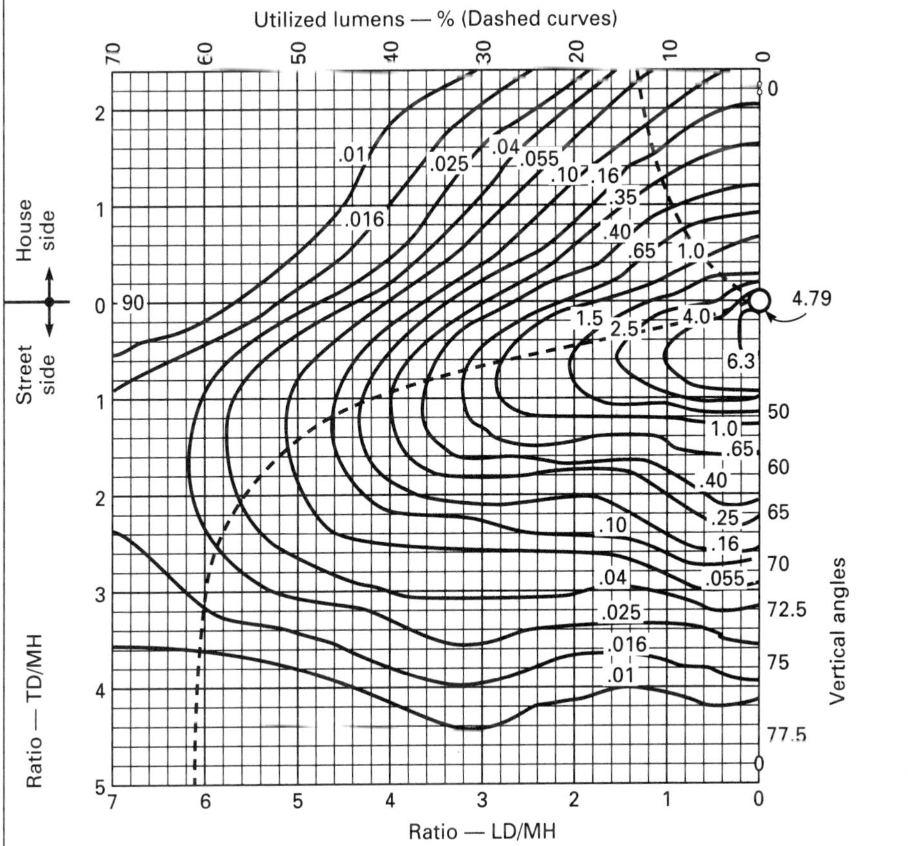

Isofootcandle & utilization curve
Footcandle data for 30' MH. Utilization curve no correction.

MH - FT.	20	21	22	23	24	25	26	27	28	29	30	31	32	33	34	35
Factor	2.25	2.04	1.86	1.70	1.56	1.44	1.33	1.24	1.15	1.07	1.0	.94	.88	.83	.78	.74

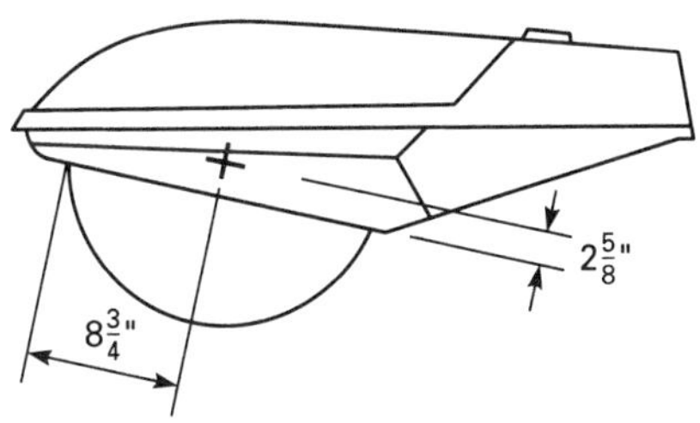

Luminaire description:
Roadway unit
TYPE M400A
Socket position 5

Lamp: 400 watt metal halide
Refractor 510
Medium semi-cutoff type III

No. 30000

Fig. 11-24. Roadway Luminaire Report Showing Isofootcandle and Utilization Curves.

Part A: Spacing

$$\text{spacing} = \frac{\text{LMS} \times \text{UF} \times \text{LF} \times \text{LLF}}{E_r \times W}$$

Street side: $\frac{\text{TD}}{\text{MH}} = \frac{18}{45} = 0.40$ $\qquad \text{UF}_A = 0.16$

$\frac{\text{TD}}{\text{MH}} = \frac{54}{45} = 1.2, \qquad \text{UF}_B = 0.46$ $\qquad \text{UF}_{BC} = \text{UF}_B - \text{UF}_C = 0.16$

$\frac{\text{TD}}{\text{MH}} = \frac{30}{45} = 0.67, \qquad \text{UF}_C = 0.30$ $\qquad \text{UF}_{SS} = 0.32$

House side: $\frac{\text{TD}}{\text{MH}} = \frac{6}{45} = 0.13$ $\qquad \text{UF}_{HS} = 0.02$

$\text{UF}_{TOT} = 0.34$

$$\text{S} = \frac{31{,}500 \times 0.34 \times \left(\frac{32.0}{31.5}\right) \times 0.70}{0.8 \times 48} = 198.3 \text{ ft} \quad \text{Use 200 ft oc.}$$

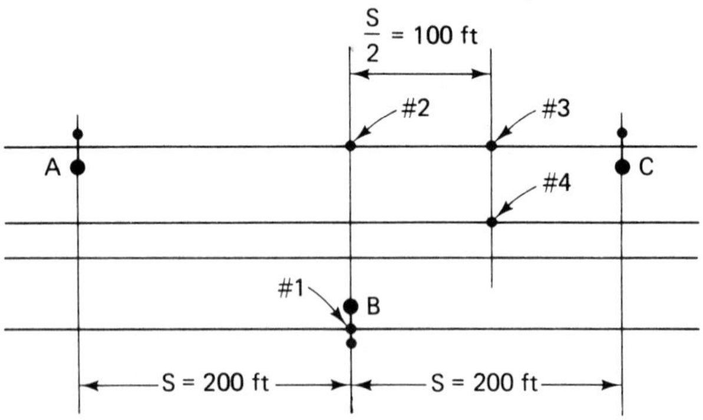

Part B: Uniformity Check

1. Find E_i at each point.

	$\frac{TD}{MH}$	$\frac{LD}{MH}$	E_i
Point 1			
Pole *A*	$\frac{54}{45} = 1.2$	$\frac{200}{45} = 4.44$	0.13
Pole *B**	$\frac{6}{45} = 0.13$	$\frac{0}{45} = 0$	4.00
Pole *C*	$\frac{54}{45} = 1.2$	$\frac{200}{45} = 4.44$	0.13
			$E_{pl_i} = 4.26$ fc

Point 2			
A^*	$\frac{6}{45} = 0.13$	$\frac{200}{45} = 4.44$	0.025
B	$\frac{54}{45} = 1.2$	$\frac{0}{45} = 0$	1.25
C^*	$\frac{6}{45} = 0.13$	$\frac{200}{45} = 4.44$	0.025
			$E_{p2_i} = 1.30$ fc
Point 3			
A^*	$\frac{6}{45} = 0.13$	$\frac{300}{45} = 6.67$	0
B	$\frac{54}{45} = 1.2$	$\frac{100}{45} = 2.22$	1.0
C^*	$\frac{6}{45} = 0.13$	$\frac{100}{45} = 2.22$	0.52
			$E_{p3_i} = 1.52$ fc
Point 4			
A	$\frac{18}{45} = 0.40$	$\frac{300}{45} = 6.67$	0.01
B	$\frac{30}{45} = 0.67$	$\frac{100}{45} = 2.22$	1.35
C	$\frac{18}{45} = 0.40$	$\frac{100}{45} = 2.22$	1.25
			$E_{p4_i} = 2.61$ fc

* indicates that distances and illuminance values are House side (HS).

$$E_{\min_i} = E_{p2_i} = 1.30 \text{ fc}$$

$$E_{\min} = E_{\min_i} \times \left(\frac{\text{test dist.}}{\text{act. dist.}}\right)^2 \times \text{LF} \times \text{LLF}$$

$$E_{\min} = 1.30 \times \left(\frac{30}{45}\right)^2 \times \left(\frac{32.0}{31.5}\right) \times 0.70 = 0.40 \text{ fc}$$

2. Check uniformity.

$$\text{criteria, } \frac{E_{\text{ave}}}{E_{\min}} \leqslant 3.0, \qquad \text{actual, } \frac{E_{\text{ave}}}{E_{\min}} = \frac{0.80}{0.40} = 2.00 < 3.00 \text{ (O.K.)}$$

Example Problem 2: Building Floodlighting

A light-white building located in a dark ambient surround is to be illuminated with ground-mounted floodlights set back 20 ft. Determine the actual CBU and check the uniformity. Use the floodlight photometric found in Figure 11-25 using old test data given.

$E_r = 5$ fc   area = $75 \times 240 = 18{,}000$ ft^2 (one side)

LLF = 0.75   assume CBU = 0.70

BL = 21,512 (photometrics from Figure 11-25)

Photometric Report Prepared for: ABC Corp.

Catalog Number: 5600 Date: 1975 Report no.: 8000

Reflector: Segmented aluminum, specular finish.

Aux. Reflector

Cover: Clear glass

Lamp data: LU 400			Test Results
Watts: 400	Fila. Type: Arc tube		Avg. Max. CP.: 59600
Volts: 120	Lumens: 42,000		Max. CP.: 93360
Bulb: E-18		Nom. Actual	Beam Lumens: 21512
Base: Mogal screw	Fila. Height:		Beam Efficiency: 51.2%
Service: General	Fila. Width:		Horiz. Beam Spread: 84°
Light Center Length: 5¾			Vert. Beam Spread: 55°

Test procedure and data form in accordance with I.E.S. and N.E.M.A. standards

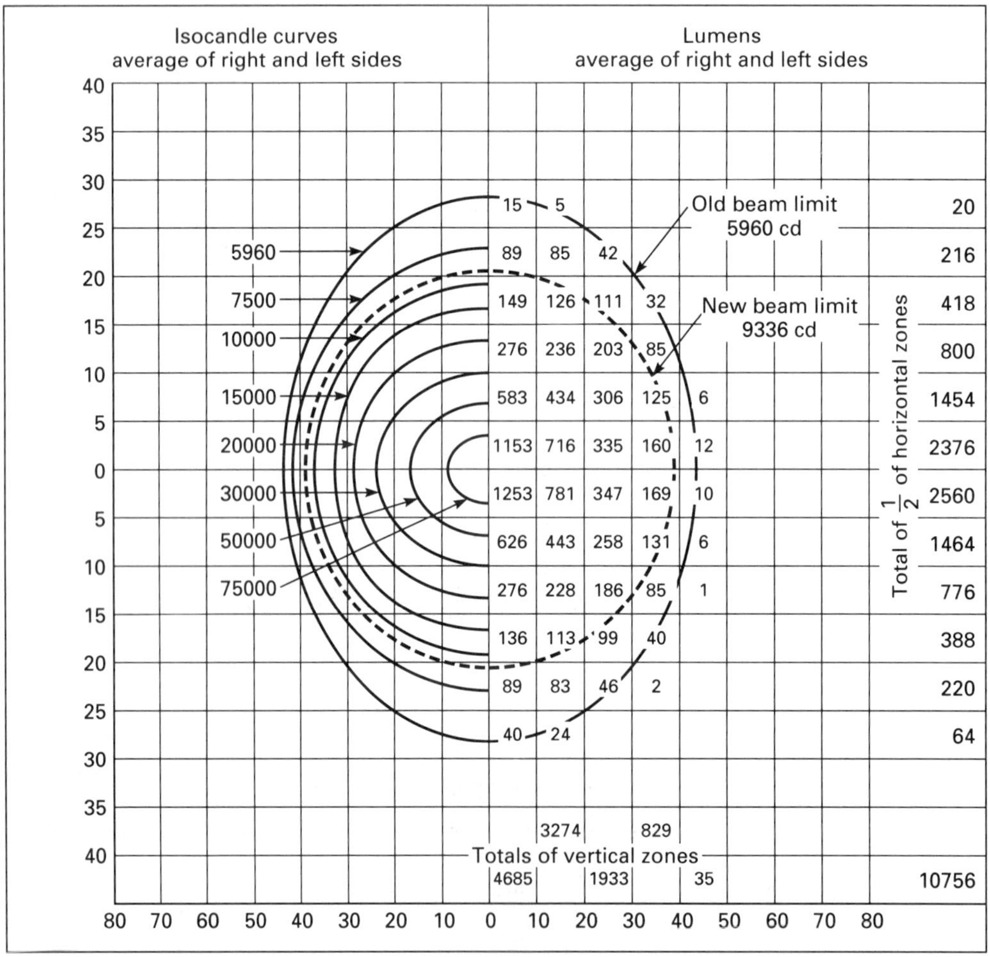

Fig. 11-25. Typical Floodlight Photometric Test Report.

Solution

1. Calculate the number of floodlights required.

$$N = \frac{E \times A}{\text{BL} \times \text{CBU} \times \text{LLF}}$$

$$= \frac{5 \times 18{,}000}{21{,}512 \times 0.70 \times 0.75} = 7.97, \qquad \text{use } 8$$

Two luminaires per ground position will be tried for a total of four ground positions. Four of the eight luminaires below will be aimed at points 40 ft above ground. The remaining four luminaires will require a lower-wattage lamp, wider distribution, and lower aiming point. For the purposes of this example, only the top four luminaires will be analyzed. The maintained illumination at points A, O, B, x, and y will be checked for uniformity. The illumination contribution from the remaining four lower-wattage luminaires should be added to the points. For this example, they will not be included.

2. Determine the actual CBU.

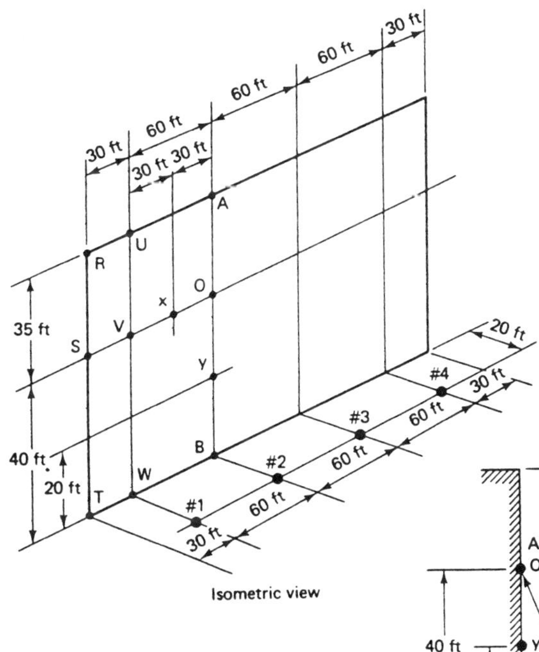

Isometric view

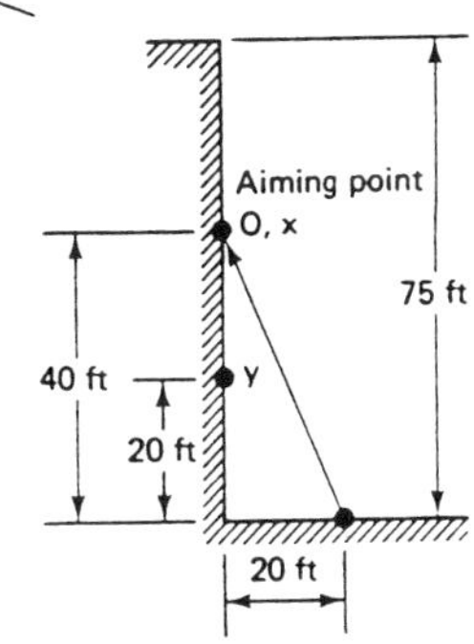

Cross section

Luminaire 1 (4)

1. Locate building on isodiagram (Fig. 11-26) for luminaire 1 (4).

$$\angle U1\ W = \arctan \frac{75}{20} = 75.1°$$

$$\text{aiming pt} < V1\ W \arctan \frac{40}{20} = 63.4°$$

$$\text{top} = 75.1° - 63.4° = 11.7°$$

$$\text{bottom} = 63.4°$$

$$\overline{U1} = \sqrt{75^2 + 20^2} = 77.6$$

$$\overline{V1} = \sqrt{40^2 + 20^2} = 44.7$$

Corner R from U:

$$\angle R1\ U = \arctan \frac{30}{77.6} = 21.1°$$

Edge S from V:

$$\angle S1\ V = \arctan \frac{30}{44.7} = 33.9°$$

Bottom corner T from W:

$$\angle T1\ W = \arctan \frac{30}{20} = 56.3°$$

2. Sum lumens on the building (columns from left to right). See Figure 11-26.

24	22	3274	4685	1933	829	35	
85	214	−5	−15	−42	−32		
104	335	−85	−89	−111	−29		
85	347	−126	−149	−124			
40	258	−156	−182				
2	186						
	99						
	46						
		2902	4250				
		× 2	× 2				sum
340	1506	5804	8500	1656	768	35 =	18,608

3. Calculate CBU for luminaire 1 (and 4).

$$\text{CBU} = \frac{18{,}608}{21{,}512} = 0.87$$

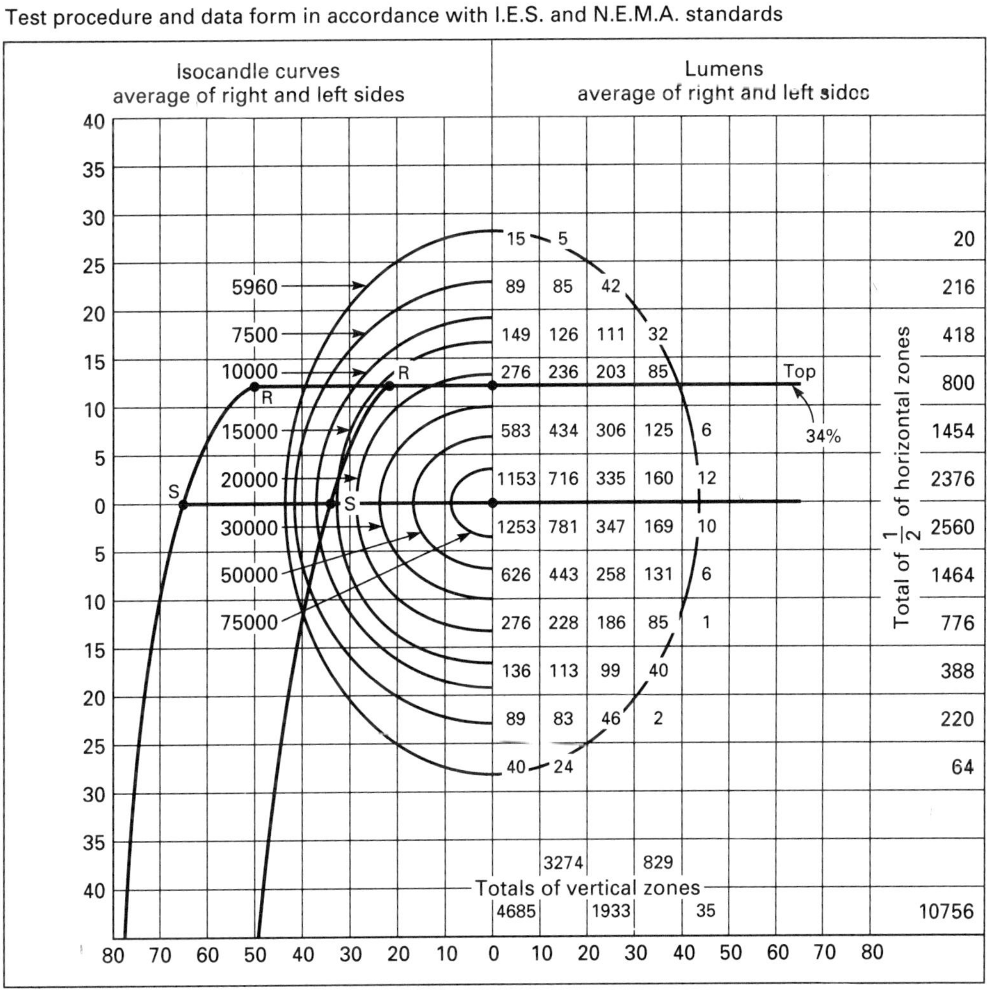

Figure 11-26. Beam Lumen Summation for Example Problem.

Luminaire 2

1. Locate building on isodiagram (Fig. 11-26) for luminaire 2 (3).
   Corner R from A:

$$\angle R2A = \arctan \frac{90}{77.6} = 49.2^\circ$$

   Edge S from O:

$$\angle S20 = \arctan \frac{90}{44.7} = 63.6^\circ$$

   Bottom corner T from B:

$$\angle T2B = \arctan \frac{90}{20} = 77.5^\circ$$

2. Sum lumens on the building (columns from left to right). See Figure 11-26.

4685	3274	1933	829	35				
−15	−5	−42	−32					
−89	−85	−111	−29					
−149	−126	−124						
−182	−156					½ sum		sum
4250	2902	1656	768	35	=	9611	× 2 =	19,222

3. Calculate CBU for luminaire 2 (and 3).

$$CBU = \frac{19{,}222}{21{,}512} = 0.89$$

   The **actual CBU** is

$$CBU = \frac{(2 \times 0.87) + (2 \times 0.89)}{4} = 0.88$$

$$\text{actual } CBU = 0.88 > \text{assumed } CBU = 0.70$$

Assumption close enough to continue example problem. (Actual CBU is high, which indicates that the solution is probably going to be spotty and nonuniform. A change in aiming point of luminaire distribution might be warranted.)

3. Uniformity Check (continued)

For luminaire 2, the contribution at points A, O, B, x, and y is as follows:

Point A

$$E_A = \frac{24{,}000}{(77.6)^2} \cos 75.1 = 1.02 \times 0.75 = 0.77 \text{ fc}$$

Point O

$$E_O = \frac{93{,}360}{(44.7)^2} \cos 63.4 = 20.92 \times 0.75 = 15.69 \text{ fc}$$

Point y

$$\angle y\,2B = \arctan \frac{20}{20} = 45^\circ$$

$$\overline{y\,2} = \sqrt{20^2 + 20^2} = 28.3$$

The angle from aiming point O down to point y is

$$\angle O2B - \angle y2B = 63.4^\circ - 45^\circ = 18.4^\circ$$

$$E_y = \frac{9000}{(28.3)^2} \cos 45^\circ = 7.95 \times 0.75 = 5.96 \text{ fc}$$

Point x

$$\angle x2O = \arctan \frac{30}{44.7} = 33.9^\circ$$

$$x2 = \sqrt{(30)^2 + (44.7)^2} = 53.8 \text{ ft}$$

$$E_x = \frac{14{,}000}{(53.8)^2} \cos 33.9^\circ \cos 63.4^\circ = 1.79 \times 0.75$$

$$= 1.34 \text{ fc}$$

Point B

The angle from the aiming point O down to the bottom of the building B is

$$\angle O2B = 63.4^\circ$$

The bottom of the building is off the diagram; therefore,

$$E_B = 0 \text{ fc}$$

This portion of the building would be illuminated by the four lower-wattage, wide-distribution luminaires (not analyzed in this example).

4. Conclusions

Concerning maintained illuminance produced by luminaire 2 (only) at points A, O, B, x, and y, it can be concluded that the uniformity ratio between the top (A) and the aiming point (O) exceeds 20 to 1, which is excessive. Examination of Figure 11-26 will show that luminaires 1 and 3 will not add significantly to point (A). Therefore, the aiming point should be moved up the building or a new luminaire should be selected. Once the layout has been modified, the maintained illuminance levels must be recalculated as shown.

TABLE 11-4. Luminance Factors, β

CIE Surface: R3

Luminance Factors: β(h, Vs)

Viewing Angle = 1°
vo = 89°

Vertical V_s	Horizontal, h 0	2	5	10	15	20	25	30	35	40	45	60	75	90	105	120	135	150	165	180
.00	.0294	.0294	.0294	.0294	.0294	.0294	.0294	.0294	.0294	.0294	.0294	.0294	.0294	.0294	.0294	.0294	.0294	.0294	.0294	.0294
14.04	.0240	.0240	.0244	.0249	.0253	.0258	.0262	.0271	.0280	.0294	.0298	.0303	.0308	.0308	.0312	.0317	.0321	.0321	.0326	.0357
26.57	.0194	.0194	.0199	.0199	.0199	.0199	.0204	.0217	.0235	.0262	.0276	.0289	.0298	.0308	.0317	.0326	.0339	.0339	.0344	.0480
36.87	.0140	.0136	.0136	.0145	.0149	.0149	.0149	.0158	.0176	.0204	.0222	.0244	.0267	.0285	.0303	.0321	.0339	.0353	.0353	.0697
45.00	.0100	.0100	.0100	.0100	.0100	.0100	.0100	.0104	.0118	.0140	.0158	.0181	.0204	.0226	.0249	.0276	.0326	.0352	.0362	.1023
51.34	.0078	.0077	.0077	.0077	.0074	.0071	.0070	.0073	.0083	.0104	.0118	.0136	.0154	.0176	.0208	.0244	.0298	.0348	.0357	.1464
56.31	.0062	.0061	.0060	.0060	.0060	.0058	.0057	.0060	.0072	.0078	.0086	.0100	.0117	.0145	.0176	.0217	.0267	.0326	.0348	.2068
60.26	.0047	.0046	.0045	.0045	.0046	.0045	.0044	.0045	.0051	.0062	.0070	.0079	.0089	.0104	.0127	.0172	.0231	.0303	.0335	.2775
63.43	.0038	.0037	.0036	.0036	.0035	.0034	.0034	.0034	.0039	.0048	.0054	.0062	.0071	.0082	.0100	.0136	.0190	.0280	.0321	.3644
68.20	.0025	.0024	.0024	.0024	.0024	.0024	.0023	.0022	.0023	.0025	.0064	.0068	.0044	.0054	.0065	.0086	.0127	.0222	.0280	.5641
71.57	.0017	.0017	.0016	.0016	.0015	.0015	.0014	.0015	.0015	.0018	.0020	.0023	.0025	.0031	.0038	.0053	.0085	.0163	.0235	.8000
74.05	.0013	.0012	.0012	.0011	.0011	.0009	.0009	.0009	.0009	.0013	.0015	.0016	.0019	.0022	.0025	.0035	.0060	.0122	.0194	1.0466
75.96	.0009	.0009	.0009	.0008	.0008	.0007	.0007	.0007	.0007	.0009	.0009	.0012	.0014	.0016	.0020	.0026	.0043	.0090	.0163	1.3317
77.47	.0008	.0008	.0007	.0007	.0006	.0005	.0004	.0004	.0005	.0007	.0008	.0009	.0009	.0012	.0015	.0020	.0031	.0073	.0136	1.5967
78.69	.0007	.0006	.0006	.0005	.0004	.0003	.0003	.0003	.0004	.0006	.0006	.0007	.0008	.0009	.0012	.0016	.0024	.0060	.0109	1.9223
79.70	.0000	.0000	.0000	.0000	.0000	.0000	.0000	.0000	.0000	.0000	.0005	.0006	.0006	.0007	.0009	.0014	.0018	.0047	.0094	2.2185
80.54	.0000	.0000	.0000	.0000	.0000	.0000	.0000	.0000	.0000	.0000	.0000	.0005	.0006	.0008	.0009	.0011	.0015	.0036	.0077	2.5432
81.25	0	0	0	0	0	0	0	0	0	0	0	0	.0004	.0005	.0006	.0008	.0011	.0030	.0068	2.9580
81.87	0	0	0	0	0	0	0	0	0	0	0	0	.0003	.0004	.0005	.0006	.0008	.0024	.0060	3.3587
82.41	0	0	0	0	0	0	0	0	0	0	0	0	0	.0003	.0004	.0005	.0007	.0021	.0053	3.7686
82.87	0	0	0	0	0	0	0	0	0	0	0	0	0	.0003	.0003	.0004	.0006	.0017	.0047	4.3495
83.29	0	0	0	0	0	0	0	0	0	0	0	0	0	.0002	.0003	.0003	.0005	.0015	.0042	4.8899
83.66	0	0	0	0	0	0	0	0	0	0	0	0	0	0	.0002	.0003	.0004	.0012	.0038	5.4205
83.99	0	0	0	0	0	0	0	0	0	0	0	0	0	0	.0002	.0003	.0003	.0010	.0034	6.0144
84.29	0	0	0	0	0	0	0	0	0	0	0	0	0	0	.0002	.0002	.0003	.0009	.0032	6.5977
84.56	0	0	0	0	0	0	0	0	0	0	0	0	0	0	.0001	.0002	.0003	.0008	.0029	7.2751
84.81	0	0	0	0	0	0	0	0	0	0	0	0	0	0	.0001	.0001	.0002	.0007	.0026	7.9504
85.03	0	0	0	0	0	0	0	0	0	0	0	0	0	0	0	.0001	.0002	.0006	.0024	8.6136
85.24	0	0	0	0	0	0	0	0	0	0	0	0	0	0	0	.0001	.0002	.0005	.0022	9.2539

12 advanced lighting calculations

Chapter 8 dealt with basic calculation procedures used in interior lighting. Chapter 11 discussed basic calculation procedures used in exterior lighting. The methods introduced in both of those chapters were variations on the point-by-point method and the lumen method. As was pointed out in Chapter 8, both methods are based on certain design assumptions that may introduce small errors in the results. The assumptions are introduced in the interest of simplicity. This chapter will present more sophisticated mathematical techniques for the lighting professional wishing to advance beyond the basic formulas introduced in Chapters 8 and 11. These procedures should be used in critical situations where the small errors introduced by the simpler methods cannot be tolerated. Understanding these procedures can also be helpful toward more fully comprehending the simpler methods that are derived from them such as the methods of Chapters 8 and 11. These mathematical techniques can and should be used to develop microcomputer software application programs and may aid in understanding commercially available lighting software.

The first topic of this chapter is a presentation of **vector algebra** and its application to exterior lighting calculations utilizing the point-by-point method. Next we introduce calculus-based methods for computing the direct component of illuminance from complex sources or for situations which violate the assumptions of the point-by-point method. Then the mathematics of light interreflecting within a space is discussed. This is known as luminous flux transfer theory. It is included because it forms the basis for the lumen method and allows a more complete understanding of the method. Finally, the theoretical and mathematical basis of the VCP method are considered. For the student not inclined to pursue the more complex mathematics involved, omitting this chapter from a course will not prevent a basic understanding of point-by-point or lumen method calculations.

VECTOR ALGEBRA

There are a number of exterior lighting applications where the point-by-point method is applied such as roadway lighting, area lighting, parking lot lighting, pedestrian lighting, and building floodlighting. Longhand calculations utilizing the point-by-point method to calculate illuminance at points on a surface is relatively simple but extremely time consuming if a large number of points are calculated. A computer program can perform the calculation much more rapidly. The development of a computer program to perform that simple calculation is complicated by the reported photometrics for a non-symmetrical, aimed luminaire. The identification of calculation points on the "house side" versus the "street side" (see Chapter 11) and the potential for horizontal or vertical rotation (area lighting or building floodlighting) of a luminaire makes the application of simple trigonometric relationship more cumbersome in a computer program. The application of vector algebra simplifies these complex interactions of photometric planes, calculation grids, and rotating planes. For a complete understanding of the basic principle of vector algebra, the reader should consult a math text or an engineering statics text. We present here a summary of the equations developed in vector algebra that are useful in the application of vector algebra to exterior point-by-point calculations.

Dot Product

The **dot product** of two vectors can be used to determine the angle between those two vectors (see Fig. 12-1).

$$\text{dot product} = \overline{V_1} \cdot \overline{V_2} = |\overline{V_1}|\ |\overline{V_2}| \cos \theta$$

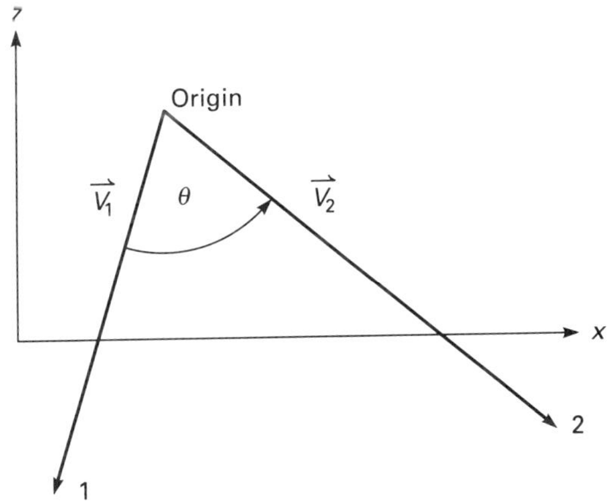

Figure 12-1. Dot Product Vectors and Terminology.

where $|\overline{V_1}|$ represents the length of vector $\overline{V_1}$.

$$DP = (V_{1i}V_{2i}) + (V_{1j}V_{2j}) + (V_{1k}V_{2k}) = |\overline{V_1}|\ |\overline{V_2}| \cos \theta$$

Therefore,

$$\cos \theta = \frac{(V_{1i}V_{2i}) + (V_{1j}V_{2j}) + (V_{1k}V_{2k})}{|\overline{V_1}|\ |\overline{V_2}|}$$

Cross Product

The cross product defines a vector that is perpendicular to a plane formed by two vectors (See Fig. 12-2.)

Vector $\overline{V_N}$ is a vector perpendicular to a plane formed by vectors $\overline{V_1}$ and $\overline{V_2}$.

$$|\overline{V_N}| = |\overline{V_1}|\ |\overline{V_2}| \sin \Gamma$$

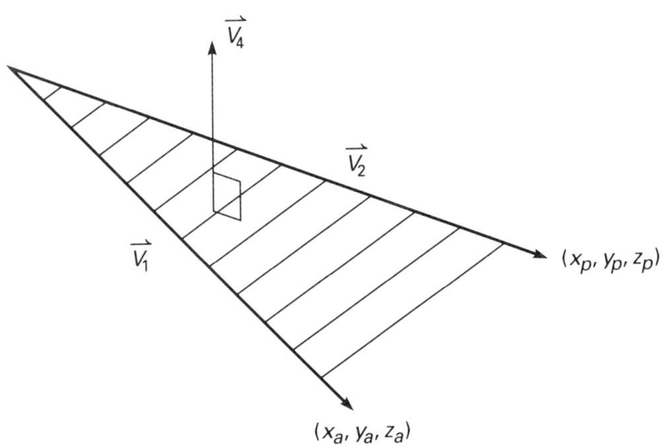

Figure 12-2 Cross Product Vectors and Terminology.

Direction Numbers

The letters u, v, and w, will be used to denote the **direction numbers** of a vector. The u, v, w direction numbers are the i, j, k components of a vector.

Application of Vector Algebra to Lighting Calculations

Photometric Angle, θ (Elevation Angle, Fig. 12-3)

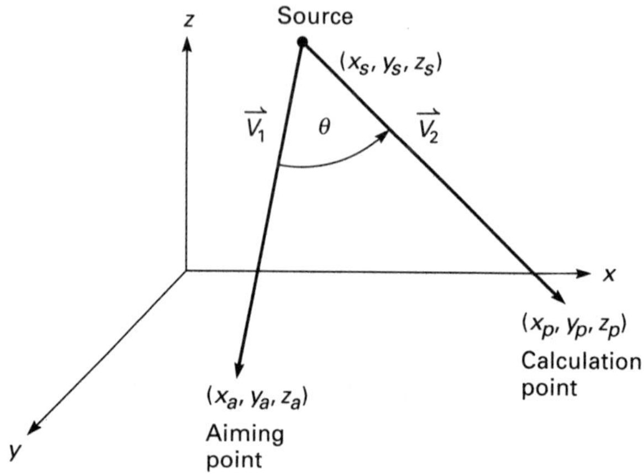

Figure 12-3. Elevation Angle Terminology and Vectors.

Calculation Vectors

Direction Numbers. The letters u_1, v_1, w_1, u_2, v_2, and w_2 are the direction numbers for the lines from the source to the aiming point and the source to the calculation point).

$\overline{V_1}$—Direction Numbers

$$u_1 = V_{1i} = (x_a - x_s) \qquad \text{source to aim, } x$$
$$v_1 = V_{1j} = (y_a - y_s) \qquad \text{source to aim, } y$$
$$w_1 = V_{1k} = (z_a - z_s) \qquad \text{source to aim, } z$$

$\overline{V_2}$—Direction Numbers

$$u_2 = V_{2i} = (x_p - x_s) \qquad \text{source to point, } x$$
$$v_2 = V_{2j} = (y_p - y_s) \qquad \text{source to point, } y$$
$$w_2 = V_{2k} = (z_p - z_s) \qquad \text{source to point, } z$$

Length of Vectors $\overline{V_1}$ *and* $\overline{V_2}$

$$|\overline{V_1}| = \sqrt{u_1^2 + v_1^2 + w_1^2}$$
$$|\overline{V_2}| = \sqrt{u_2^2 + v_2^2 + w_2^2}$$

Dot Product

$$\cos\theta = \frac{(u_1u_2) + (v_1v_2) + (w_1w_2)}{\sqrt{u_1^2 + v_1^2 + w_1^2}\sqrt{u_2^2 + v_2^2 + w_2^2}}$$

Elevation Angle, θ

$$\theta = \arccos\left(\frac{(u_1u_2) + (v_1v_2) + (w_1w_2)}{\sqrt{u_1^2 + v_1^2 + w_1^2}\sqrt{u_2^2 + v_2^2 + w_2^2}}\right) \tag{12.1}$$

Photometric Angle, α

Photometric Plane of Symmetry. The reference vector $\overline{V_R}$ and the aiming axis vector $\overline{V_1}$ define the **photometric plane of symmetry.**

Roadway Lighting

Reference Vector. We define a reference vector $\overline{V_R}$ as a unit vector that is coplanar with the photometric plane of symmetry and perpendicular to the aiming axis. (See Fig. 12-4.) $\overline{V_R}$ coincides with the $\alpha = 0°$ photometric angle.

1. No Luminaire Tilt. If the luminaire face is parallel with the roadway (no tilt), the reference vector is coplanar with the $\alpha = 0°$ photometric plane. (See Fig. 12-5.)

2. Luminaire with Tilt. If the lens of the roadway luminaire is not parallel with the ground, the tilt angle must be specified. For roadway luminaires, the tilt angle

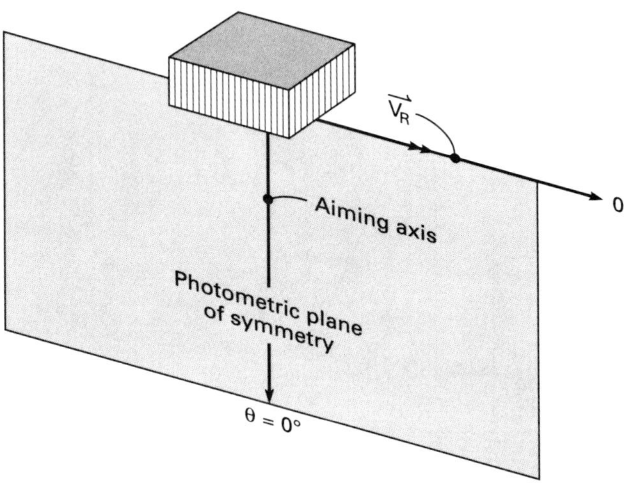

Figure 12-4. Roadway: Reference Vector and Plane of Symmetry.

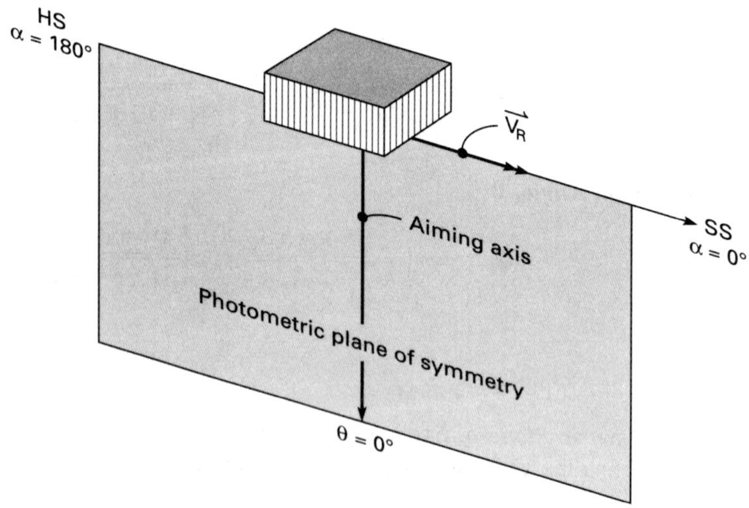

is measured from a horizontal line parallel with the ground up to the reference vector. (See Fig. 12-6.)

3. Luminaire Orientation (Lateral Rotation). **Xangle** is the rotation angle of the roadway luminaire with the $\alpha = 0°$ plane (terminal side) pointing in a $+x$ direction (initial side), which results in an xangle $= 0°$.

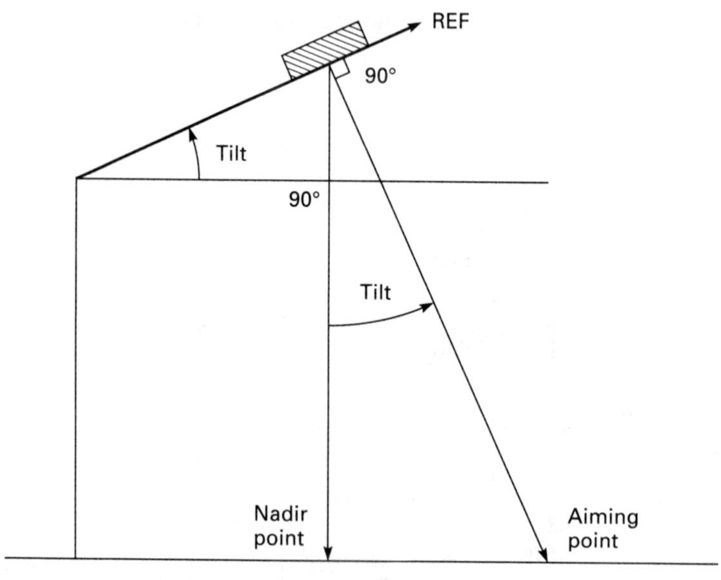

Figure 12-6. Tilt and the Plane of Symmetry.

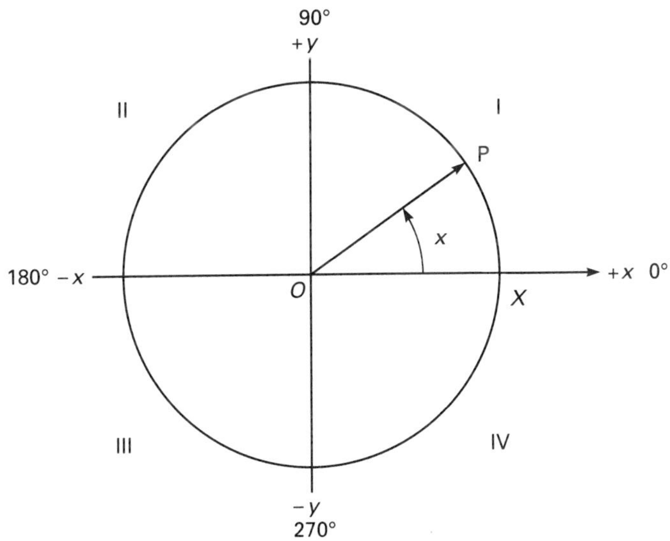

Figure 12-7. Trigonometric Relationship by Quadrant.

In Figure 12-7, the line OX is called the initial side of the angle, and the line OP (REF vector) is called the terminal side of the angle (xangle). The distance of any point P from the x-x axis is called the ordinate and the distance from the y-y axis is called the abscissa.

Signs of Coordinates in the Four Quadrants

Quadrant I: abscissa = (+), ordinate = (+), all functions (+)
Quadrant II: abscissa = (−), ordinate = (+), sin (+)
Quadrant III: abscissa = (−), ordinate = (−), tan (+)
Quadrant IV: abscissa = (+), ordinate = (−), cos (+)

For a unit circle with a radius = 1,

$$\sin x = \frac{\text{ordinate}}{\text{radius}} = \text{ordinate} = y \text{ distance or } y \text{ component}$$

$$\cos x = \frac{\text{abscissa}}{\text{radius}} = \text{abscissa} = x \text{ distance or } x \text{ component}$$

Example:

For a total of four luminaires on a single pole, the xangles and direction numbers would be as shown in Figure 12-8.

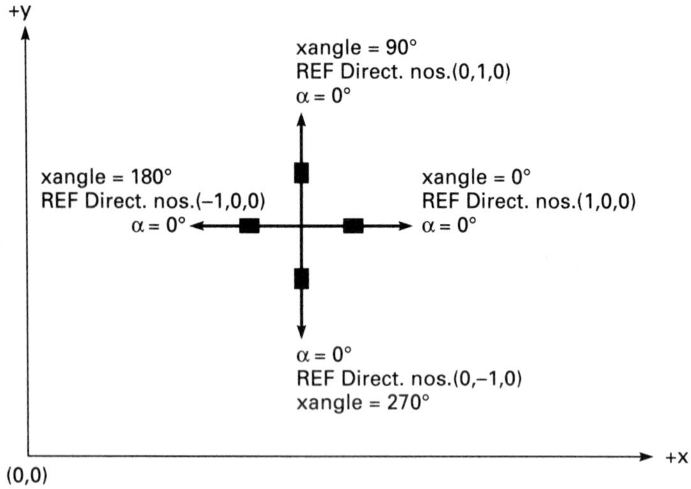

Figure 12-8 Xangle and Direction Numbers Defined.

$\overline{V_R}$—Direction Numbers

$$\text{REFA} = x \text{ component of } \overline{V_R} = \text{abscissa} = \cos(\text{xangle})$$

$$\text{REFB} = y \text{ component of } \overline{V_R} = \text{ordinate} = \sin(\text{xangle})$$

$$\text{REFC} = z \text{ component of } \overline{V_R} = \sin(\text{tilt})$$

$$V_{Ri} = u_r = \text{REFA} = \cos(\text{xangle})$$

$$V_{Rj} = v_r = \text{REFB} = \sin(\text{xangle})$$

$$V_{Rk} = w_r = \text{REFC} = \sin(\text{tilt})$$

Floodlighting

Reference Vector. The floodlight reference vector $\overline{V_R}$ is defined as a vector straight up along the face of the unit which is perpendicular to a horizontal aiming axis vector, $\overline{V_1}$ (Fig. 12-9). The reference vector $\overline{V_R}$ and the aiming axis vector $\overline{V_1}$ define the **photometric plane of symmetry.**

With the reference vector $\overline{V_R}$ pointing straight up and the aiming axis vector $\overline{V_1}$ pointing horizontal, the floodlight is defined as *no tilt*. The tilt angle for a floodlight is measured from a straight up, vertical line to the reference vector. If the aiming axis vector $\overline{V_1}$ moves above the horizontal reference line (up), the reference vector $\overline{V_R}$ will have a *backward tilt*. If the aiming axis vector $\overline{V_1}$ moves below the horizontal reference line (down), the reference vector $\overline{V_R}$ will have a *forward tilt*.

1. *Lateral Rotation*: *Xangle* is a horizontal angle measured counterclockwise (CCW) from the $+x$ reference axis to the vertical plane containing the aiming vector, $\overline{V_1}$.

2. *Back Tilt of the Floodlight*: Ground-mounted floodlighting equipment will nor-

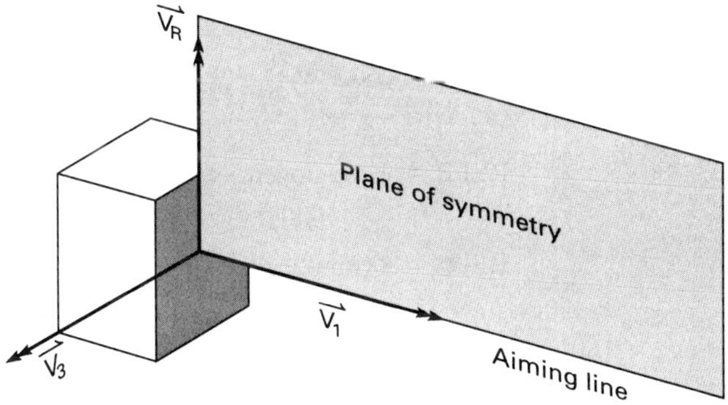

Figure 12-9. Floodlighting: Reference Vector and Plane of Symmetry.

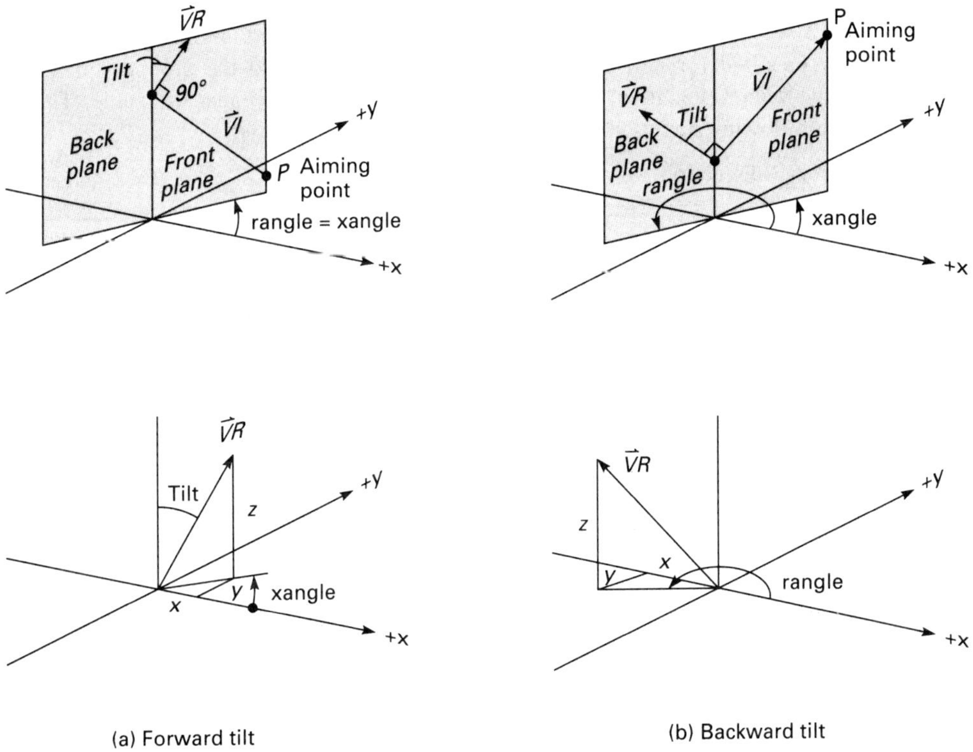

Figure 12-10 Backward and Forward Tilt with Xangle.

mally be tilted backward to move the aiming axis up the building surface. For a backward tilt the vector $\overline{V_R}$ is in the back plane; hence, "rangle" is greater than 180° as shown in Figure 12-10. This results in the following direction numbers:

$\overline{V_R}$—Direction Numbers

$$\begin{aligned}\text{REFA} &= x \text{ component of } \overline{V_R} = \text{abscissa}\\ &= \cos\ (\text{rangle})\end{aligned}$$

$$\begin{aligned}\text{REFB} &= y \text{ component of } \overline{V_R} = \text{ordinate}\\ &= \sin\ (\text{rangle})\end{aligned}$$

$$\begin{aligned}\text{REFC} &= z \text{ component of } \overline{V_R}\\ &= \cos\ (\text{tilt})\end{aligned}$$

where range = 180° + xangle.

$$V_{Ri} = u_r = \text{REFA} = \cos(180° + \text{xangle})$$

$$V_{Rj} = v_r = \text{REFB} = \sin(180° + \text{xangle})$$

$$V_{Rk} = w_r = \text{REFC} = \cos(\text{tilt})$$

3. Forward Tilt of the Floodlight: Pole-mounted floodlighting equipment will normally be tilted forward to move the aiming axis down to the ground. For a forward tilt, the vector, $\overline{V_R}$, is in the front plane, hence, rangle is equal to xangle. This results in the following direction numbers:

$\overline{V_R}$—Direction Numbers

$$\begin{aligned}\text{REFA} &= x \text{ component of } \overline{V_R} = \text{abscissa}\\ &= \cos\ (\text{xangle})\end{aligned}$$

$$\begin{aligned}\text{REFB} &= y \text{ component of } \overline{V_R} = \text{ordinate}\\ &= \sin\ (\text{xangle})\end{aligned}$$

$$\begin{aligned}\text{REFC} &= z \text{ component of } \overline{V_R}\\ &= \cos\ (\text{tilt})\end{aligned}$$

$$V_{Ri} = u_r = \text{REFA} = \cos\ (\text{xangle})$$

$$V_{Rj} = v_r = \text{REFB} = \sin\ (\text{xangle})$$

$$V_{Rk} = w_r = \text{REFC} = \cos\ (\text{tilt})$$

Normal Vector to the Plane of Symmetry. The cross product of the two lines ($\overline{V_1}$ and $\overline{V_R}$) results in the direction numbers (u_3, v_3, w_3) of a vector $\overline{V_3}$ normal to the plane of symmetry.

$$\overline{V_3} = \overline{V_1} \times \overline{V_R}$$

Right-Hand Rule. The perpendicular direction of vector $\overline{V_3}$ is determined by applying the right-hand rule (Fig. 12-9). The vector $\overline{V_3}$, is normal to the plane found by crossing $\overline{V_1}$ with $\overline{V_R}$. The fingers of the right-hand curl in the direction from $\overline{V_1}$ to $\overline{V_R}$ and the extended thumb points in the direction of the vector $\overline{V_3}$ (normal to the plane formed by $\overline{V_1}$ and $\overline{V_R}$).

Cross Product

$$\begin{vmatrix} i & j & k \\ V_{1i} & V_{1j} & V_{1k} \\ V_{Ri} & V_{Rj} & V_{Rk} \end{vmatrix}$$

$$CP = \hat{\imath}\begin{vmatrix} V_{1j} & V_{1k} \\ V_{Rj} & V_{Rk} \end{vmatrix} - \hat{\jmath}\begin{vmatrix} V_{1i} & V_{1k} \\ V_{Ri} & V_{Rk} \end{vmatrix} + \hat{k}\begin{vmatrix} V_{1i} & V_{1j} \\ V_{Ri} & V_{Rj} \end{vmatrix}$$

$$CP = \hat{\imath}\,[V_{1j}V_{Rk} - V_{1k}V_{Rj}] - \hat{\jmath}\,[V_{1i}V_{Rk} - V_{1k}V_{Ri}] + \hat{k}\,[V_{1i}V_{Rj} - V_{1j}V_{Ri}]$$

$\overline{V_3}$—Direction Numbers

$$u_3 = V_{3i} = (V_{1j}V_{Rk}) - (V_{1k}V_{Rj}) = \text{PLNSYM } i$$

$$v_3 = V_{3j} = -[(V_{1i}V_{Rk}) - (V_{1k}V_{Ri})] = (V_{1k}V_{Ri}) - (V_{1i}V_{Rk}) = \text{PLNSYM } j$$

$$w_3 = V_{3k} = (V_{1i}V_{Rj}) - (V_{1j}V_{Ri}) = \text{PLNSYM } k$$

$$u_3 = (v_1w_r) - (w_1v_r)$$

$$v_3 = (w_1u_r) - (u_lw_r)$$

$$w_3 = (u_1v_r) - (v_1u_r)$$

Plane of Rotation

The vector from the source to the aiming point $\overline{V_1}$ and the vector from the source to the point of calculation $\overline{V_2}$ defines the **plane of rotation** (Fig. 12-11).

Perpendicular to the Plane of Rotation. The cross product of the two vectors $\overline{V_1}$ and $\overline{V_2}$ results in the direction numbers (u_4, v_4, w_4) of a vector ($\overline{V_4}$) normal to the plane of rotation.

$$\overline{V_4} = \overline{V_1} \times \overline{V_2}$$

Right-Hand Rule. The perpendicular direction of vector V_4 is determined by applying the right-hand rule. The vector $\overline{V_4}$ is normal to the plane found by crossing $\overline{V_1}$ with $\overline{V_2}$. The fingers of the right-hand curl in the direction from $\overline{V_1}$ to $\overline{V_2}$ and the extended thumb points in the direction of the vector $\overline{V_4}$.

Cross Product. The vector, $\overline{V_4}$, is the cross product of $\overline{V_1}$ with $\overline{V_2}$ and it defines a perpendicular vector to the plane of rotation.

$$\begin{vmatrix} i & j & k \\ V_{1i} & V_{1j} & V_{1k} \\ V_{2i} & V_{2j} & V_{2k} \end{vmatrix}$$

$$CP = \hat{\imath}\begin{vmatrix} V_{1j} & V_{1k} \\ V_{2j} & V_{2k} \end{vmatrix} - \hat{\jmath}\begin{vmatrix} V_{1i} & V_{1k} \\ V_{2i} & V_{2k} \end{vmatrix} + \hat{k}\begin{vmatrix} V_{1i} & V_{1j} \\ V_{2i} & V_{2j} \end{vmatrix}$$

$$CP = \hat{\imath}\,[V_{1j}V_{2k} - V_{1k}V_{2j}] - \hat{\jmath}\,[V_{1i}V_{2k} - V_{1k}V_{2i}] + \hat{k}\,[V_{1i}V_{2j} - V_{1j}V_{2i}]$$

$\overline{V_4}$—Direction Numbers

$$u_4 = V_{4i} = (V_{1j}V_{2k}) - (V_{1k}V_{2j}) = \text{PLNROT } i$$

$$\begin{aligned} v_4 = V_{4j} &= -[(V_{1i}V_{2k}) - (V_{1k}V_{2i})] \\ &= (V_{1k}V_{2i}) - (V_{1i}V_{2k}) = \text{PLNROT } j \end{aligned}$$

$$w_4 = V_{4k} = (V_{1i}V_{2j}) - (V_{1j}V_{2i}) = \text{PLNROT } k$$

$$u_4 = (v_1w_2) - (w_1v_2)$$

$$v_4 = (w_1u_2) - (u_1w_2)$$

$$w_4 = (u_1v_2) - (v_1u_2)$$

Photometric Angle, α. The dot product of the two vectors $\overline{V_3}$ and $\overline{V_4}$, which are normal to the two planes is used to determine the angle α (Fig. 12-11) between the photometric plane of symmetry and the plane of rotation. The angles θ and α are used to interpolate the spherical coordinate luminous intensity value from the source to the calculation point.

The vector $\overline{V_3}$ is the cross product of $\overline{V_1}$ with $\overline{V_R}$, and it defines a perpendicular vector to the plane of symmetry.

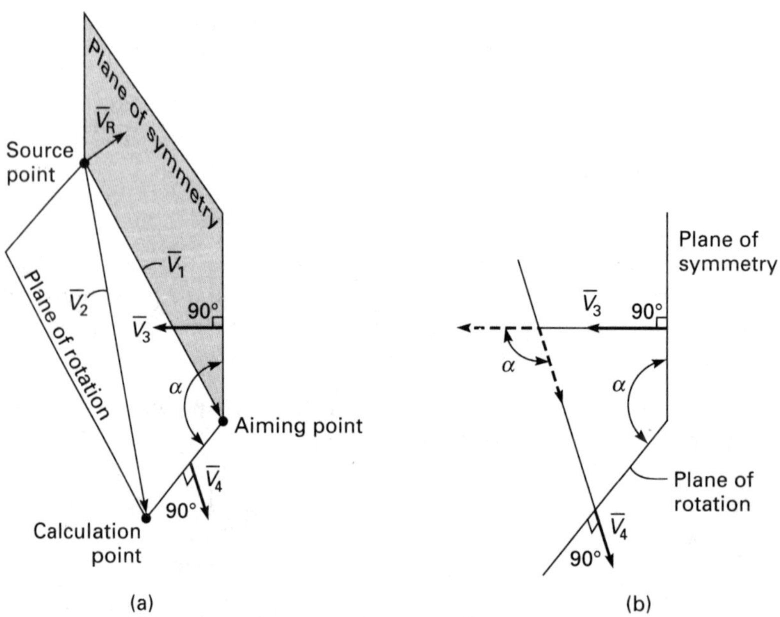

Figure 12-11. Rotational Angle, α.

$\overline{V_3}$—Direction Number (see above)

$$u_3 - V_{3i} = (V_{1j}V_{Rk}) - (V_{1k}V_{Rj}) = \text{PLNSYM } i$$
$$v_3 = V_{3j} = (V_{1k}V_{Ri}) - (V_{1i}V_{Rk}) = \text{PLNSYM } j$$
$$w_3 = V_{3k} = (V_{1i}V_{Rj}) - (V_{1j}V_{Ri}) = \text{PLNSYM } k$$

$$u_3 = (v_1 w_r) - (w_1 v_r)$$
$$v_3 = (w_1 u_r) - (u_1 w_r)$$
$$w_3 = (u_1 v_r) - (v_1 u_r)$$

Length of the Vectors $\overline{V_3}$ *and* $\overline{V_4}$:

$$|\overline{V_3}| = \sqrt{u_3^2 + v_3^2 + w_3^2}$$
$$|\overline{V_4}| = \sqrt{u_4^2 + v_4^2 + w_4^2}$$

Dot Product

$$\text{dot product} = V_3 \bullet V_4 = |\overline{V_3}|\,|\overline{V_4}| \cos \alpha$$
$$DP = (V_{3i}V_{4i}) + (V_{3j}V_{4j}) + (V_{3k}V_{4k}) = |\overline{V_3}|\,|\overline{V_4}| \cos \alpha$$

Therefore,

$$\cos \alpha = \frac{(u_3 u_4) + (v_3 v_4) + (w_3 w_4)}{\sqrt{u_3^2 + v_3^2 + w_3^2}\sqrt{u_4^2 + v_4^2 + w_4^2}}$$

Photometric Angle, α (Fig. 12-11)

$$\alpha = \arccos\left[\frac{(u_3 u_4) + (v_3 v_4) + (w_3 w_4)}{\sqrt{u_3^2 + v_3^2 + w_3^2}\sqrt{u_4^2 + v_4^2 + w_4^2}}\right] \qquad (12.2)$$

Illuminance Component Perpendicular to a Surface

Calculation Angle, Φ. The angle Φ (Fig. 12-12) is the angle between the vector $\overline{V_N}$ and the vector $\overline{V_6}$ which is a line from the calculation point to the source.

Normal to the Calculation Surface. The normal vector $\overline{V_N}$ is defined as a vector normal to the calculation surface and pointing *away* from the surface at the point of calculation.

Since vectors are defined as diverging from the origin of two vectors, the vector is *from* the calculation point *to* the source.

Calc. Value	Calc. Plane	Direction Numbers
E_h	XY	(0, 0, 1)
E_v	XZ	(0, 1, 0) or (0, −1, 0)
E_v	YZ	(1, 0, 0) or (−1, 0, 0)

For illuminance (Fig. 12-12) on a horizontal surface in the *XY* plane:

Normal Vector $\overline{V_N}$—Direction Numbers

$$u_5 = V_{Ni} = 0$$
$$v_5 = V_{Nj} = 0$$
$$w_5 = V_{Nk} = 1$$

Calculation Vector. The calculation vector is from the calculation point to the source which is the *negative* of $\overline{V_2}$.

$\overline{V_6}$—Direction Numbers. (From the calculation point to the source)

$$u_6 = V_{6i} = (x_s - x_p) = -(x_p - x_s) = -u_2$$
$$v_6 = V_{6j} = (y_s - y_p) = -(y_p - y_s) = -v_2$$
$$w_6 = V_{6k} = (z_s - z_p) = -(z_p - z_s) = -w_2$$

Length of the Vectors $\overline{V_N}$ *and* $\overline{V_6}$

$$|\overline{V_N}| = \sqrt{u_5^2 + v_5^2 + w_5^2}$$
$$|\overline{V_6}| = \sqrt{u_6^2 + v_6^2 + w_6^2}$$

Dot Product. Dot product $= \overline{V_N} \bullet \overline{V_6} = |\overline{V_N}|\,|\overline{V_6}| \cos \Phi$

$$DP = (V_{Ni}V_{6i}) + (V_{Nj}V_{6j}) + (V_{Nk}V_{6k}) + |\overline{V_N}|\,|\overline{V_6}| \cos \Phi$$

Therefore,

$$\cos \Phi = \frac{(V_{Ni}V_{6i}) + (V_{Nj}V_{6j}) + (V_{Nk}V_{6k})}{|\overline{V_N}|\,|\overline{V_6}|}$$

$$\cos \Phi = \frac{(u_5u_6) + (v_5v_6) + (w_5w_6)}{\sqrt{u_5^2 + v_5^2 + w_5^2}\sqrt{u_6^2 + v_6^2 + w_6^2}}$$

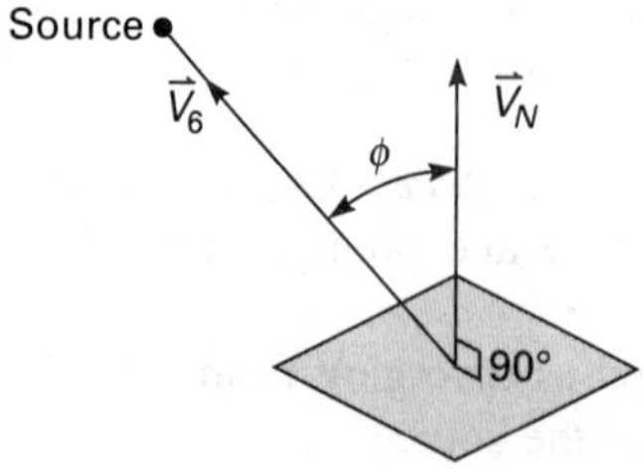

Figure 12-12. Illuminance Component Perpendicular to the Horizontal Roadway, Angle Φ.

***Perpendicular Angle* Φ**

$$\Phi = \arccos\left[\frac{(u_5u_6) + (v_5v_6) + (w_5w_6)}{\sqrt{u_5^2 + v_5^2 + w_5^2}\sqrt{u_6^2 + v_6^2 + w_6^2}}\right] \tag{12.3}$$

Distance,* D, *from the Calculation Point to the Source

$$D^2 = |\overline{V_2}|^2 = u_2^2 + v_2^2 + w_2^2$$
$$D^2 = (x_p - x_s)^2 + (y_p - y_s)^2 + (z_p - z_s)^2$$

Calculation of Illuminance on a Horizontal Surface

The formula (Cosine Cubed Law of Illuminance; see Chapter 8) for calculating the illuminance at a point on a horizontal surface becomes

$$E_h = \frac{I_{(\theta,\alpha)}}{H^2} \times \cos^3 \Phi \tag{12.4}$$

Calculation of Roadway Luminance

The calculation of roadway luminance which is described in Chapter 11 requires the selection of a luminance factor and the calculation of the "normal" component of illuminance.

$$\beta_{(h,vs)} = r_{(h,vs)} = q \times \cos^3 v_s \times 10^4$$

therefore

$$L_p = \beta_{(h,vs)} \times \frac{I_{(\theta,\alpha)}}{H^2}$$

The luminance design method section of Chapter 11 explains the following formula for calculating the luminance at a point.

$$L_p = \sum_1^n \frac{\beta_{(h,vs)} \times I_{(\theta s,\alpha)}}{H^2} \tag{12.5}$$

where h = the horizontal angle from the β reference plane to the plane containing the calculation line.

v_s = the vertical angle between a perpendicular to the roadway and the vector from the calculation point to the source. (See Fig. 1-8, p. 12, luminance factor for a diagram showing h and v_s.)

H = the mounting height or the vertical distance between the luminaire and the roadway surface.

The calculation of the angles θ_s and α were given in the roadway illuminance calculation earlier in this chapter. The only remaining unknown value in equation 12.5 is the β (Fig. 12-13) value which is a function of v_s and h.

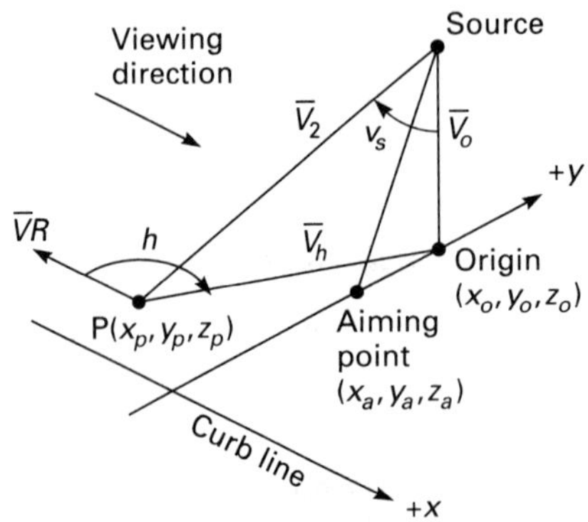

Figure 12-13. Geometrical Relationships for Finding β.

Photometric Angle, *h* (Horizontal Angle)

$\overline{V}_h$—Direction Numbers

$$u_h = V_{hi} = (x_o - x_p) \quad \text{point to origin, } x$$

$$v_h = V_{hj} = (y_o - y_p) \quad \text{point to origin, } y$$

$$w_h = V_{hk} = (z_o - z_p) \quad \text{point to origin, } z$$

$\overline{V}_R$ (Beta)—Direction Numbers

$$u_b = V_{Ri} = -1 \quad \text{beta reference, } x$$

$$v_b = V_{Rj} = 0 \quad \text{beta reference, } y$$

$$w_b = V_{Rk} = 0 \quad \text{beta reference, } z$$

Length of Vectors $\overline{V}_h$ and $\overline{V}_R$

$$|\overline{V}_h| = \sqrt{u_h^2 + v_h^2 + w_h^2}$$

$$|\overline{V}_R| = \sqrt{u_b^2 + v_b^2 + w_b^2}$$

Dot Product

$$\cos h = \frac{(u_h u_b) + (v_h v_b) + (w_h w_b)}{\sqrt{u_h^2 + v_h^2 + w_h^2}\sqrt{u_b^2 + v_b^2 + w_b^2}}$$

$$h = \arccos\left[\frac{(u_h u_b) + (v_h v_b) + (w_h w_b)}{\sqrt{u_h^2 + v_h^2 + w_h^2}\sqrt{u_b^2 + v_b^2 + w_b^2}}\right]$$

Photometric Angle, v_s (Vertical Angle)

$\overline{V}_o$—Direction Numbers

$$u_o = V_{oi} = (x_o - x_s) \quad \text{source to origin, } x$$

$$v_o = V_{oj} = (y_o - y_s) \quad \text{source to origin, } y$$

$$w_o = V_{ok} = (z_o - z_s) \quad \text{source to origin, } z$$

$\overline{V_2}$—Direction Numbers (see above)

$$u_2 = V_{2i} = (x_p - x_s) \qquad \text{source to point, } x$$
$$v_2 = V_{2j} = (y_p - y_s) \qquad \text{source to point, } y$$
$$w_2 = V_{2k} = (z_p - z_s) \qquad \text{source to point, } z$$

Length of Vectors $\overline{V_o}$ and $\overline{V_2}$

$$|\overline{V_o}| = \sqrt{u_o^2 + v_o^2 + w_o^2}$$
$$|\overline{V_2}| = \sqrt{u_2^2 + v_2^2 + w_2^2}$$

Dot Product

$$\cos v_s = \frac{(u_o u_2) + (v_o v_2) + (w_o w_2)}{\sqrt{u_o^2 + v_o^2 + w_o^2}\sqrt{u_2^2 + v_2^2 + w_2^2}}$$

$$v_s = \arccos\left[\frac{(u_o u_2) + (v_o v_2) + (w_o w_2)}{\sqrt{u_o^2 + v_o^2 + w_o^2}\sqrt{u_2^2 + v_2^2 + w_2^2}}\right]$$

ILLUMINANCE AT A POINT—GENERALIZED PROCEDURE

In Chapter 8 we saw that the Inverse-Square Cosine Law of Illumination (ISL) could be used in many situations to calculate the direct component of illuminance at a point. One restriction on using the ISL is that the source must be far enough away that it approximates a point source. In cases where this condition is not met (a very large source, or a calculation point close to an area source such as a fluorescent troffer), we can return to fundamental principles and develop a calculus-based method for determining the exact value of the direct component of illuminance.

Consider the source of arbitrary three-dimensional shape shown in Figure 12-14. If the source is far enough away from point P on the receiving surface, then the illuminance at point P can be calculated using the ISL:

$$E_p = \frac{I \cos \theta}{R^2} \tag{12.6}$$

If the source cannot be considered a point source, it could be discretized into many point–source–sized segments and the approximate illuminance at P found by summing all the illuminances (found using the ISL) from the individual segments. The smaller the discretizations, the more accurate would be the approximation. In the limit, we take an infinite number of discrete elements of differential area, dA, and sum the illuminances from these elements, in other words we integrate over the surface of the source. The illuminance at P due to one of these discrete elements would be

$$dE_p = \frac{dI \cos \theta}{R^2} \tag{12.7}$$

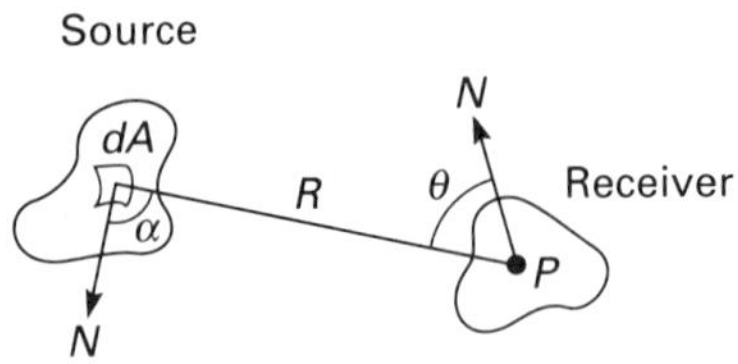

Figure 12-14. Arbitrary Source and Receiver with Normal Vectors.

Recalling that intensity is equal to luminance times projected area we have

$$dI = LdA \cos \alpha \tag{12.8}$$

where L is the luminance normal to the differential element. The $\cos \alpha$ term accounts for the fact that the element is "viewed" obliquely from P, and therefore the area of the element appears smaller by a factor of $\cos \alpha$.

Inserting equation 12.8 into equation 12.7 yields

$$dE_p = \frac{L \cos \alpha \cos \theta \, dA}{R^2} \tag{12.9}$$

Integrating this expression over the entire source gives us

$$E_p = \int \frac{L \cos \alpha \cos \theta \, dA}{R^2} \tag{12.10}$$

Equation 12.10 is perfectly general. It can be applied to any size or shape of source, at any distance from the calculation point, and for any distribution of luminance across the source. It remains to put the integrand into an integrable form appropriate for the specific application considered. A few cases are considered here to illustrate the general procedure.

Hemisphere

Consider a point on a flat plane covered by a luminous hemisphere. This could represent, for example, the sky dome. In Chapter 13 we find that the luminance of the overcast sky can be approximated by the following:

$$L_p = \frac{L_z(1 + 2 \cos \theta)}{3} \tag{12.11}$$

where L_p is the luminance of the sky at a point and L_z is the luminance at the zenith, or directly overhead. Zenith luminance is a constant for a given calculation.

From this expression we see that luminance is independent of which direction around the dome we face; it depends only on the angle from the zenith, θ. Therefore, any differential band around the hemisphere of height dh will have the property of constant luminance around the band. To apply the general method for calculating illuminance at a point, we can discretize the sky dome into differential bands and integrate. The situation is depicted in Figure 12-15.

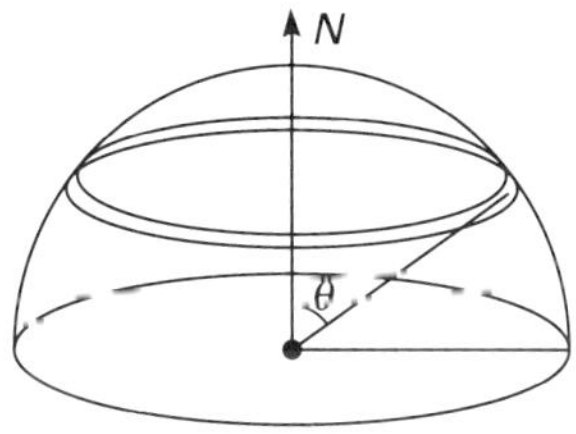

Figure 12-15. A Luminous Hemisphere.

Since the vector normal to the hemisphere is collinear with the line from the hemisphere to the calculation point P, angle α is always zero. Thus in equation 12.10, the cos α term is unity. This gives

$$dE_{oc} = \frac{L_p \cos\theta \, dA}{R^2} \tag{12.12}$$

where E_{oc} is the illuminance at a point on the horizontal plane under an overcast sky.

Next an expression for dA, the area of the differential circular band must be found. Refer to Figure 12-16. The area of the band is its circumference times its height, dh. The circumference is given by

$$C = 2\pi r$$

but,

$$r = R \sin\theta$$

so,

$$C = 2\pi R \sin\theta$$

To find an appropriate expression for dh we make use of the small-angle approximation $\tan\mu = \mu$. Thus $dh/R = \tan d\theta = d\theta$, and $dh = R\, d\theta$. Therefore,

$$dA = C\, dh = (2\pi R \sin\theta)(R\, d\theta) = 2\pi R^2 \sin\theta \, d\theta \tag{12.13}$$

substituting into equation 12.12 gives

$$dE_{oc} = 2\pi L_p \cos\theta \sin\theta \, d\theta \tag{12.14}$$

The R has conveniently dropped out, convenient because we would be hard pressed to determine the distance to the sky dome! Inserting the luminance from equation 12.11

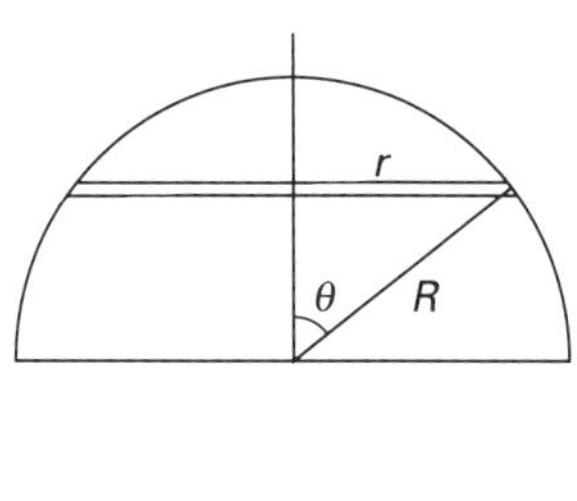

Figure 12-16. A Vertical Section through the Hemisphere.

and integrating from zero to $\pi/2$ results in the following equation for illuminance at a point on a horizontal plane from a standard overcast sky:

$$E_{oc} = \frac{2\pi L_z}{3}\int_0^{\pi/2} \cos\theta \sin\theta\,(1 + 2\cos\theta)d\theta \tag{12.15}$$

which, when evaluated, yields

$$E_{oc} = \frac{7\pi L_z}{9} \tag{12.16}$$

It is interesting to note that, if we had assumed constant luminance instead of the luminance distribution of equation 12.11, the result would have been

$$E = \pi L \tag{12.17}$$

In other words, the illuminance at a point on a horizontal plane surrounded by a hemisphere of constant luminance is equal to the exitance of the hemisphere. The same result is found if we replace the hemisphere with an infinite plane of constant luminance. In some situations, the ceiling of a very large room can approximate this condition.

Flat Strip

Another practical example of the generalized procedure considers a flat luminous strip located in a plane that is parallel to the plane in which lies the calculation point. This configuration is shown in Figure 12-17. It could represent, for example, a row of 1 × 4 fluorescent troffers with acrylic prismatic lenses.

Equation 12.10 is used to find the illuminance at point P. Note that the angles α and θ are always equal, and that $\cos\alpha = \cos\theta = Z/R$. Note also that $dA = W\,dy$. Thus equation 12.10 becomes

$$E = \int \frac{LZ^2W\,dy}{R^4} \tag{12.18}$$

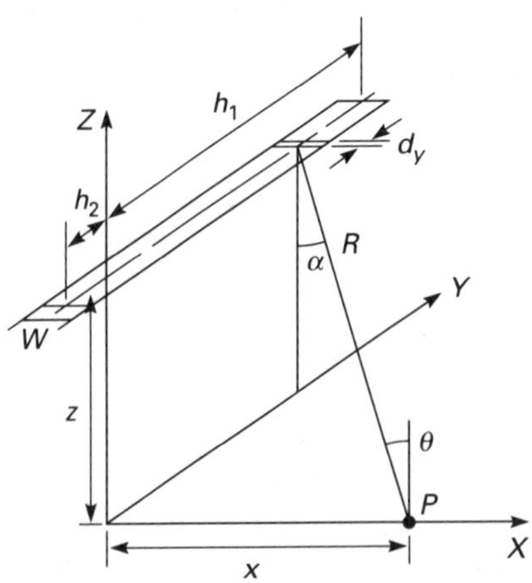

Figure 12-17. Geometry for Finding Illuminance at a Point from a Flat Strip Source.

If luminance is assumed constant (a dubious assumption for a real-world luminaire), then luminance can be moved outside the integral sign. The limits of integration will be from *negative* h_2 to h_1. Since $R^2 = X^2 + y^2 + Z^2$, the integral becomes

$$E = LZ^2W \int_{-h_2}^{h_1} \frac{dy}{(X^2 + y^2 + Z^2)^2} \qquad (12.19)$$

which, when evaluated, yields

$$E = \left(\frac{LZ^2W}{X^2 + Z^2}\right)\left[\frac{h_1}{X^2 + Z^2 + h_1^2} + \frac{\tan^{-1} h_1/(X^2 + Z^2)^{1/2}}{2(X^2 + Z^2)^{1/2}} + \frac{\tan^{-1} h_2/(X^2 + Z^2)^{1/2}}{2(X^2 + Z^2)^{1/2}}\right] \qquad (12.20)$$

In the more realistic case where luminance is not constant, an analytic expression for luminance could be determined from regression analysis of the photometric distribution of the source. Moeck and Belcher reported on a project where this was done for a prism light guide having a flat aperture 20 ft long by 8 in wide and a single 250 W metal halide lamp at one end. Their regression analysis yielded the following expression for luminance:

$$L = 0.821\, Z/R[(512 + 34.3h_2 + 1.12h_2^2) + (2.24h_2 - 34.3)y + 1.12y^2] \qquad (12.21)$$

When equation 12.21 is substituted into equation 12.19, we get

$$E = 0.547Z^3 \int_{-h_2}^{h_1} \frac{(512 - 34.3h_2 + 1.12h_2^2) + (2.24h_2 - 34.3)y + 1.12y^2\, dy}{(X^2 + y^2 + Z^2)^{5/2}} \qquad (12.22)$$

The nontrivial process of evaluating the integral in equation 12.22 results in

$$E = 0.547Z^3\left[\left(\frac{(512 - 34.3h_2 + 1.12h_2^2)y}{3ab}\right)\left(\frac{1}{b^2} + \frac{2}{a}\right) - \frac{(2.24h_2 - 34.3)}{3b^3} + 0.373y\left(\frac{1}{ab} - \frac{1}{b^3}\right)\right]_{-h_2}^{h_1} \qquad (12.23)$$

where

$$a = X^2 + Z^2$$
$$b = (X^2 + y^2 + Z)^{1/2}$$

Moeck and Belcher then used equation 12.21 to calculate illuminance at 20 reference points. These values were compared with measured values at the same points and with results of ISL calculations. They estimate that the predictions from this method were, on average, about 4% more accurate than those from the ISL method.

Flat Rectangular Source

Many large sources in architectural lighting are rectangular (troffers, windows, etc.). To develop expressions for dealing with these sources when they cannot be considered a point source we start again with equation 12.9. Figure 12-18 illustrates the configuration.

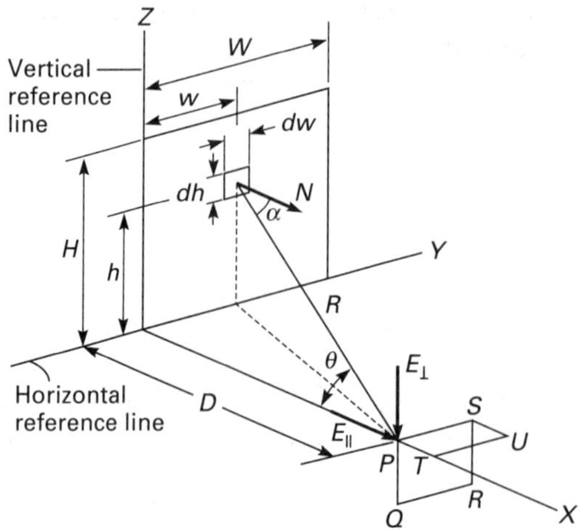

Figure 12-18. A Rectangular Luminous Source.

The differential area dA can be written as the product of the differential distances dh and dw. Cos α and cos θ can be written in terms of D, h, and w, but these expressions will depend on whether the calculation point is located in a plane perpendicular to the source (PSTU) or in a plane parallel to the source (PQRS).

Parallel plane

For the case of a point in a plane parallel to the source plane, the two angles α and θ are equal. Therefore, $\cos\alpha = \cos\theta = D/R$. We note that $R^2 = D^2 + w^2 + h^2$. Substituting into equation 12.9 and integrating from zero to W and from zero to H gives

$$E_{\parallel} = \int_0^W \int_0^H \frac{LD^2\, dw\, dh}{(D^2 + w^2 + h^2)^2} \qquad (12.24)$$

Perpendicular plane

For the case of a point in a plane perpendicular to the source plane, we still have $\cos\alpha = D/R$, but angle θ has changed so that $\cos\theta = h/R$. The relationship $R^2 = D^2 + w^2 + h^2$ still holds, and so we again substitute into equation 12.7 and integrate from zero to W and from zero to H to yield

$$E_{\perp} = \int_0^W \int_0^H \frac{LDh\, dw\, dh}{(D^2 + w^2 + h^2)^2} \qquad (12.25)$$

If luminance of the source varies across its surface, and an expression can be determined for L in terms of the other variables, then that expression can be introduced into equation 12.24 or 12.25, and the resulting integral evaluated. In the special case where we can assume constant luminance of the source, the L in equations 12.24 and 12.25 moves outside the integral signs, and upon evaluation yields equations 12.26 and 12.27.

$$E_{\|} = \frac{L}{2}\left[\frac{H}{\sqrt{D^2 + H^2}} \times \sin^{-1}\left(\frac{W}{\sqrt{D^2 + W^2 + H^2}}\right)\right] + \left[\frac{W}{\sqrt{D^2 + W^2}} \times \sin^{-1}\left(\frac{H}{\sqrt{D^2 + W^2 + H^2}}\right)\right] \tag{12.26}$$

$$E_{\perp} = \frac{L}{2}\left\{\tan^{-1}\left(\frac{W}{D}\right) - \left[\left(\frac{D}{\sqrt{D^2 + H^2}}\right) \times \sin^{-1}\left(\frac{W}{\sqrt{D^2 + W^2 + H^2}}\right)\right]\right\} \tag{12.27}$$

If luminance is expressed in cd/ft^2 in equations 12.26 and 12.27 then illuminance will be in footcandles.

It is important to note the coordinate system employed in Figure 12-18. The point for which illuminance is being calculated lies on a line perpendicular to one corner of the luminous source. Frequently we wish to calculate illuminance at other points not on this line. The method for doing this is illustrated in Figure 12-19. The illuminance at P due to source A can be expressed as

$$E_A = E_{ABCD} - E_{CD} - E_{BC} + E_C \tag{12.28}$$

Note that the contribution from C was subtracted twice and therefore must be added back once.

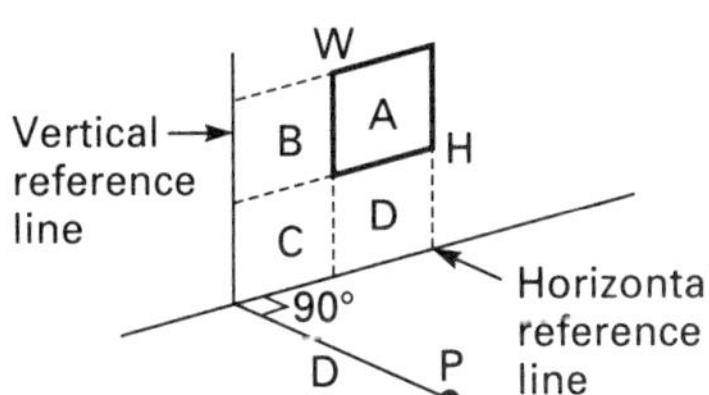

Figure 12-19. Illuminance from a Rectangular Source.

LUMINOUS FLUX TRANSFER

When an area source emits light in a perfectly diffuse fashion, that is, according to Lambert's cosine law (Chapter 1), we can evoke radiative flux transfer theory in order to understand movement of light from that source to other surfaces or to points in space. The student of heat transfer will recognize in the following that light behaves precisely as heat radiating from an area source; in each case we are dealing with flux transfer, only the type of flux has changed from heat to light.

In lighting we are concerned with two situations where flux transfer is helpful. The first involves predicting illuminance on a surface due to the flux emitted from a diffuse area source. We multiply the exitance of the source by a **form factor** to obtain the illuminance on the surface. The second involves predicting the illuminance at a point due to the flux emitted from a diffuse area source. We multiply the exitance of the source by a **configuration factor** to obtain the illuminance at the point. We will consider each of these in turn.

Illuminance on a Surface

Figure 12-20 shows two arbitrary surfaces, *a* and *b*, with areas A_a and A_b, respectively. A form factor f_{a-b} is defined as the fraction of flux leaving surface *a* that arrives at surface *b*. If surface *a* has an exitance of M_a lumens per unit area, then the flux leaving *a* is simply A_aM_a. Therefore, the flux arriving at surface *b* from surface *a* is

$$\phi_{a-b} = A_aM_af_{a-b} \tag{12.29}$$

Since illuminance of a surface is defined as the flux arriving divided by the area of the surface, the illuminance of surface *b* due to source *a* is given by

$$E_{b(a)} = \frac{A_aM_af_{a-b}}{A_b} \tag{12.30}$$

Intuitively we feel that there should be some sort of reciprocal relationship between the form factor from *a* to *b* and the form factor from *b* to *a*. Further, it seems reasonable that this reciprocal relationship should involve the two areas. Without offering a formal proof then, we think the reader will have little trouble accepting the following reciprocity relationship:

$$A_af_{a-b} = A_bf_{b-a} \tag{12.31}$$

or

$$f_{a-b} = \frac{A_bf_{b-a}}{A_a} \tag{12.32}$$

Substituting Equation 12.32 into Equation 12.30 yields

$$E_{b(a)} = M_af_{b-a} \tag{12.33}$$

Equation 12.33 has been called the "wrong-way law" of flux transfer since the form factor subscripts seem backward. They seem backward since f_{b-a} is the fraction of flux *leaving* *b* that arrives at *a*, however, when calculating the illuminance at *b*, we are interested in the fraction leaving *a* that arrives at *b*. The student should review the sequence of equations 12.30–12.33 to be convinced that equation 12.33 is indeed the "right way."

The following formula may be used to calculate form factors (using the terminology of Fig. 12-20):

$$f_{a-b} = \frac{1}{A_a}\int_{A_a}\int_{A_b} \frac{\cos\theta_a \cos\theta_b}{\pi R^2}\, dA_a\, dA_b \tag{12.34}$$

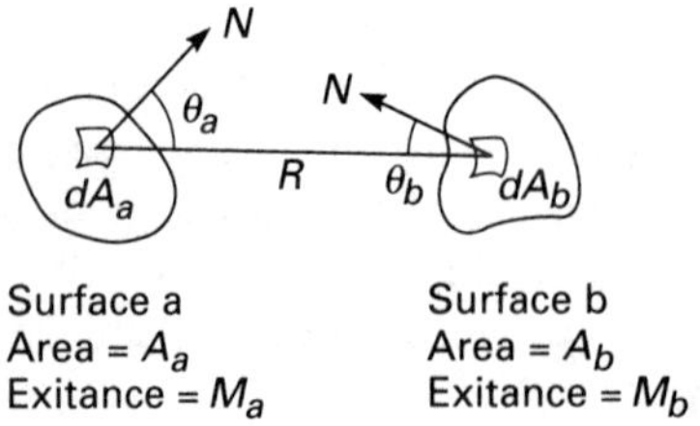

Figure 12-20. Geometry for Form Factors.

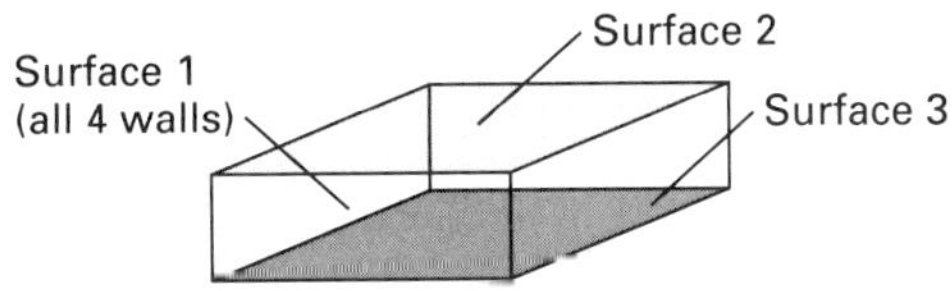

Figure 12-21. Three Surface Flux Transfer Model.

Since solving double surface integrals is not the way most lighting professionals want to spend their spare time, it is useful to develop formulas based on the specific case of the right rectangular parallelepiped. This shape is chosen since it accounts for the vast majority of rooms encountered.

Consider the room in Figure 12-21. We have divided the room into three surfaces—the walls collectively which we call surface 1, the ceiling (surface 2), and the floor (surface 3). If we want to know the form factor from the ceiling to the floor, it can be calculated by evaluating equation 12.32 over the appropriate limits. This results in

$$f_{2-3} = \frac{2}{\pi xy}\left\{ \ln\left[\frac{(1+x^2)(1+y^2)}{1+x^2+y^2}\right]^{1/2} + y(1+x^2)^{1/2}\tan^{-1}\left[\frac{y}{(1+x^2)^{1/2}}\right] + x(1+y^2)^{1/2}\tan^{-1}\left[\frac{x}{(1+y^2)^{1/2}}\right] - y\tan^{-1}(y) - x\tan^{-1}(x)\right\} \qquad (12.35)$$

where $x = L/H$ and $y = W/H$.

The other form factors, f_{2-1}, f_{3-1}, f_{3-2}, f_{1-2}, f_{1-3}, and f_{1-1} (the last of which must be included since part of the flux leaving 1 strikes 1 directly), could be computed directly from equation 12.34. However, by observing a few relationships we can avoid a lot of work. First, by symmetry, $f_{2-3} = f_{3-2}$, $f_{3-1} = f_{2-1}$, and $f_{1-2} = f_{1-3}$. By noting that all of the flux leaving the ceiling must initially strike the wall or floor, we infer the relationship $f_{2-1} + f_{2-3} = 1$. Similarly, $f_{1-1} + f_{1-2} + f_{1-3} = 1$. By the reciprocity relationship, $f_{1-2} = (A_2/A_1)f_{2-1}$. From these relationships, all the form factors for a rectangular room can be expressed in terms of the previously calculated f_{2-3} as shown in Table 12-1.

The form factors can now be used to solve the problem of interreflectance of light within a space. Assume that each of the three surfaces has an initial exitance, which we will call M_{01}, M_{02}, and M_{03}. These initial values could be due to light passing through the surface, light generated by the surface, or, more typically, by the light which directly strikes the surfaces from a luminaire in a room. In the latter case, we would require the density of luminaires in the ceiling be constant across the ceiling. This is necessary so that all four walls have about the same average exitance and so can be grouped together as one surface. Constant luminaire density could be accomplished by a uniform symmetric layout; however, there is no reason why the pattern should not be random. The important thing to keep in mind about the initial exitances is that they are the values *before* light interreflects in the space. The reflectances of the surfaces will be denoted ρ_1, ρ_2, and ρ_3. After light is allowed to interreflect within the room, each surface will have a terminal value, M_1, M_2, and M_3. Since all the surfaces are

TABLE 12-1 Form Factors for Three Surface Model in Terms of f_{2-3}

To surface \ From surface	1	2	3
1	$1-2\frac{A_2}{A_1}(1-f_{2-3})$	$1-f_{2-3}$	$1-f_{2-3}$
2	$\frac{A_2}{A_1}(1-f_{2-3})$	—	f_{2-3}
3	$\frac{A_2}{A_1}(1-f_{2-3})$	f_{2-3}	—

assumed to be completely diffuse (which is not a bad approximation for typical room finishes), the final exitance of a surface will equal the initial exitance plus that fraction of lumens per unit area that reflects off of the surface, or the product of reflectance and illuminance. For example, for surface 1;

$$M_1 = M_{01} + \rho_1 E_1 \tag{12.36}$$

The illuminance of any surface will be the sum of the illuminances due to flux coming from each of the other surfaces. Therefore, $E_1 = E_{1(1)} + E_{1(2)} + E_{1(3)}$. Substituting this expression and Equation 12.33 into 12.36 yields:

$$M_1 = M_{01} + \rho_1(f_{1-1}M_1 + f_{1-2}M_2 + f_{1-3}M_3) \tag{12.37}$$

and similarly for the other surfaces:

$$M_2 = M_{02} + \rho_2(f_{2-1}M_1 + f_{2-3}M_3) \tag{12.38}$$

$$M_3 = M_{03} + \rho_3(f_{3-1}M_1 + f_{3-2}M_2) \tag{12.39}$$

Equations 12.37–12.39 are a system of three simultaneous equations in three unknowns, M_1, M_2, and M_3. Dividing any of these final exitances by the reflectance of that surface gives us the illuminance of that surface. The lumen method for calculating average illuminance on the workplane from the zonal cavity system is based upon the preceding. In that method, the coefficient of utilization subsumes the form factors as well as the photometric distribution of the luminaires.

Illuminance at a Point

To determine illuminance at a point from a diffuse area source, configuration factors are used. Analogous with equation 12.33 we may write

$$E_{2(1)} = M_1 C_{2-1} \tag{12.40}$$

where $E_{2(1)}$ is the illuminance at point 2 due to surface 1, M_1 is the exitance of source 1, and C_{2-1} is the configuration factor between surface 1 and point 2. Actually, configuration factors were introduced earlier in the chapter, although that term was not used. When discussing calculation of illuminance at a point using the generalized procedure, the assumption was sometimes made that the area source being considered had constant luminance. This is equivalent to saying that it is a perfect Lambertian emitter. Thus, by replacing L by M/π in equations 12.17, 12.20, 12.26, and 12.27, those expressions become configuration factors.

VISUAL COMFORT PROBABILITY

In Chapter 8 we saw that the VCP metric provides information about occupant comfort in a room containing direct glare sources. In that chapter the student learned how to use VCP tables that manufacturers provide for each luminaire. Several assumptions must be made in order to use the VCP tables, however. In this section we examine how to assess VCP under circumstances where those assumptions are inappropriate.

First, let us consider what constraints the tables place on VCP analyses.

1. Tables consider an assumed "worst case" location in the room. It is not possible to calculate from the tables the VCP at a specific point of interest.
2. Tables assume a uniform layout of luminaires. If your design concentrates light sources at the task, the tables will not be usable.
3. Tables are based on a single luminaire. If there is more than one type of luminaire in a space, VCP analysis using tables is not possible.

In order to overcome these problems we must use the original VCP formulation from which the tables are derived.

The first step is to find the **index of sensation** for each glare source (luminaire) in the field of view. This index is traditionally given the symbol M; do not confuse this with the photometric quantity of exitance which has also, unfortunately, the symbol M. The index of sensation is calculated by

$$M = \frac{0.5 L_s Q}{P\, F^{0.44}} \tag{12.41}$$

In equation 12.41, L_s is the luminance of the glare source in cd/m^2. Q is a function of the solid angle subtended by the glare source, ω_s (in steradians). It is calculated by

$$Q = 20.4\ \omega_s + 1.52\ \omega_s - 0.075 \tag{12.42}$$

The average luminance of the entire field of view, F, or **field luminance** is found using the expression

$$F = \frac{L_w\omega_w + L_f\omega_f + L_c\omega_c + \sum_{i=1}^{n} L_{si}\omega_{si}}{5} \tag{12.43}$$

All the luminances in equation 12.43 are in cd/m^2, the subscripts w, f, c, and s refer to walls, floor, ceiling, and source, respectively. Each term in the numerator takes the luminance of a portion of the visual field and weights it according to the amount of the visual field it subtends. The summation term in the numerator adds up the weighted luminances for all of the n luminaires in the visual field. The denominator is 5 because it is assumed that the entire visual field comprises 5 steradians. Since F is a weighted average of the luminances in the field of view, the sum of the weighting factors in the numerator (the ω's) must also be 5 steradians.

The solid angle subtended by a surface can be calculated from the expression: $\omega = A_p/D^2$, where A_p is the projected area of the surface and D is the distance from observer to the surface. The projected area is found by multiplying the area of the surface by the cosine of the angle formed by the normal to the surface and a line from the observer to the surface.

The average field luminance appears in the denominator of equation 12.41. This is intuitively correct since it represents adaptation level. For a given glare luminance, the higher the adaptation luminance, the less annoying is the glare source. So, the greater the F in the denominator, the smaller the M.

The **position index** P takes into account that one's sensitivity to glare varies with location in the visual field. In general, glare sources near the axis of sight are more annoying than ones on the periphery. P is inversely related to glare sensitivity: a large value of P indicates that a glare source in that position will not cause much discomfort. Therefore, P also appears in the denominator of equation 12.39 and a large P gives a small M. Position index is determined from

$$P = \exp[(35.2 - 0.31889\tau - 1.22e^{-2\tau/9})10^{-3}\sigma + (21 + 0.26667\tau - 0.002963\tau^3)10^{-5}\sigma^2] \tag{12.44}$$

The angles τ and σ are the angular displacements (in degrees) of the position from the line of sight. These angles are shown in Figure 12-22.

Once indices of sensation are calculated for all of the glare sources in the field, their cumulative impact is determined. This combined effect is called the **discomfort glare rating,** or **DGR.** It is calculated according to

$$DGR = \left(\sum_{i=1}^{n} M_i\right)^{n^{-0.0914}} \tag{12.45}$$

This nonlinear relationship tells us that two equal glare sources are not twice as discomforting as one, that four sources aren't twice as bad as two, and so on. In other words, the effect of each additional glare source is of decreasing importance.

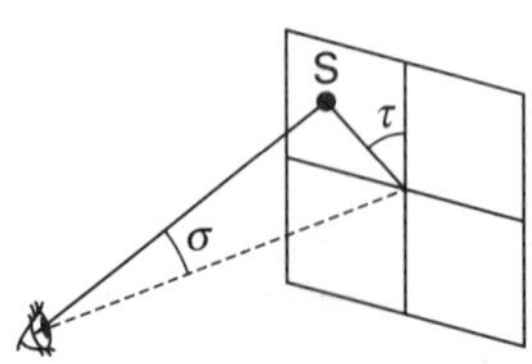

Figure 12-22. Angular Displacement for VCP Position Index.

The final step is to translate the *DGR* into a *VCP*, the probability that the space will not be objectionably uncomfortable from the point of view of glare. If we consider a large number of people, this translates into the percentage of people who will find the space acceptable. *VCP* may be found from Figure 12-23 or by evaluating the integral

$$VCP = \frac{100}{\sqrt{2\pi}} \int_{-\infty}^{6.374 - 1.3227 \ln (DGR)} e^{(-t^2)/2}\, dt \qquad (12.46)$$

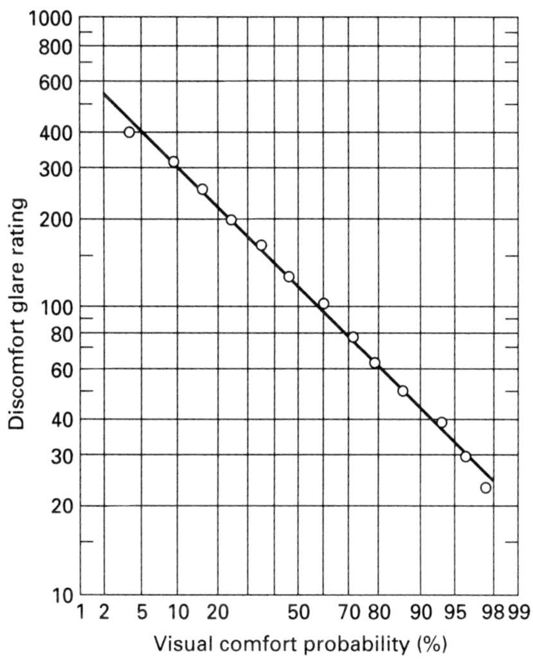

Figure 12-23. Plot of VCP as a Function of DGR (eq. 12.45). (Courtesy of the Illuminating Engineering Society of North America.)

13 daylighting design

When a tomb of an ancient Egyptian king is discovered one of the most exciting aspects of the find is the splendid polychrome drawings and decorations on the walls of the burial vault, perfectly preserved through the centuries in the dry desert air. One thing is conspicuous in its absence in these windowless rooms, though. The ceilings and walls are not smudged with soot as we would expect if they had been painted by torchlight. How did the ancient artists see to create their masterpieces?

The answer to the puzzle is the subject of this chapter—daylight. The ancient Egyptians had learned to harness daylight to work for them by positioning sol-

380 - 402

diers holding polished bronze shields to reflect the sun's rays deep into the tombs and monuments. In this chapter we present tools and ideas that will equip the student to use that same daylight resource to produce pleasing, effective, energy-efficient spaces.

A cursory look at people's involvement with daylight through the centuries will serve to heighten our awareness of the unique position in which we, at the end of the twentieth century, find ourselves. Centuries before the Egyptians reflected light into tombs and calculated the position a shaft of light would penetrate on a certain date, the druids of Stonehenge were building their monument to the sun's position. This mysterious observatory/temple indicates that societal interest in the daylight resource spans at least four millennia. Nearly an entire volume of Roman law dealt with light from the sun. Through the Christian era, light has held an important symbolic position (Jesus himself claimed to be the light of the world). This importance can be seen in development through time of the fenestration of church architecture. From the early Christian period to romanesque to gothic, the maximum possible fenestration was used in churches; the ratio of fenestration to solid wall continually grew, hampered only by building technology constraints.

Daylight conscious design may have reached its peak, however, around the turn of the twentieth century at the hands of the great masters Louis Sullivan and especially Frank Lloyd Wright. In the work of these men we see daylighting design as it must be, neither ignored nor tacked on as an afterthought, but as an integral element of the functional and aesthetic character of the building.

The international style changed all of that, however. With the glass boxes of the 1950s, 1960s, and early 1970s came an almost total insensitivity to the exterior environment. These buildings rarely responded to site-specific or climatological information; instead they created their own thermal and luminous environments, fueled by massive infusions of cheap, petroleum-based energy.

The house of glass blocks finally began to crumble when the OPEC oil embargo caused energy prices to soar in the early 1970s. Architects and engineers rushed to implement any half-baked lighting solution which someone claimed would conserve energy, regardless of impact on visual performance or aesthetics. Fortunately, by the mid-1980s and into the 1990s, calm reflection based on solid research and experience took the place of panic reaction. Today there is an exhilarating proliferation of books, data, methods, articles, in short, interest in using natural light to its fullest advantage; not just from the economic standpoint, but aesthetically, visually, and in response to the conservation ethic. This is truly a renaissance period for daylighting.

WHY DAYLIGHT?

What motivates this resurgent interest in natural lighting? Though responses will vary and the answer is probably complex, a major reason involves the potential for **energy conservation.** Successfully admitting renewable natural light into a space reduces the need to supply electric light, which is usually generated at the expense of a nonrenewable resource. Using daylight can therefore be considered good stewardship of the vast natural wealth with which we have been blessed.

Another reason for advocating daylighting is closely associated with, but distinctly different from, energy conservation. **Energy cost savings** can accrue because of lower electric lighting bills. We say *can* accrue because electric lighting is not the only building energy cost potentially affected by the introduction of natural light. Turning off the electric lights means a reduction of heat gain to the space from those lights. This implies that more energy must be used in that space during the heating season (and less in the summer if the space is air conditioned). To complicate the issue further, admitting daylight to a space requires some form of fenestration. This means that heat loss/gain characteristics of the building skin may be altered, which in turn would affect energy costs.

Though the final economic result of all the above is not intuitively obvious, one thing is certain; if the electric lights are not switched off in response to the presence of adequate daylight, no energy will be saved. Any building with windows can be said to be daylit, but only when we turn off the electric lights can we save energy or money. Switching the lights off can be done in a variety of ways using switches, dimmers, or a combination of the two. However accomplished, control of the electric lights must be done automatically. This is because people simply cannot be relied upon to turn off lights in response to sufficient daylight. This phenomenon has been borne out in numerous case studies. We can speculate on the reasons for this. They may include laziness, desire not to be interrupted, inability to remember to do it, and undesirability of a sharp drop in light level.

This last reason deserves a closer look. Imagine arriving at your daylit office at 8:00 A.M. on a winter day. In most parts of the world there will be insufficient light to perform office tasks comfortably. As an example, let's say 10 footcandles are available, while 50 fc is the recommended level. So you turn on the electric lighting system which provides 50 fc. By midmorning, daylight availability has increased to the point where 50 footcandles of the light on your desk is due to daylight. Therefore, you now have 100 fc (50 fc from the electric lights, 50 fc from natural light). At this point you could turn off the electric lights and still have sufficient light to perform the visual task. However, this requires cutting your light level in half. Most people, adapted to the higher level, are very reluctant to turn off the lights and plunge themselves into relative obscurity.

The possibility of **peak demand reduction** can be another strong incentive to use natural light. Most electric utility companies charge their commercial, institutional, and industrial customers a monthly fee in addition to the price they charge per kilowatt-hour of electric energy consumed. This additional fee is based on the maximum rate at which the customer consumed electricity during the month. It is a per kilowatt fee: 60 kW of peak demand will incur a demand charge of twice that assessed the user who demands a peak of only 30 kW. Thus the customer who uses a large amount of electricity for a short period of time will be billed more than the customer who uses the same number of kilowatt-hours, but stretches out that usage over the entire month. The utility's rationale for this rate structure is that they are required to provide all of the electricity their customers demand. As the utility reaches its generating capacity, peak demand charges act as disincentives for their customers to demand power, or viewed another way, revenues generated by peak demand charges help finance additional generating capacity.

Daylighting can help reduce peak demand because the daylight resource is most plentiful at precisely the time when commercial installations demand the most electric power (viz., early afternoon in the summer). Right after lunch everyone is in the office, and all of the computers, copiers, typewriters, coffee pots, and fax machines are running. More important, the air conditioning system is receiving its maximum load because, in addition to dissipating heat from these electrical loads, heat gain from the summer sun is maximum. Thus if we can turn off some of the lights in response to daylight availability, we may be able to avoid some of the peak demand charge.

For many designers, the **quality of light** produced by the sun and sky is a very attractive reason for daylighting. Daylight is dynamic. It changes in intensity from moment to moment on a partly cloudy day. Every day the color, directionality, and intensity vary from morning to midday to evening. The sun's low position in the winter sky causes deeper penetration through vertical glazing and causes longer shadows to be cast than in the summer. This variability can bring a certain richness to many designs and add interest to a space. It is this same variability that provides the designer with a great challenge when the intent is to provide consistent light for visual performance.

In Chapter 5 we examined many of the nonvisual affects of light on humans. Many of these **physiological** and **psychological benefits** are good arguments for using natural light. Examples include beneficial effects on calcium absorption, metabolism, and hormone secretion to mention a few. Intuitively, it seems we should function better, emotionally and physically, under the natural light in which our bodies were designed to operate.

Lighting designer and educator William Lam adds to the case for daylighting by describing it as a **formgiver for architecture.** For Lam, appropriate daylighting is an integral part of a design, reinforcing the architectural concepts (4). This becomes obvious when carried to its absurd extreme: when no light is present, architectural forms cannot be seen. In this very real sense daylight can become part of the architecture.

Daylight must enter through an aperture. In many cases that same aperture affords the occupants of the space a **view** out. Though a view may or may not be beautiful, its presence is generally considered beneficial because it allows us to keep in touch with what is happening in the exterior environment: weather, time of day, seasons. On the other side of the coin from view out, however, is view in, which raises the issue of privacy. When view windows are used in daylighting design, privacy concerns must be addressed. These include the type of behavior that is observable, the frequency with which it can be observed, the regularity with which it can be observed, and interactions of these. For example, privacy will not be a problem if slightly embarrassing behavior is observable even frequently, but the potential for even rarely having very private behavior observed will raise strong privacy objections from occupants.

Finally, one argument for daylighting that is very hard to resist is the fact that people like it; it is **desired by building occupants.** Perhaps people like natural light for some combinations of the other reasons cited, but survey-type studies reveal a desire for daylight among the large majority of office workers. To ignore this desire would be a serious programming flaw when planning a space.

Why Daylight?

- Energy conservation
- Energy cost savings
- Peak demand reduction
- Quality of light
- Physiological benefits
- Psychological benefits
- Design formgiver
- View
- Desired by building occupants

ANALYSIS OF DAYLIGHTING

Daylighting is analyzed in terms of the three primary aspects: (1) exterior environment, (2) interior environment, and (3) interface medium (window, skylight, etc.). The analysis of these components should be used to establish design parameters and techniques.

The primary difficulty in daylighting is the variability of daylight with respect to the time of day and year, location (longitude and latitude), and environmental conditions. These variations in the quality and quantity of daylight (exterior environment) result in variations in the interior environment.

Exterior Environment

The exterior environment produces and influences the daylight that becomes the source of light for interior daylighting calculations. The constantly changing exterior environment is the key factor in the analysis of daylight. The daylight resource is appropriately thought of as not one, but three sources of light. These three sources are the sun, the sky, and the ground.

Sun

Though approximately 93 million miles away, the sun is the primary source of heat and light for the earth. Its diameter is more than 100 times that of the earth The radiant energy or solar radiation received from the sun is transmitted in the form of short waves, including ultraviolet, visible, and infrared energy. Solar radiation is the most important source of heat, which is produced by the absorption of most of this radiant energy including light.

Only 50 to 60% of the radiant energy reaching the outer limits of the earth's atmosphere actually reaches the earth's surface. The percentage of energy reaching the earth varies with latitude, season, and cloud cover. Losses in radiant energy are due to selective scattering, diffuse reflection, and absorption in the atmosphere. Scattering and reflection account for approximately 70% of the loss; the remainder is due to absorption by water vapor.

The rotation of the earth on its polar axis causes daily changes, and the revolution of the earth around the sun causes seasonal changes (Fig. 13-1). The two **equinoxes** are the times when the sun's noon rays are directly vertical at the equator. The vernal equinox occurs in the spring (March 21 in the northern hemisphere, NH). The autumnal equinox occurs in the fall (September 23, NH).

The **solstices** are the times when the sun's noon rays are directly overhead at latitude $23\frac{1}{2}°$ north and south. The summer solstice in the northern hemisphere occurs on June 22 when the sun is directly overhead at the Tropic of Cancer. The winter solstice occurs on December 22 (NH) when the sun is directly overhead at the Tropic of Capricorn. For latitudes greater than $23\frac{1}{2}°$, the sun reaches its highest point at noon on the summer solstice.

The position of the sun at any given instant of time is expressed in terms of two angles. **Altitude** or **solar altitude** is the vertical angle (or elevation) of the sun above the horizontal plane. **Azimuth** or **solar azimuth** is an angle measured in the horizontal plane. The horizontal angle is measured from a reference direction (such as due south) to the vertical plane through the sun (see Fig. 13-2).

Sky

Small obscuring particles (such as dust) and water vapor act to diffuse and scatter the radiant energy as it passes through the atmosphere. This results in what is referred to as **sky luminance.** The belt of maximum radiant energy moves back and forth across the equator as the seasons change and from east to west as the earth rotates. The luminance of the sky forms a vault or dome of nonuniform luminance. Sky luminance is a function of cloud cover, thus we speak in terms of **overcast** sky or **clear** sky luminance.

Overcast sky luminance varies as a function of location, time, density of cloud cover, and uniformity of cloud cover. The luminance pattern is not uniform and will have a luminance of roughly three times greater at zenith (directly overhead) than on the horizon. This luminance is approximately independent of azimuth, in other words, for a given altitude above the horizon the overcast sky has about the same luminance

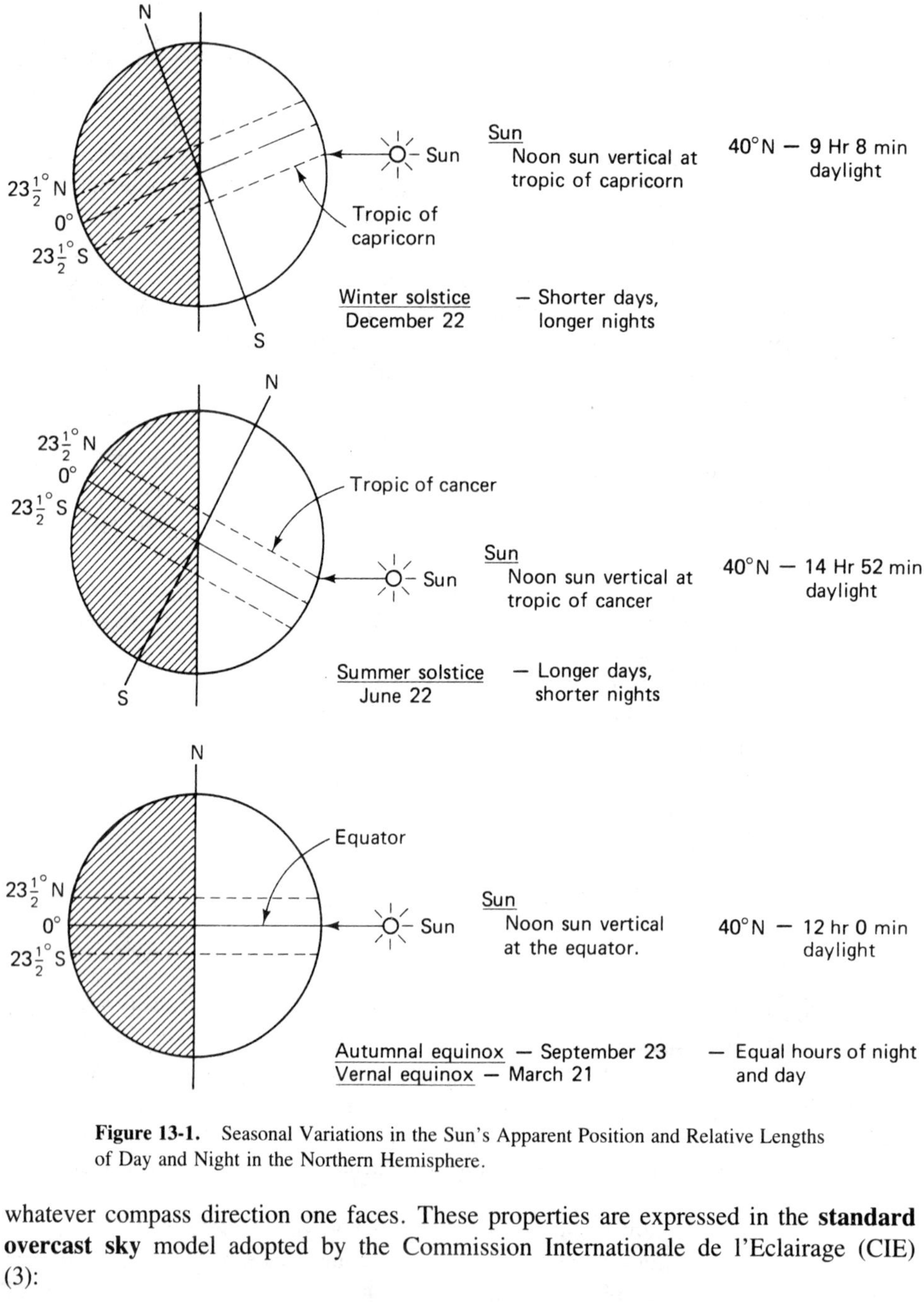

Figure 13-1. Seasonal Variations in the Sun's Apparent Position and Relative Lengths of Day and Night in the Northern Hemisphere.

whatever compass direction one faces. These properties are expressed in the **standard overcast sky** model adopted by the Commission Internationale de l'Eclairage (CIE) (3):

$$\text{luminance at a point in the sky} = L_p = L_z \frac{(1 + 2 \cos \theta)}{3}$$

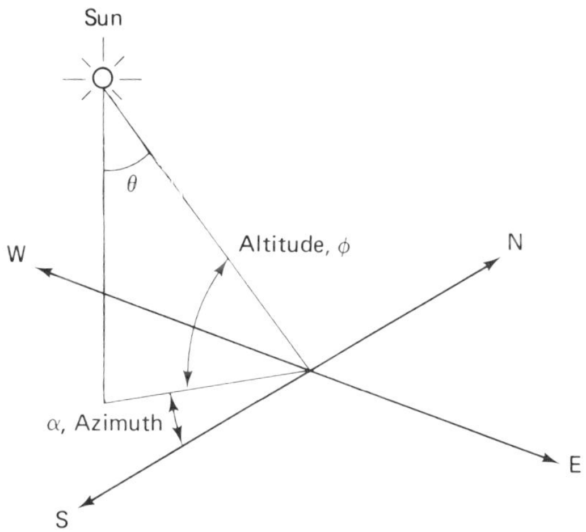

Figure 13-2. The Sun's Apparent Position Is Defined by Two Angles: Altitude and Azimuth.

where L_z is the luminance at zenith of the overcast sky and the angle θ is measured down from the zenith to the point of interest.

Clear sky luminance depends on the location of the sun, location (both altitude and azimuth) of the point of interest, and the amount of atmospheric haze or dust. The luminance of a clear sky is generally several times greater near the horizon than at zenith except in the vicinity of the sun, where sky luminance increases. The minimum clear sky luminance is found 90° opposite the position of the sun. The CIE **standard clear sky** model (3) can be used to approximate the luminance of a point in the clear sky:

$$L_e = L_z \frac{(1 - e^{-0.32 \sec(a)})(0.91 + 10e^{-3b} + 0.45 \cos^2 b)}{0.274\,(0.91 + 10e^{-3c} + 0.45 \cos^2 c)}$$

where a = angular displacement of the point of interest from the zenith
b = angular displacement of the point of interest from the sun
c = angular displacement of the sun from the zenith

A **partly cloudy condition** is said to exist if the sky is not completely overcast, but more than 30% of the sky dome is covered by clouds. When the sky is partly cloudy, the following expression can be used to approximate the luminance of a point in the sky (3):

$$L_\theta = L_z \frac{(1 - e^{-0.8/\cos a})(0.526 + 5e^{-3b/2})}{0.551\,(0.526 + 5e^{-3c/2})}$$

Theoretically, any of the foregoing expressions for sky luminance could be inserted into the integral form of the inverse square law (see Chapter 11) and the illuminance at a point determined by integrating over that portion of the sky "visible" to the receiving point. Practically, however, the formulae for clear sky and partly cloudy sky are much too complex for integral analysis. The simple form of the overcast sky model, on the other hand, makes it suitable to such analysis. An example is presented in Chapter 11.

Ground

Light reflected from the ground depends on the reflective properties of the ground cover. The effectiveness of ground light is also dependent on the orientation (exposure to direct sun) of the window, obstructions between sun and ground and/or ground and window, and the height of the window above the ground.

As an approximation, we assume the ground plane extends away from the vertical plane of the window an infinite distance. This simplifies calculations because the configuration factor (see Chapter 11) between a point on a vertical plane and an infinite horizontal plane is 0.5. This means that half of the light leaving the ground arrives at the vertical external surface of the window.

Obstacles

Obstacles in the exterior environment may obscure contributions or increase reflected light from any one or more of the sources of light. Obstacles would include adjacent buildings or structures and landscape. High landscape may obscure contributions from the ground, sun, and/or sky. The blocking effect of high landscape, such as deciduous trees, may vary from season to season. Lower landscape or ground cover may obscure contributions from the ground and may also be seasonal, depending on the type of landscaping material.

Interior Environment

The daylight entering a space must be analyzed in terms of the quantity and quality of the light. Daylight may be sufficient in quantity to reduce the artificial lighting level needed, but result in false energy conservation if the quality of the light is not analyzed. A poor quality of daylight may result in discomfort and a loss in visibility, which may cause a decrease in human performance and productivity. This, in turn, may cause increased use of the space, resulting in additional energy consumption.

Quantity

The footcandle (or lux) is the unit associated with the quantity of daylight produced in a space (i.e., illuminance). An illuminance criterion is used in most methods of daylighting design because of the availability of instruments to measure footcandles or lux and because of the relative ease of calculating illuminance. Illuminance levels may vary throughout the space at any given time depending on the number, location, and orientation of windows, the season, and exterior conditions. Currently, illuminance or illuminance ratio criteria are the most popular calculation procedures available to the designer. The lighting professional must be aware, however, of the limitations of methods that deal only with the quantity of light. That is, illuminance has little or no bearing on one's ability to perform a visual task (see Chapter 8).

Quality

The quality of daylighting in a space affects its physical and psychological impact on the human occupants. To deal with the quality aspects of daylight properly, the human occupant must be included. Because people are such complex creatures, the analysis of quality is almost as variable as the stimulus (daylight). The physical effects of daylighting can be investigated in terms of the visual, thermal, and acoustical environment. The psychological effect of daylight is investigated in terms of the subjective response to the environment. The emotional response to daylight is investigated in terms of very basic *like–dislike* and *need–don't need* modes of response.

Physical Effects

Visual Environment. The visual quality of the daylit environment is a very complex problem. One must deal with the response of the human visual system to daylight. The quality of the visual stimulus is dependent on many factors. We will concentrate on the three most important factors that affect quality: (1) glare, (2) luminance ratios, and (3) color.

Glare can be defined as any excessively bright source of light within the visual field that creates discomfort and/or a loss in visibility. This discomfort is associated with pain, fatigue, strain, or increased tension. Loss of visibility is, as the term implies, a complete or partial loss in the ability to see the task or object of interest. As a glare source moves closer to the line of sight, both discomfort and loss of visibility increase as an exponential function. For example, as the bright lights of an oncoming automobile approach the driver's line of sight, the effect on visibility and comfort will become greater. Therefore, glare is a function of the source, location, intensity, surrounding luminance, and direction of view. If the occupant is involved in heads-up tasks, direct glare will be the phenomenon of concern; occupants involved in heads-down tasks will be influenced by reflected glare or veiling reflections (see Chapter 8).

If the window is serving one of its primary psychological and emotional functions, that is, visual relief, it will be on the line of sight, which will compound the problems of direct glare. Direct glare from a source on the line of sight may cause discomfort depending on the differences in luminance between the window and the interior environment. The greater the luminance difference or luminance ratio, the greater the discomfort.

Loss in visibility due to direct glare will be due to transient adaptation. Transient adaptation is a measure of the adaptation state of the visual system as the eye moves about the nonuniform environment. It takes time for the visual system to adapt to the variation in luminances within the luminous environment. Relative to daylight, the primary concern would be readaptation to the lower luminance of a visual task or object in the environment after subjecting the visual system to the potentially high luminance of the window. Adaptation from light to dark situations is referred to as **dark adaptation.** The luminance values in most interior environments would not be thought of as dark. However, relative to the potentially high levels of luminance of the window, the luminance difference could vary from a ratio of 1 to 1 to more than 10,000 to 1. Fortunately, this adaptation for most interior environments involves the cone receptors, which have a much quicker recovery time than the rods. The time it takes to readapt to the interior luminances is a function of the luminance difference between the window and the visual task, the length of exposure to the window luminance, the magnitude of the window luminance, and the size or visual angle subtended by the window. Readaptation time will increase with an increase in luminance difference, exposure, window luminance, and a decrease in window size.

During the readaptation time, there is a temporary loss in the ability to see. Most occupants will compensate for this loss in visibility by sacrificing speed or accuracy. Loss in speed and/or accuracy will result in a reduction in performance, and hence a potential loss in productivity. Although these losses may be small for a single glance at an excessively bright window, the accumulated losses over an 8 hr day could be significant. When evaluating energy reduction by replacing artificial light by daylight, one must look at the potential increase in work time and energy consumed to compensate for the possible loss in productivity due to direct glare, transient adaptation, and readaptation.

As we saw in Chapter 8, reflected glare can result from the reflection of the image of the window off a specular surface. The problem of reflected glare involves source location and orientation of the task. Care in analyzing the location of the offending zone and proper orientation of tasks relative to windows will result in little or no problems with reflected glare.

Veiling reflections result in a loss in visibility due to a reduction in contrast. Because of the strong unidirectional (unilateral placement) quality of most windows, the relationship between source and task location can be optimized. If the guidelines in Chapter 8 are followed, daylighting can result in increased visibility values. A computer program could be used to study the visibility of various tasks under different task-window orientations.

Luminance ratio is a ratio of the luminance of a task to the luminance of the area surrounding the task. In daylighting, the primary concern is between the luminance of the window and its immediate surrounding area of walls and/or mullions and frame. The visual system experiences increasing discomfort as the ratio between the window luminance and the surrounding area increases. Typically, the maximum luminance ratio anywhere in the visual field should not exceed 40 to 1. This means that the ratio between the highest and lowest luminance level that can be seen from one location as an occupant casually looks around a space should be less than about 40 to 1. However, if the occupant is working on a critical visual task, luminance ratios in the field of view should not be greater than 5 to 1. If the ratio exceeds 5 to 1, a loss in visibility will occur as described previously.

Color is an important element in the consideration of any light source, fabricated or natural. Object color is seen by means of a phenomenon known as selective absorption, that is, some of the wavelengths of light incident on an object are absorbed while others are reflected. This means that object color is dependent on both the pigment characteristics *and* the spectral distribution characteristics of the light source.

The spectral distribution characteristics of daylight (3) depend on the mixture of direct sunlight and diffuse sky light present. In general, the diffuse sky creates more short-wavelength energy (blue, indigo, violet) than does direct sunlight. The ratio of direct sunlight to diffuse skylight varies as a function of the solar altitude and azimuth, latitude, and atmospheric density and transmission. The thicker the atmosphere, the greater the effect on short-wavelength transmission through the atmosphere.

The spectral distribution characteristics of daylight may also be modified by transmission through different types of glass. Since the spectral distribution of daylight plus the spectral transmission of the window can vary over quite a range, one must be careful when selecting building finishes and furnishings. Since most artificial light sources used for interior applications have vastly different spectral power distribution characteristics than sunlight, colors samples should be viewed under the electric light source to be used as well as under the specific conditions of daylight plus glazing that will be encountered in the space. In this way the designer can avoid unpleasant surprises after the building is occupied.

Acoustical Environment. Fenestration affects the acoustical environment in terms of sound transmission through openings around the glass or through the glass. The sound transmission loss of glass varies from roughly 20 to 40 decibels (dB) for a frequency range of 125 to 4000 hertz (compare with 50 dB for a brick wall with plaster). This represents a large span, ranging from a very poor sound barrier (at 20 dB) to a fairly good one (at 40 dB). However, the transmission of noise from the exterior environment into the interior environment is not only dependent on the transmission loss of the glazing, but also on the air tightness of the building. Street or traffic noise transmitted through and around glass can be annoying and distracting. In addition to exterior noise transmission, the window and frame combination may itself create noise due to the vibration or rattling of improperly fitted materials. An additional acoustical concern stems from the fact that smooth, hard surfaces such as glass can reflect sound, which may add to the ambient interior noise level.

Thermal Environment. The thermal aspects of the interior environment as affected by the window are dealt with in detail later in this chapter. In general, fenestration affects the heat transfer characteristics between the interior and exterior environments, and it affects heat exchange and distribution within the interior environment.

Psychological and Emotional Effects. In Chapter 5 we looked at some of the psychological and emotional aspects of lighting. Most of those studies must be considered preliminary; few of them dealt specifically with daylighting. This type of research needs to be extended and expanded to include the effects of daylight on attitude, well-being, and motivation. The basic question is whether the space should have windows or be windowless. The problem is more than simply balancing physical aspects of quality (visual, thermal, acoustic) against artificial lighting to produce energy conservation. One must be concerned with the way that daylight affects the user's spatial perception and behavior. If daylighting or the absence of daylighting has an adverse effect on attitude and motivation, it can have a detrimental effect on performance and productivity. Loss in performance and productivity may result in increased use of energy, since people working inefficiently will have to work longer to perform the same task.

One aspect of the effects of a fenestration can be seen from a 1962 study of the effect of windowless classrooms on the learning process (1). In that study, the Architectural Research Laboratory of the University of Michigan investigated daylight in two elementary schools of the same type of construction. Mann Elementary School was used as the control situation with windows. Hoover Elementary was modified to be windowless for 1 year and a follow-up of 1 year with windows was included. Both elementary schools included kindergarten through third grade. Performance testing of the children in the windowless classroom indicated no effect on the learning process when compared to students in the control school.

The study of windowless versus windowed environments needs to be extended. In most cases where windows exist, occupants find that excessive direct sun and glare result in the use of a covering or shielding system, which turns the space into a windowless environment. A covered window seems to be less disturbing than a truly windowless room, since the occupant knows that the windows are there and contact with the external world is possible if desired.

Fenestration

The interface between the exterior environment and the interior environment is the **fenestration.** The fenestration or openings are analyzed in terms of their application. The term **window** applies to all openings in the sidewalls, or sidelight; the term **toplight** applies to all systems utilizing an opening in the ceiling-roof plane, which provides overhead light. This section concentrates on the analysis of the physical characteristics of the most common interface material, glass. Some important characteristics of plastics are described as they apply to the interface for toplight.

Terminology and Units

1. *Transparent*:—property of a material that transmits light without any apparent change in direction or color. Objects can be seen clearly through the material in either direction.
2. *Translucent*:—property of a material that transmits light but diffuses the light as it passes through. Objects cannot be seen clearly through the material.
3. *Opaque*:—property of a material that will not transmit light.
4. *Reflective glass*:—a glass material that is coated on the outside with a transparent metallic oxide coating. During the daytime, when viewed from the inside the material appears transparent, and when viewed from the outside it appears to be opaque and acts as a mirror surface. The reverse is true at night.
5. *Tinted glass*:—glass that contains additives that change the color and appearance, and reduce transmission.
6. *U-coefficient of heat transmission* (*U-value*):—the number of British thermal units per hour (Btuh) that passes through 1 ft^2 of interface material when the temperature difference between interior air and exterior air is 1°F for a steady rate of heat flow.

7. *Conductivity*, *k*:—the number of Btuh that passes through 1 ft^2 of interface material 1 in. thick when the temperature difference (ΔT) between interior and exterior is 1°F (ΔT = 1°F) for steady rate of heat flow.
8. *Conductance*, *C*:—same as conductivity except for a specific thickness (x) or a thickness other than 1 in. ($C = k/x$).
9. *Thermal resistance*, *R*:—the reciprocal of conductivity ($R = 1/k$); it is the number of hours required for 1 Btu to flow through interface material of k conductivity. For thicknesses other than 1 in. expressed as conductance, C, the thermal resistance would be x/k.
10. *Air-space conductance*, *a*:—the rate of heat flow (Btuh) between the bounding surfaces through 1 ft^2 of area for ΔT = 1°F. It is affected by orientation and the emissivity, E, of the bounding surfaces.
11. *Emissivity*, *E*:—the effective thermal absorption of the bounding surfaces of the air space.
12. *Surface film conductance*, *h*:—conductivity dependent on the speed at which the air strikes the interface material. It is the rate of heat flow (Btuh) through 1 ft^2 at ΔT = 1°F due to air motion across the surface:

    outside: h_o = 4.0 Btuh per ft^2 per °F for a 7.5 mph wind

    inside: h_i = 1.46 Btuh per ft^2 per °F
13. *Infiltration*:—movement of air between interior and exterior environments due to cracks around window. The air flow is usually from cracks on the windward side to cracks on the leeward side.
14. *Thermal pressure*:—the difference (ΔT) between interior (t_i) and exterior (t_o) temperature.
15. *Greenhouse effect*:—trapping of solar radiation by the conversion of absorbed radiation by surfaces within a room into long-wavelength radiation. Most glass is opaque to long-wavelength radiation beyond 3 micrometers (μm).
16. *Solar optical properties*:—transmittance (τ), reflectance (ρ), and absorptance (α). The total solar radiation striking a surface must equal the sum of the transmitted, reflected outward, and absorbed energy (Figure 13-3): $\tau + \rho + \alpha = 1.00$. The

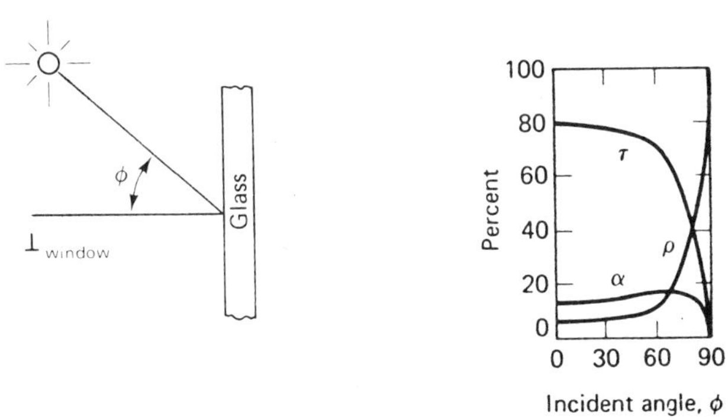

Figure 13-3. Variations in Optical Properties for Typical Quarter-inch Plate Glass.

magnitude of each of the three optical properties is a function of the thickness of the glass, physical properties of the glass, surface properties (coatings or film), and the angle of incidence, ϕ.

17. *Shading coefficient* (*SC*):—a ratio of the solar heat gain of a sheet of glass to the solar heat gain of the reference glass. The reference glass is double-strength sheet glass with $\tau = 0.87$, $\rho = 0.08$, and $\alpha = 0.05$. Shading coefficient can be calculated as a ratio of the solar heat gain coefficient (F) of a sheet of glass to the F value for the reference glass. For single glass, $F = \tau + (U\alpha/h_o)$; for the reference glass, $F = 0.87$. Therefore,

$$SC = \frac{F \text{ of glass}}{0.87}$$

Physical Characteristics of the Material

Both single glass and double or insulating glass may be clear, tinted, or reflective. Clear glass is an optically transparent material while tinted glass contains additives that reduce the visual transmittance and shading coefficient and modify the color appearance. Reflective glass consists of a transparent metallic oxide coating applied to the surface of the clear or tinted glass. For single-sheet glass the reflective coating is normally placed on the outside surface. For double or insulating glass, the reflective coating can be placed on the outside surface of the outdoor glass, on the air-space surface of the outdoor glass, or on the air-space surface of the indoor glass. The outdoor glass and indoor glass can both be clear or the outdoor glass can be tinted while the indoor glass is clear.

Glass has a specific gravity of 2.40 to 2.60 and weighs approximately 156 to 158 lb/ft^3. This is heavy compared to most building material. For example, concrete has a weight of 144 lb/ft^3, and common brick weighs 120 lb/ft^3. A more meaningful way to express the weight of glass is in terms of its weight per square foot of surface area. For example, $\frac{1}{4}$ in.-thick glass will weigh between 3.25 and 3.29 lb/ft^2.

The strength of glass is a key consideration in its selection. The strength attributes are expressed in terms of the thickness of glass required. The required thickness is a function of the external pressure loading (wind, sonic, etc.), the surface area of the glass, strength properties of the glass, surface characteristics of the glass, and the support conditions.

Environmental Impact of the Window

Thermal Characteristics. The thermal characteristics of glass in combination with shading devices (if any) are an important property that must be considered. The heat gain and/or heat loss properties of glass will have an influence on the initial and operating cost of the mechanical conditioning system, which in turn will influence the level of human comfort and performance. Increases in heat gain (or loss) will occur from an increase in the shading coefficient. U-value, and air-to-air temperature difference.

The U-value represents the reciprocal of the sum of the thermal resistance values

TABLE 13-1 Typical *U*-Values

	Summer*,‡	Winter†,‡
Single glass	1.06	1.16
Double glass	0.61	0.65
Triple glass	0.45	0.47

* Outside air film, summer at $7\frac{1}{2}$ mph; $h_o = C = 4.0$.

‡ Outside air film, winter at 15 mph; $h_o = C = 6.0$.

† Inside air film, still air, $h_i = C = 1.46$.

Source: Stein, Reynolds, and McGuinnes (1986).

of each material, including air spaces and surface films. Table 13-1 shows *U*-values of glass for various conditions. Note that the lower the *U*-value, the lower the heat gain or heat loss. The argument is often presented that the use of double or triple glazing improves the energy characteristics of a window. While this is true, it must be kept in mind that even the best glazing configuration is substantially worse than even a simple stud wall from the perspective of reducing heat loss/gain. To appreciate this, compare the *U*-values in Table 13-1 to the approximate *U*-value of 0.20 for a typical 2 x 4 wood stud wall with wood siding, vegetable fiber board sheathing, and gypsum wallboard (9).

Maintenance Characteristics. Maintenance of the glass is important to maintain the level of light transmission and desirable visibility. The length of time between cleaning periods is a function of the internal and external environmental conditions. Maintenance of the interior surface of the glass will usually be more convenient, which should result in a more repetitive cleaning cycle. For multistory, fixed window systems, exterior cleaning is more difficult and costly resulting in longer cleaning cycles. With longer cleaning cycles for exterior glass surfaces, loss of light transmission can vary substantially depending on the characteristics of the exterior environment. The percent of transmission loss for one study indicated a variation in light loss of from 73% for a "typical clean location" to 55% for a "typical dirty location" after 6 months.

This means that, for glass having a transmittance of 85% when clean

$$
\begin{aligned}
t_{\text{glass}} &= 85\% \\
&= 0.85 \times 0.73 = 0.62 = 62\%, \text{ clean location} \\
&= 0.85 \times 0.55 = 0.47 = 47\%, \text{ dirty location}
\end{aligned}
$$

It is obvious that this loss in transmission can have quite an effect on the quantity of light reaching the interior space. Dirt or film deposits on the exterior of the glass surface will also affect the *quality* of the light and the mode of heat transmission. The quality of the light will probably be more diffuse in nature, resulting in more nondirectional scatter of the light entering the space. Heat transmission will be affected due to the change in the surface air film characteristics, which will affect the h_o value.

To prevent scratches, abrasive cleaners should be avoided when cleaning glass. Fingerprints, grease, scum, and glazing putty can be removed with a wash, rinse, and dry technique recommended by the manufacturer. A mild soap, detergent, or slightly acidic cleaning solution can be used to wash most glass surfaces. For some types of glass, mild commercial solvent can be used to remove surface grime. Glass with reflective coatings can be damaged or scratched if not cleaned properly. Alkaline or fluorine material that is associated with concrete, masonry, or decorative crushed rock can stain or etch the glass. Oxide deposits from weathering steel or rust can also stain or etch glass.

The maintenance of toplight or skylight systems can create additional problems of dirt accumulation and accessibility. The horizontal orientation of a toplight system will cause dirt to accumulate more rapidly than on a vertical surface. Cleaning of the toplight from the outside may be easier than cleaning the outside of vertical windows; however, accessibility to interior toplight surfaces may be more difficult and require ladders or scaffolding to be brought into the space, which may discourage cleaning. Deposits of dirt and surface film materials such as smoke present in the environment will adversely affect the efficiency of transmission and the distribution characteristics. Static charges, which are common with plastic surfaces found in toplight systems, may compound the maintenance problem.

Control Elements

Control elements are introduced to prevent the direct transmission of the sun, to reduce glare, and to reduce heat gain and heat loss. The control elements can be internal or external. They can be an integral part of the window or a separate element from the window.

Direct transmission of radiant energy from the sun must be prevented in any space where critical visual tasks are anticipated so as to reduce glare and luminance ratio problems. Note that this does not apply to circulation spaces and other areas where the visual tasks are not demanding. In these spaces, direct sunlight may even be desirable to create visual interest and sparkle. An additional benefit of occluding direct sunlight is that it prevents color fading.

Both interior and exterior shielding elements can be used. Exterior shielding elements would include overhangs, fins, louvers, building elements or projections, and landscaping. Interior shielding usually consists of shades, blinds (vertical and horizontal), and draperies. Interior shielding has the disadvantage of allowing the penetration of the radiant energy into the space, which will be converted to heat and result in heat gain due to the greenhouse effect. Interior shielding is used more often than exterior shielding because of its accessibility and the ease of maintenance. To be effective as a shielding device, the material must be opaque or of very low transmittance.

An external overhang or awning is perhaps the simplest shading device. Its functioning is elegantly ingenious: in the summer when the sun's rays are more nearly vertical, the overhang casts a long shadow and effectively shades the window. In the winter, the low-angle rays of the sun are not blocked by the overhang but can penetrate and warm the interior. Though effective for regulating heat gain, if control of luminance ratios is desired, the overhang may not be ideal since it still allows a bright sun patch to be traced through the space on sunny winter days.

Glare control can be achieved by reducing the relative luminance of the window surface or by moving the glare source out of the direct field of view. Reductions in window luminance can be achieved by using tinted or low-transmission glass or by using low-transmission translucent shades, blinds, or draperies. This is not an ideal method, however, because reducing the window luminance simultaneously reduces the amount of light entering the space. The result could be an overall dimmer environment in which the luminance level of the window is still excessive for the new lower adaptation level. People tend to suggest using ''fill'' light in such situations to raise light levels in the dark areas. The result could be a system that uses *more* energy than the nondaylit alternative.

Two preferred solutions for excessive window brightness are shown in Figures 13-4 and 13-5. Figure 13-4 illustrates use of a splayed window recess. The angled portion will have a luminance that is intermediate between the bright window and the dark interior wall, making for an easier visual transition. In Figure 13-5, the view function of the window is separated from the light admitting fenestration. The small lower window is heavily tinted and thus not excessively bright. The upper clear glass window is shielded from view by a light shelf, the primary function of which is to reflect sunlight onto the ceiling and deeper into the space. Therefore, use of a light shelf enhances uniformity of illuminance in the space and can also be designed to interrupt our line of sight toward bright windows.

Some types of glare control are not applicable in some situations since they adversely affect the ability of occupants to look out of the window and may produce unacceptable color distortion because of selective transmission of tinted glazing. Direct glare can be reduced by moving the daylight source overhead, as in toplighting. However, direct solar transmission must also be controlled when using toplighting. Also, because of the strong directional quality of the light from toplight, reflected glare and veiling reflections may be problems.

Heat gain can be reduced by using tinted and/or reflective thermal glass. Double and triple glass systems utilizing reflective coatings and tinted glass have been introduced to reduce heat loss and heat gain. Visibility through the glass and color perception should be considered when selecting thermal glass. The consequences of reflecting radiant

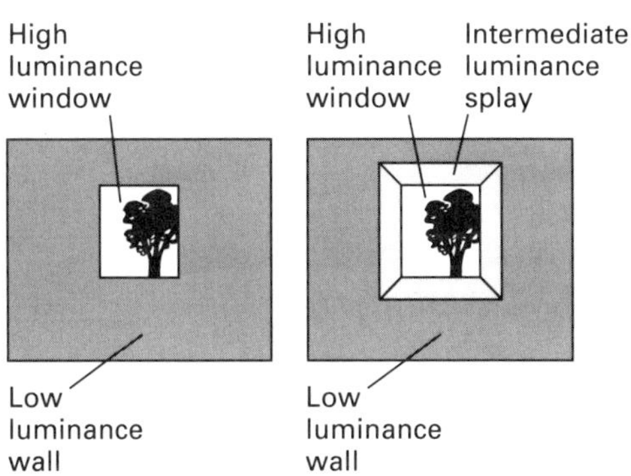

Figure 13-4. Reducing the Effects of Excessive Window Brightness with the Use of a Splayed Recess.

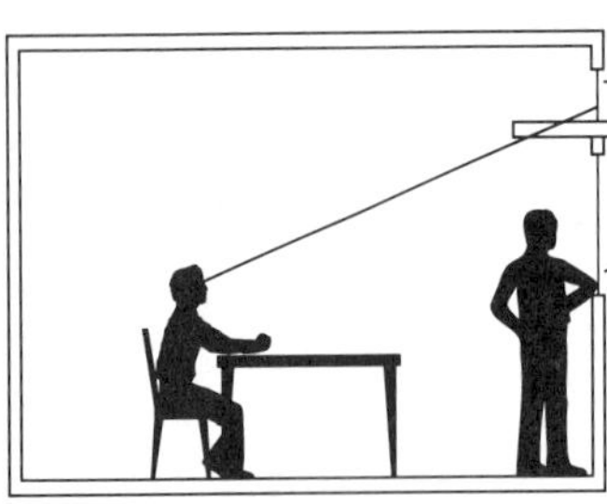

Figure 13-5. Separating the View Function from the Daylight Function can Reduce Glare.

energy from the building to adjacent property owners must be investigated as this has been the grounds for many lawsuits.

A dual-mode shading device has been investigated by Silverstein (8). The system consists of a reversible roller shade that has a dark-colored absorbing side while the other side is a solar reflector. The system provides for the control of direct sunlight, the control of glare, and the control of heat gain and heat loss. The dark side is turned toward the sun during winter months to act as a solar collector, that, when combined with the proper natural ventilation, allows the heat to be circulated into the space. The air space created by the shade adds to the insulating capabilities of the system during winter evenings. The reflective side of the shade can be used to reject solar heat gain during the summer. To increase its efficiency during the summer, an interior-exterior air circulation system should be provided.

Rosenfeld has suggested a modification to the Silverstein system (5). He has suggested using venetian blinds (see Figure 13-6) constructed of low-transmission gray plastic with a metallic reflective coating on one side. The primary advantage of the venetian blind is that the blinds can be opened slightly to allow for more natural ventilation during the summer, and on cloudy days they can be partially opened to admit more daylight.

A beam-daylighting system has been investigated by Rosenfeld and Selkowitz to increase the effective utilization of daylighting by increasing daylighting penetration in the space (6). The system utilizes a venetian blind with a metallic reflective coating on one side. The silvered beam blind is mounted above eye level and utilizes direct solar radiation incident on the top 1 to 2 ft of the window (see Figure 13-6). The beam blind is separate from the lower window shading system to allow for independent action. The purpose is to reflect the solar radiation from the beam blind to the diffuse white ceiling plane. This indirect lighting system provides greater daylight penetration with more uniform diffusion. The sun orientation, blind tilt, and reflective properties of the ceiling are critical to the efficient utilization of the daylight. Since this system allows for the penetration of direct solar radiation (redirected to the ceiling) into the space, it will result in direct solar heat gain to the interior environment. If the system is not performing at its optimum, sufficient daylighting may not be present to allow for a sufficient quantity of artificial lights to be switched off to offset the heat gain produced by the direct solar radiation.

Photochromic and **electrochromic materials** are being considered for use as

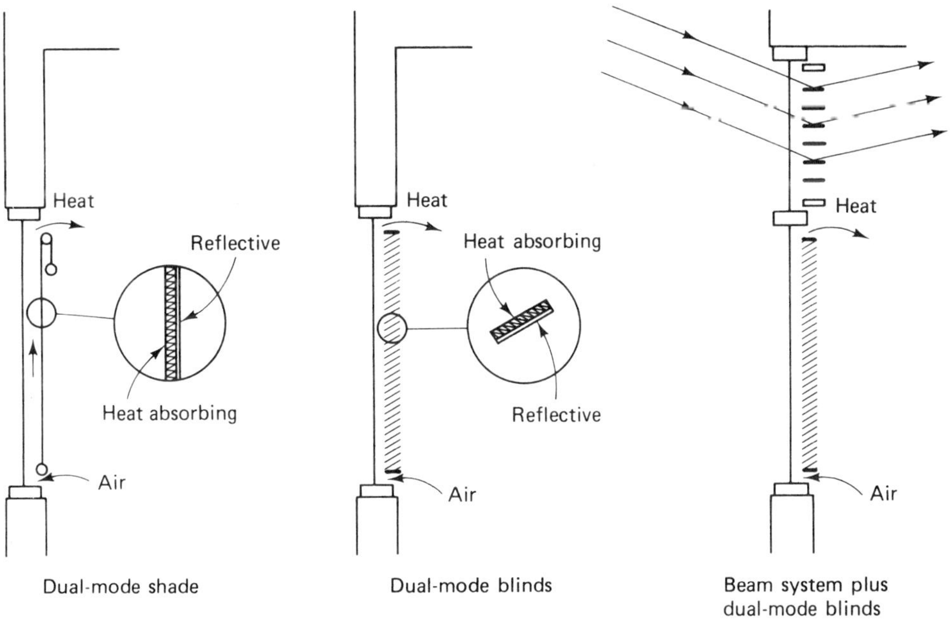

Figure 13-6. Examples of Daylight and Heat Control Devices.

daylight control devices. Photochromic refers to the property of a material whose transparency decreases with increased incident light level (available in some sunglasses). A window constructed of photochromic glass would be clear when only skylight is present, but would darken in direct sunlight. This would be very beneficial for excluding radiant heat in the summer, but it would also block the desirable warming rays of the low winter sun.

Electrochromic is the ability to alter the transmittance of a material by passing an electric current through it (example: liquid crystal displays). An electrochromic window could be designed with addressable pixels similar to a computer monitor. A microprocessor could control which pixels were clear and which ones were dark at any given moment. Several effects would be possible: a uniform pattern could be created allowing various degrees of shading, a decorative pattern could be displayed, or dark part of the window could move through the day in response to the sun's path so as to cast a consistent shadow in the room, on the visual task for example. A major automobile manufacturer plans to introduce an electrochromic sunroof; however, there are no electrochromic or photochromic products currently on the market for architectural fenestration.

A control device of a different nature is the **heliostat.** Rather than attempting to block the sun, the heliostat collects and redirects the suns rays. A motor-driven mirror (the primary mirror) tracks the sun's position through the sky and reflects sunlight onto the stationary secondary mirror. From the secondary mirror, light is directed (usually vertically) into the interior space where it can be effectively redistributed. The advantage is that, under clear sky conditions, more light is admitted to the interior and this amount is more constant through the day than with fenestration without a heliostat.

DESIGN OF DAYLIGHTING

An analysis of the important aspects of the exterior environment, interior environment, and fenestration is essential to the beginning of the design process. Each opening placed in the exterior wall of a building should be evaluated in terms of its total impact on the interior environment and the human occupants. The decision to place fenestration in a building should not be based on arbitrary, superficial reasoning. Table 13-2 lists the advantages and disadvantages of placing fenestration in the exterior walls.

The list is not complete. Each designer may not agree with all factors and should compile a personal list. However, this type of listing procedure accompanied by a study of the analysis section should give the designer sufficient information to make a decision for or against placing fenestration in a room. This decision may vary from room to room, depending on the function of the room, its orientation, and its occupants. This statement implies a potentially drastic change in the outward appearance of buildings to more appropriately express the inward function, while weighting the impact on energy conservation.

Once the decision has been made to provide fenestration in a space, the size, proportion, and placement of the opening must be decided by the designer. The size

TABLE 13-2 Advantages and Disadvantages of Placing Fenestration in the Exterior Walls

Advantages	
Interior/exterior visual communication	Indoor plant growth
Design enhancement: rhythm, relief drama, pattern, etc.	Fire escape
Ventilation*	Security: easier to observe interior when entering the space
Opens the space, feeling of enlargement	Awareness of exterior environmental conditions
Heat gain in winter	Visual and psychological relief: varying source of light, changing color, changing mood
Potentially good visibility with proper task placement	
Disadvantages	
Unpleasant view	Condensation
Glare source	Undependable, constantly varying light source
Potential color fading and ultraviolet damage	Shadowing from strong direction source
Heat loss in winter	Sound transmission*
Heat gain in summer	Security: easily penetrated
Bodily discomfort: radiation loss due to surface temperature	Air pollution, if operable for ventilation*
Increased maintenance: cleaning and breakage	Limits circulation
Higher cost than conventional wall materials	

* Ventilation will increase sound transmission and the amount of air pollution entering the space.

may be a compromise between the thermal impact of the opening and the daylighting potential. The proportions and placement should be based on the quality and quantity of daylight entering the space. The following factors should be considered in window design:

1. Avoid the direct penetration of sunlight if critical visual tasks are present.
2. For low buildings of one or two stories, use light-reflecting ground cover to improve light penetration into the space.
3. Provide openings in more than one wall if possible to improve penetration, uniformity, and reduce harsh shadows.
4. Use high, clear glazing; the head of the window is most effective in admitting light to the inner portion of a space.
5. Two rules of thumb:
   a. Optimum daylighting penetration and uniformity will be achieved if the window height is at least one-half the room depth.
   b. Glass area should be approximately equal to 25% of the floor area to optimize the uniformity of illumination in a space.
6. The reflectance properties of external shielding devices can be critical to daylight penetration.

The following factors should be considered in toplight or skylight design:

1. Avoid the direct penetration of sunlight if critical visual tasks are present.
2. For monitor roof, clerestory, and sawtooth configurations, the use of light-reflecting roof materials will increase the quantity of light entering the space.
3. Rules of thumb:
   a. Maximum spacing between skylights should be 1.5 times the ceiling height.
   b. Skylights should occupy approximately 5 percent of the ceiling area; this percentage is recommended to minimize ceiling clutter when skylights (5%) are combined with ceiling-mounted luminaires (10 to 20% coverage) to allow for a 75 to 85% ceiling surface area, which is important to the interreflection of energy.
4. Openings through a flat roof plane create flashing problems that are difficult to solve because of extreme temperature variations, causing differential expansion and contraction.
5. Heat buildup due to natural heat convection currents in a room may result in excessive heat loss.
6. Use moisture seals for condensation prevention and drainage.

Assuming suitable sun control, a southern exposure (continental United States) is preferred to optimize the daylight contribution into a space. Sufficient daylighting must be provided to replace artificial lighting if energy conservation is to be realized. Northern exposure may not provide a sufficient quantity of illumination to allow for a significant reduction in artificial lighting. In general, east-west orientations should be avoided since they present the most difficult problems in daylight control and create complicated, extreme heating and cooling problems.

Daylight Modeling

Two modeling techniques are commonly used for assessing the design of a daylit space. Often the designer will build a physical scale model which can be observed under conditions simulating the daylight environment which will be available at the actual building site. The other possibility is to build a mathematical model, that is, to use a calculation procedure for predicting illuminances in the space. As shown in Table 13-3, each of these methods has its own strengths and weaknesses so the designer will usually apply both methods in order to have a complete understanding of the project.

Scale Models

Though frequently used to help *visualize* the daylit appearance of a space, scale models also lend themselves to analysis by photometric measurements. Since the sun's rays arrive virtually parallel and the sky dome is effectively infinitely far away, scale models can be perfect representations of the actual space, as far as the illuminance meter is concerned. In other words, no scale factor is necessary; the illuminance level read from the meter is the illuminance level expected in the full-scale room. Accuracy of this prediction will depend solely on the accuracy with which the model is constructed. Two common trouble spots warrant discussion: the scale of the sensor, and the scale of textures. Many illuminance meters have fairly large sensors—an inch or more tall by 2 inches or more in diameter. If the scale used in construction of the model is too

TABLE 13-3 Strengths and Weaknesses of Scale and Mathematical Models

Advantages	Disadvantages
Scale models	
Helps to visualize the daylit appearance of a space	Quantitative data can be difficult to obtain
Simple model is easily constructed	Requires special modeling skills, materials, and equipment
Elaborate model can be presentation tool	Light leaks can erroneously influence conclusions
Photographs of results can be very realistic	Textures difficult to model
Mathematical models	
Gives quantitative results	Difficult to visualize appearance based on numbers
Hand calculations yields quick approximations	Some software not very user-friendly
Computer simulations can be done on inexpensive microcomputers	Results can vary widely depending on algorithm used
Computer simulations allow quick investigation of numerous alternatives	

small, the sensor may dominate it and alter the interreflectance properties. A minimum scale of $\frac{3}{4}$ in. equals 1 ft is recommended in such situations. Using actual samples of floor and wall covering materials can cause a texture scale problem. The threads of a carpet sample, for example, will be many times too large when placed in a scale room. This can cause alteration of the light reflecting properties of the space. Therefore, it is generally safer to model reflectances only and not textures. The serious modeler will find the Schiller reference a valuable source of information about daylight modeling (7).

Calculation Methods

Numerous calculation procedures for predicting illuminance from daylight have been developed over the last few decades. Many of them are sufficiently complex and sophisticated as to require computer solutions. We will investigate here two methods that are simple enough to work through longhand, yet provide a reasonable level of accuracy. These two are the Illuminating Engineering Society of North America's lumen method and the Commission Internationale de l'Eclairage's daylight factor method. They were chosen because they are sanctioned by the world's leading lighting institutions and thus have the greatest acceptance and application.

IES Lumen Method

The calculation method approved by the Illuminating Engineering Society of North America is one which permits calculation of illuminance at each of three specific points within a room. Both the direct and interreflected components of illuminance are accounted for, but the individual values are not available, only the total is calculated. The specific points that can be calculated are on a line perpendicular to the center of the window at workplane height (2.5 feet above the floor). One point is 5 ft from the window, another in the center of the room, and the third is 5 ft from the "rear wall" or the wall opposite the window. These are referred to as the max, mid, and min points, respectively.

The following equation defines illuminance at a point in the IES method:

$$E_p = (E_i)(A_w)(\tau)(LLF)(K)(C)$$

where

E_i = the vertical illuminance incident on the window

A_w = the area of the glazing

τ = the transmittance of the glazing

LLF = the light loss factor due to dirt accumulation on the glazing

K = a coefficient of utilization that accounts for ceiling height and width

C = a coefficient of utilization that accounts for room length and width

The procedure can be thought of as a four-step process:

1. Determine the vertical illuminance on the exterior of the window.
2. Multiply the vertical illuminance by the area of the glazing. Recalling that average illuminance is flux divided by area, we see that this step gives us the flux on the exterior surface of the window.
3. Determine the flux transmitted by the window. This will be the product of the flux determined in step 2 and the loss factors (viz., τ and *LLF*).
4. Finally, we multiply the flux entering the room by factors that account for the interreflective properties of the room (geometry and reflectances) to tell us how much of the flux arrives at the workplane per unit area, in other words, illuminance.

One complication that arises in implementing this procedure is that the *C* and *K* factors are very sensitive to the direction from which the flux is arriving. A much different percentage of the flux coming from the sky will arrive on the workplane than will flux reflected into the room from the ground. Thus for each of the three points in the room, an illuminance calculation is performed for the contribution due to sky light and a separate calculation is performed for the contribution due to ground light. The two values are then added to yield the total illuminance at that point.

The component of vertical illuminance on the exterior of the window due to ground light is found by multiplying the horizontal illuminance from Table 13-4 by the reflectance of the ground plane and by 0.5, the configuration factor between a point on a vertical surface and an infinite horizontal plane. The vertical illuminance on the exterior of the window due to sky light is found from Table 13-5. *C* and *K* factors for ground light and sky light are given in Table 13-6. Note that for these tables, width is defined as the room dimension perpendicular to the window wall and length is the dimension along the window wall. These definitions can result in the surprising situation of the width being longer that the length. This should not be cause for alarm as the terms are merely convention. Light loss factors can be taken from Table 13-7. A sample problem is given in the data section at the end of the chapter. Following is a list of additional constraints and assumptions of the IES method that must be kept in mind when analyzing the results of such calculations.

- Window is assumed to extend to the ceiling.
- Direct sunlight on window is not permitted.
- Only certain room dimensions permitted (20, 30, 40 ft).
- Assumes unilateral window configuration.
- Method has been extended to include toplighting and the effects of shading devices (not covered here).

CIE Daylight Factor Method

The **daylight factor (DF)** for a point is a ratio, expressed as a percentage or decimal fraction, of the interior horizontal illuminance at the point to the *simultaneous* external horizontal illuminance. The method was developed in Europe where overcast conditions are much more prevalent than in the United States. It is not surprising, therefore, that the method is only strictly applicable for overcast sky conditions (though adaptations of the method to clear sky conditions are in use). The beauty of the method lies in the fact that, since it is based on the CIE uniform overcast sky, the daylight factor is a constant that is independent of time of day or day of the year. This does not mean that interior illuminance is constant, merely that it is a constant fraction of the (changing) exterior illuminance.

One of the most salient factors in dealing with natural light is its variability. At times during the day, and for certain days of the year, the amount of light available is greatly in excess of requirements. Rarely, however, is the amount of light available sufficient to light the workplace during the entire working day. The daylight factor method attempts to provide information about the percentage of time during which acceptable conditions exist. These are referred to as the **tolerable minimum conditions.**

Figure 13-10 in the Data Section is a key element of the CIE method. It allows the designer to answer the question ''what is the value of external illuminance which will be equaled or exceeded for the required percentage of normal working hours throughout the year. For example, at 30° latitude, the exterior illuminance on a horizontal plane due to an overcast sky will be 950 footcandles or greater for 90% of the hours between 9:00 A.M. and 5:00 P.M. throughout the year. If a different work day is assumed, the percentages can be adjusted using the values in Table 13-7.

Since the daylight factor compares interior illuminance to exterior illuminance, when any two of the three quantities (interior illuminance, exterior illuminance, DF) are known, the third is easily found. For example, if the exterior illuminance is known from Figure 13-10 and the desired internal illuminance needed to perform a specific task is known (say, from the IES illuminance selection procedure), then the ratio of the two can be taken to determine the DF for which the space must be designed. Given the geometry of a space, daylight factors can be determined from Figure 13-11 for rooms lit by vertical windows on one wall or from Figure 13-12 for rooms lit by a uniform layout of skylights in a flat roof. When the sky is partly obstructed by adjacent buildings, Figure 13-13 can be used to adjust the DF to account for the reduction in interior light that will occur. Note that in the case of unilateral sidelighting, the DF is for the worst point in the space, or the minimum DF. However, in the case of skylights, an average DF is determined. Following is a list of additional restrictions and conventions that must be considered when using the CIE method. A worked example appears in the data section.

- Window head is assumed to be 12 in. below the ceiling.
- The sill height is assumed to be 36 in.
- Window width is expressed as a percentage of room length such as 30%, 60%, or 90%.
- The ceiling height is assumed to be 9 to 15 ft.
- Room length is measured parallel to the window wall.
- Room reflectances (in percent) are assumed as 70/50/15 (ceiling/walls/floor).

- Room depth is expressed in window heights.
- Glazing transmittance is assumed as 85%.

COMBINED ARTIFICIAL/DAYLIGHTING CONTROL SYSTEMS

Basic Principles

The control of the electric lighting system to optimize the utilization of daylight is essential if energy savings are to be realized. If artificial lighting levels are not reduced as daylight contribution increases in a space, the overall energy consumption will increase due to the added burden on the mechanical system. That is, the increase in daylighting must be offset by a decrease in power consumption in the artificial lighting to produce meaningful energy reductions.

The addition of daylighting to artificial lighting can result in excessively high lighting levels near the window (Figure 13-7). Flexible control systems must be installed

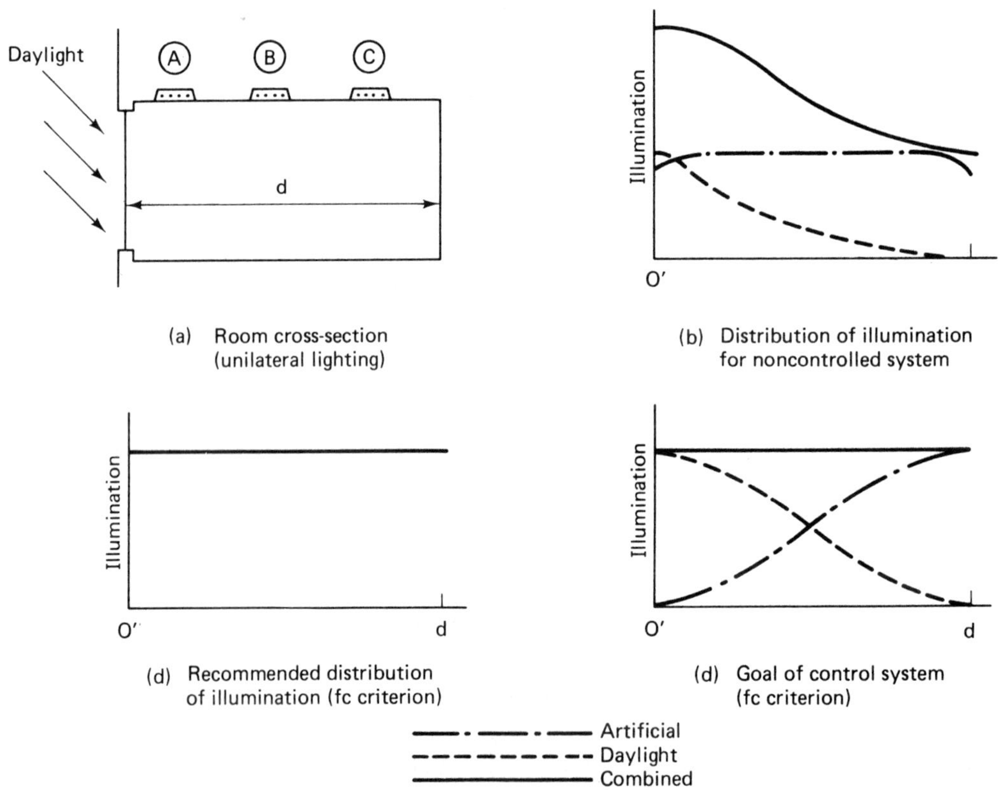

Figure 13-7. Combining the Contributions of Daylight and Electric Lighting.

to maintain a more uniform lighting level throughout the space. Figure 13-7 illustrates the problem.

The simplest, least expensive, and most obvious solution would be to provide multiple on/off switching within each room. The cross section in Figure 13-7 shows three rows of luminaires. One switch could be provided to control all the lamps in row *A*; a second switch could be used to turn off half the lamps in the second row *B*; a third switch would provide on/off night control to row *C* and the other half of the lamps in row *B*. When a sufficient daylighting contribution exists in the space, switches *A* and *B* could be turned off to approximate the distribution in Figure 13-7. If the daylighting contribution is inadequate (overcast or nighttime), or if opaque shielding devices are used to control direct light, all three switches *A*, *B*, and *C* could be turned on. The switching could be either a low- or normal-voltage system.

However, the control system is only as good as its weakest link. In this case, the weakest link is the human operator. As variations in the daylighting contribution occur during a normal working day, the user may have to adjust the switching. Experience indicates that unless the user of the space is also the one who pays the utility bills, the user will grow tired of making the appropriate adjustments and leave all the artificial lights on. To conserve energy and optimize the use of daylight, the control of the artificial lighting must be taken out of the hands of the user.

Energy Conservation and Daylighting

The user of a space should be given only one control option; that is, the lighting-level control system is either on or off. This is called the **user system.** The changes that occur during the on period to compensate for variations in the daylighting contribution must be completely automatic. This is known as the **compensating system.** The compensating system could involve two or more levels of switching or a continuously varying mode (dimming). In either mode, the compensating system should have a delay circuit that would prevent response to momentary daylight reductions caused by cloud movement. Of the two compensating systems, the continuously varying mode would be preferred. The cycling effect of the on/off mode would be less expensive and better suited for retrofit applications, but it could also be quite distracting and annoying. The user system, which is the on/off control of the compensating system, should be controlled with a timer or other device that would automatically turn off the compensating system at the end of the day. A short-period (maximum 3 hr) manual override would have to be available for those who must work overtime or come in at night. In larger rooms, multiple dimmers and photocell sensors may be required to control separate rows of luminaires to assure a more uniform lighting level throughout the space.

Figures 13-8 and 13-9 are the result of daylighting computer simulation studies (7) and provide a great deal of insight into the energy effectiveness of various control strategies. They represent the cumulative energy usage of a typical building through an entire year. Figure 13-8 compares the *lighting* energy savings only for one-step switching, two-step switching, and continuous dimming. In all cases, as the amount of fenestration or aperture size is increased (to the right along the abscissa), less lighting energy is

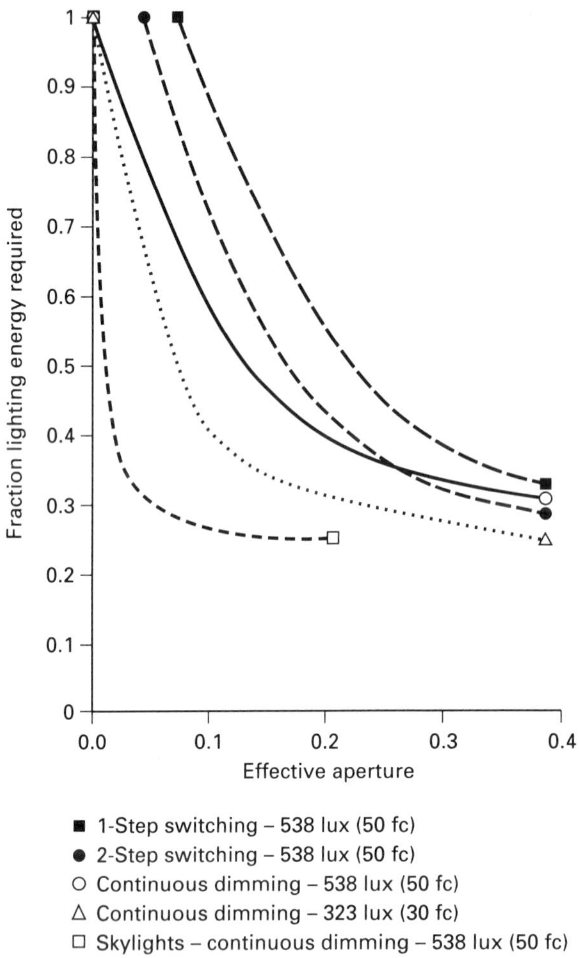

Figure 13-8. Electric Lighting Energy Requirements from a Daylight Simulation Study. (Courtesy of Lawrence Berkeley Laboratory.)

required. In general, for a given size aperture, continuous dimming saves the most lighting energy and one-step switching the least, with two-step switching being intermediate between the two. Exceptions occur toward the right and left sides of the graph. On the left side, there is an amount of fenestration for each of the switching alternatives below which no lighting energy savings occurs. This is because, at that level, insufficient daylight is present to permit turning off electric lights and still maintain the prescribed task illuminance (50 fc in the case of this study). Continuous dimming saves lighting energy all the way to the extreme left of the graph (progressively smaller apertures) because no matter how little daylight is available, some electric light can be deleted. It is interesting to note that two-step switching is more effective than continuous dimming for very large amounts of fenestration. This is because, for very large fenestration area, all of the lights can frequently be turned off. If the lights are *switched* off, no

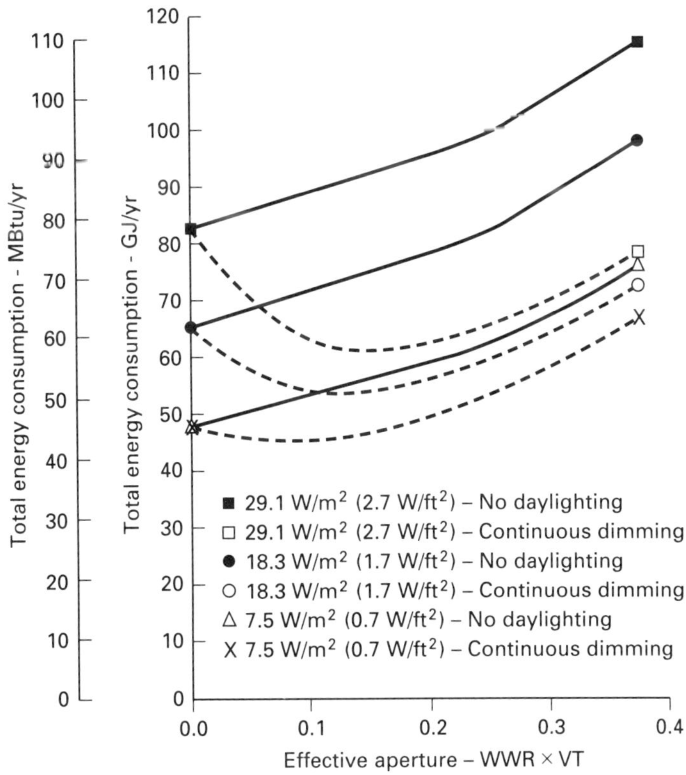

Figure 13-9. Total Building Energy Consumption from a Daylight Simulation Study. (Courtesy of Lawrence Berkeley Laboratory.)

energy is being used. However, if the lights are *dimmed* to the lowest setting, the dimmer/ballast/lamp combination still draws some energy.

Lighting energy is only part of the story, though. Figure 13-9 graphs *total* energy consumption (lighting plus HVAC) for three different power densities of electric light, with and without daylighting (that is, continuous dimming vs. ignoring daylighting and leaving all the lights on continually). This graph indicates that there is an optimum amount of daylight that should be admitted to a space (corresponding to the aperture size where the dimming curves are at minima). Using less fenestration than the optimum means that extra energy is consumed for electric lighting. Using more fenestration than the optimum causes excessive heat loss/gain and the associated energy consumption, without an offsetting reduction in lighting energy.

Studies of this type help point out the complexity of the daylighting issue. Energy gains are possible, but it is also possible to hurt the energy use profile of a building through careless or uninformed application of daylighting principles. A complete daylighting design procedure will involve an energy and economic analysis of the building's performance over time.

REFERENCES

1. Architectural Research Laboratory, "The Effect of Windowless Classrooms on Elementary School Children," Department of Architecture, University of Michigan, Ann Arbor, 1965.
2. Johnson, R., D. Arasteh, D. Connell, and S. Selkowitz, "The Effect of Daylighting Strategies on Building Cooling Loads and Overall Energy Performance," *Proceedings of Thermal Performance of the Exterior Envelopes of Buildings III*, Clearwater Beach, Florida, 1985.
3. Kaufman, John E. and Jack F. Christensen, eds., *IES Lighting Handbook,* Reference Volume, 1984. New York: Illuminating Engineering Society, 1984.
4. Lam, William M. C., *Sunlighting as Formgiver for Architecture*. New York: Van Nostrand Reinhold, 1986.
5. Rosenfeld, A. H., "Some Comments on Dual Solar-Control Venetian Blinds," Lawrence Berkeley Laboratory, Department of Physics, University of California, Berkeley.
6. Rosenfeld, A. H., and S. E. Selkowitz, "Beam and Diffuse Daylighting, and Peak Power," *Proceedings: The Basis for Effective Management of Lighting Energy Symposium*, Federal Energy Administration, Washington, D.C., October 29, 30, 1975.
7. Schiller, M. ed., *Simulating Daylight with Architectural Models*. Daylight Network of North America, undated.
8. Silverstein, S. D., "Efficient Energy Utilization in Buildings: The Architectural Window," *Proceedings 10th Intersociety Energy Conversion Engineering Conference*, 1973, p. 685.
9. Stein, B., J. Reynolds, and W. McGuinnes, *Mechanical and Electrical Systems for Buildings*, 7th ed. New York: John Wiley, 1986.

EXAMPLE PROBLEMS

IES lumen method example

Consider a school located in Los Angeles, CA (34° north latitude). Find the max, mid, and min illuminances at 10:00 A.M. on June 21 for a room having a 30 ft long south-facing window and a room width of 20 ft. The sky is clear, the sill height is 3 ft, the transmittance of the glazing is 85%, and the reflectance of the surrounding ground is 30%.

1. Find the two components (sky and ground) for the vertical illuminance on the exterior of the window.
   a. From Table 13-5 we find that the sky contributes 800 fc.
   b. From Table 13-4 the total horizontal illuminance on the ground is 10,100 fc. This value must be multiplied by the reflectance of the ground and the configuration factor of 0.5 in order to obtain the contribution from ground light to the vertical illuminance on the window:

$$(10{,}100)(0.30)(0.5) = 1515 \text{ fc}$$

2. Find the flux on the exterior of the window by multiplying illuminance by window area:

   a. sky: (800 fc)(30 ft)(5 ft) = 120,000 lm
   b. ground: (1515)(30 ft)(5 ft) = 227,250 lm

3. Find the flux transmitted through the window by multiplying by the transmittance (given as 0.85) and the LLF (found from Table 13-7 as 0.9)

a. sky: (120,000)(0.85)(0.9) = 91,800 lm
b. ground: (227,250)(0.85)(0.9) = 173,846 lm

4. Multiply the flux by the appropriate C and K factors from Table 13-6:

Illuminance due to light from the sky:

$$E_{max} = (91{,}800)(0.0143)(0.145) = 190 \text{ fc}$$
$$E_{mid} = (91{,}800)(0.0100)(0.100) = 101 \text{ fc}$$
$$E_{min} = (91{,}800)(0.0079)(0.105) = 76 \text{ fc}$$

Illuminance due to light reflected from the ground:

$$E_{max} = (173{,}846)(0.0102)(0.124) = 220 \text{ fc}$$
$$E_{mid} = (173{,}846)(0.0094)(0.123) = 201 \text{ fc}$$
$$E_{min} = (173{,}846)(0.0082)(0.0994) = 142 \text{ fc}$$

Total illuminance

$$E_{max} = 190 + 220 = 410 \text{ fc}$$
$$E_{mid} = 101 + 201 = 302 \text{ fc}$$
$$E_{min} = 76 + 142 = 218 \text{ fc}$$

CIE daylight factor method example

Consider an office building located in Topeka, KS (40° north latitude). It is desired that for 70% of the hours between 9:00 A.M. and 5:00 P.M., 30 fc be available at the worst location in the space under overcast conditions. What is the maximum depth a 35 ft long office could be if the ceiling height is 9 ft, 90% of the wall is glazed, and an adjacent building obstructs the sky at an angle of 10° above the horizon?

a. From Figure 13-10 we find the horizontal external illuminance to be 1400 fc (this value will be met or exceeded during 70% of working hours).
b. This yields a minimum DF of 30/1400 or 2.1%.
c. The dirt correction from Table 13-7 is 0.9.
d. The design DF is then 2.1/.9 = 2.33.
e. From Figure 13-11 this gives a maximum permissible room depth to prevent the daylight factor from being less than 2.33 of $3H$.
f. Entering Figure 13-13 with this value gives a correction factor due to a 10° occlusion of 0.88.
g. The eventual DF used will have to be multiplied by the correction factor in part **f** to find the design DF. Thus a DF of 2.65 = 2.33/0.88 is required.
h. A DF of 2.65 implies (from Figure 13-11) a room depth of $2.7H$.
i. $2.7H$ in turn requires (from Figure 13-13) a correction factor of 0.89. Then, if the 2.65 DF is used and corrected for a 10° occlusion, the result is

$$2.65 \times 0.89 = 2.36 \text{ DF}$$

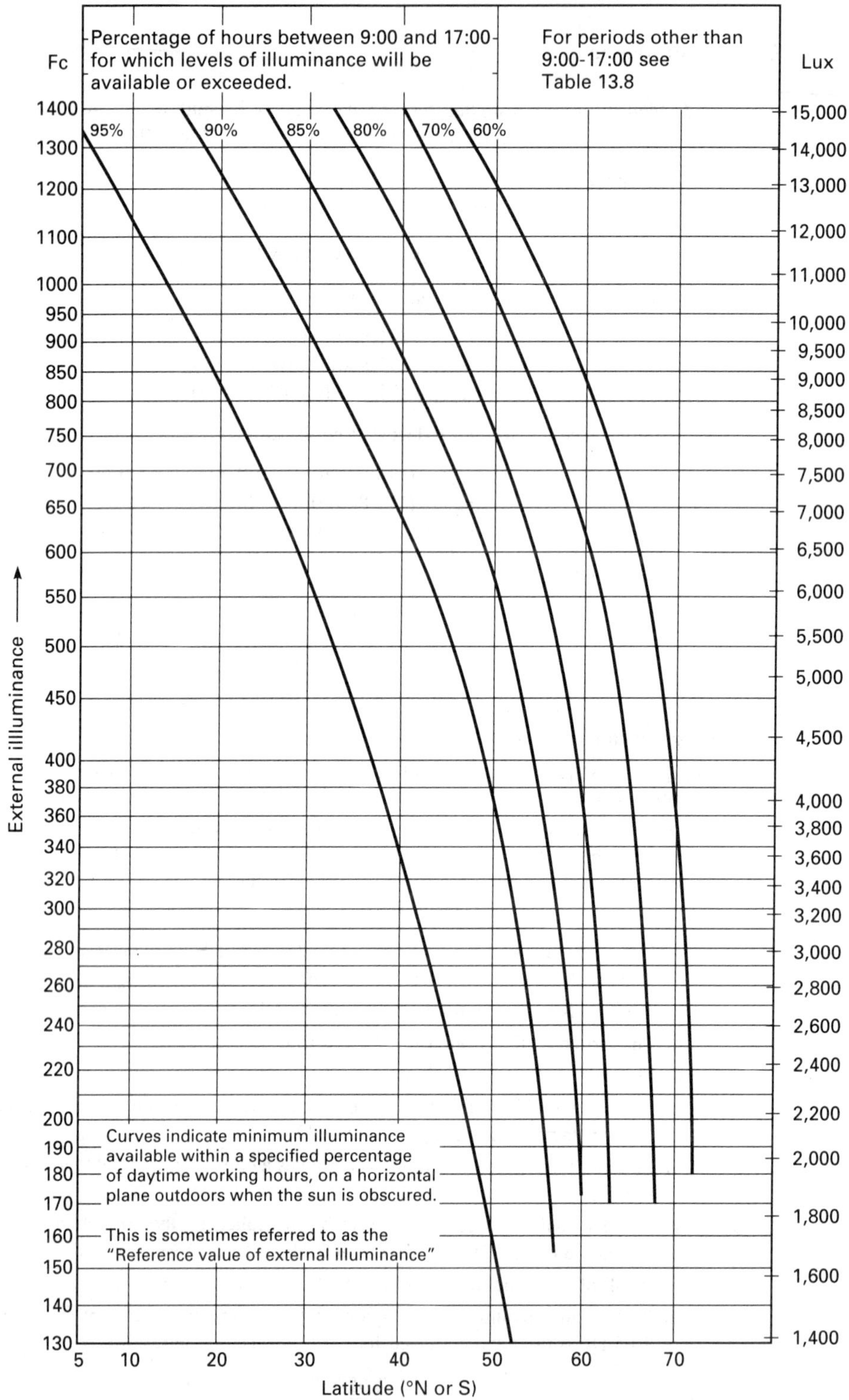

Figure 13-10. External Illuminance that Will Be Met or Exceeded for Various Percentages of the Day between 9:00 AM and 5:00 PM. (Courtesy of the Commission Internationale de l'Eclairage.)

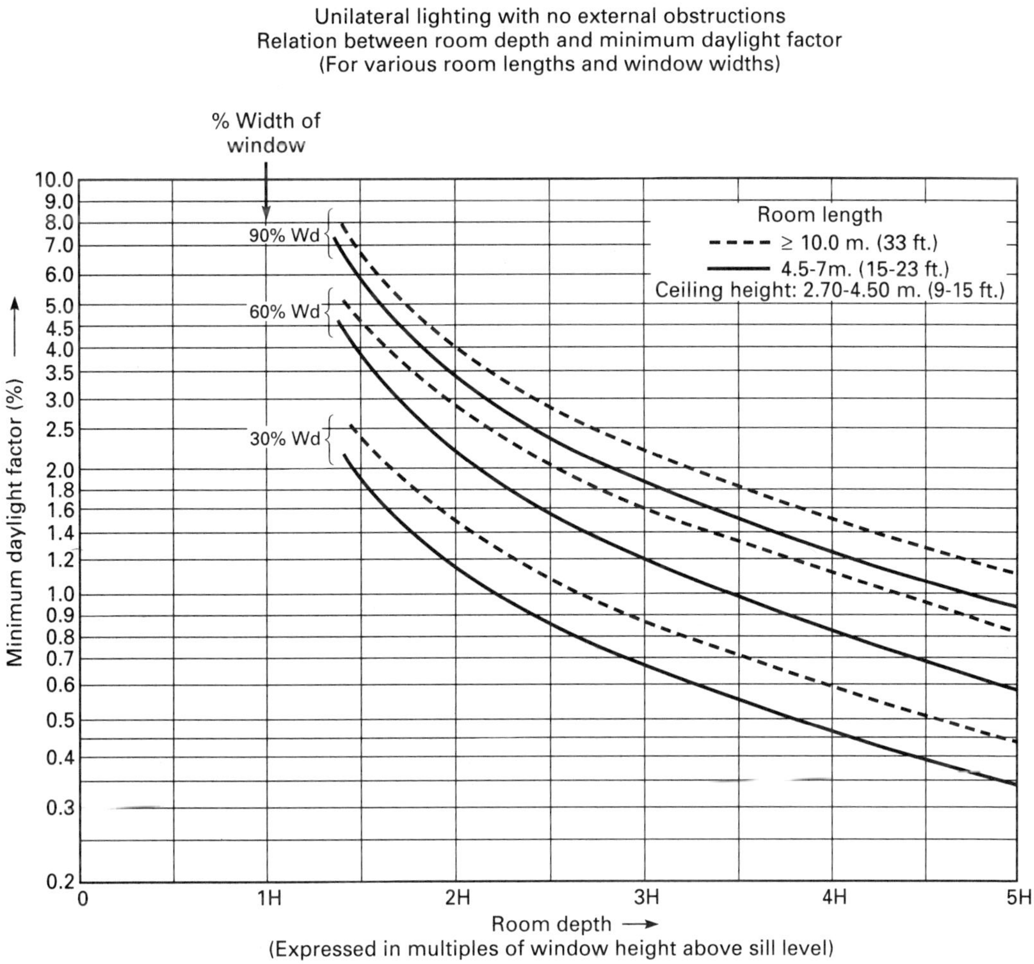

Figure 13-11. Daylight Factors for Vertical Windows in One Wall. (Courtesy of the Commission Internationale de l'Eclairage.)

This DF is reasonably close to the 2.33 design DF we determined was necessary in step **d.**

j. The height of the window is determined by subtracting the sill height and head space from the room height. So the window height is

$$9 - 3 - 1 = 5 \text{ ft}$$

k. When multiplied by the $3H$ of step e, the maximum permissible room depth for the conditions given becomes

$$3 \times 5 \text{ ft} = 15 \text{ ft}$$

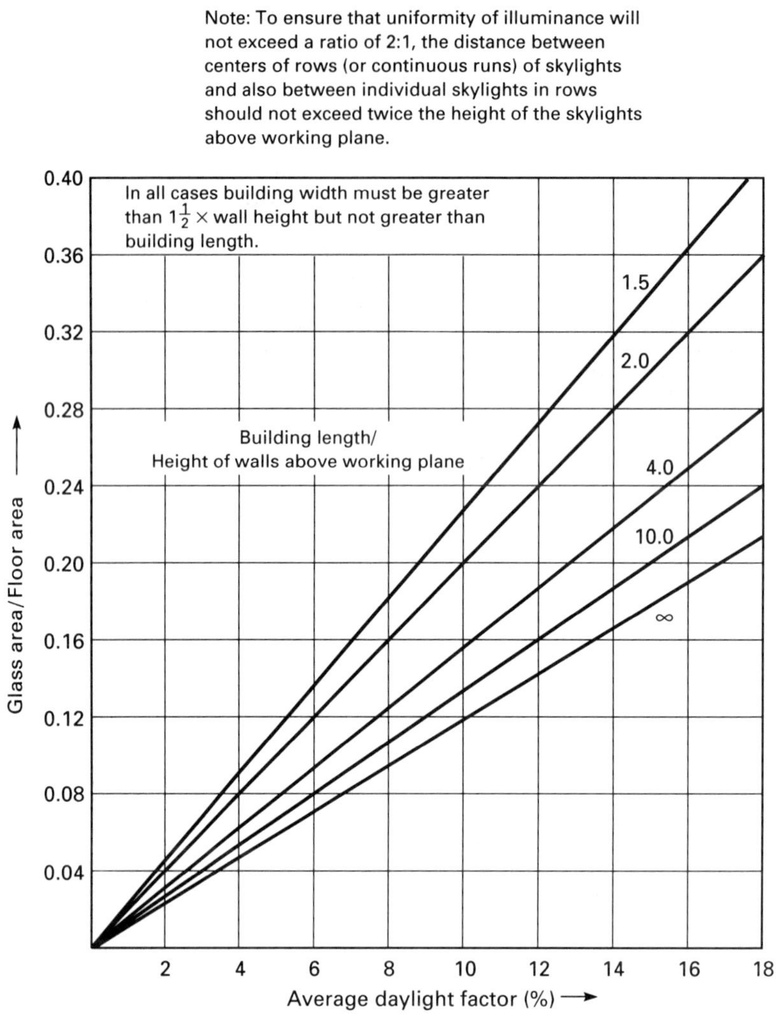

Figure 13-12. Daylight Factors for Skylights in a Flat Roof. (Courtesy of the Commission Internationale de l'Eclairage.)

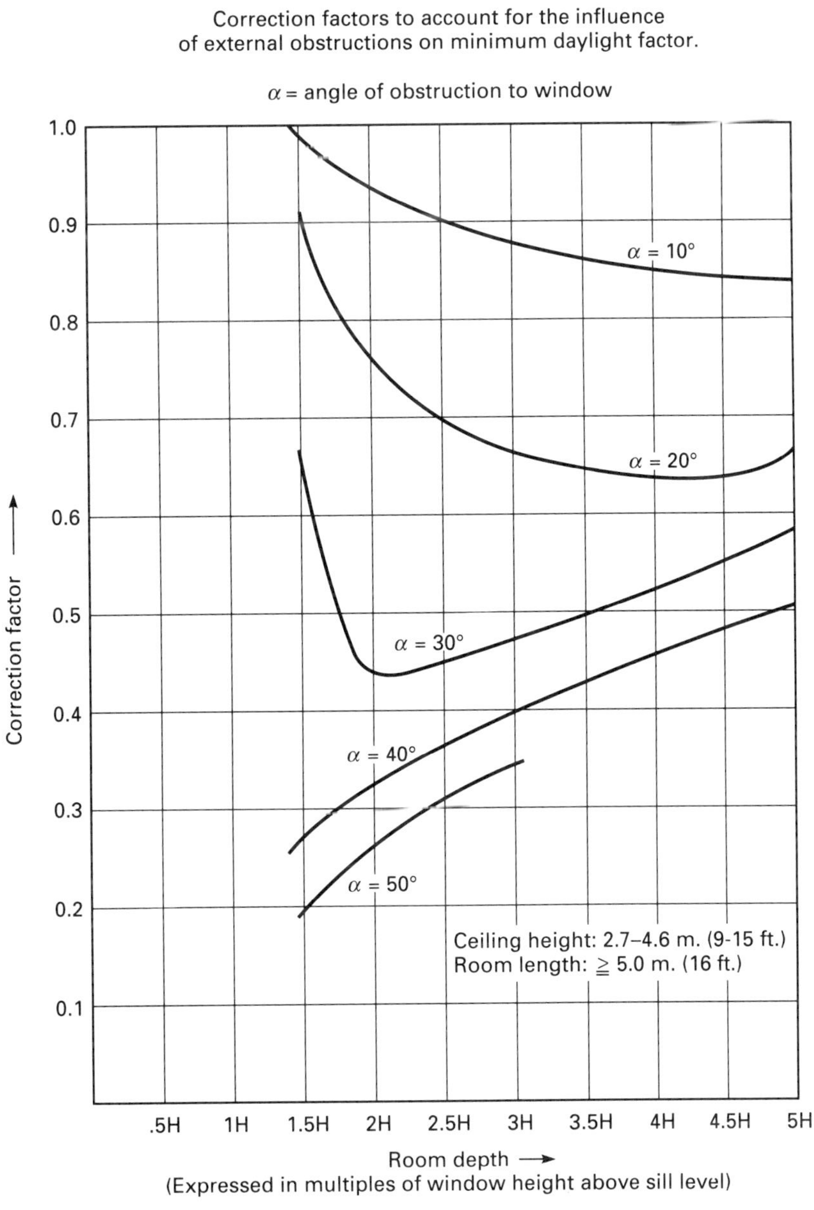

Figure 13-13. Correction Factors to Account for External Obstructions. The angle is the angle formed by a horizontal line through the window and a line from the window to the top of the adjacent building. (Courtesy of the Commission Internationale de l'Eclairage.)

DATA SECTION

TABLE 13-4 Exterior Horizontal Illuminance

		Exterior Horizontal Illuminance in Kilolux (Multiply by 100 for Footcandles*)								
		December 21			March and September 21			June 21		
Latitude (degrees north)	Component	8 A.M. 4 P.M.	10 A.M. 2 P.M.	Noon	8 A.M. 4 P.M.	10 A.M. 2 P.M.	Noon	8 A.M. 4 P.M.	10 A.M. 2 P.M.	Noon
					Clear Day†					
30	Direct	9	42	55	34	72	87	52	86	99
	Diffuse	8	12	13	11	14	15	13	15	16
	Total	17	54	68	45	86	102	65	101	115
34	Direct	5	35	48	32	68	82	52	86	98
	Diffuse	7	11	12	11	14	15	13	15	16
	Total	12	46	60	43	82	97	65	101	114
38	Direct	3	29	41	29	64	77	53	84	96
	Diffuse	6	10	12	11	14	15	13	15	16
	Total	9	39	53	40	78	92	66	99	112
42	Direct	1	22	33	27	59	71	53	83	94
	Diffuse	5	10	11	10	13	14	13	15	16
	Total	6	32	44	37	72	85	66	98	110
46	Direct	0	16	25	24	54	65	53	80	91
	Diffuse	4	9	10	10	13	14	13	15	16
	Total	4	25	35	34	67	79	66	95	107
50	Direct	0	10	18	21	49	59	52	78	87
	Diffuse	0	8	9	10	12	13	13	15	15
	Total	0	18	27	31	61	72	65	93	102

Partly Cloudy Day‡

30	Direct	0	13	20	9	33	44	19	44	55
	Diffuse	9	22	27	20	34	39	27	40	45
	Total	9	35	47	29	67	83	46	84	100
34	Direct	0	9	16	8	30	40	20	44	54
	Diffuse	7	20	25	19	33	38	27	40	45
	Total	7	29	41	27	63	78	47	84	99
38	Direct	0	6	12	7	27	36	20	43	52
	Diffuse	6	17	22	18	31	36	28	39	44
	Total	6	23	34	25	58	72	48	82	96
42	Direct	0	4	8	6	24	32	20	42	50
	Diffuse	4	15	19	17	29	34	28	39	43
	Total	4	19	27	23	53	66	48	81	93
46	Direct	0	2	5	4	20	28	20	40	48
	Diffuse	2	12	16	16	27	32	28	38	42
	Total	2	14	21	20	47	60	48	78	90
50	Direct	0	1	2	3	17	24	19	38	45
	Diffuse	0	10	13	15	25	29	27	37	41
	Total	0	11	15	18	42	53	46	75	86

TABLE 13-4 (continued)

Exterior Horizontal Illuminance in Kilolux (Multiply by 100 for Footcandles*)										
Latitude (degrees north)	Component	December 21			March and September 21			June 21		
		8 A.M. 4 P.M.	10 A.M. 2 P.M.	Noon	8 A.M. 4 P.M.	10 A.M. 2 P.M.	Noon	8 A.M. 4 P.M.	10 A.M. 2 P.M.	Noon
Overcast Days§										
30	Direct	0	0	0	0	0	0	0	0	0
	Diffuse	4	11	13	9	16	18	13	19	21
	Total	4	11	13	9	16	18	13	19	21
34	Direct	0	0	0	0	0	0	0	0	0
	Diffuse	4	9	12	9	15	18	13	19	21
	Total	4	9	12	9	15	18	13	19	21
38	Direct	0	0	0	0	0	0	0	0	0
	Diffuse	3	8	10	9	15	17	13	19	21
	Total	3	8	10	9	15	17	13	19	21
42	Direct	0	0	0	0	0	0	0	0	0
	Diffuse	2	7	9	8	14	16	13	18	20
	Total	2	7	9	8	14	16	13	18	20
46	Direct	0	0	0	0	0	0	0	0	0
	Diffuse	1	6	8	8	13	15	13	18	20
	Total	1	6	8	8	13	15	13	18	20
50	Direct	0	0	0	0	0	0	0	0	0
	Diffuse	0	5	6	7	12	14	13	17	19
	Total	0	5	6	7	12	14	13	17	19

* More exact multiplier is 92.9.

† Atmospheric extinction coefficient = 0.21. Atmospheric extinction coefficient is the reciprocal of the optical air mass times the natural logarithm of the ratio of the extraterrestrial direct normal solar illuminance to the sea level direct normal solar illuminance. Optical air mass is the ratio of the sea level path length through the atmosphere toward the sun to the sea level path toward the zenith.

‡ Atmospheric Extinction Coefficient = 0.80.

§ Typical nonprecipitative minimum.

Courtesy of the Illuminating Engineering Society of North America.

TABLE 13-5A Exterior Vertical Illuminance on a Clear Day† in Kilolux. (For approximate footcandle values, multiply by 100.*)

Latitude (degrees north)	Component	December 21					March and September 21					June 21				
		8 A.M.	10 A.M.	Noon	2 P.M.	4 P.M.	8 A.M.	10 A.M.	Noon	2 P.M.	4 P.M.	8 A.M.	10 A.M.	Noon	2 P.M.	4 P.M.
							Facing North									
30	Direct	0	0	0	0	0	0	0	0	0	0	10	0	0	0	10
	Diffuse	3	4	4	4	3	6	6	5	6	6	8	7	7	7	8
	Total	3	4	4	4	3	6	6	5	6	6	18	7	7	7	18
34	Direct	0	0	0	0	0	0	0	0	0	0	6	0	0	0	6
	Diffuse	3	4	4	4	3	5	5	5	5	5	8	7	6	7	8
	Total	3	4	4	4	3	5	5	5	5	5	14	7	6	7	14
38	Direct	0	0	0	0	0	0	0	0	0	0	3	0	0	0	3
	Diffuse	3	4	4	4	3	5	5	5	5	5	8	7	6	7	8
	Total	3	4	4	4	3	5	5	5	5	5	11	7	6	7	11
42	Direct	0	0	0	0	0	0	0	0	0	0	0	0	0	0	0
	Diffuse	2	3	3	3	2	5	5	5	5	5	7	6	5	6	7
	Total	2	3	3	3	2	5	5	5	5	5	7	6	5	6	7
46	Direct	0	0	0	0	0	0	0	0	0	0	0	0	0	0	0
	Diffuse	2	3	3	3	2	5	5	5	5	5	7	6	6	6	7
	Total	2	3	3	3	2	5	5	5	5	5	7	6	6	6	7
50	Direct	0	0	0	0	0	0	0	0	0	0	0	0	0	0	0
	Diffuse	0	3	3	3	0	5	5	4	5	5	7	6	6	6	7
	Total	0	3	3	3	0	5	5	4	5	5	7	6	6	6	7
							Facing South									
30	Direct	26	64	74	64	26	20	42	50	42	20	0	5	11	5	0
	Diffuse	8	12	14	12	8	8	11	11	11	8	7	8	8	8	7
	Total	34	76	88	76	34	28	53	61	53	28	7	13	19	13	7
34	Direct	20	63	75	63	20	21	46	55	46	21	0	11	18	11	0
	Diffuse	8	12	14	12	8	8	11	12	11	8	7	8	8	8	7
	Total	28	75	89	75	28	29	57	67	57	29	7	19	26	19	7
38	Direct	13	61	75	61	13	23	50	60	50	23	0	17	25	17	0
	Diffuse	7	12	14	12	7	8	11	12	11	8	7	8	9	8	7
	Total	20	73	89	73	20	31	61	72	61	31	7	25	34	25	7
42	Direct	5	57	72	57	5	24	53	64	53	24	1	23	31	23	1
	Diffuse	6	12	14	12	6	8	12	13	12	8	7	9	9	9	7
	Total	11	69	86	69	11	32	65	77	65	32	8	32	40	32	8

TABLE 13-5A (continued)

Latitude (degrees north)	Component	December 21					March and September 21					June 21				
		8 A.M.	10 A.M.	Noon	2 P.M.	4 P.M.	8 A.M.	10 A.M.	Noon	2 P.M.	4 P.M.	8 A.M.	10 A.M.	Noon	2 P.M.	4 P.M.
46	Direct	0	51	68	51	0	25	56	68	56	25	5	28	38	28	5
	Diffuse	5	11	13	11	5	8	12	13	12	8	8	9	10	9	8
	Total	5	62	81	62	5	33	68	81	68	33	13	37	48	37	13
50	Direct	0	41	60	41	0	25	58	70	58	25	8	34	44	34	8
	Diffuse	0	10	12	10	0	8	12	14	12	8	8	10	11	10	8
	Total	0	51	72	51	0	33	70	84	70	33	16	44	55	44	16
Facing East																
30	Direct	36	39	0	0	0	68	48	0	0	0	69	45	0	0	0
	Diffuse	10	10	7	5	3	13	11	7	5	4	14	11	7	6	4
	Total	46	49	7	5	3	81	59	7	5	4	83	56	7	6	4
34	Direct	28	37	0	0	0	67	48	0	0	0	69	45	0	0	0
	Diffuse	9	10	7	5	2	13	11	7	5	3	14	11	7	6	4
	Total	37	47	7	5	2	80	59	7	5	3	83	56	7	6	4
38	Direct	17	35	0	0	0	65	47	0	0	0	69	44	0	0	0
	Diffuse	8	9	7	5	2	13	11	7	5	3	14	11	7	6	4
	Total	25	44	7	5	2	78	58	7	5	3	83	55	7	6	4
42	Direct	6	32	0	0	0	63	46	0	0	0	69	44	0	0	0
	Diffuse	7	9	7	4	2	13	11	8	5	3	14	11	7	6	4
	Total	13	41	7	4	2	76	57	8	5	3	83	55	7	6	4
46	Direct	0	27	0	0	0	60	45	0	0	0	69	44	0	0	0
	Diffuse	5	8	6	4	1	12	11	8	5	3	14	11	7	6	4
	Total	5	35	6	4	1	72	56	8	5	3	83	55	7	6	4
50	Direct	0	22	0	0	0	57	44	0	0	0	69	44	0	0	0
	Diffuse	0	8	6	4	0	12	11	7	5	3	14	11	7	6	4
	Total	0	30	6	4	0	69	55	7	5	3	83	55	7	6	4

Facing West

30	Direct	0	0	0	39	36	0	0	0	48	68	0	0	0	45	69
	Diffuse	3	5	7	10	10	4	5	7	11	13	4	6	7	11	14
	Total	3	5	7	49	46	4	5	7	59	81	4	6	7	56	83
34	Direct	0	0	0	37	28	0	0	0	48	67	0	0	0	45	69
	Diffuse	2	5	7	10	9	3	5	7	11	13	4	6	7	11	14
	Total	2	5	7	47	37	3	5	7	59	80	4	6	7	56	83
38	Direct	0	0	0	35	17	0	0	0	47	65	0	0	0	44	69
	Diffuse	2	5	7	9	8	3	5	7	11	13	4	6	7	11	14
	Total	2	5	7	44	25	3	5	7	58	78	4	6	7	55	83
42	Direct	0	0	0	32	6	0	0	0	46	63	0	0	0	44	69
	Diffuse	2	4	7	9	7	3	5	8	11	13	4	6	7	11	14
	Total	2	4	7	41	13	3	5	8	57	76	4	6	7	55	83
46	Direct	0	0	0	27	0	0	0	0	45	60	0	0	0	44	69
	Diffuse	1	4	6	8	5	3	5	8	11	12	4	6	7	11	14
	Total	1	4	6	35	5	3	5	8	56	72	4	6	7	55	83
50	Direct	0	0	0	22	0	0	0	0	44	57	0	0	0	44	69
	Diffuse	0	4	6	8	0	3	5	7	11	12	4	6	7	11	14
	Total	0	4	6	30	0	3	5	7	55	69	4	6	7	55	83

* More exact multiplier is 92.9.

† Atmospheric extinction coefficient = 0.21.

Courtesy of the Illuminating Engineering Society of North America.

TABLE 13-5B Exterior Vertical Illuminance on a Partly Cloudy Day† in Kilolux. (For approximate footcandle values, multiply by 100.*)

Latitude (degrees north)	Component	December 21					March and September 21					June 21				
		8 A.M.	10 A.M.	Noon	2 P.M.	4 P.M.	8 A.M.	10 A.M.	Noon	2 P.M.	4 P.M.	8 A.M.	10 A.M.	Noon	2 P.M.	4 P.M.
							Facing North									
30	Direct	0	0	0	0	0	0	0	0	0	0	4	0	0	0	4
	Diffuse	3	7	8	7	3	8	12	13	12	8	14	17	17	17	14
	Total	3	7	8	7	3	8	12	13	12	8	18	17	17	17	18
34	Direct	0	0	0	0	0	0	0	0	0	0	2	0	0	0	2
	Diffuse	3	6	7	6	3	8	11	12	11	8	14	16	17	16	14
	Total	3	6	7	6	3	8	11	12	11	8	16	16	17	16	16
38	Direct	0	0	0	0	0	0	0	0	0	0	1	0	0	0	1
	Diffuse	2	5	6	5	2	7	10	11	10	7	13	15	16	15	13
	Total	2	5	6	5	2	7	10	11	10	7	14	15	16	15	14
42	Direct	0	0	0	0	0	0	0	0	0	0	0	0	0	0	0
	Diffuse	1	4	5	4	1	7	9	10	9	7	13	15	15	15	13
	Total	1	4	5	4	1	7	9	10	9	7	13	15	15	15	13
46	Direct	0	0	0	0	0	0	0	0	0	0	0	0	0	0	0
	Diffuse	1	4	4	4	1	6	9	9	9	6	13	14	14	14	13
	Total	1	4	4	4	1	6	9	9	9	6	13	14	14	14	13
50	Direct	0	0	0	0	0	0	0	0	0	0	0	0	0	0	0
	Diffuse	0	3	3	3	0	6	8	8	8	6	12	13	13	13	12
	Total	0	3	3	3	0	6	8	8	8	6	12	13	13	13	12
							Facing South									
30	Direct	1	19	28	19	1	5	19	25	19	5	0	3	6	3	0
	Diffuse	8	19	24	19	8	12	21	26	21	12	12	18	21	18	12
	Total	9	38	52	38	9	17	40	51	40	17	12	21	27	21	12
34	Direct	0	16	25	16	0	5	20	27	20	5	0	6	10	6	0
	Diffuse	6	18	22	18	6	12	21	26	21	12	12	19	22	19	12
	Total	6	34	47	34	6	17	41	53	41	17	12	25	32	25	12

38	Direct	0	13	22	13	0	5	21	28	21	5	0	9	14	9	0
	Diffuse	5	16	21	16	5	11	21	26	21	11	13	20	23	20	13
	Total	5	29	43	29	5	16	42	54	42	16	13	29	37	29	13
42	Direct	0	9	17	9	0	5	21	29	21	5	0	11	17	11	0
	Diffuse	3	14	19	14	3	11	21	26	21	11	13	20	24	20	13
	Total	3	23	36	23	3	16	42	55	42	16	13	31	41	31	13
46	Direct	0	5	13	5	0	5	21	29	21	5	2	14	20	14	2
	Diffuse	2	12	17	12	2	11	21	25	21	11	14	21	25	21	14
	Total	2	17	30	17	2	16	42	54	42	16	16	35	45	35	16
50	Direct	0	2	8	2	0	4	20	28	20	4	3	16	23	16	3
	Diffuse	0	10	14	10	0	10	20	25	20	10	14	21	25	21	14
	Total	0	12	22	12	0	14	40	53	40	14	17	37	49	37	17
							Facing East									
30	Direct	2	12	0	0	0	17	22	0	0	0	26	23	0	0	0
	Diffuse	9	15	13	8	3	19	23	17	11	6	24	26	19	13	8
	Total	11	27	13	8	3	36	45	17	11	6	50	49	19	13	8
34	Direct	1	10	0	0	0	16	21	0	0	0	26	23	0	0	0
	Diffuse	7	14	12	7	2	18	22	17	11	5	24	25	18	13	8
	Total	8	24	12	7	2	34	43	17	11	5	50	48	18	13	8
38	Direct	0	7	0	0	0	14	20	0	0	0	26	23	0	0	0
	Diffuse	6	12	11	7	2	17	21	16	10	5	24	25	18	13	8
	Total	6	19	11	7	2	31	41	16	10	5	50	48	18	13	8
42	Direct	0	5	0	0	0	13	18	0	0	0	26	22	0	0	0
	Diffuse	4	11	10	6	1	16	20	15	10	5	24	25	18	13	8
	Total	4	16	10	6	1	29	38	15	10	5	50	47	18	13	8
46	Direct	0	3	0	0	0	11	17	0	0	0	26	22	0	0	0
	Diffuse	2	9	8	5	0	16	19	15	9	4	24	24	18	13	8
	Total	2	12	8	5	0	27	36	15	9	4	50	46	18	13	8
50	Direct	0	1	0	0	0	9	15	0	0	0	26	21	0	0	0
	Diffuse	0	7	7	4	0	15	18	14	9	4	24	24	17	12	8
	Total	0	8	7	4	0	24	33	14	9	4	50	45	17	12	8

TABLE 13-5B (continued)

Latitude (degrees north)	Component	December 21					March and September 21					June 21				
		8 A.M.	10 A.M.	Noon	2 P.M.	4 P.M.	8 A.M.	10 A.M.	Noon	2 P.M.	4 P.M.	8 A.M.	10 A.M.	Noon	2 P.M.	4 P.M.
							Facing West									
30	Direct	0	0	0	12	2	0	0	0	22	17	0	0	0	23	26
	Diffuse	3	8	13	15	9	6	11	17	23	19	8	13	19	26	24
	Total	3	8	13	27	11	6	11	17	45	36	8	13	19	49	50
34	Direct	0	0	0	10	1	0	0	0	21	16	0	0	0	23	26
	Diffuse	2	7	12	14	7	5	11	17	22	18	8	13	18	25	24
	Total	2	7	12	24	8	5	11	17	43	34	8	13	18	48	50
38	Direct	0	0	0	7	0	0	0	0	20	14	0	0	0	23	26
	Diffuse	2	7	11	12	6	5	10	16	21	17	8	13	18	25	24
	Total	2	7	11	19	6	5	10	16	41	31	8	13	18	48	50
42	Direct	0	0	0	5	0	0	0	0	18	13	0	0	0	22	26
	Diffuse	1	6	10	11	4	5	10	15	20	16	8	13	18	25	24
	Total	1	6	10	16	4	5	10	15	38	29	8	13	18	47	50
46	Direct	0	0	0	3	0	0	0	0	17	11	0	0	0	22	26
	Diffuse	0	5	8	9	2	4	9	15	19	16	8	13	18	24	24
	Total	0	5	8	12	2	4	9	15	36	27	8	13	18	46	50
50	Direct	0	0	0	1	0	0	0	0	15	9	0	0	0	21	26
	Diffuse	0	4	7	7	0	4	9	14	18	15	8	12	17	24	24
	Total	0	4	7	8	0	4	9	14	33	24	8	12	17	45	50

* More exact multiplier is 92.9.

† Atmospheric extinction coefficient = 0.80.

Courtesy of the Illuminating Engineering Society of North America.

TABLE 13-5C Exterior Vertical Illuminance on an Overcast Day†, Any Orientation. (For approximate footcandle values, multiply by 100.*)

Latitude (degrees north)	Component	December 21					March and September 21					June 21				
		8 A.M.	10 A.M.	Noon	2 P.M.	4 P.M.	8 A.M.	10 A.M.	Noon	2 P.M.	4 P.M.	8 A.M.	10 A.M.	Noon	2 P.M.	4 P.M.
30	Direct	0	0	0	0	0	0	0	0	0	0	0	0	0	0	0
	Diffuse	2	4	5	4	2	4	6	7	6	4	5	8	8	8	5
	Total	2	4	5	4	2	4	6	7	6	4	5	8	8	8	5
34	Direct	0	0	0	0	0	0	0	0	0	0	0	0	0	0	0
	Diffuse	1	4	5	4	1	4	6	7	6	4	5	7	8	7	5
	Total	1	4	5	4	1	4	6	7	6	4	5	7	8	7	5
38	Direct	0	0	0	0	0	0	0	0	0	0	0	0	0	0	0
	Diffuse	1	3	4	3	1	3	6	7	6	3	5	7	8	7	5
	Total	1	3	4	3	1	3	6	7	6	3	5	7	8	7	5
42	Direct	0	0	0	0	0	0	0	0	0	0	0	0	0	0	0
	Diffuse	1	3	4	3	1	3	5	6	5	3	5	7	8	7	5
	Total	1	3	4	3	1	3	5	6	5	3	5	7	8	7	5
46	Direct	0	0	0	0	0	0	0	0	0	0	0	0	0	0	0
	Diffuse	0	2	3	2	0	3	5	6	5	3	5	7	8	7	5
	Total	0	2	3	2	0	3	5	6	5	3	5	7	8	7	5
50	Direct	0	0	0	0	0	0	0	0	0	0	0	0	0	0	0
	Diffuse	0	2	2	2	0	3	5	5	5	3	5	7	8	7	5
	Total	0	2	2	2	0	3	5	5	5	3	5	7	8	7	5

* More exact multiplier is 92.9.

† Typical nonprecipative minimum.

Courtesy of the Illuminating Engineering Society of North America.

TABLE 13-6 Coefficients for use with the IES Lumen Method

A. Illuminance from an overcast sky, without window controls

C

Room Length			6.1 M (20 FT)		9.1 M (30 FT)		12.2 M (40 FT)	
Wall Reflectance (%)			70	30	70	30	70	30
	Room Width M	FT						
MAX	6.1	20	.0276	.0251	.0191	.0173	.0143	.0137
	9.1	30	.0272	.0248	.0188	.0172	.0137	.0131
	12.2	40	.0269	.0246	.0182	.0171	.0133	.0130
MID	6.1	20	.0159	.0117	.0101	.0087	.0081	.0071
	9.1	30	.0058	.0050	.0054	.0040	.0034	.0033
	12.2	40	.0039	.0027	.0030	.0023	.0022	.0019
MIN	6.1	20	.0087	.0053	.0063	.0043	.0050	.0037
	9.1	30	.0032	.0019	.0029	.0017	.0020	.0014
	12.2	40	.0019	.0009	.0016	.0009	.0012	.0008

K

Ceiling Height			2.4 M (8 FT)		3 M (10 FT)		3.7 M (12 FT)		4.3 M (14 FT)	
Wall Reflectance (%)			70	30	70	30	70	30	70	30
	Room Width M	FT								
MAX	6.1	20	.125	.129	.121	.123	.111	.111	.0991	.0973
	9.1	30	.122	.131	.122	.121	.111	.111	.0945	.0973
	12.2	40	.145	.133	.131	.126	.111	.111	.0973	.0982
MID	6.1	20	.0908	.0982	.107	.115	.111	.111	.105	.122
	9.1	30	.156	.102	.0939	.113	.111	.111	.121	.134
	12.2	40	.106	.0948	.123	.107	.111	.111	.135	.127
MIN	6.1	20	.0908	.102	.0951	.114	.111	.111	.118	.134
	9.1	30	.0924	.119	.101	.114	.111	.111	.125	.126
	12.2	40	.111	.0926	.125	.109	.111	.111	.133	.130

B. Illuminance from a clear sky without window controls

C

	Room Width M	Room Width FT						
MAX	6.1	20	.0206	.0173	.0143	.0123	.0110	.0098
	9.1	30	.0203	.0173	.0137	.0120	.0098	.0092
	12.2	40	.0200	.0168	.0131	.0119	.0096	.0091
MID	6.1	20	.0153	.0104	.0100	.0079	.0083	.0067
	9.1	30	.0082	.0054	.0062	.0043	.0046	.0037
	12.2	40	.0052	.0032	.0040	.0028	.0029	.0023
MIN	6.1	20	.0106	.0060	.0079	.0049	.0067	.0043
	9.1	30	.0054	.0028	.0047	.0023	.0032	.0021
	12.2	40	.0031	.0014	.0027	.0013	.0021	.0012

K

	Room Width M	Room Width FT								
MAX	6.1	20	.145	.155	.129	.132	.111	.111	.101	.0982
	9.1	30	.141	.149	.125	.130	.111	.111	.0954	.101
	12.2	40	.157	.157	.135	.134	.111	.111	.0964	.0991
MID	6.1	20	.110	.128	.116	.126	.111	.111	.103	.108
	9.1	30	.106	.125	.110	.129	.111	.111	.112	.120
	12.2	40	.117	.118	.122	.118	.111	.111	.123	.122
MIN	6.1	20	.105	.129	.112	.130	.111	.111	.111	.116
	9.1	30	.0994	.144	.107	.126	.111	.111	.107	.124
	12.2	40	.119	.116	.130	.118	.111	.111	.120	.118

C. Illuminance from a uniform ground without window controls

C

	Room Width M	Room Width FT						
MAX	6.1	20	.0147	.0112	.0102	.0088	.0081	.0071
	9.1	30	.0141	.0112	.0098	.0088	.0077	.0070
	12.2	40	.0137	.0112	.0093	.0086	.0072	.0069
MID	6.1	20	.0128	.0090	.0094	.0071	.0073	.0060
	9.1	30	.0083	.0057	.0062	.0048	.0050	.0041
	12.2	40	.0055	.0037	.0044	.0033	.0042	.0026
MIN	6.1	20	.0106	.0071	.0082	.0054	.0067	.0044
	9.1	30	.0051	.0026	.0041	.0023	.0033	.0021
	12.2	40	.0029	.0018	.0026	.0012	.0022	.0011

K

	Room Width M	Room Width FT								
MAX	6.1	20	.124	.206	.140	.135	.111	.111	.0909	.0859
	9.1	30	.182	.188	.140	.143	.111	.111	.0918	.0878
	12.2	40	.124	.182	.140	.142	.111	.111	.0936	.0879
MID	6.1	20	.123	.145	.122	.129	.111	.111	.100	.0945
	9.1	30	.0966	.104	.107	.112	.111	.111	.110	.105
	12.2	40	.0790	.0786	.0999	.106	.111	.111	.118	.118
MIN	6.1	20	.0994	.108	.110	.114	.111	.111	.107	.104
	9.1	30	.0816	.0822	.0984	.105	.111	.111	.121	.116
	12.2	40	.0700	.0656	.0946	.0986	.111	.111	.125	.132

Courtesy of the Illuminating Engineering Society of North America.

TABLE 13-7 Typical Light Loss Factors for Daylighting Design

	Light loss factor glazing position		
Locations	Vertical	Sloped	Horizontal
Clean areas	0.9	0.8	0.7
Industrial areas	0.8	0.7	0.6
Very dirty areas	0.7	0.6	0.5

Courtesy of the Illuminating Engineering Society of North America.

TABLE 13-8

Percentages to Use When Figure 13-10 Curves Are Applied to Periods Other than 09:00–17:00

Curve in Fig. 13-10	95%	90%	85%	80%	70%	60%
Alternative Period	Percentage of Alternative Period					
07.00–15.00	95	90	85	80	70	60
08.00–16.00	100	100	95	85	70	60
07.00–17.00	95	85	75	65	55	45
06.00–18.00	75	70	65	60	50	40

Courtesy of the Commission Internationale de l'Eclairage.

14 lighting economics

Not one single person can make a luminaire. Surprising statement? Not really. Not when you consider that, as simple a device as it is, a luminaire comprises five or six major components (lamps, lampholders, reflector, housing, ballast, and lens/diffuser) and that each major component consists of perhaps hundreds of subcomponents, and that each subcomponent consists of sub-subcomponents, and that each sub-subcomponent (such as a screw) requires the know-how and labor of hundreds, even thousands of people around the world in order for it to be produced.

How can a simple screw involve the efforts of

thousands? Well, consider that the ore for the metal has to be mined. In order for each miner to do his job, he must be outfitted with the right equipment. Each piece of equipment has myriad components, each one requiring the work of hundreds to bring it into being. Even the cup of coffee the miner drinks on his break (a requisite ingredient in the performance of his job) required not only Juan Valdez and the television camera crew, but longshoremen, roasters, merchants, and so on, to bring it to the miner. And mining the ore is certainly not the end of it. The ore must be refined, the refined metal must be fabricated into a screw, and the screw must be packaged, marketed, and transported to the company that manufactures the luminaire subcomponent.

Considering that the entire economy of an industrialized nation consists of thousands upon thousands of products of greater or lesser complexity than a luminaire, it boggles the mind that the entire system can function with any degree of efficiency. How do the manufacturers of luminaire screws know just how many luminaire screws to produce to keep all of the luminaire manufacturers supplied? Even if the total number of luminaire screws required in a year is known, how does a manufacturer know how many luminaire screws the competing luminaire screw manufacturers plan to produce in the coming year? Similar questions could be asked for the millions of elements that make up those thousands upon thousands of products within an economy. It's the proverbial can of worms on an astronomical scale.

In a socialist country the government takes a proactive stance in sorting through the gargantuan complexities of supply and demand. The government sets quotas and production limits on luminaire screws as well as toilet paper, stereos, dog food, automobiles, rutabagas, computers, and every other product conceivable. The government dictates who may sell to whom. Bureaucrats determine when a new factory is needed, and functionaries decide when a plant can be shut down. Every detail of the intricate workings of the economy is planned by the government. This is referred to as a **managed economy.**

But what's wrong with this picture? As the decade of the 1990s opens, Soviet bloc countries are falling all over each other to see who can abandon their managed economy the fastest. Faster than Michail Gorbachev can say *peristroika*, socialist economies are rushing to adopt capitalist concepts. It would seem a paradox that in the face of the intricate interdependencies on a colossal scale that make up a national economy, the governments that in the past have tried to actively manage those interdependencies are feeling defeated in their efforts and have in fact begun to emulate those governments which give free reign to their economies. If a country's economy is left unmanaged to wander at will, why does it not rapidly dissolve into chaos?

THE PRICE SYSTEM

The answer to this puzzle has sometimes been called the *invisible hand*. It is this "invisible hand" that guides the free market economy and ensures that just the right quantity of luminaire screws is produced each year. It is this "invisible hand" that ensures that the miners mine just the right amount of ore for those luminaire screws and that the coffee roasters roast just the right amount of coffee for the miners that mine the ore for luminaire screws. This ''invisible hand'' is the **price system.**

We are each intimately familiar with the price system whether we realize it or not. Suppose you are interested in having rutabaga for dinner. You go to your local supermarket and find it priced at $3.89 a pound. You decide that you didn't want rutabaga that badly and settle for green beans instead. A while later you are in the store and notice that the rutabaga is now priced at 29 cents a pound. Its not very fresh, but you decide that old rutabaga at 29 cents a pound is better than no rutabaga so you buy the last of it. The price system has worked. There was just the right amount of rutabaga, though no central planner told the rutabaga farmer how much rutabaga to plant.

Or suppose you manufacture luminaire screws. You figure that a dollar a screw is just compensation for your efforts, so you quote that price to the luminaire manufacturers. After getting laughed at several times, and when no orders come in for months, you decide that perhaps a more reasonable price would be 29 cents a pound. Now the luminaire manufacturers are more interested, and you find that, though they won't buy as many as you hoped, you can efficiently scale back production to where they will buy them just about as fast as you can make them. Again, chalk up a success to the price system.

We commonly refer to the working of the price system as the **law of supply and demand.** Traditionally, it is illustrated as shown in Figure 14-1. The abscissa indicates the total quantity of a particular item; the ordinate shows the price of the item. The curve labeled *D-D* indicates the locus of all points at which buyers are willing to purchase the product. This is the demand curve. Note, that as price increases, purchasers are less and less willing to buy—as indicated by the fact that the quantity demanded decreases. *S-S*, on the other hand, the supply curve, indicates the locus of all points at which sellers are willing to sell their product. As price increases, so does the quantity sellers are willing to supply. This is due, in part, to the fact that, as the price goes up, it becomes possible for even an inefficient producer to make a profit in this market. This encourages new producers to enter the market and so the total quantity available on the market goes up.

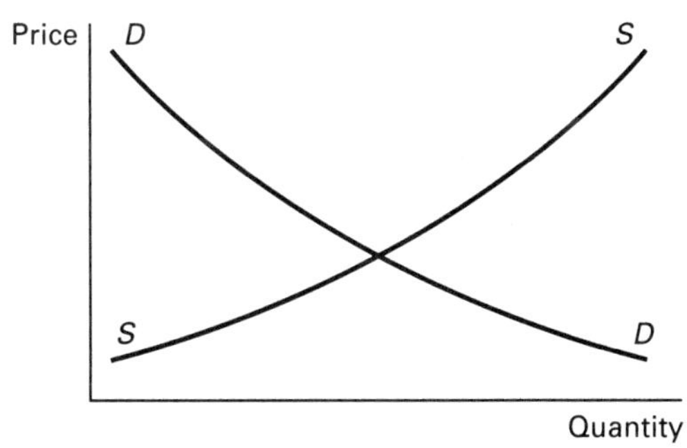

Figure 14-1. The Supply and Demand Curves.

To relate our example to the law of supply and demand, consider two prices, P_1 and P_2 shown in Figure 14-2. P_1 corresponds with our attempt to sell luminaire screws for $1 a piece. At this price we were perfectly willing to provide a quantity of screws equal to Q_s, but luminaire manufacturers are only willing to buy Q_d. By lowering our price to P_2, however, we find that the amount we are willing to provide just corresponds to the amount manufacturers want to buy at that price. This is the equilibrium price. Thus the price system has sent us a signal. By telling us the price P_2, it has told us how many screws to make, Q_e. We could ignore that signal if we chose, but only to our own detriment. In that sense, the free economy is like the managed economy—you can ignore the signal sent to you by the central planner in a socialist state, but it will be to your detriment to do so. Here ignoring an economic signal is financially costly. There, ignoring an economic signal wins you an all-expenses-paid trip to the gulag.

There is another difference. In a planned economy, imperfect humans use incomplete information to make supply and price decisions. The socialist system makes many mistakes. In a free market, an efficient impersonal system uses an abundance of price data to make those decisions. By and large, the price system makes few mistakes. The notable exception is what economists call **externalities.** These are the costs of creating goods or services which are not borne by the producer. An obvious example of an externality is pollution. When our luminaire screw manufacturer pollutes the air, society pays the cost. The cost to the manufacturer of producing the screws does not accurately reflect the total cost of production. Laws requiring the producer to install smokestack scrubbers and the like cause the manufacturer to incur the total cost. The price of the screw is increased to partially offset this, demand decreases, and a new equilibrium is reached. Greed, materialism, and depersonalization of the marketplace are other externalities. It is doubtful that laws can be enacted to cure these externalities. This may prove to be the downfall of capitalism.

So what does all of this have to do with the lighting professional? Precious little, perhaps, were it not for the facts that Figures 14-1 and 14-2 apply to all scarce resources, that money is a scarce resource, and that the governing factor in most lighting designs is the consideration of money. Thus, economics is of central importance to the lighting professional. The following should help to assist in making sound economic judgments about light.

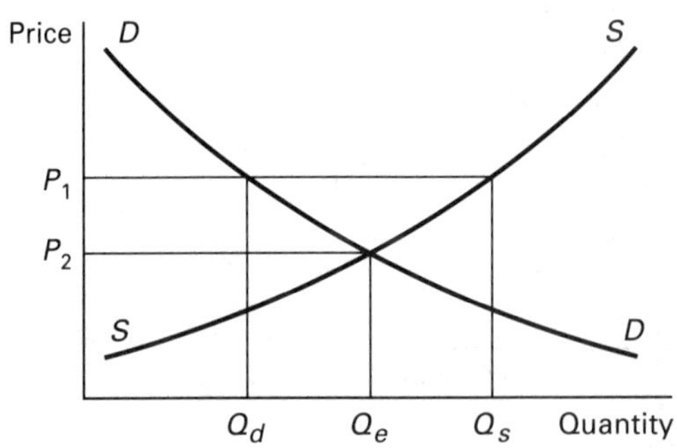

Figure 14-2. Lowering the Price from P_1 to P_2 Brings Quantity Supplied and Quantity Demanded into Equilibrium.

Let's consider first the notion of money as a scarce resource. This idea comes as no surprise to most students. It is hard to find enough money to both buy textbooks and go to Padre for spring break. When demand for money exceeds the supply of money, we are sometimes willing to pay a higher price for money. In other words, we take out a loan. So, for each dollar the travel agent wants for sending us to Padre, we spend, say, \$1.10; \$1.00 to the travel agent now, and 10¢ interest later to the banker who loans us the money. Interest is the fee charged for using someone else's money.

Note that interest has a time component. The ten cents on the dollar we pay our banker assumes that we pay back the loan at the end of a stipulated period. If we want that period to be longer, the interest is higher. (If the period is too long, the banker has no interest in making the loan in the first place, but that's a different issue.) We refer to the time component associated with interest as the **time value of money.**

To see that money has a "time value," consider the answer to this question: which would you prefer, \$1000 today or \$1000 a year from now? We can feel pretty confident that the person who answers "a year from now" didn't understand the question. In general, we prefer to consume today rather than tomorrow. But even if we are willing to delay consumption until a year from now, we generally prefer more to less. This means we still prefer the \$1000 today because, though we don't want to spend it on things to consume today, we can put it in the bank and accrue interest on it (that is, have someone pay us for the use of it) for one year. At the end of the year we will have more than the \$1000 which we would possess had we accepted the other side of the offer.

TIME EQUIVALENCES

The preceding gives a qualitative understanding of the time value of money. In order to assess the economic aspects of lighting systems, a quantitative understanding is necessary. Insight can be gained by answering a series of questions such as

Which do you prefer, \$1000 today or \$1000 a year from now?
Which do you prefer, \$1000 today or \$1010 a year from now?
Which do you prefer, \$1000 today or \$1020 a year from now?
Which do you prefer, \$1000 today or \$1030 a year from now?
Which do you prefer, \$1000 today or \$1040 a year from now?
Which do you prefer, \$1000 today or \$1050 a year from now?
Etc. . . .

Going through the list sequentially, one eventually arrives at a question where it is very difficult, impossible even, to choose between the two options. Let's say this occurs at \$1100. This means that you are indifferent as to whether you receive \$1000 today or \$1100 a year from now. Either one is equally acceptable. In this sense, the two values are equal. You have found a **time equivalence.** We infer that the gratification you receive from having the money now rather than later is worth \$100 to you. Or you believe that you can invest the \$1000 in such a way that it will grow to \$1100 in a year. Another way of stating this is that you require \$100 in interest to persuade you to wait a year to receive your money.

In calculating equivalences, then, four elements are involved: (1) The amount you are considering today, called the **present value** and frequently denoted by PV; in mortgage lending, this value is called the principal; (2) the **interest rate,** i expressed as a percent or decimal fraction; (3) the period of time being considered, n; (4) the value being considered in the future, or **future value,** FV. In our example $PV = \$1000$, $n = 1$ year, $FV = \$1100$. From this information we can deduce that $i = 10\%$ since the present value grows by 10% from \$1000 to \$1100 over the year. We would like a general formula involving these four variables so that, given any three, the fourth can be calculated. The following is a derivation of such a formula.

Consider investing our \$1000 for *three years* at 10% interest, compounded. Compounding means interest is paid on interest as well as the principal. At the end of one year, our investment has grown to

$$FV_1 = \$1000 + \$1000 \times 10\% = \$1100$$

after two years

$$\begin{aligned} FV_2 &= FV_1 + FV_1 \times 10\% && \text{(what we had at the beginning} \\ &= \$1100 + \$1100 \times 10\% && \text{of year two, plus interest)} \\ &= \$1210 \end{aligned}$$

after three years

$$\begin{aligned} FV_3 &= FV_2 + FV_2 \times 10\% && \text{(beginning value plus interest)} \\ &= \$1210 + \$1200 \times 10\% \\ &= \$1330 \end{aligned}$$

Therefore, we say that \$1330 three years hence is equivalent to \$1000 today, *at 10% interest compounded annually*. Let's generalize this to n years at interest rate i (decimal fraction):

At the end of one year:

$$\begin{aligned} FV_1 &= PV + \text{PV} \times i \\ &= PV(1 + i) \end{aligned}$$

after two years

$$\begin{aligned} FV_2 &= FV_1 + FV_1 \times i \\ &= PV(1 + i) + PV(1 + i) \times i \end{aligned}$$

which, after factoring out $PV(1 + i)$ yields

$$= PV(1 + i)(1 + i)$$

after three years

$$\begin{aligned} FV_3 &= FV_2 + FV_2 \times i \\ &= PV(1 + i)(1 + i) + PV(1 + i)(1 + i) \times i \\ &= PV(1 + i)(1 + i)(1 + i) \end{aligned}$$

which we can then generalize for n years:

$$FV_n = PV(1 + i)^n \tag{14.1}$$

Equation 14.1 gives us the future equivalent of any present value for given conditions of interest and time. If, for example, you put \$100 in the bank today at 10% interest and would like to know how much this will be worth when you retire in 44 years, the formula is

$$FV_{44} = \$100(1 + 0.10)^{44} = \$6626$$

Frequently, we know a future value and would like to calculate the present equivalent. By rearranging equation 14.1, we find

$$PV = FV_n/(1 + i)^n \tag{14.2}$$

As an example, how much would a father need to invest in a 9% certificate of deposit today in order to have \$30,000 on hand to pay his 5-year-old child's college tuition 13 years from now? Using equation 14.2 gives

$$PV = \$30{,}000/(1 + 0.09)^{13} = \$9785$$

For completeness, let's solve equation 14.1 for the interest rate:

$$i = (FV_n/PV)^{1/n} - 1 \tag{14.3}$$

Suppose the father in the previous example only has \$8700 today. If he still wants to have \$30,000 13 years hence, what interest rate must he be able to get on his \$8700 investment? From equation 14.3,

$$i = (30{,}000/8700)^{1/13} - 1 = 0.0999 \text{ or about } 10\%$$

The concept of equivalence as expressed by these equations, particularly equation 14.2, is central to the economic analysis of lighting systems. As seen in the following, when people try to ignore this concept, very unsatisfactory results can ensue.

TOOLS FOR DECISION MAKING

What is meant by the term "economic analysis of lighting"? Certainly we are interested in such things as the cost of equipment, the cost to install that equipment, the energy costs associated with providing light, the cost of extracting the heat the lights generate, and the like. All these things play a role in the analysis, but the essence of an economic analysis is that it is a tool to help make a decision or form an opinion. There are many possible solutions to a given lighting problem—which one is the cheapest? Should source A be specified in favor of B, even though it has lower efficacy? Which of three systems provide the best economic benefits? Is one luminaire type always preferred to another? In the following we examine the tool kit to see what methods are available to assist in making these decisions.

First-Level Analysis Methods

First-level analysis is performed using simple methods, generally capable of being computed "by hand," that is, without resorting to a computer. Since you get what you pay for, however, the minimal effort required for these techniques yields only minimal information. Worse still, the information they do provide may be misleading or simply wrong. This is because these tools usually don't take into consideration such things as the time value of money, tax effects, influences of the lighting system on the cost of other systems, and the like. For these reasons, the authors don't advocate their use. However, because many uninformed people still use these methods, descriptions are included here to help the student avoid being misled upon encountering these people in professional practice.

Some people claim that these methods are helpful in removing alternatives from consideration. In the preliminary stages in the design process, perhaps less damage can be done. But certainly by the time the decision is narrowed down to the final few alternatives, more rigorous methods are required. Stated another way, the decision of whether or not to use a first-level method can be framed in terms of the question "What is the cost of a wrong answer?" If it is minimal, then these methods can do little harm. If it is significant, then a more thorough analysis is called for.

The Cost of Light

The most simplistic economic analysis consists of a rule such as compare initial cost and buy the cheapest. Thus if lamp A costs \$1.20 and lamp B costs \$1.00, we would buy lamp B. If the lamps are identical in the service they provide, then this may be a sufficient analysis. However, if lamp A produces 1000 lumens but lamp B produces only 800 lumens, then we might choose A based on a cost-per-lumen comparison (A = .12 cents/lumen vs. B = .125 cents/lumen). But lamp A may only have a rated life of 1000 hours as compared with 1100 hours for lamp B, and a further refinement to the calculation would be necessary.

The need to consider simultaneously lumens, life, and lamp cost was recognized early in the history of lighting. A basic measure of lighting value was developed using the idea of cost per unit of light delivered (9). This method has traditionally been labeled the "cost of light." Because the lumen is a relatively small quantity, the cost per lumen-hour of general lighting systems is typically a fraction of a cent per lumen-hour. Therefore, the unit of dollars per million lumen-hours is usually used.

$$U_1 = (10/Q) \times [(P + h)/L + (W \times R)] \tag{14.4}$$

where

U_1 = unit cost-of-light for a lamp in dollars per million lumen-hours

Q = mean lamp lumens

P = lamp price in cents

h = cost of labor to replace one lamp in cents

L = average rated lamp life in thousands of hours

W = mean input watts per lamp (lamp + losses)

R = energy cost in cents per kilowatt-hours

The cost of light is calculated by dividing the product of average light output and the life of the lamps into the total owning and operating costs for the same time period. Equation 14.4 shows the most simple formulation. Note that equation 14.4 contains no information about the luminaire in which the lamp is housed. It tells us nothing about *usable* light, only about generation of light by the lamp. Therefore it is suitable only for use in a crude comparison of light sources, not for analyzing lamp/luminaire systems.

To help appreciate this concept, consider a lamp manufacturer who promotes a new lamp as being an efficient replacement for metal halide. By performing the calculation of equation 14.4 twice, once for a metal halide lamp and once for the new lamp, we could approximately compare how much it costs each of the two lamps to produce 1 million lumen-hours. If the lamps differed appreciably in physical characteristics such as size, shape, and/or operating temperature, the luminaires that house them might also differ appreciably in efficiency, cleaning requirements, intensity distribution, and/or maintenance characteristics. Each of these differences could affect the cost of owning and using the lamp/luminaire system. These differences could also result in different intensity distributions for the two luminaires, with one being more efficient at putting light where we want it, thus causing a discrepancy in the number of units required and consequently in the cost of the systems. In order to account for these cost differences, the cost-of-light equation must be modified to encompass the entire lamp/luminaire system as indicated in equation 14.5.

$$U_s = [10/(Q \times CU \times D)] \times [(P + h)/L + (W \times R) + (F + M)/H] \qquad (14.5)$$

where all quantities are as for equation (14.1) except

U_s = unit cost of light for a lamp/luminaire system in dollars per million lumen-hours

P = lamp price in cents, total for all lamps in the luminaire

h = cost of labor to relamp one luminaire in cents

W = mean input watts to luminaire (lamps + losses)

F = fixed or owning costs in cents per luminaire-year

M = cleaning costs in cents per luminaire-year

H = annual hours of operation in thousands of hours

D = product of all luminaire-specific depreciation factors such as dirt depreciation, ballast factor, temperature factor

CU = an appropriate coefficient of utilization

Equation 14.5 provides the cost in dollars for a specific lamp/luminaire system to supply a million lumen-hours to the workplane. If the calculation is for a specific space, then the appropriate coefficient of utilization is the CU based on the actual reflectances and room cavity ratio of that space. If a more generic comparison is desired, then certain assumptions must be made. For example, the *CU* could be based on 80/50/20 reflectances and a *RCR* of 1. These assumptions should be clearly stated in reporting results of the calculation.

In addition to the problem that it doesn't consider the time value of money, this type of calculation requires implicit assumption of one of two things: either (1) all systems under consideration are capable of providing precisely the amount of light required in a space, or (2) any increase (decrease) in illuminance above (below) the target value carries with it a commensurate increase (decrease) in the value of the benefits of the lighting system to the owner. For example, consider luminaires A and B, which have identical owning and operating costs. Though the initial cost for A is higher than for B, the cost of light for A is cheaper by virtue of the additional assumption that A has greater efficiency. A lumen method illuminance calculation reveals that, for a certain room, 3.6 type A luminaires are required to achieve an average maintained target illuminance of 50 fc on the workplane. A similar calculation for type B indicates a need for 4.3 luminaires. Obviously, four luminaires of either type will be installed.

The cost-of-light calculation indicated that type A was cheaper. However, by installing 4 type A's rather than the 3.6 actually needed, the efficiency advantage (which is what caused the cost-of-light calculation to be in A's favor in the first place) is lost, and assumption 1 is violated. What about assumption 2? Will the future benefits of that extra light equal or exceed the additional cost? At present, there are no objective measures for predicting the answer. All that we can do is make an inference from an owner's actions: if the owner is willing to pay for the extra light, then the owner must believe that the benefit of the extra light exceeds its cost.

The presence of these issues greatly compromises the usefulness of the cost-of-light method. But in realizing the shortcomings of this method a valuable lesson is learned. We see that concentrating on one element of a system (the lamp) can give misleading information. It is necessary to consider the entire system in an economic analysis. Another of the first-level methods, the annual cost method, has the advantage that it does consider the entire system (though it still fails to consider the time value of money).

Annual Cost Method

A typical example of this method is found in Figure 14.3. Simply stated, this model attempts to assign to each lighting system an equivalent yearly cost of ownership. To find this value, one adds to the annual maintenance and energy costs the annualized capital investment and other one-time costs. This annualized cost is the sum of all one-time costs divided by the useful life of the system. Thus a system requiring a total of \$20,000 in capital investment over the course of its 20-year life would have a \$1000 annual cost. In comparing several proposed systems, the one with the lowest annual cost would be preferred.

Simple Payback

How long will it take to recoup the initial investment in a system? This is the question which the simple payback method attempts to answer. It is defined as the initial cost of a system divided by the annual cash flow or savings which the system engenders.

$$P = \frac{I}{A}$$

$$\text{payback} = \frac{\text{incremental investment}}{\text{incremental annual cash flow}} \tag{14.6}$$

The term "incremental" in equation 14.6 indicates a comparison of one system to another. Two types of comparisons are possible: comparing a proposed replacement system to an existing system, or comparing two design alternatives for a new space in which there is no "existing" system. The former is the more common usage of the simple payback concept in lighting.

Frequently questions arise concerning the desirability of replacing existing lighting with a new technology. Will it pay for itself? Or, more appropriate to the payback method, how long will it take for it to pay for itself? For example, an owner of an office building is considering replacing 1000 standard fluorescent ballasts with a new energy efficient version. It is estimated that the ballasts will save about \$2000 in energy costs yearly. The expense associated with purchasing and installing the new ballasts is estimated at \$11,000. Then the payback would be P = 11,000/2000 or 5.5 years. (Note: These numbers are totally fictitious and should in no way be used to justify a specific decision concerning ballast replacement.)

Using the payback method to compare alternatives involves applying the same formula to *differences* in investments and cash flows as the following illustrates. Consider four lighting systems compared in Table 14-1. The three alternates, A, B, and C, are compared with the "base system." Column I shows the initial investments in thousands of dollars. In column II the first cost of each alternate is compared with the first cost of the base and is recorded as an increase above the base cost. Annual operating and maintenance costs are shown in III, and column IV gives differences in annual cost between each alternate and the base case. In column V the payback is calculated as the quotient of II divided by IV.

In traditional use of the payback method, alternate C is removed from consideration since it does not pay back, and alternate B is preferred to alternate A since (although the initial cost is higher) it pays back more quickly. Either B or the base case is selected depending on whether 2.86 years is considered rapid enough payback.

	Item	Lighting Method #1	Lighting Method #2
A. Installation data			
Type of installation (office, industrial, etc.)	1	....	....
Luminaires per row	2	....	....
Number of rows	3	....	....
Total luminaires	4	....	....
Lamps per luminaire	5	....	....
Lamp type	6	....	....
Lumens per lamp	7	....	....
Watts per luminaire (including accessories)	8	....	....
Hours per start	9	....	....
Burning hours per year	10	....	....
Group relamping interval or rated life	11	....	....
Light loss factor	12	....	....
Coefficient of utilization	13	....	....
Illuminance, maintained	14	....	....
B. Capital expenses			
Net cost per luminaire	15	....	....
Installation labor and wiring cost per luminaire	16	....	....
Cost per luminaire (luminaire plus labor and wiring)	17	....	....
Total cost of luminaires	18	....	....
Assumed years of luminaire life	19	....	....
Total cost per year of life	20	....	....
Interest on investment (per year)	21	....	....
Taxes (per year)	22	....	....
Insurance (per year)	23	....	....
Total capital expense per year	24	....	....
C. Annual operating and maintenance expenses			
Energy expense			
Total watts	25	....	....
Average cost per kWh (including demand charges)	26	....	....
Total energy cost per year*	27	....	....
Lamp renewal expense			
Net cost per lamp	28	....	....
Labor cost each individual relamp	29	....	....
Labor cost each group relamp	30	....	....
Percent lamps that fail before group relamp	31	....	....
Renewal cost per lamp socket per year†	32	....	....
Total number of lamps	33	....	....
Total lamp renewal expense per year	34	....	....
Cleaning expense			
Number of washings per year	35	....	....
Worker-hours each (est.)	36	....	....
Worker-hours for washing	37	....	....
Number of dustings per year	38	....	....
Worker-hours per dusting each	39	....	....
Worker-hours for dustings	40	....	....
Total worker-hours	41	....	....
Expense per worker-hour	42	....	....
Total cleaning expense per year	43	....	....
Repair expenses			
Repairs (based on experience, repairman's time, etc.)	44	....	....
Estimated total repair expense per year	45	....	....
Total operating and maintenance expense per year	46	....	....
D. Recapitulation			
Total capital expense per year	47	....	....
Total operating and maintenance expense per year	48	....	....
Total lighting expense per year	49	....	....

$$\text{* total energy cost per year} = \frac{\text{total watts} \times \text{burning hours per year} \times \text{cost per kWh}}{1000}$$

Figure 14-3 Annual Cost Method. (Courtesy of the Illuminating Engineering Society of North America.)

TABLE 14-1 Four Lighting Systems

	I Initial Investment	II Increase Above Base	III Annual Costs	IV Annual Saving Comp. w/Base	V Payback
Base	110		20		
A	120	10	17	3	10/3 = 3.33
B	130	20	13	7	20/7 = 2.86
C	140	30	21	−1	Does not pay back

Those who like the simple payback method argue that it is easy to use and simple to understand. Closer inspection reveals that, like all first-level methods, it is seriously flawed. In fact, it is actually a risk assessment tool posing as a profitability metric. This is seen from analysis of the question the method seeks to answer. It does not answer the question, "Is a certain investment profitable?" Rather, it assuages the concerns of the person who is unsure about the future and hopes to recoup the investment as soon as possible. It is intended to answer the question, "If I buy this system designed to save me money, how fast can I get my money back?" If getting money back is the primary concern, why invest at all? If no investment is made, then that money is available immediately with no risk incurred.

Another problem with this method is that it fails to consider what happens after the investment is paid back. For example, if alternate A continues to save money for 10 years, but the savings of system B decline sharply each year after the payback period, the payback method has led to the choice of an inferior alternative. A similar situation would be encountered if the systems had different lengths of economic life. Therefore the payback method cannot be used when the alternatives have nonuniform cash flows; it is inappropriate for options having different lives and should be used with skepticism in cases of equal lives.

Simple Rate of Return

The simple rate of return is just the inverse of the payback:

$$\text{simple ROR} = \text{incremental annual cash flow/incremental investment} \quad (14.7)$$

Thus if a system saves $20,000/yr and requires an initial investment of $100,000 then its simple return is 20/100 or 20% (or a 5 yr payback). Advantages and disadvantages are identical to the payback method. It is simple to apply and understand, but it cannot deal with nonuniform savings streams or unequal lives.

Advanced Tools

The most serious flaw with the first level methods is their failure to consider the time value of money. We saw earlier the importance of time when dealing with equivalents. Let's consider now a tool for economic decision making which accounts for time. Then we will be in a position to compare that method with the first-order methods to demonstrate the kinds of problems caused by failure to consider time.

Discounted Cash Flow Analysis

The time value of money is troublesome in lighting economics because various events occur at different times for each lighting system. In one system, the ballasts may need to be replaced in the tenth year after it is installed, whereas in another system they may last 15 years. Different types of lamps have different expected lives, thus replacement costs are incurred at different times. Cleaning schedules may vary and so the timing of costs associated with cleaning.

What is needed is a way to remove the effects of time from the various costs incurred during the life of a system. But of course we already have the tools to do this. Equations 14.1 through 14.3 give us equivalences such that any sum may be compared with another regardless of when they occur. Specifically, equation 14.1 allows us to find a present equivalent of any future sum of money. Finding the present value of a future sum using equation 14.1 is known as **discounting.** In a **discounted cash flow analysis,** all the cash flows (that is, economic events which cause the owner out-of-pocket expense or savings) which occur throughout the life of a lighting system are discounted to the present. This ensures that all the cash flows are on an equal basis. They may then be directly compared with one another.

A common discounted cash flow analysis is called the **present value** or **present worth model.** Here, the discounted cash flows for a given system or alternative are summed (remember, we can do this because they are on an equal basis—today). Then this present worth can be directly compared with the present worth of other systems under consideration. The one with the lowest cost is preferred.

Consider an example in which two alternative lighting systems are being considered for a building:

> *System A* uses high-quality, efficient luminaires. It initially costs $48,000 for equipment and installation. Energy and maintenance costs total $5000/yr. At the end of the economic life of the system, the components can be salvaged (scrapped) for $4000.
>
> *System B* costs $40,000 installed. It is less efficient than A and so annual energy costs are greater—$5900. These lower quality luminaires are deemed worthless at the end of their 20 year life.

Table 14-2 shows the cash flows and their present (discounted) values assuming an 8% discount rate. That is, equation 14.2 is applied to each of the values in the "Future Amount" columns to yield the amounts in the "Present Value" columns. Year zero is the time of the initial investment. Note that the sign of the salvage value for system A is negative, or opposite all the other signs. This is because all other items represent outflows of cash or costs, whereas the salvage value is a cash inflow, or a revenue. It is crucial in a present worth analysis that one be consistent with signs.

TABLE 14-2 Cash Flows and Their Present (Discounted) Values

	System A		System B	
Year	Future Amount	Present Value	Future Amount	Present Value
0	$48,000	$48,000	$40,000	$40,000
1	5,000	4,630	5,900	5,463
2	5,000	4,287	5,900	5,058
3	5,000	3,969	5,900	4,684
4	5,000	3,675	5,900	4,337
5	5,000	3,403	5,900	4,015
6	5,000	3,151	5,900	3,718
7	5,000	2,917	5,900	3,443
8	5,000	2,701	5,900	3,188
9	5,000	2,501	5,900	2,951
10	5,000	2,316	5,900	2,733
11	5,000	2,144	5,900	2,530
12	5,000	1,986	5,900	2,343
13	5,000	1,838	5,900	2,169
14	5,000	1,702	5,900	2,009
15	5,000	1,576	5,900	1,860
16	5,000	1,459	5,900	1,722
17	5,000	1,351	5,900	1,595
18	5,000	1,251	5,900	1,476
19	5,000	1,159	5,900	1,367
20	5,000	1,073	5,900	1,266
	−4,000	−858		
		$96,231		$97,921

In this simple comparison, system A is preferred to system B because its cost is lower (on a present value basis). Note that looking only at first costs would have given the false impression that B is cheaper. Note also the relatively small effect which the $4000 salvage value has on the present worth of system A. This illustrates the fact that events in the distant future have only slight relevance to today's decisions. For this reason, the economic life, or analysis horizon, is rarely taken as longer than 20 years.

In a present worth analysis of lighting system alternatives, it is imperative to quantify *all* cash flows (out-of-pocket costs, revenues, savings, etc.) which implementation of the proposed system would engender. The omission of an item (for example, the cost of replacing ballasts in the tenth year) could cause one to select an inferior option. Most of the cash flows are obvious and not likely to be omitted: initial equipment cost, relamping costs, energy costs, ballast replacement, salvage value, and so on. Slightly more subtle are the costs incurred due to the impact of the lighting system on other systems. For example, it is encouraging to note that most present worth models currently in use explicitly consider the impact of lighting on HVAC costs—initial equip-

TABLE 14-3 Items That Should Be Included in a Discounted Cost Analysis

Initial Costs
Luminaires, lamps, ballasts, poles, mounting hardware
Conductors, conduits
Control devices, switches, dimmers, occupancy sensors, photocells, etc.
Sales tax
Labor to install the system
Design fees
Incremental effect on HVAC equipment
Energy Costs
Energy cost of lighting
Peak demand charge
Effects of lights on HVAC energy cost
Maintenance Costs
Relamping—lamps and labor
Cleaning—materials and labor
Replacing damaged or worn-out components such as lenses, diffusers, ballasts, reflectors, control devices
Other Recurring Costs
Insurance
Taxes
Salvage Value
Revenue from sale of equipment
Cost of selling or disposing of equipment

ment costs as well as energy consumption (2). Other costs are easy to overlook. Table 14-3 lists many of the obvious, as well as subtle, items which should be included in a discounted cost analysis. The following discusses some of the major items.

Initial Cost. Lighting design professionals encounter many pressures to keep initial costs of the lighting system as low as possible. This frequently results in the specification of low-quality equipment or insufficient light for task performance. What are these pressures? Why are some members of the design team motivated to look, for example, at the first costs in the previous example and automatically prefer system B?

Pressures to keep first-cost down generally fall into three categories. (1) The owner is shortsighted. (2) The person paying the initial costs is not the same person as the one who pays the maintenance and energy costs. (3) The contractor hopes to increase his or her profit. Let's consider each of these separately.

1. Maintenance and energy costs are directly tied to the design of the lighting equipment. Durable, well-constructed equipment with fast, easy access will help to minimize maintenance costs. Such features, of course, add to initial equipment costs. Well-designed reflectors efficiently put light where it is needed, thus reducing energy costs. But well-designed reflectors are expensive. Cheaper luminaires sacrifice reflector efficiency, thus increasing energy costs. The shortsighted owner will ignore these effects and insist that initial costs be minimized.

In view of what we have learned about time equivalents, owners' myopia is a curious phenomenon. Since future cash outflows can be translated into present values, it is possible to determine the most efficient way of spending one's money over time. Placing total emphasis on first cost does not maximize economic efficiency. Discounted cash flow analyses do.

But what about the owner/developer who plans to sell the building in three years? Can emphasizing first cost at the expense of operating costs be justified in this case? Not if the present worth analysis is properly performed. One prominent element of a short-term (such as three year) present worth analysis will be the salvage value. How much does the developer hope to receive when the building is sold? If the buyer has life-cycle costs in mind, he or she will be willing to pay more for a building if it has low energy and maintenance bills. Therefore the "salvage" value of an energy-efficient building at the end of three years would be much higher than what could be obtained for a building that consumes much energy. Though the first cost may be higher for an energy-efficient building, the resale will also be higher. So even if the owner isn't planning to keep the building long, this most emphatically does not imply that a total emphasis on first cost is warranted.

2. Many building owners pass on energy and maintenance costs to their tenants, either through appropriate metering methods or by prorating energy costs based on leased square footage. An owner who practices this may be inclined to say "I want to keep the first cost cheap, even if that means making the building energy inefficient. After all, I won't be paying the energy bill, the tenants will." Is this owner justified in minimizing first cost of the lighting, even though it raises the energy bill? The answer is the same as for the developer in case 1 who plans to sell the building in a short time. The owner should consider this question: "Is a tenant willing to pay more rent for a space with higher or lower electric bills?" An energy-efficient building may be highly attractive to tenants. When this concept is appropriately factored into the analysis, first costs will not be overemphasized.

3. Contractors, by virtue of greater hands-on experience, can frequently find less expensive ways to solve construction problems than can designers. When they are able to do so, they should share in the savings. However, when the lighting professional specifies a high-quality luminaire because it optimizes the economics of maintenance and energy costs, he or she should disregard the contractor's change order that claims to "save a few bucks up front by using cheaper fixtures."

Cleaning and Maintenance. A well-planned maintenance program can actually reduce life cost of a system. This is because proper cleaning and so-called preventative maintenance will minimize light loss due to dirt accumulation and surface deterioration. These factors are typically accounted for in the design by installing more light in the space than is necessary for the task. This way, when the system has fully deteriorated, the designer can still be assured of sufficient light levels. If all other aspects of the system are held constant, this increase in light level means, of course, an increase in installed watts, which will increase the operating costs.

Energy Costs. Electric utilities have several rate structures for different classes of customers. The residential customer typically pays a flat rate per kilowatt hour of energy consumed. Commercial and industrial users also pay a certain fee per kilowatt, but this energy fee may vary depending on the level of usage. In addition to the energy fee, commercial customers may also be assessed a peak demand charge. This demand charge is based on the maximum or peak kilowatts the customer requires at some point during a billing cycle (typically monthly). Refer to Chapter 10 for a complete treatment of electric utility rates.

Impact on Other Systems. All the electric energy used to light a building is ultimately converted into heat energy. In the winter this heat is welcome. Its presence reduces the need for heating energy from other sources such as a boiler or furnace. In the summer, however, this heat-from-light is detrimental. Extra energy must be input to the cooling system to extract it. In general, cooling is more energy intensive than heating and so the net result over the year is that the presence of heat-from-light causes extra energy to be used in the building. These additional HVAC energy costs must be included as annual costs in order for an economic analysis of a lighting system to be complete.

However, if only incremental HVAC energy costs are included, an important element of the cost impact of lighting is omitted. The HVAC equipment must be appropriately sized to account for the presence of heat-from-light. The heating plant can be smaller, but the cooling equipment must be larger due to the heat which electric lights introduce. This will affect the initial cost of the building. Note, though, that it is usually inappropriate to assume a price per Btu hr of heating or cooling and then multiply that price by the various heat energy inputs of the lighting systems under consideration. This is because heating and particularly cooling systems are available only in certain standard sizes, with fairly large increments between sizes. For example, if 23.5 tons of cooling are required to dissipate all the heat loads in a building including the heat from lighting system A and 23.7 tons of cooling are required to dissipate all the heat loads in that building including the heat from lighting system B, then 25 tons of cooling will be installed regardless of which lighting system is chosen. However, if lighting system C is large enough to require 26 tons of cooling, then a 30-ton chiller will be used and the additional cost of this equipment must be considered in the economic analysis of the lighting system.

Income Tax. The impact of income taxes is one item frequently overlooked in discussions of present worth analyses (a notable exception is DeLaney (5). Income taxes are major disbursements that often have (or should have) marked influence on decisions. This is clearly seen when it is recognized that corporate income taxes are frequently in the 50% range and that taxable income is reduced by the purchase of any depreciable asset (such as a lighting system). For tax purposes, the yearly depreciation of an asset is subtracted from a company's reported income, thus lowering the income and consequently the income *tax* the company must pay. For example, equipment costing \$100,000 and depreciated evenly over an assumed 20-year life would create a tax deduction of \$5000 per year. For a firm in the 50% tax bracket, this results in a sizable tax savings of \$2500 per year.

Depreciation. The depreciation method used plays a significant role in this analysis. The example assumed a "straight-line" depreciation of equal yearly write-offs. Many times the tax code allows larger deductions to be taken early in the asset's life, declining toward the end of its economic life. This results in larger tax savings earlier, a distinct advantage when one recalls that a dollar sooner is preferred to a dollar later. The depreciation rules also vary for different types of assets with portable equipment such as relocatable wiring and furniture-mounted luminaires receiving what amounts to preferential treatment over fixed assets such as traditional hard-wired systems.

Productivity. Another area not currently included in present worth calculations is the relationship between lighting and productivity. This is because of ambiguities that still exist in that area. There is a strong intuitive urge to say that better lighting results in economic gains through increased worker productivity. A plethora of studies purport to show such a relationship (see, for example, references 1 and 7 and references they cite). However, much of that line of research to date lacks rigorous experimental control. As noted recently by Rea (10), further complications arise if the effects on visual performance are not segregated from other factors affecting productivity such as nonvisual effects of light and other environmental factors. Whereas great progress is being made on the visual performance side of the issue, knowledge concerning the role played by behavioral responses to light lags far behind.

Once the total (visual plus nonvisual) effect on productivity of the luminous environment is understood, it will still be necessary to make economic assessments. It will still be necessary to show, for example, that \$1.00 worth of lighting improvement yields \$1.10 in productivity increase and not merely a \$0.90 increase.

Example. To conclude this discussion of the present worth model, let's consider an example which demonstrates how using more simplistic methods can give erroneous results. Table 14-4 shows the yearly expenditures required for two hypothetical lighting systems proposed for the same space. The initial cost for system A is significantly less than for system B; however, system A requires a significant overhaul (costing \$3500) in the twelfth year whereas system "B" does not. Annual maintenance and operating costs of the two alternatives are the same: \$1200. Summing the total expenditures required for system A and annualizing (dividing by the assumed life of 20 years) indicates an annual cost of \$1775. Similarly, the annual cost model yields a yearly cost of \$1700 for alternate B. By this method, then, B appears to be the more attractive solution.

The present worth model, when applied to this example, takes into consideration that money has a time value. It explicitly accounts for the fact that, if system B is chosen, one forgoes the opportunity of investing the \$2000 premium in initial cost (as compared to A). If the opportunity rate of interest is taken as 8%, then the \$1200 stream of annual expenses in each case has a present value of \$11,782. At the same

TABLE 14-4 Example Comparing the Present Worth and Annual Cost Methods

Year	System A		System B	
0 (Initial Cost)	$ 8,000	{8,000}	$10,000	{10,000}
1	1200	{1111}	1200	{1111}
2	1200	{1029}	1200	{1029}
3	1200	{953}	1200	{953}
4	1200	{882}	1200	{882}
5	1200	{817}	1200	{817}
.	.		.	
.	.		.	
.	.		.	
11	1200	{515}	1200	{515}
12	1200	{477}	1200	{477}
	3500	{1390}		
13	1200	{441}	1200	{441}
.	.		.	
.	.		.	
.	.		.	
20	1200	{257}	1200	{257}
Totals	$35,500	{21,172}	$34,000	{21,782}
Annual Cost:	$35,500 ÷ 20 = $1775		$34,000 ÷ 20 = $1700	

8%, the twelfth year overhaul of system A has a present value of $1390. Summing the present value of all expenditures for each system yields a difference of $610 in favor of system A ($21,172 versus $21,782). Choosing system A over system B is equivalent to pocketing nearly 8% of the initial cost. Recall, however, that use of the annual cost model would lead to selection of system B, or a loss of $610.

Resistance to Use of Present Worth. Clearly, as shown by this example and the previous discussion, the present worth method is indicated as the only model appropriate for making unambiguous decisions in the economic analysis of lighting systems. Though examples have been suggested for which present worth may not be warranted (e.g., any project with an intentionally short life of, say, less than a year, such as Christmas decorations), these are surely the exceptions rather than the rule. If present worth is the rule, why then is there still resistance to the use of such a method in the routine of design decision making? Two objections are frequently raised: (1) it's too complex and (2) clients don't ask for it.

In dealing with the first objection, it must be admitted that a present value analysis, when properly performed, can be complicated. For at least two reasons though, the difficulty of the subject must not be allowed to be an impediment to its widespread use. First, in this age of microcomputer proliferation, computational difficulties have been virtually eliminated, as indicated by the presence on the market numerous cost analysis software packages (8). Conceptual difficulties are all that remain, though these can be great. However, the second reason for forging ahead, overcoming these difficulties, and implementing present value methods is that professional ethics demand it. Whenever one procedure is shown to be clearly superior to all others, the lighting design professional owes it to his or her clients and to the profession to adopt that procedure. This is particularly true in this case where use of any other method can systematically lead to erroneous results. It is incumbent upon the lighting practitioner, then, to either seek training in these methods or obtain the services of a consultant when questions of economic analysis arise.

The argument that clients don't request present value analyses can be refuted on similar ethical grounds as well as from a more pragmatic perspective. Educating the client has always been the duty of the illumination professional. There is no reason why such education should be limited to visual and aesthetic considerations; informing clients as to how to make good economic decisions should also be considered part of the lighting specialist's role. Practically speaking, it is always in the designer's own best interest to provide the best service possible to his or her clients. In the economic realm as well as the technical, if the client is convinced that the most up-to-date procedures have been used, he or she is more likely to be satisfied with the job and thus become a repeat customer.

Sensitivity Analysis

One problem incurred with even a good model (such as present worth) is that, once selected as being the most appropriate, there is a temptation to adhere to the results as though they were chiseled in stone. Constant reminders are necessary that output from any model can be no better than the input. This leads directly to the suggestion that sensitivity analysis should be performed to determine the "stability" of the model's output for a given comparison.

In sensitivity analysis, questions such as the following are asked: How good are the assumptions serving as input into the model? Is the opportunity rate of interest really 8%? Or could it be 7%? Or 70%? What if ballasts aren't replaced in the tenth year, but in the twelfth? Would it matter? Would such changes make a previously unacceptable alternative emerge as the most attractive?

Figure 14-4 illustrates the effect of considering different possible opportunity rates for two alternatives. System A involves expenditures of \$5000 today plus \$2000 10 years from now. In B, all expenditures are delayed until year 10 in which \$12,000 must be paid. This could represent, for example, an existing system which could either be (1) upgraded today at a cost of \$5000 and overhauled 10 years from now at a cost of \$2000 or (2) completely replaced in 10 years at a cost of \$12,000.

At a zero percent opportunity rate, A is preferred (since it is cheapest: \$8000). At an opportunity rate approaching infinity, B is preferred since it is free (as compared with the \$5000 outlay required for A). Somewhere in between (at an opportunity rate of approximately 7.177%) one is indifferent—the alternatives are equally attractive since they both cost \$6000 on a present worth basis. Thus if a manager felt strongly that the firm's cost of capital fell in the 8.0% to 9.0% range, he would perform a sensitivity

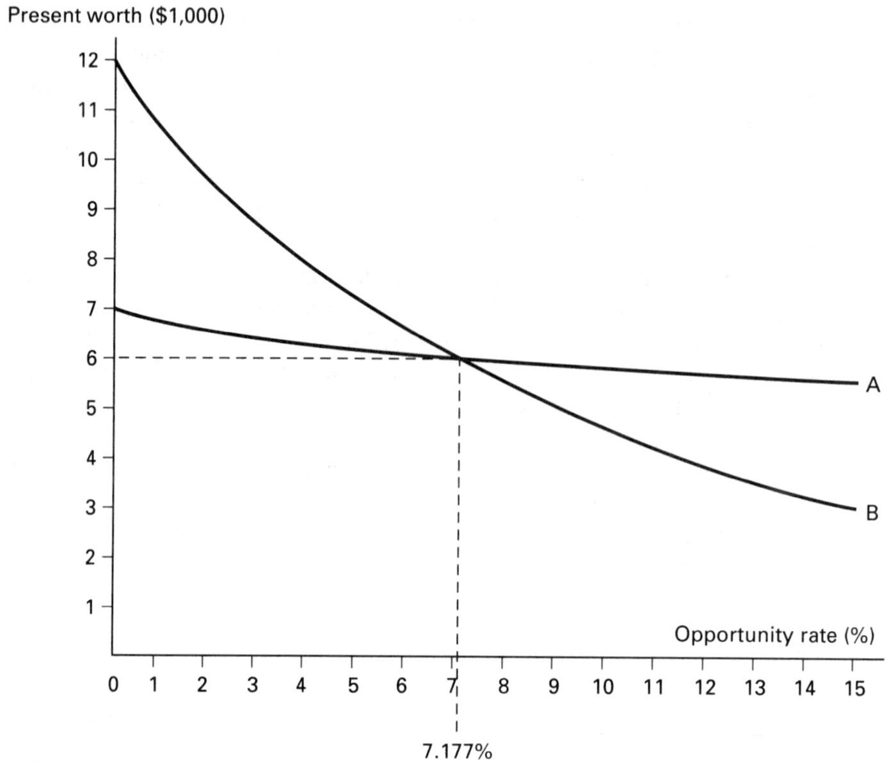

Figure 14-4. Opportunity Rate Sensitivity Analysis.

analysis comparing the alternatives at several interest rates over that interval. He would find that the alternatives were insensitive to interest rates in that range in the sense that B is always preferable. The manager with a cost of capital in the 7.0% to 8.0% range is not so lucky, however. Her sensitivity analysis would tell her that the opportunity rate does matter. If her cost of capital is really 7.0%, then A is preferred, but if it is actually 8.0%, then B is the proper choice. When such a situation arises, you can either ''sharpen your pencil'' with regard to estimating the cost of capital (if the decision is important), or else conclude that the alternatives are roughly equal and it doesn't matter which one is chosen.

Similar analyses should be performed to determine if the decision is sensitive to any of the other input variables for which the value or timing is uncertain. However, accountants warn that *all* costs and their timing, even those that have already occurred, are approximations! This opens quite a can of worms because it indicates that for an economic analysis to be rigorous requires simultaneously taking into account all the uncertainties in all the inputs. There are many ways to approach this sticky problem, indeed volumes have been written on the subject of decision making under uncertainty (for example, references 4 and 11). The most beneficial course for the lighting industry to pursue is likely to be one that continues to tap the power of the microcomputer. For example, Belcher has developed a program for the IBM PC environment which demonstrates how Monte Carlo simulation could be used to consider the effects of ranges of input values on present worth of lighting systems (3).

In that program the designer inputs ranges for costs, probability distributions for opportunity rates, ranges for timing of cash flows, and so on. The strength of the computer then comes into play as it begins making hundreds of random selections per minute of the input variables based on the given parameters. For example, on the first pass it might select an opportunity cost of 8.0%, ballast replacement in year 11, energy rate of $0.087/kWh, and so on, and from these compute a present worth of the alternative. The second time through it might randomly pick a 7.6% interest rate, ballast replacement in the tenth year, $0.092/kWh, and so on, and again calculate the present worth. After numerous repetitions of such computations, a very clear picture emerges of the probability distribution of the present value of the alternative under consideration. This allows construction of confidence intervals and inferences such as the following could be drawn: one can be 95% sure that the present worth of system A is between $8500 and $8900. Figure 14-5 illustrates the results of a Monte Carlo analysis.

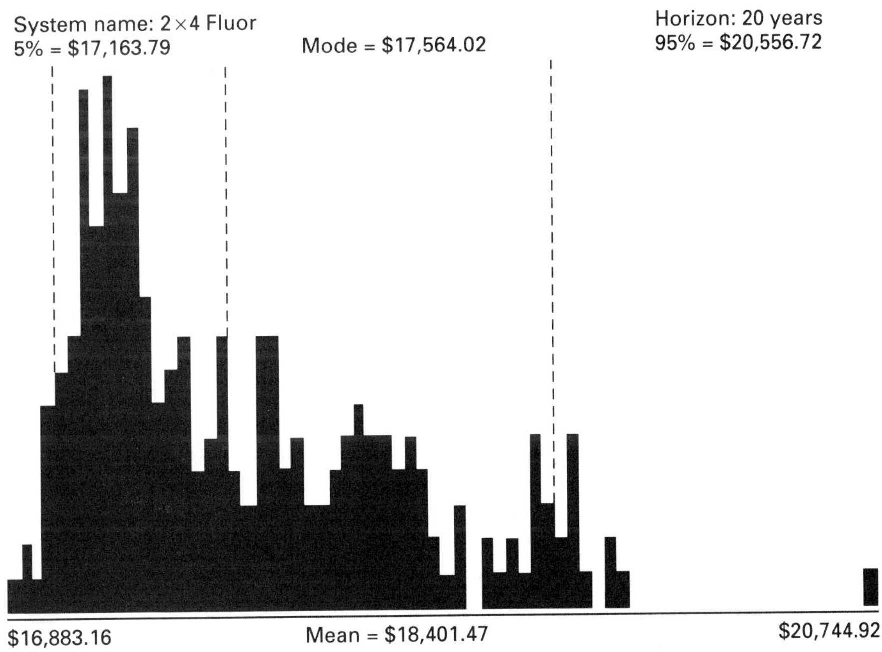

Figure 14-5. Computer Output from Monte Carlo Analysis.

Benefit Analysis

In most of the preceding we have discussed quantification of the various costs that will occur over the life of a lighting system. However, if only costs are considered, we ignore an important aspect of the situation, namely, benefits. A lighting system is purchased for the benefits it gives. If we make a choice from among several alternative lighting systems based solely on cost, the benefit side of the equation has been ignored, or at best, some critical (often dubious) assumptions have been made. Usually the implicit assumption is that that all systems provide equal benefits, therefore the one that provides these benefits at the lowest cost is the best.

The assumption of equal benefits is rarely true. Even if two systems are equivalent on one scale of benefits measurement (for example, they provide equal illuminance), it is likely that they will differ on others (such as glare, color rendering, aesthetic appeal of the luminaire, etc.). If we could assign economic values to these benefits, then the analysis would be straightforward; dollar amounts for benefits would be included in the present worth analysis in the same way that we include other revenues, such as salvage values. However, many of the benefits of light are intangible and difficult to quantify. This is complicated by the fact that many of the benefits are complexly interrelated (e.g., pleasantness of a lighting system may involve a complex interaction of color, illuminance, and the activity being performed under the lighting). For these reasons, the benefits of light are rarely (if ever) explicitly accounted for in lighting economics. Indeed, a research effort aimed at untangling all of the relationships among these issues would be astronomical and almost certainly, at the current level of knowledge, impossible.

Therefore, a radically different approach to the problem is required. One proposed method would take advantage of the fact that people regularly make decisions about the benefits of light, decisions which have monetary consequences. The goal is to then make explicit that information which lighting designers and consultants already use implicitly in their decision-making process. For example, in choosing from between two lighting systems, a lighting engineer may select system A even though it has a higher life-cycle cost than system B. The "nonquantifiable" factor leading to this choice may be that system A has superior color rendering properties. The designer was willing to incur, say, a 10% greater cost for the benefit of better color. Would B still have been chosen had system A been 20% less expensive? 30%? Through a series of questions of this nature, a point of indifference could be determined, a point at which the designer would not care whether the cheaper system was chosen or the more beneficial one. From this process, certain inferences could be drawn regarding the dollar value of color rendering.

Color is but one example of a possible dimension which may enter into the decision process. In defining this quantification method, other dimensions will need to be determined. From there, a framework will be developed within which the designer can quickly determine weighting factors for various benefits on a particular project. It is important to note that the eventual goal for this line of research is not to assign monetary values to benefits once for all, but to develop a methodology or framework with which to make those assignments for a particular project. Lighting professionals should keep abreast of developments on this topic so that they may provide their clients with state-of-the-art economic analysis.

REFERENCES

1. Barnaby, J. F., "Lighting for Productivity Gains." *Lighting Design + Application*, Vol. 10, No. 2 (Feb. 1980), p. 20.
2. Belcher, M. C., "Lighting Cost Analysis: Is There a Better Way?" *Lighting Design + Application*/Vol. 19, No. 6 (June 1989), p. 14.
3. Belcher, M. C., "MCSEALS: A Monte Carlo Simulation for Economic Analysis of Lighting Systems." *JIES*, Vol. 18, No. 2 (1989), p. 40.
4. Bodily, S. E., *Modern Decision Making*. New York: McGraw-Hill, 1985.
5. DeLaney, W. B., "How Much Does a Lighting System Really Cost?" *Lighting Design + Application*, Vol. 3, No. 1 (Jan. 1973), p. 22.
6. Horngren, C. T., *Cost Accounting*: *A Managerial Emphasis*. Englewood Cliffs, NJ: Prentice-Hall, 1977.
7. Jacob, E., "Lighting and Productivity." *International Lighting Review*, Vol. 103 (1966).
8. McCloskey, M. M. (chairman, computer committee), "Available Lighting Computer Programs. A Compendium and a Survey." *Lighting Design + Application*, Vol. 16, No. 9 (Sept. 1986), p. 40.
9 Merrill, G. S., "The Economics of Light Production with Incandescent Lamps." *Trans. Illum. Eng. Soc.*, Vol. XXXII (December 1937), p. 1077.
10. Rea, Mark S., "Toward a Model of Visual Performance: Foundations and Data. *Journal of the Illuminating Engineering Society*, Vol. 15, No. 2 (1986), pp. 41–57.
11. Schlaifer, R., *Analysis of Decisions Under Uncertainty*. New York: McGraw-Hill, 1969.

index

C

E

F

G

M

S

T

Y

Z